RESOURCES FOR THE FUTURE LIBRARY COLLECTION
WATER POLICY

Volume 1

Technology in American Water Development

Full list of titles in the set
WATER POLICY

Technology in American Water Development

Edward A. Ackerman and George O.G. Löf

Washington, DC • London

TECHNOLOGY IN AMERICAN WATER DEVELOPMENT

TECHNOLOGY
IN
AMERICAN
Water
DEVELOPMENT

by **Edward A. Ackerman** *and* **George O. G. Löf**

with the assistance of CONRAD SEIPP

PUBLISHED FOR
Resources for the Future, Inc.
BY
The Johns Hopkins Press, BALTIMORE

RESOURCES FOR THE FUTURE, INC., Washington, D. C.

Resources for the Future is a nonprofit corporation for research and education to advance development, conservation, and use of natural resources, primarily in the United States. It was established in 1952 with the cooperation of The Ford Foundation and its activities have been financed by grants from that Foundation. The main research areas of the resident staff are Water Resources, Energy and Mineral Resources, Land Use and Management, Regional Studies, and Resources and National Growth. One of the major aims of Resources for the Future is to make available the results of research by its staff members and consultants. Unless otherwise stated, interpretations and conclusions in RFF publications are those of the authors. The responsibility of the organization as a whole is for selection of significant questions for study and for competence and freedom of inquiry.

This book is one of Resources for the Future's studies in water resources, which until the fall of 1958 were directed by Edward A. Ackerman. The present director of water resource studies is Irving K. Fox. George O. G. Löf, a chemical engineer, is research associate with Resources for the Future.

Preface

No consistent factual background has been available in the United States for relating technology to water development and water policy decisions. For this reason the research of the water resources program of Resources for the Future has included studies of technical events against a background of water development, economics, and administration.

Technology in American Water Development is an exploratory study. The objectives of the study are: (1) a review of the past technical changes important in water development, and the construction of a meaningful classification of such technical events; (2) appraisal of the socially and economically relevant technical subjects suitable for more detailed study; and (3) appraisal of the manner of studying the relation between technology and the administration of water development in the United States.

The greater part of the study treats the proven technology which has influenced water development in the past or which still is effective in shaping its course. However, the study also treats the emerging technology of today. This may have only modest immediate significance, but its potential importance is large. What this emerging technology may mean in terms of the evolving technological structure of water development in the United States is considered in the last chapter of this study.

No attempt has been made to provide an exhaustive description of all the technical phenomena that have had some bearing upon water development. Had this been the objective, obviously, thousands of technical events could have been considered relevant. Instead, technical phenomena have been classified as they affect the development and operational management of water resources, and illustrations for each class have been provided. The heart of the study therefore is in the thirty-one case descriptions selected from both the emerging and proven technology. Each case description has been treated as briefly as is consistent with clarity, and nontechnical language has been used wherever possible.

Responsibility for the research and writing of the study was divided as follows: Ackerman—planning and organization of the study, writing of chapters 1-5, 11, 17-22, and parts of chapters 6-10; Löf—technical review and technical studies, writing of chapters 12-16 and parts of chapters 6 and 9; Seipp—research and original drafts for a number of the case descriptions included in chapters 6, 7, 8, 9, 10, 15 and 16, photograph selection, and compilation of map and chart illustrations. Dana Little of Brunswick, Maine, also assisted in the compilation of maps and charts. Dale Fester of Denver, Colorado, assisted in the compilation of bibliography and references. The data on ground-water development and exploration techniques included within chapters 6, 8, 10, and 15 were derived in part from a report prepared for Resources for the Future by Louis Koenig of San Antonio, Texas: "Ground Water—The Impact of Recent Technological Advances on its Utilization."

Both senior authors wish to note the value of Mr. Seipp's assistance in preparation of the study. His painstaking research, accurate reporting, and ready comprehension of the complex tasks involved in the study made its completion possible much earlier than otherwise would have been the case. Also, the first draft benefited from his imagination and critical suggestion.

The authors are grateful for guidance and research assistance which have been given from many quarters. The following have been especially helpful in providing data or answering professional questions: R. R. Bennett, W. Durum, W. B. Langbein, and K. A. MacKichan, United States Geological Survey; M. M. Moore, Federal Power Commission; L. F. Warrick, Sanitary Engineering Service, United States Department of Health, Education, and Welfare; C. H. Wadleigh, Agricultural Research Service, and Bernard Frank, Forest Service, United States Department of Agriculture; and R. A. Canham, Federation of Sewage and Industrial Wastes Associations. Detailed and valuable critical review on the first draft was given by Maynard Hufschmidt and Harold S. Kemp, Harvard University; Frank L. Weaver, Federal Power Commission; Edwin Wilson, Bureau of Reclamation, Billings, Montana; Charles McKinley, Reed College; and Blair Bower, Delaware River Basin Advisory Committee.

Painstaking research and careful review of the final draft by Donald Patton, consultant for Resources for the Future, contributed much to its accuracy. Many valuable suggestions also were contained in comments which came from second readings by Charles McKinley and Blair Bower. The careful attention of Roy Bessey, Portland, Oregon, and Stephen C. Smith of the Giannini Foundation of Agricultural Economics, and of Reed Elliot and others of the Division of Water Control Planning, Tennessee Valley Authority, also was most helpful at this stage. Further and also valuable comment on the final draft came to the authors from R. A. Canham, and Frank L. Weaver, Washington, D. C.; Milton Heath, Jr., of the University of North Carolina; Sheppard T. Powell, and Abel Wolman, Baltimore, Maryland; Hubert G. Schenck, Stanford University; Gilbert White, University of Chicago, and the Technical Association of the Pulp and Paper Industry, New York City.

Finally, both authors found the task of writing lightened by the patient and devoted secretarial assistance of Doris Stell.

It is hoped on the one hand that *Technology in American Water Development* may bring the technical horizon into better focus for businessmen, social scientists, administrators, and lawyers interested in water development and management. On the other hand, it also is hoped that the study may help to introduce engineers and physical scientists to the administrative problems and opportunities which stem from their works.

Edward A. Ackerman
George O. G. Löf
December 1, 1958

Contents

LIST OF TABLES

LIST OF FIGURES

LIST OF PLATES

PART I

THE PHYSICAL AND ECONOMIC
ENVIRONMENT OF WATER DEVELOPMENT

CHAPTER 1

The Relation of Advancing Technology
to Problems of Public
Administration and Organization

As long as men have been concerned with problems of political and administrative organization, they have been faced with the influences technologic change exerts on their organization. This has been well illustrated where the military arts of war are concerned. From the days of the earliest weapon improvements, technologic advance has meant eventual disaster for outmoded policies and organization. Throughout history the competition among politically organized groups, both large and small, has been intense enough to alter existing organizational and administrative forms when new ways of waging war emerged. Often the administrative change has come in the form of adjustments within an existing pattern, as for the recent administrative and policy adaptations in many countries to the use of nuclear explosives. Often the change has been revolutionary, as where technically inferior peoples have been conquered by more powerful groups and thereupon subjected to outside control. In whatever form, the influence of technologic change in the military arts on political organization and administrative institutions has been direct and inescapably manifest.[1]

The influence of technology on the more peaceful aspects of administrative organization is less spectacular, but still clear. Some advances, like the Roman art of road making, or modern radio communication,

[1] See Henry Adams, *The Education of Henry Adams,* for an interesting general statement which is one of many on this subject. (Modern Library edition, New York, 1931), pp. 482-88.

3

have affected large areas and have remained valid for long periods of time. The Roman road extended the geographical grasp of administrative devices, introduced diversity of responsibilities, promoted centralization, and permitted a degree of administrative specialization which was previously impossible. The effect of radio communication on the organization of many public activities of the present day, from police protection to the conduct of diplomacy, is common knowledge.

More often, the technical influence on public administrative organization has been selective or specialized. The important technical changes of the past century have all been followed by administrative responses in western or westernized countries. The development of the long-distance, high-speed railroad resulted either in the establishment of governmental agencies for construction, management, and operation (as in a number of European countries), or in the creation and maintenance of agencies for governmental regulation of private enterprise in the public interest (as in the United States). The generation and distribution of electricity for widespread consumption and the development of telephone and telegraph communication were followed by similar administrative responses. In all instances, examples can be found of state-owned and -managed systems or state regulatory agencies where development and operation is undertaken by private management. In some instances, as for electric power in the United States, both publicly managed and privately managed systems (in combination with regulatory public supervision) are found side by side.

These familiar examples of public administrative changes which eventually have followed technical developments may be supplemented by many others. The automobile has resulted in the organization of highway departments, road administrations, vehicular inspection bureaus, safety bureaus, and other public bodies; progress in pharmaceutical therapy and chemical food preservation techniques has been followed by the establishment of public agencies for food and drug inspection or assay; and the widespread peacetime use of aircraft has brought a new set of public agencies to carry on essential regulation, surveillance, or operation. The freshest example stems from the introduction of atomic energy. The administrative changes resulting from this striking addition to our technical knowledge are undoubtedly not yet complete, but the creation of a major independent agency in the Executive Branch of the United States Federal Government is one immediate consequence. This administrative change has been paralleled in other countries, and also in international organization.

The general correlation between important technical developments and significant change in the form of public administration is clear. Technical innovations that have changed military power or economic strength, economic growth, and economic institutions inevitably have required adjustments in the form and technique of public administration.

The Technology of Water Development

From prehistory the technology of water development and use has exerted a strong influence in human society. Theories as to the origin and growth of those enduring early states which progressed beyond tribal organization frequently mention the influence of irrigation upon the rise of well-organized public administration.[2] The arid and semi-arid Near and Middle East furnish a number of examples, from the Indian peninsula to Egypt. The Far East and even the pre-Columbian Western Hemisphere also are not without their cases.[3] The organized Indian communities of the Southwest, Central America, and Peru were supported by irrigation agriculture.

There have been equally interesting examples of the effects of inadequate water-use technology upon the welfare of a society and its form of organization. It seems safe to assume that the decline of closely organized, cohesive political units based on irrigation agriculture in the Near East and Mediterranean area sometimes was preceded by the ravages of lowland sedimentation resulting from improperly managed upstream watersheds.[4] Some interesting evidence on this point is provided by rare observational controls, like the provincial organization which attends the use of the Min River delta in Szechwan, China. There, despite many periods of external confusion, violence, and national political fragmentation, the early development of simple techniques for controlling sedimentation of irrigation works has permitted

[2] Cf., for example, James Fairgrieve, *Geography and World Power* (7th ed.; London: University of London Press, 1932).

[3] Karl A. Wittfogel and Chiasheng Feng (eds.), *History of Chinese Society* (New York: Macmillan, 1949); K. S. Latourette, *The Chinese, Their History and Culture* (3rd ed.; New York: Macmillan, 1946); George B. Sansom, *Japan* (London: Oxford University Press, 1944); Joseph Needham, *Science and Civilization in China* (Cambridge, England: University Press, 1954), I.

[4] H. H. Bennett, *Soil Conservation* (New York: McGraw-Hill, 1939); William Willcocks, *Egyptian Irrigation* (2nd ed.; London: E. and F. M. Spon, 1899).

the survival of a coherent provincial administration for two thousand years.

The early and patent influences of water technology on the organization of public administration were not followed by a consistent advance through the centuries. For many Mediterranean lands the high point of water-use techniques was reached under the Romans, who had mastered long-distance water transportation in a remarkable manner.[5] After the end of the Roman era, only in a few atypical situations, mainly the Netherlands, was there important technological advance in water control. From the sixteenth century onward, during a time when much of the remainder of the world was technologically dormant in this field, Dutch hydraulic engineers continued to develop the techniques of water control. But it is only in the twentieth century that water-use and water-control technology again has commenced to move ahead in a manner which re-establishes it as a significant influence on the content of governmental responsibility and the form of public administration in advanced countries.

Within the last sixty years the technique of making cement and using concrete has been much improved, the hydraulic turbine has been developed, means of large-scale water storage were designed for the first time, practical methods of transmitting the energy of falling water long distances from streams were found, the suction dredge was invented and put to use, new methods of applying irrigation water were invented, and a number of other important new techniques were applied to water use and control. Not the least of these changes has been the application of multiple-purpose design to stream control. This in itself has generated a still unsettled controversy within the United States and elsewhere as to form of public administration.

The technical changes that have already been applied have resulted in certain changes in the form and function of government administration. In the United States they at least partly helped to shape the form and functions of the Tennessee Valley Authority, the rise and continuing growth of the civil-functions divisions of the United States Army Corps of Engineers, the growth and influence of the Bureau of Reclamation, and other administrative changes like the establishment of pollution control functions of the state health services and the federal Department of Health, Education, and Welfare.

[5] E. C. Semple, *The Geography of the Mediterranean Region: Its Relation to Ancient History* (New York: Holt, 1931).

It is probable that advances in the technology of water development in this modern period are far from complete. Technical influences on the need for water control and development, like the problem of nuclear waste removal and other industrial waste disposal, are likely to increase in importance. It also seems reasonable to say that the effects of past technical changes have not yet fully run their course as far as adjustments in public administration are concerned. The stresses which today are evident in United States organization for the development, regulation, and management of water use and control may have technical origins not yet fully reflected in political action. We are in the midst of the first period since ancient times in which there has been a relatively rapid change in the technical basis for such development. A careful appraisal of the technical background of modern water development, therefore, is requisite for understanding present and future public administrative problems and organizational structure related to water use and control. In surveying the past, it is clear that water problems become more intense as settlement increases in density. This has been taking place rapidly in the United States in the last three decades. Along with increases in population density, some hitherto sporadic uses of water have become widespread, like hydroelectric generation and recreational activities.

Technology and administrative organization in a sense are directed toward the same ends in water development. Both are concerned with the reconciliation of demand for water and the natural supplies, whatever the physical location of the specific demand or supply. Technology in this instance is concerned with the application of scientific knowledge and material techniques to the alteration or channeling of water demand, to the production, storage, or transportation of water. Administrative organization is concerned with the application of social techniques to the same ends; in parts of the United States, for instance, it prescribes the priority of use by type of use and specific user.

Accordingly, it seems obvious that technology and administrative organization of water development must have relations which profoundly affect the course that water and later economic development follow. Organizational form and technological status exhibit patterns of compatibility or incompatibility from time to time. If organizational form remains static while technology is advancing, administrative and management stresses may be created. There seem to be logical as well as empirical reasons for examining the nature of modern water development technology and its relation to administrative organization.

Technology, Administrative Organization, and the Political Process

The connection between technology and administrative organization is not always a direct one. Usually, in fact, the connection is indirect, operating in any society largely through economic and political processes. Administrative organization implies a political structure, and technical events have always occurred within and must be viewed in their relation to some political structure. The larger political adaptations have an important bearing on the administrative responses to technology.

Charles McKinley, an experienced observer and outstanding student of American resource administration, has described this relation thus:

> Technological change generally causes modifications in the number of social interests and relations affected by technical consequences. Hence political structures change and reallocations in legal responsibilities take place which in turn become the new foundations for administrative programs and sub-policy decision making. . . . New social interests are either generated or aroused by new relations created as people seize upon the advantages implicit in technological changes. Even as some are advantaged by the latter, it often happens that other groups are injured—at least in the short run. Even if no one is disadvantaged differential benefits create competition, and stresses result. If those who suffer (or think they suffer) cannot find satisfactory solutions through private dealings, or through the old political-administrative mechanisms and the legal assignment of functions, these stresses induce pressures for legal or administrative changes or both.[6]

This general sequence of events is well illustrated in the resource development history of the United States. Public organization for irrigation development, for electric generation and distribution, for navigation, and for pollution control have all had a history illustrating the principle expressed by Professor McKinley.

The present federal organization for the "reclamation" (irrigation) of western arid lands had its origin in the late nineteenth century desire of this country to open and settle the West. However, the continuance and greatly increased size of the federal reclamation organization in the last three or four decades can not be related to the opening of the West in any literal sense. Improvements in refrigerated transportation and in techniques of storing and transporting water gave enticing economic opportunities for some favorably situated western lands, as in California.

[6] Personal communication to the authors, July 1957.

Extensive development accordingly proceeded on some western lands within the existing politico-economic structure, locally organized and privately financed. Repayment often was guaranteed through local taxing power as in the case of irrigation districts. At the same time, a great many of the lands physically adaptable to irrigation could not be developed successfully if a cost-benefit reckoning were to be made within the manner of accounting and financing afforded by private enterprise, even where supported by local government taxing powers. Political pressure for federal financing and entrepreneurship to irrigate the less favorably situated lands therefore gathered momentum, and was applied successfully. The program and the administrative organization of the Bureau of Reclamation is the result. It is probably the largest, most elaborate, and technically one of the most competent irrigation services in the world.

The story, with variations, is repeated for electrical distribution. After the development of central station generation of electricity in this country, and the equipment for household lighting, the electrification of cities and other areas of consolidated settlement went ahead rapidly under the existing political and economic structure. However, relatively little attention was paid to the potential rural consumer or his needs. Rural electrification appeared uneconomic in all except some high-income areas. Political pressure gathered and finally found expression in the program and organization of the Rural Electrification Administration, the Tennessee Valley Authority, the Bureau of Reclamation, and the regional power administrations of the Department of the Interior, in local public utility districts, and in other ways.

Similar events can be traced in the history of the federal navigation facilities program, and in the stream pollution control program, with differences in degree.

While forces other than technical are to be reckoned with in these events of the past, technical stimulus leading to political expression was not far from the surface.

There are also technical answers to the problems which technology creates. Thus the turbine pump has been an important contribution to western irrigation. No political process was resorted to in the application of this important device to the irrigation problems of millions of acres. On the other hand, the application of the large-scale concrete dam, and techniques of long-distance water transportation did depend on the activation of a political procedure. A classic example in the United States was the creation of the Metropolitan Water District of

Southern California to develop interbasin water transportation. In general we can say that the broader and more complicated aspects of technology, whether they have created the problem or are applied as answers to a problem, are ultimately likely to involve some political adjustment.

The origins of problems, and even cause-and-effect relations, tend to become clouded when the political process is part of a solution. This in part may explain some of the present obscurity in our understanding of the relation between technology and administration. At the same time it may be helpful to remember that both technology and administrative devices are instruments to be used in the development and management of a resource. If they are understood as instruments which can be coordinated in their use, economic efficiency and the welfare of the social group as a whole can gain.

CHAPTER 2

The Geographical Nature of Water Occurrence

Regional Characteristics of Distribution in the United States—soil moisture; distribution of other water sources; inherent physical and chemical properties; organic and microbiotic content. Summary of the Physical Nature of Water Supply.

The effects of technology upon the development of water resources must be considered in terms of the physical environment in which water occurs. Water is to be found everywhere on earth, and men require it daily. Yet water is supplied by nature in no two localities exactly in the same manner as to quality, amount, timing, and mode of occurrence. Furthermore, men's requirements are specialized, their general uses having a relatively low tolerance of mineralization or other impurity. The greatest bodies of water on earth, the seas and oceans, thus far have been the least versatile for man's use.

The oceans are, by an overwhelming margin, the dominant water feature of the earth. They probably contain something on the order of 1,060 trillion acre-feet of water. This water is usable, as it always has been, as a transportation medium, as a natural source of food and industrial plants and animals, and as an economic source of certain minerals (sodium chloride, potash, and magnesium). It may gain value as a culture medium for managed production of plants and animals in the future, although it has only minor importance in this respect at present. But the oceans are at present of little direct value for consumptive use of water. Indirectly they have great value as a principal source of atmospheric moisture for the hydrologic cycle. Because of its practically inexhaustible bulk, sea water presents an important technologic challenge—that of broadening its utility into the consumptive uses for which water is most commonly demanded.

Far less of the earth's total supply of water is contained in the atmosphere, on the surface of the land, or within the rocks of the mantle. However, this has been the most valuable water because most of it is "fresh," or much less mineralized than the water of the oceans. The total estimated supply on the land surface, in the atmosphere, or in the rocks to a depth of 12,500 feet is only about 3 per cent of that in the oceans, or 33 trillion acre-feet. Almost 75 per cent of this is immobilized in polar ice and glaciers, and a small amount in the hydrated earth minerals (table 1). The most widespread sources for human use (the

TABLE 1. *Estimated Relative Quantities of Water Available within the Earth's Hydrosphere*

Item	Million acre-feet	Index of amount relative to soil moisture	Per cent of total estimated fresh water present
1. Oceans	1,060,000,000	51,960	—
2. Atmosphere, earth's crust,[1] fresh water bodies	33,016,084	1,618	100
a. Polar ice and glaciers	24,668,000	1,209	74.72
b. Hydrated earth minerals	336	0.16	0.001
c. Lakes	101,000	5	0.31
d. Rivers	933	0.046	0.003
e. Soil moisture	20,400	1	0.01
f. Ground water:			
(1) Fissures to 2500 ft.	3,648,000	179	11.05
(2) Fissures 2500 ft. to 12,500 ft.	4,565,000	224	13.83
g. Plants and animals	915	0.045	0.003
h. Atmosphere	11,500	0.56	0.035
3. Hydrologic cycle (annual):			
a. Precipitation on land	89,000	4.4	—
b. Stream runoff	28,460	1.4	—

[1] To 12,500 ft. depth only.

Source: Based upon data given in C. S. Fox, *Water* (New York: Philosophical Library, 1952); R. L. Nace, *Water Management, Agriculture and Ground Water Supplies,* U. S. Geological Survey, Mimeo., Washington 1958; and R. L. Nace, personal communication, 1959.

atmosphere, streams, and soil moisture) comprise only about 0.39 per cent of the estimated available remaining 25 per cent of the fresh water. Neither the atmosphere nor the soil zone is a major point of storage for the earth's fresh water. For this purpose, lakes are much more important than all of the three together (about 1.2 per cent of the fluid supply).

The importance of the hydrologic cycle is brought out in some additional estimates. The total annual flow of streams is estimated at about 31 times their average water content at any given time. The average annual rainfall over the land surface is about 7.7 times the moisture content of the entire atmosphere at any given time. It is 4.4 times the total moisture content held in the soil. On the average, five or more times as much moisture falls on land in the hydrologic cycle as appears in stream runoff. Evapotranspiration therefore returns about four-fifths of precipitation to the atmosphere. Thus the major sources of fresh water for human use—streams and soil moisture—are in reality momentary storage vehicles for water supplied to the earth's surface through the hydrologic cycle. The high ratio of annual precipitation to standing stream water content, to soil moisture content, and to average atmospheric moisture content has this meaning: any significant general increase in the moisture available from atmospheric sources will have to come from an acceleration of the hydrologic cycle.

One of the notable features of the hydrosphere is the great quantity of water stored on or within accessible parts of the earth's crust. Very large amounts are stored in the rocks below the root zone of the soil. Nearly three-quarters of all the fresh water thought to exist in the hydrosphere is stored in the outer crust of the earth to a depth of about 12,500 feet. It may amount to more than 8 trillion acre-feet. Finding how to use this tremendous amount of stored water is no less a challenge to science and engineering than the ultimate greater use of the oceans and the alteration of the hydrologic cycle.

Regional Characteristics of Distribution in the United States

Considered in the narrow range of man's general needs, the outstanding physical feature of water occurrence is its regionalization, or localization. Disregarding the oceans as a transportation medium, water resources exhibit strongly differentiated regional and local patterns of availability. While the dynamics of water supply are highly dependent on the character of precipitation and evapotranspiration on the earth's surface, local and regional availability is profoundly altered by the configuration of the earth's surface, and by the differential storage capacity of the rock interstices in the earth's crust.

SOIL MOISTURE[1]

The most widely employed water source on the surface of the earth unquestionably is soil moisture. In this, the United States is no exception. All unirrigated agriculture depends on soil moisture, and irrigation, considered simply, is nothing more than an attempt to create optimum soil moisture conditions for the growth of crops. Soil moisture varies on the earth according to the balance of precipitation and evaporation in any region or locality; according to the type of soil;[2] and, to a lesser extent, according to the configuration of the land surface.

The amount of moisture in storage as soil moisture at any one time is relatively small. Over the earth as a whole it has been estimated as equivalent to a layer about 4.6 inches thick averaged for 57 million square miles of land surface on the earth.[3] As any farmer knows, this amount of water would provide little plant growth. The significant data on soil moisture therefore relate to the frequency with which moisture in the soil is replenished and the length of time it remains available for plant growth. This information is given in precipitation, evaporation, transpiration, runoff, and water balance data. The relative importance of viewing soil moisture in terms of the moisture cycle is shown by the fact that annual average precipitation over the earth's land surface as a whole is four times the estimated soil moisture storage at any given moment. Because significant variations in amount of precipitation are the rule, even within broad climatic regions, and because soil types have wide local variations, soil moisture conditions have great variety in the

---

[1] Soil moisture may be defined here as the moisture available to plant growth. It is mostly capillary water, but includes some gravitational water (see glossary). Capillary water is that held by surface tension in the capillary spaces, and in a continuous film around soil particles, and which is free to move in response to capillary forces. See R. K. Linsley, M. A. Kohler, and J. L. H. Paulus, *Applied Hydrology* (New York: McGraw-Hill, 1949), p. 289.

[2] Some soils have very high moisture retention capacities (loams), while others are low (e.g. sands, or grainy textured). Also some soils have high absolute water retention capacities (e.g. clays), but have very little available to plants. According to the above definition their moisture content (available to plants) is low.

[3] C. S. Fox, *Water* (New York: Philosophical Library, 1952), p. 36.

United States and elsewhere. Nevertheless, it is possible to make some generalizations for the United States. For instance, soil moisture supply is very low over most parts of the Great Basin[4] nearly all of the time. But it almost always is high on the windward slopes of the Cascade Range in Washington, Oregon, and northern California.

Like most attributes of water, soil moisture supply varies not only geographically, but also seasonally, weekly, and even daily. Its availability to plants can be directly correlated with weather conditions, as it is influenced by precipitation, insolation, soil and air temperature, and wind.[5] The details of this relationship are outlined in chapter 16.

While the many different crops have different moisture requirements, continuously optimum soil moisture conditions are rare on the earth's surface. A most important disruption of soil moisture availability occurs in freezing soil temperatures, which characterize this continent north of 40° N. latitude for extended periods each year. Significant intermittent fluctuations in soil moisture also occur during the growing season on nearly all soils in all climates. They are especially significant on the soils of low moisture retention capacity, like the Atlantic and Gulf coastal plains sands of the United States, and generally over the long-growing-season area of southeastern United States. Few of the crops grown in the eastern United States are produced under optimum soil moisture conditions at all times during their development. Considering the great variety of local soil classes, local surface configuration conditions, and the equally great variety in the local course of precipitation, the pattern of soil moisture occurrence over both space and time is extremely complex. This is a most important pattern to understand because the largest present use of water is for crop growth, drawn from soil moisture, and because one large potential demand for water development is likely to be an amelioration of soil moisture conditions. Yet we know remarkably little about this problem; it is only within the last ten years that meaningful concepts and practical means of measuring soil moisture have been devised.[6] We now know that the concept of

---

[4] See map on p. 666 for location of regions or localities mentioned by name in the text and not shown in specific maps elsewhere in this study.

[5] See C. W. Thornthwaite and J. R. Mather, *The Water Balance,* Laboratory of Climatology Publications, Vol. VIII, No. 1 (Centerton, N. J., 1955).

[6] The determination of soil moisture conditions has been handicapped in the past by the great variation to be found in soil materials, and by the difficulties of instrumentation for accurate measurement. These problems have been studied through a number of different disciplines, including agronomy, plant physiology,

"humid" and "arid" climates is somewhat misleading. Every climate, as Thornthwaite has shown, has both moisture and aridity indices, with very little of the area of the United States covered by *perhumid* (constantly humid) climates.[7] Soil moisture conditions closely parallel the moisture factor in climate.

The natural supply of soil moisture for probably 95 per cent of the farmed area of the United States in an average year provides less than optimum soil moisture for crop growth.[8] And there is relatively little precise information on the seasonal water supplies necessary to obtain optimum production for individual crops. Yet knowledge of the extent of water demands, and the economics of artificial supply are dependent on this information. These facts alone give some indication of the potentially enormous size and the comprehensiveness of water development problems in this country and elsewhere. The dimensions of the field of water development appear to be limited only by our abilities to perceive and our technical capacities to treat these problems.

DISTRIBUTION OF OTHER WATER SOURCES

Although soil moisture is the most widespread form of water supply, it long has been supplemented by other water sources for man's use. Where concentrated supplies are required for withdrawal—as for livestock, human domestic consumption, or industrial processing—streams, lakes, and ground water are essential resources. With widespread realization of the almost universal deficiency of soil moisture when compared to the optimum amounts usable under cultivation, these supplies also may be widely used for soil moisture supplements.

ecology, soil science, and others. One recent contribution to soil moisture measurement has come from climatology. C. W. Thornthwaite has developed the concept of potential evapotranspiration; his calculations of water balance appear to have opened the road to a clearer understanding of soil moisture conditions, although the extent of applicability of his method is not yet known. See Thornthwaite and Mather, *op. cit.* Other research workers also are continuing to conduct research in this area of study. See also chapter 16, pp. 442-51.

[7] "Perhumid" here is used to indicate maintenance of a soil moisture condition in which the wilting point is never reached for plants on the prevailing soils of the given area. Adequacy of moisture supply at critical periods in the growing season may be important determinants of crop growth.

[8] This statement is not to be interpreted as a reference to the *economic* desirability of making up the deficit between natural soil moisture supply and the optimum supply.

While soil moisture conditions vary over the earth's surface, the surface water supplies are still more unevenly distributed. These supplies essentially are the excess from the moisture cycle operation which cannot be stored naturally in soil or rock and which do not re-enter the atmosphere through evaporation or plant and animal transpiration. The same is true for ground water, although its distribution is much wider than that of concentrated surface waters. Streams and other fresh surface waters probably occupy less than a hundredth of the surface of the earth's land masses, and 4 per cent of United States territory. Their distribution therefore is spotty in comparison to all the localities of potential need.

The distribution of ground water has somewhat different characteristics. As shown by the general estimates, all land masses are thought to have great amounts of water stored in the interstices of the rocks which lie below the soil zone of suitable depth for cultivation. However, the many different types of rock to be found in the lithosphere have varying capacities to store and release the water which may be sought in them. Some sandstones may have large storage capacities, with relatively free movement of water within; other standstones may be so consolidated as to release little water. Some shales may be nearly impervious to water movement and store very little that is not contained in their hydrated minerals. Siliceous limestones and many igneous and metamorphic rocks contain stored water which moves only along erratically placed jointing, fault zones, or, for the limestone, along solution channels. Concentrations of water accordingly are localized below the earth's surface also, although minor deposits of water are much more widely distributed over the zone below root penetration (on the average, about six feet, with many exceptions for specific plants).

Within the United States, as elsewhere, concentrated water supplies display a wide variety in their amount, seasonal time of availability, frequency of occurrence within areas of given size, and other characteristics. Of the average precipitation received annually on continental United States (probably about 4.75 billion acre-feet), less than thirty per cent (1.37 billion acre-feet) may be available as runoff (table 2). The occurrence and concentration of this vast amount of water follow the precipitation pattern, as it is modified by topography, latitude, and the character of the soil and country rock. A large part of western United States contains almost no permanent runoff, except in the few places where exotic streams originating in mountain areas may cross

TABLE 2. *Summary Data Concerning Water Resources of Continental United States*

	Square miles	Acre-feet
Gross area of continental United States[1]	[a] 3,080,809	
Land, area, excluding inland water[2]	[b] 2,974,726	
Volume of average annual precipitation		[c] 4,750,000,000
Volume of average annual runoff (discharge to the sea)		[d] 1,372,000,000
Estimated total usable ground water		[e] 47,500,000,000
Average amount of water available in the ground as soil moisture		[f] 635,000,000
Estimated total of lake storage		[g] 13,000,000,000
Total storage of reservoirs (with a capacity of 5,000 acre-feet or more)[3]		[h] 365,000,000

[1] Includes that part of the Great Lakes within United States territory.
[2] As defined by the U. S. Bureau of the Census.
[3] Includes reservoirs under construction as of January 1, 1954.

Sources:

[a] U. S. Department of Commerce, Bureau of the Census, *Area of the United States: 1940*, Washington, 1942.

[b] *Idem, Statistical Abstract of the United States, 1955*, Washington, 1955, p. 7.

[c] U. S. Department of Agriculture, *Water, Yearbook of Agriculture, 1955*, Washington, 1955, p. 36.

[d] See table 11. See also W. E. Wrather, "A Summary of the Water Situation with Respect to Annual Runoff in the United States," in *The Physical and Economic Foundation of Natural Resources: II, The Physical Basis of Water Supply and Its Principal Uses* (Washington: U. S. Government Printing Office, 1952), p. 36.

[e] A. M. Piper, "The Nation-Wide Water Situation" in *The Physical and Economic Foundation of Natural Resources: IV, Subsurface Facilities of Water Management and Patterns of Supply—Type Area Studies* (Washington: U. S. Government Printing Office, 1953), p. 15.

[f] Based upon the estimate of C. W. Thornthwaite. Cf., for example, p. 353 of *Water, Yearbook of Agriculture, 1955*.

[g] Estimate for that part of the Great Lakes which is within United States territory: C. L. McGuinness, *The Water Situation in the United States with Special Reference to Ground Water* (U. S. Geological Survey Circular No. 114, Washington, 1951), p. 10. Figure for remaining waters not included elsewhere, estimated on the basis of available information.

[h] Estimate based upon information presented in N. O. Thomas and G. E. Harbeck, Jr., *Reservoirs in the United States* (U. S. Geological Survey Water Supply Paper No. 1360-A, Washington, 1956).

the land. Some parts of the country may experience no runoff, even intermittent, for years on end. Death Valley, California, is such a place. On the other hand, sections of the Olympic Mountains, in the state of Washington, average about six feet of runoff in a typical year. On the whole, mountain regions are the most productive of runoff. Flat areas, particularly those of moderate to scant rainfall, do not favor runoff.

The United States may be divided into three major runoff regions (figure 1). One is the moist area generally east of the 95th meridian, most of which has an average annual runoff of ten inches or more, and nearly all of which has a well-developed stream network.[9] Within this eastern region nearly all streams of any size are permanent. Major concentrations of flowing water are to be found within this area in the Mississippi-Ohio and the Great Lakes-St. Lawrence systems, but many lesser rivers, like the Hudson, the Susquehanna, the Connecticut, and the Alabama have substantial flows.

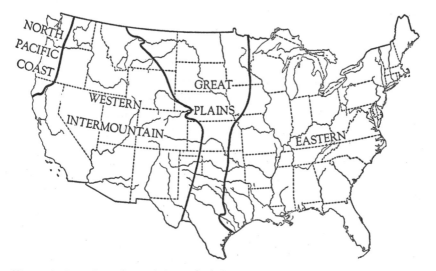

Figure 1. Runoff Regions of the United States.

The second runoff region is that of the Great Plains. It has relatively little continuous runoff because of low precipitation, low relief, and high evapotranspiration. However, during periods of thaw (in the

[9] Some local exceptions occur in karst formations on limestone, as in Indiana, Kentucky, Tennessee, and Florida.

north) or during intense storms, heavy runoff may be experienced for limited periods. Some runoff is contributed by mountains bordering the region.

The remainder of the country is comprised mainly of alternating mountain and intervening dry lowland or plateau. Only the North Pacific Coast region, which is one of the most humid parts of the United States, is an exception to this pattern. In the Western Intermountain region substantial runoff from the mountains is naturally favored, and this runoff is the source of the only important supplies of surface water in the region. The pattern of the Intermountain region generally is one in which the lowland or plateau provides the land, and the mountains the water which is used on the flat land. With few exceptions, little runoff originates in any lowland or plateau area of this region.

For all major runoff regions there is a pronounced concentration of water in one season of the year. The only subregional exception is in Florida, which has a more evenly distributed runoff. Even in Florida, however, 30 per cent of the annual runoff is concentrated within three months of an average year. In arid and semi-arid parts of the country, as much as 75 per cent of the year's runoff comes as a "wave" within a few weeks, the product of snow melt on the upper watersheds. This pattern is profoundly influenced by the seasonal course of temperature in regions or localities that have snow cover or frozen soil, but it also is affected by summer maxima of precipitation in all but the Pacific Coast states. Thus, not only is runoff distribution geographically diverse in the country, but it is unevenly distributed seasonally (figure 2). The importance of artificial storage for the best use of water in the United States therefore becomes clear.

The daily pattern of flow in many streams also is highly uneven. In humid regions between 50 and 70 per cent of total annual runoff occurs as intermittent stream rises or freshets which come from rainstorms or melting snow or ice. Vegetative water consumption also may cause diurnal fluctuation.[10] The smaller and steeper the profile of a stream, the greater the fluctuation in runoff tends to be. All streams tend to have some short period fluctuation, but many of the smaller streams of the country have runoff patterns with peak flows several hundred

---

[10] W. E. Wrather, "A Summary of the Water Situation with Respect to Annual Runoff in the United States," *The Physical and Economic Foundation of Natural Resources: II, The Physical Basis of Water Supply and its Principal Uses* (U. S. Congress, House Committee on Interior and Insular Affairs [Washington: U. S. Government Printing Office, 1952]), p. 41.

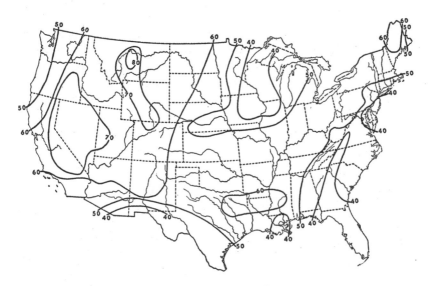

Figure 2. Seasonality of Runoff—per cent median discharge in three consecutive wettest months.

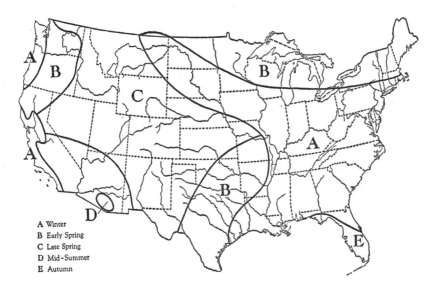

A Winter
B Early Spring
C Late Spring
D Mid-Summer
E Autumn

Figure 3. Seasons of Highest Flow.

times the average flow over the greater part of the year (figure 3). Thus the streams which would be normal surface supplies for the smaller cities and other communities over most of the country have potential capacities under storage far beyond the usual course of unregulated flow. Other things being equal, the larger and climatically more diverse the watershed of a stream, the less violent its fluctuations in flow are apt to be. Geological conditions that favor infiltration also will produce locally smoother runoff curves. The influence of extensive tree cover also is clearly perceptible in runoff.

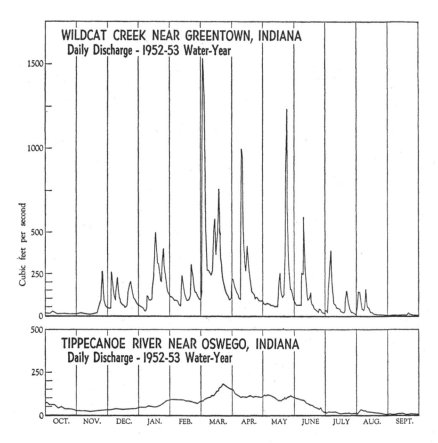

Figure 4. Variation in Stream Flow in Basins of Close Proximity. Drainage area illustrated: Wildcat Creek—166 square miles; Tippecanoe River—126 square miles. Average elevation above sea level: Wildcat Creek near Greentown—811-812 feet; Tippecanoe River near Oswego—832 feet.

The same is true for the manner of cultivation or land use on farmed areas. Infiltration rates are notably greater for grass-covered lands than for cropped areas. Close-growing crops are associated with higher infiltration rates than row crops. Runoff fluctuations are reduced where infiltration rates are increased (figures 4 and 5).[11]

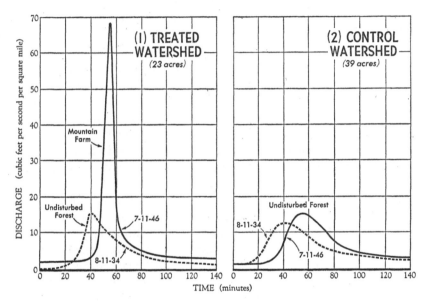

Figure 5. Influence of Forest Cover on Stream Flow. Note on cover: (1) treated watershed (1946 only)—26 per cent too rough for agriculture; (2) control watershed (1934 and 1946)—entirely wooded.

A special characteristic of runoff conditions in the United States is the size of the drainage basins. The basin of the Colorado covers 246,000 square miles, and that of the Columbia, 259,000. Each is considerably larger than the total land area of any nation in western Europe. The drainage basin of the Missouri, only a tributary of the Mississippi, is 530,000 square miles. The Mississippi Basin as a whole

[11] See H. L. Cook, "The Effects of Land Management upon Runoff and Ground Water," *Proceedings of the United Nations Scientific Conference on the Conservation and Utilization of Resources,* Vol. IV, Water Resources (New York: U. N. Department of Economic Affairs, 1951), pp. 193-202.

covers 1,250,000 square miles, its main stream extending over 3,860
miles and carrying a third of the entire United States runoff. The tributaries of the Mississippi touch thirty-one different states.

Even secondary basins of the country are of considerable size. The
basins of the Tennessee, the Trinity, or the Brazos in Texas, for
example, all cover 30,000 square miles or more. In view of the notably
close relation between hydrologic events, not only upstream and downstream in a basin but also between surface and ground water, the size
of unified drainage areas in the United States is most significant.

Ground-water distribution. The pattern of ground-water availability
reflects in part, but not entirely, the pattern of precipitation and
evapotranspiration. Indeed, ground water reflects a pattern of availability
in some respects the obverse of the runoff pattern in the United States
(figure 6). The pattern of ground-water distribution can be clearly
described only with an understanding that there are two component
parts of present ground-water supplies. One part is included in the
atmosphere-ground-surface-atmosphere circulation (hydrologic cycle)
continually taking place. Another much larger part has been trapped
within the earth from past precipitation and no longer is part of circulation within the hydrologic cycle, except insofar as man brings it to the
earth's surface. Together they are enormous, comprising perhaps 48
billion acre-feet of "usable water",[12] thirty-five times the annual runoff,
and ten times the total precipitation received on United States territory
in a year. In addition there are further great quantities of salt water
stored underground, particularly in the formations favorable to petroleum
occurrence. These latter deposits do not enter discussion here, except
insofar as they may have created local water supply problems where
their waters have been released.

Geological structure is an important determinant in the location of
the ground-water resources considered usable in the United States.
Most localities in the eastern part of the country, underlain by unconsolidated sediments of moderate depth, contain some usable ground

[12] H. E. Thomas, "Underground Sources of Water," *Water, The Yearbook of
Agriculture, 1955,* U. S. Department of Agriculture, Washington, 1955, p. 62.
"Usable water" means that which is of sufficiently low mineralization to be employed under existing technology, if it can be lifted to the surface at economically
feasible cost. However, great quantities would not be available at present costs
of lifting and present levels of demand.

Figure 6, opposite. Major Aquifers in the United States.

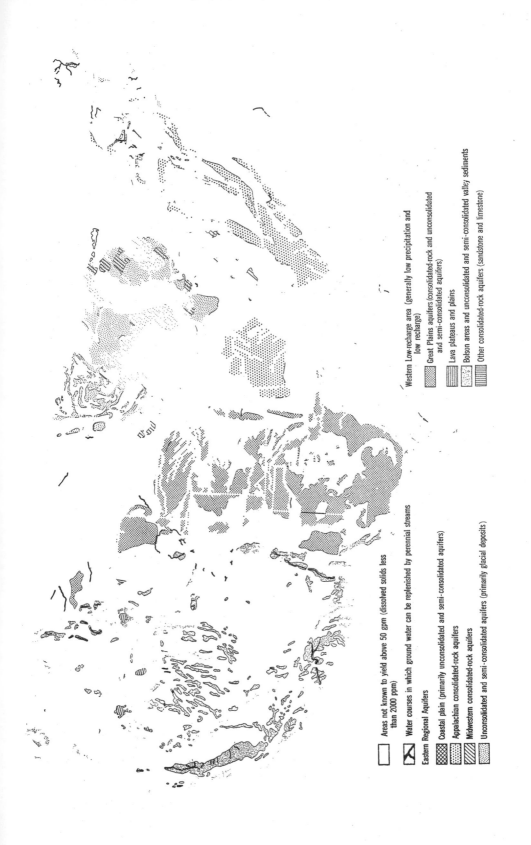

Areas not known to yield above 50 gpm (dissolved solids less than 2000 ppm)

Water courses in which ground water can be replenished by perennial streams

Eastern Regional Aquifers

Coastal plain (primarily unconsolidated and semi-consolidated aquifers)

Appalachian consolidated-rock aquifers

Midwestern consolidated-rock aquifers

Unconsolidated and semi-consolidated aquifers (primarily glacial deposits)

Western Low-recharge area (generally low precipitation and low recharge)

Great Plains aquifers (consolidated-rock and unconsolidated and semi-consolidated aquifers)

Lava plateaus and plains

Bolson areas and unconsolidated and semi-consolidated valley sediments

Other consolidated-rock aquifers (sandstone and limestone)

water. However, productive deposits which might be used as sources
of supply for irrigation, for domestic purposes, or for an industrial
plant are much more limited. The formations capable of yielding 50
gallons per minute or more to an individual well with total dissolved
solids 2,000 ppm or less, have been mapped by the United States
Geological Survey (figure 6). Their distributional pattern on first
glance appears disordered. However, several clear geological correlations can be discerned. One is the association of ground water with
the flood plain alluvium of major streams, especially in the eastern
part of the country. Another is the presence of ground water in the
bolsons and other detritus-filled depressions lying adjacent to mountains
in the western states. A second type of western aquifer is the flat-lying
volcanic rocks, or lava beds, found in Oregon, California, Washington,
Idaho, Wyoming, Arizona, and New Mexico. Other types of aquifers
are much more regionalized. They include the sedimentary rocks of
the Great Plains, Appalachian Valley, and the Midwest; the glacial till
deposits of the Lakes States; and the great aquifer in the unconsolidated
sediments of the Atlantic and Gulf coastal plains and the lower Mississippi Valley. With certain minor exceptions, it can be seen that the
mountain regions of both East and West are not the sites of significant
ground-water supplies;[13] neither are the northern Great Plains; nor are
the predominantly granitic and metamorphic rock areas of New England and the southern Piedmont.

These several aquifer types have varying capacities for water production related to their extent, their rates of recharge, and their vulner-
ability to the incursion of low-quality water. Roughly, one may classify
the aquifers into four groups:

    1) Those with close intercommunication with surface-water sup-
    plies, in which the aquifer is fed by gravitational water[14] and releases

[13] Because of rapid runoff, massive impermeable formations, and elevated position which discourages retention of ground waters.

[14] Water beneath the earth's surface is found in two distinct zones divided by
a hypothetical surface called the water table. As commonly observed, the water
table is the surface represented by the levels to which free water will rise in
unpumped wells. Beneath the water table all permeable formations are saturated
with water. The percentage of water contained may vary at different depths due
to differing porosities of the rocks, but beneath the water table every interstitial
space in rock is filled with water. A "zone of aeration" exists above the water
table, where saturation with water is usually not encountered, and interstitial
space in soil or rock may be filled with air or water. Where water receipts from
the atmosphere are sufficient, water may percolate through the zone of aeration
to reach the zone of saturation. Such percolating water is gravitational water.

water freely to surface flow. Flood plain and river valley gravels are good examples. The characteristics of these aquifers are very directly dependent upon the size and seasonality of the surface flows with which they are related. These aquifers are most numerous and most productive in the eastern United States, although some are found in the Pacific Northwest. In their management needs, they must be considered as almost coextensive with surface flows. Tributaries of the Ohio, the lower Missouri and the lower Mississippi are related to some of the more extensive flood plain deposits of this type.

2) The more extensive "regional" aquifers which occur in the United States east of the 100th meridian. They have moderate to high rates of recharge and over extended areas offer some of the largest permanent ground-water yields in the country. One of the best of these is the potentially very productive aquifer associated with the sands and other formations of the Atlantic and Gulf coastal plains.

3) The aquifers in the low recharge area between the 100th and the 120th meridians, which includes much of the West. Relative to the draft which could be or has been placed upon them, these aquifers have very little inflow. They may contain large volumes of water, but this water must be considered basically a minable material. With a few exceptions, the water in these deposits is not the "renewable resource" which we usually consider water to be. Experience in southern Arizona and California, and the probable course of events in the High Plains of Texas, suggest that use of these deposits very quickly makes them more expensive to exploit, and places them in a "limited life" category. They offer special problems of management and exploitation policy which have not as yet been treated adequately. The nature of their exploitation may help to determine the extent and type of western development, and the creation of water development problems with repercussions on the nation.[15]

4) Within each of the three foregoing ground-water deposit types, areas are found which possess special limitations on exploitation. They are those which, as they are exploited, may be made either temporarily or permanently unusable, by the incursion of highly mineralized water normally confined to adjacent strata or water bodies. Most typically, these susceptible areas are adjacent to the sea, where heavy use of ground water, or artificial drainage may cause a backflow of sea water into the aquifer. This has happened in New Jersey, New York, Florida, and elsewhere on the Atlantic Coast; in Louisiana and elsewhere on the Gulf Coast; and in California. Susceptible aquifers also have been found in Arizona, and other parts of the dry West may be affected if exploitation of local aquifers becomes heavy (figure 7).

---

[15] It may be noted that almost any aquifer may be depleted if draft upon it is heavy enough. A good example occurs in the Chicago area, where the relatively productive aquifers have been much depleted by heavy pumping.

Figure 7. Areas of Known Excessive Perennial Ground-water Withdrawal.

Some of the most extensive susceptible areas are in neither of these situations. Relatively large sections of the country are underlain by beds containing vast quantities of salt water. These occur in oil fields and elsewhere, and probably are found more commonly beneath the surface in eastern United States than in the West. Wherever they occur there is danger of contaminating fresh-water aquifers. This has been widely experienced in Kansas, and currently presents problems in Oklahoma, Michigan, Louisiana, Texas, Arkansas, Indiana, Kentucky, New Mexico, and South Dakota.[16]

INHERENT PHYSICAL AND CHEMICAL PROPERTIES

In noting the unusability of sea water, the existence of vast underground salt-water deposits, and the reduced value of fresh-water aquifers that lie adjacent to salt-water bodies, some important variations among different aquifers, streams, and water bodies, of significance in water use and development, have been implied.

The principal variations in water found on the earth's surface are

[16] Gerald G. Parker, "The Encroachment of Salt Water into Fresh," in *Water, The Yearbook of Agriculture, 1955,* pp. 615-35.

due mainly to its superior capacity as a solvent and as a medium for transporting solids in suspension. Water in different sections of the United States (or elsewhere) may vary in "quality" according to the amount of solids which it carries either in suspension or in solution. Solids also are carried as bed load by surface streams.

The suspended materials, mainly sediment of soil mineral origin[17] are a property of surface water only. Variation in mineral sediment content is roughly a function of the vegetative cover of the watershed contributing runoff to a stream or lake, the profile of a stream, the geological formations over which the water flows, and the presence or absence of meteorological conditions that cause freshets. Especially in this century, cultivation systems have so greatly altered the natural vegetative cover that they also must be considered among the influential factors determining the mineral sediment content of water. In general, the mineral sediment content of surface waters in the United States is greatest in the arid or semi-arid sections. It also tends to be relatively large where water flows over loosely consolidated sedimentary rocks, as in the Missouri Basin, or over unconsolidated marine sediments, as on the Atlantic and Gulf coastal plains (figure 8). Finally, mineral sediment content recently has become large where cultivation or grazing has removed a large share of the natural vegetation, and some of the natural conditions favoring sedimentation of water are found, such as intense rainfall and sloping terrain. These conditions are present in the southern Piedmont area, the valley of the Tennessee, parts of the Midwest, the Rio Grande Valley, and elsewhere. From the point of view of the development and control of water, it is significant that mineral sediment contributions to stream flow or natural and artificial reservoir storage may move over great distances. The sediment content of a stream may be dependent upon physical conditions a hundred or more miles away.

The *dissolved* mineral solids are of even greater importance in the use of water. They affect the quality of water wherever it is found on or within the earth. Both ground and surface water vary in their dissolved solids content. No fewer than fifty elements other than oxygen and hydrogen (and the isotopes deuterium and tritium) have been identified in sea water.[18] The oceans, which are great repositories

---

[17] Hereinafter called "mineral sediment."

[18] Aluminum, antimony, arsenic, barium, bismuth, boron, bromine, cadmium, caesium, calcium, carbon, cerium, chlorine, chromium, cobalt, copper, fluorine,

Figure 8. Average Suspended Sediment Load of Selected Streams in the United States. [Ratio of total annual load (tons) to total discharge (second-foot-days).]

of dissolved elements moved by water from the rocks of the land, have a relatively high dissolved mineral content, about 3.5 per cent or 35,000 parts per million. Of the component elements in the dissolved solids, sodium and chlorine are by far dominant (10,769 ppm,

gallium, germanium, gold, iodine, iron, lanthanum, lead, lithium, magnesium, manganese, mercury, molybdenum, nickel, nitrogen, phosphorus, potassium, radium, rubidium, scandium, selenium, silicon, silver, sodium, strontium, sulfur, thorium, tin, titanium, uranium, vanadium, yttrium, zinc, zirconium (H. W. Harvey, *The Chemistry and Fertility of Sea Waters* [Cambridge, England: Cambridge University Press, 1955], pp. 140-41; H. V. Sverdrup, M. W. Johnson, and R. H. Fleming, *The Oceans: Their Physics, Chemistry, and General Biology* [New York: Prentice-Hall, 1942], p. 220). It is not unlikely that all elements are present in sea water, at least in trace amounts.

and 19,353 ppm respectively). Magnesium (1,350 ppm), sulfur (901 ppm), calcium (408 ppm), and potassium (387 ppm) are of secondary incidence, and all other elements are present in small or minute quantities.[19] The dissolved solids content of the oceans is exceeded or equaled only by bodies of surface water on the continents. Some tributaries of the Pecos in Texas are reported as having a chloride content almost two and one-half times that of sea water,[20] and various analyses of Great Salt Lake water have indicated a dissolved solids content as high as 277,200 ppm, and never less than 137,900 ppm.[21] One of the saltiest bodies of water on earth, the Dead Sea, at its surface has a saline content six times (210,000 ppm) as concentrated as the ocean. Its salinity increases with depth.[22] Brines contained in large quantities within underground strata, like those underlying part of the Michigan lower peninsula, or associated with southwestern oil fields, are like the saltiest surface waters.[23] Brines brought to the surface incidental to Kansas oil production are reported to run as high as 248,000 ppm total solids, containing 125,000 ppm chlorides.[24] In South Dakota brines containing 140,000 ppm total solids, including 83,000 ppm chlorides, have been reported.[25] In very localized situations, like hot springs, unusually high mineral content may be present, although not necessarily dominated by chlorides. Sulfur hot springs (which usually contain dissolved hydrogen sulfide and the sulfides of other elements) are probably the most frequent example. In places, underground brines or salt deposits are near enough the surface to affect the salinity of surface-water supplies. In a past analysis, the dissolved solids content of Saline River water in Kansas, for instance,

[19] Sverdrup, Johnson, and Fleming, *ibid.*

[20] Parker, *op. cit.*, p. 626.

[21] F. W. Clarke, *The Data of Geochemistry* (Washington: U. S. Government Printing Office, 1908), pp. 170-71.

[22] P. H. Kuenen, *Realms of Water: Some Aspects of Its Cycle in Nature* (New York: Wiley, 1955), p. 230.

[23] The total dissolved solids of Michigan brines are frequently found to be in excess of 200,000 ppm. C. W. Cook, *Brine and Salt Deposits of Michigan, Their Origin, Distribution and Exploitation* (Michigan Geological and Biological Survey, Pub. No. 15), Lansing, Mich., 1914.

[24] U. S. Public Health Service, *Kansas River Basin Water Pollution Investigation*, Washington, 1949, p. 44.

[25] South Dakota School of Mines and Technology, Engineering and Mining Experiment Station, *Inventory of Published and Unpublished Data on the Characteristics of Saline Surface and Ground Waters of South Dakota*, 1954, p. 32.

was found to be 2,624 ppm, while that of other Kansas streams ranged between 340 and 880 ppm.[26]

On the whole, waters found on land, or in the shallow-lying aquifers underground, have a relatively low dissolved solids content. This also may be true for substantial amounts of deep-lying water. "Hard" water in the Great Lakes region has a dissolved solids content of about 100-150 ppm. The Mississippi River is within the same range (166 ppm).[27] A relatively saline major stream by United States standards, the Colorado, has about 700 ppm. Higher analyses, of course, are found. The Rio Grande at Fort Quitman, Texas, in a recent year averaged about 1,770 ppm,[28] and small streams in the Great Basin may be found with even higher dissolved solids content. In 1956 the Red River in Texas, from which the city of Dallas obtains part of its water supply, was reported as having a content of 3,696 ppm.[29] At the other extreme is water like that in Lake Superior, which has a solids content of 50 ppm, a very pure water from a large concentrated supply. With some exceptions, water in the United States east of the 100th meridian, and particularly from the Mississippi eastward, has a low dissolved solids content. The exceptions are the saline underground waters, such as underlie Kentucky and Michigan, and certain industrially caused saline conditions, as on the Holston River downstream from Saltville, Virginia, and on Ohio's Muskingum River over most of its length.

On the other hand, high dissolved minerals content tends to characterize streams west of the 100th meridian, particularly after they leave their mountain sources. Exceptions to this pattern occur in the mountain sections themselves, and in the Pacific Coast north of the San Francisco Bay area and west of the Cascades and Sierra Nevada (figure 9 and table 3).

Compared to the remarkable uniformity of the oceans' mineral content, the waters of the land show great variation. The principal differences center on the content of sodium, chlorine, calcium, carbon, magnesium and sulfur. Minor but locally important differences may

[26] Clarke, *op. cit.*, p. 85.

[27] Fox, *op. cit.*, p. 14; Sverdrup, Johnson, and Fleming, *op. cit.*, p. 216.

[28] M. E. Fireman and H. E. Hayward, "Irrigation Water and Alkaline Soils," *Water, The Yearbook of Agriculture, 1955*, p. 321.

[29] Letter from Ben L. Grimes, Supervisor, Water Purification Plants, City of Dallas Department of Water Works, October 5, 1956. See also n^3, page 336 for acceptability of saline waters for various uses.

be caused by content of iron, manganese, boron, silicon, arsenic, zinc, selenium, lithium, and fluorine. Surface waters and shallow-lying ground water in arid and semi-arid sections of the United States in general are characterized by higher sodium, chlorine, magnesium, and sulfur content than waters in more humid sections of the country. The presence of these elements in arid-region water is of more than casual interest because if they exist in excessive quantities they may determine the usability of water by plants, animals, and man. Saline water[30] occurs

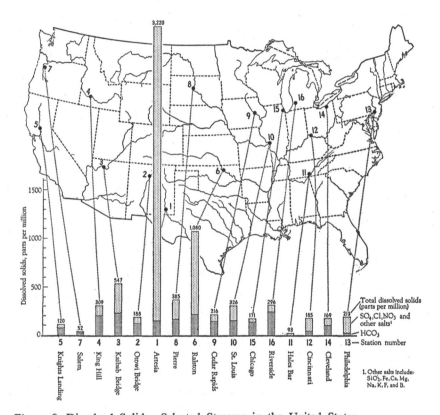

Figure 9. Dissolved Solids—Selected Streams in the United States.

[30] In general, water with a dissolved solids content above 1,000 ppm may be described as saline, particularly if sodium and sulfates are important elements among the dissolved solids. (See note [3], chapter 13.) "Hard" water, on the other hand, is caused principally by the presence of calcium and magnesium ions, which combine with carbon and oxygen to form soluble bicarbonates. "Hard" water may have as low a dissolved solids content as 150 ppm.

TABLE 3. *Water Quality: Chemical Analysis of Representative Surface Water Supplies in the United States*

(in parts per million)

Station numbers on figure 9	1	2	3	4	5	6	7
Station location	Pecos River at Artesia, N. M.	Rio Grande, at Otowi Bridge, near San Ildefonso, N. M.	Colorado River, at Kaibab Br., near Bright Angel Creek, Ariz.	Snake River, at King Hill, Idaho	Sacramento River, at Knight's Landing, Calif.	Arkansas River, at Ralston, Okla.	Willamette River, at Salem, Ore.
pH factor	n.d.	n.d.	n.d.	n.d.	n.d.	n.d.	n.d.
Silica (SiO_2)	16.0	21.0	15.0	31.0	22.0	n.d.	16.0
Iron (Fe)	n.d.	n.d.	n.d.	n.d.	n.d.	n.d.	n.d.
Calcium (Ca)	472.9	39.1	84.0	46.1	13.0	101.0	4.8
Magnesium (Mg) ...	108.0	5.6	23.0	19.0	7.8	28.9	1.9
Sodium (Na)	438.2	14.0	63.9	29.0	12.0	}239.2{	3.7
Potassium (K)	n.d.	n.d.	5.5	4.7	1.6		1.2
Bicarbonate (HCO_3) .	158.0	125.0	233.0	208.0	81.7	219.0	26.8
Sulfate (SO_4)	1,429.0	41.8	178.1	49.9	13.9	204.0	2.9
Chloride (Cl)	675.9	4.3	46.2	24.1	7.5	350.4	2.5
Fluoride (F)	n.d.	n.d.	0.38	0.57	0.19	n.d.	0.38
Nitrate (NO_3)	3.7	1.9	3.1	3.1	0.6	3.7	0.6
Total dissolved solids ..	3,220	188	547	309	120	1,080	52
Total hardness as $CaCO_3$	n.d.	n.d.	n.d.	n.d.	n.d.	n.d.	n.d.
Specific conductance (micromhos at 25°C)	4,250	816	295	486	175	1,800	54
Period (calendar year except where otherwise stated)	◄——————————— Oct. 1951–Sept. 1952[a] ———————————						

n.d.=no data.

Sources:

[a] U. S. Geological Survey, *Quality of Surface Waters for Irrigation, Western United States, 1952* (Water Supply Paper No. 1362), Washington, 1955.

8	9	10	11	12	13	14	15	16
Missouri River, at Pierre, S. D.	Cedar River, at Cedar Rapids, Iowa	Mississippi River, at St. Louis, Mo., 5 miles below confluence with Missouri River	Tennessee River, at Hales Bar Reservoir, near Chattanooga, Tenn.	Ohio River, at Cincinnati, Ohio	Schuylkill River, at Philadelphia, Pa.	Lake Erie, at municipal water intake for Cleveland, Ohio	Lake Michigan, at intake for South District Filtration Plant of Chicago, Ill.	Paw Paw River, near Riverside, Mich.
n.d.	n.d.	7.9	7.6	7.0	7.2	6.6	8.2	7.6
n.d.	12.0	13.0	6.0	7.0	8.3	2.4	2.3	10.0
n.d.	0.07	n.d.	0.4	0.04	0.05	0.12	0.09	0.45
50.9	41.0	50.0	19.2	31.0	31.0	39.0	32.0	54.0
16.1	13.0	14.0	4.8	8.0	14.0	7.3	10.0	20.7
52.0	8.8	{35.0}	5.3	12.0	{9.0}	8.7	3.5	{13.7}
n.d.	2.4		n.d.	3.5		1.3	1.0	
170.2	151.0	156.0	n.d.	46.0	46.0	103.0	138.0	244.0
151.7	31.0	97.0	n.d.	81.0	95.0	30.0	17.0	36.0
6.7	12.0	16.0	9.0	14.0	9.8	20.0	6.5	10.0
n.d.	0.20	n.d.	n.d.	0.20	0.10	0.10	0.10	0.0
1.9	6.9	4.6	1.8	2.4	6.2	1.5	n.d.	none
385	216	326	93	185	213	169	171	296
n.d.	156	183	66	110	135	128	121	220
583	338	n.d.	n.d.	304	332	286	263	500
⟶ Oct. 1948–Sept. 1949[b]		1950[c]	1947[d] 1951[c]	1951[c]	1951[c]	1951[c]	1952[c]	1954[c]

[b] U. S. Geological Survey, *Quality of Surface Waters of the United States, 1949* (Water Supply Paper No. 1162), Washington, 1954.

[c] U. S. Geological Survey, *Industrial Utility of Public Water Supplies in the United States 1952* (Water Supply Papers Nos. 1299 and 1300), Washington, 1954.

[d] Tennessee Valley Authority, *Industrial Water Supplies of the Tennessee Valley Region,* Knoxville, Tenn., 1948 (mimeo).

[e] State of Michigan, Water Resources Commission, *Report on Water Resource Conditions and Uses in the Paw Paw River Basin,* Lansing, Mich., 1955 (mimeo).

sporadically within the Great Basin and other arid areas of the United States.

Water which is high in bicarbonate content may be found in both humid and arid sections of the country; this condition depends upon the physical relation of the water to calcium- or magnesium-bearing rocks like limestone or dolomite, and the presence of carbonic acid. Relatively high calcium content is found in surface and ground water over a large section of the Midwest and the interior Southeast, as well as being frequently encountered in the arid and semi-arid West. It probably is the geographically most frequent cause of "hard" water, reducing the water's value for the many industrial processes affected by lime, and for household solvent uses such as bathing and laundering. Its agricultural effects, however, are favorable rather than unfavorable.

The remaining elements which in solution may impart other qualities to water are not as widespread in their occurrence as the foregoing.

Iron occurrence of harmfully high concentration in water is geographically sporadic, although known in all major sections of the country. It is dependent upon local geological conditions. The same may be said for each of the remaining secondary elements which may be found in water, like boron, manganese, zinc, or fluorine. Each has some significance for man's use, like iron, which is undesirable in the food processing industries; or manganese, undesirable in textile processing; and fluorine, lately discovered as a desirable dental caries inhibitor[31] (considered harmful when present in amounts above 4 ppm).

There are still other physical properties which vary with the nature of the source and the geographical origin of a body of water. Water varies in its content of dissolved gases, in the hydrogen ion concentration, in natural radioactivity, and in temperature.

The most important gases present in water are oxygen and carbon dioxide, essential respectively to animal life and to plant life within the water. They vary according to opportunity for aeration, frequency of precipitation, organic content, presence of carbonates, and in response to other environmental conditions. Hydrogen ion concentra-

[31] While great differentials can be noted within various regions of the United States in the prevailing standards regarding water quality, the above observations would probably command consensus in this field as broadly applicable. Cf., for example, Edwin W. Taylor, *The Examination of Waters and Water Supplies* (6th ed.; Philadelphia: Blakiston, 1949), p. 282.

tion (acidity or alkalinity) varies in the presence of sodium, magnesium, calcium and other elements, and with conditions favoring or discouraging the formation of carbonic acid.

Temperature is a physical property in which ground water and surface waters are distinguished from each other. Shallow surface waters follow the adjacent atmosphere in their temperatures, although without the same sharp fluctuations over most of the country; ground water in general is much more nearly isothermal. In summer ground-water temperatures are generally cooler than surface-water temperatures; in winter the opposite is true. These differences between surface and ground water are significant wherever water is used as a medium of heat exchange. Some ground waters are of high temperature. Where found they have been used for medical, recreational, and even industrial power purposes. This is true particularly in European countries, Iceland, Japan, and New Zealand.

ORGANIC AND MICROBIOTIC CONTENT

Waters also vary in their content of living microorganisms, and in the content of both suspended and dissolved substances which are waste from life processes.

It is far more difficult to appraise the organic and microbiotic character of the surface-water supplies of the United States than to summarize their mineral sediment or chemical quality. Water's organic attributes are transitory and are constantly undergoing a dynamic process of change. These aspects of water quality accordingly are highly localized in character. Analysis is complicated because no single measure exists which adequately reflects the organic condition of water.

No naturally occurring water is completely pure. Rainfall contains certain dissolved atmospheric gases and it gathers other impurities, including floating dust, the air-borne debris of combustion, bacterial life, and the spores of fungi, before it reaches the earth's surface. Water thereupon begins to carry in suspension or solution not only inorganic materials but also organic residues of plant and animal life. As it moves over land surfaces it may pick up the waste products from vegetative processes and become host to water-loving organisms.

Finally, it becomes altered with varying loads of organic and inorganic wastes which are supplied to it through the activities of man.

The suspended materials found in water thus have a content of living microorganisms along with the mineral sediment. Bacteria and other plankton, both plant and animal, are present in most water. They vary widely in amount and, in the case of algal growth, they may reach great concentrations within specific waters. Although normally absent or present in very small numbers, bacteria harmful to man may and do find water a favorable medium for survival when introduced in large quantities from "polluting" sources. The capacity of water to carry bacteria, algae, and other planktonic plants, as well as mineral material in solution or suspension, is obviously of considerable significance in exploiting surface and ground waters for human consumption.

Organic matter present in water undergoes decomposition and putrefaction through the action of bacteria. Bacteria, in decomposing organic impurities, use for their respiratory and metabolic activities some of the atmospheric oxygen which is dissolved and held in solution in water. Aquatic life similarly draws upon this supply of oxygen for respiration. Green plants and planktonic animals cannot long survive in water devoid of a minimum supply of oxygen, although wherever green plant growth occurs oxygen is liberated as a product of the photosynthetic process. There also is a tendency for the oxygen supply to be renewed by aeration at the water surface. In this respect, water can be viewed as a biochemical medium for the disposal of organic waste materials. In general, as the amount of organic matter present in water increases, the metabolic activities and the growth of bacteria are intensified. However, because oxidation is then accelerated, the supply of dissolved oxygen is depleted more rapidly. Only a fixed amount of oxygen can be dissolved by water at a given temperature and atmospheric pressure through the process of aeration. The greater the deficiency of dissolved oxygen, the more rapidly does the oxygen supply of water tend to be renewed. Within certain clearly defined limits, a balance is thus struck. Oxygen depletion in the process of oxidizing organic matter creates conditions encouraging compensating oxygen renewal in the water.

It is therefore possible to evaluate the physical and chemical effects of the presence of varying amounts of organic matter upon the condition of water by determining this oxygen balance. The ratio of the amount of dissolved oxygen which water possesses to the saturated amount, expressed as the per cent of saturation, suggests the extent to which its oxygen supply has been depleted, which in turn is of significance in appraising its ability to handle further loads of organic matter. Generally any value below 75 per cent suggests the existence of "pollution" which makes the water undesirable for human consumption. The lower the percentage the more unsuitable the water is.

One of the most useful measures in appraising water quality is the Biochemical Oxygen Demand. B.O.D. indicates the amount of oxygen required in the future for the biochemical oxidation of the organic matter present in a sample of water. This measure is generally standardized to express the oxygen demand that will be experienced when the water is incubated for a five-day period at 68°F. Water not absorbing more than 1 ppm of oxygen in five days can generally be considered very pure, 3 ppm suggests fairly clean water, while water absorbing 5 ppm or more of oxygen is of doubtful purity. In general, the more oxygen required, the more serious the "pollution" indicated.[31] The considerable variation of B.O.D. at selected stations scattered over the United States is illustrated in table 4. Variations occur at the same station over time, and geographical variation among different stations also is evident.

The most probable number of coliform organisms, another measure included in table 4, is useful in detecting the presence and estimating the extent of sewage pollution. These bacteria, derived from the fecal wastes of humans and animals, serve as effective indicators of the possible presence of disease-producing bacteria and other pathogens. The greater their number, the higher is the degree of contamination that is indicated. The density of coliform organisms is relevant in ascertaining the sanitary hazard of water supplies for domestic purposes and for bathing and recreational uses. This measure also suggests the extent of the need for purification. As illustrated in table 4, great geographical and temporal variation also is experienced in this report among different waters in the United States.

The temperature of water must also be considered when evaluating its organic condition. The warmer the water, the more rapidly does the process of oxidation take place; on the other hand, water at 32°F. can

---

[31] While great differentials can be noted within various regions of the United States in the prevailing standards regarding water quality, the above observations would probably command consensus in this field as broadly applicable. Cf., for example, Edwin W. Taylor, *The Examination of Waters and Water Supplies* (6th ed.; Philadelphia: Blakiston, 1949), p. 282.

TABLE 4. *Organic Pollution of Surface Water at Selected Stations*

River, location where sampled, and date	(¹)	Number of samples	Discharge in cfs	Temperature (°F)	Dissolved oxygen (per cent saturation)	Five-day biochemical oxygen demand (ppm)	Coliform organisms (most probable number per 100 ml)
a. Kennebec River, at Gardiner, Me., 1945-47	Average	n.d.	n.d.	72	34	4.2	24,000
b. Housatonic River below Great Barrington, Mass., August 16, 1935 to June 10, 1936	Average	7	n.d.	n.d.	74	2.2	n.d.
	Maximum		n.d.	n.d.	83	3.2	n.d.
	Minimum		n.d.	n.d.	67	0.9	n.d.
c. Yadkin River, N. C., at Station No. 6; 195 miles from N.C.-S.C. line, June 19, 1952 to August 28, 1952	Average	9	594	79	87	1.3	36,200
	Maximum		870	84	98	3.0	46,000
	Minimum		480	68	86	0.9	24,000
Same at Station No. 49, main-stream 11 miles from N.C.-S.C. line, October 1, 1952 to November 13, 1952	Average	7	5,236	50	81	0.9	220
	Maximum		7,730	75	88	1.3	9,100
	Minimum		2,410	34	74	0.6	14
d. Tennessee River at Hales-Bar Reservoir above Chattanooga, Tenn., summer of 1936	Average	22	15,300	79	83	1.1	1,400
	Maximum		29,200	88	90	1.8	7,000
	Minimum		9,400	68	69	0.4	20
Same, winter of 1936-37	Average	13	44,700	52	89	1.2	3,100
	Maximum		160,000	66	91	2.4	11,000
	Minimum		11,000	43	85	0.5	160
e. Ohio River, at Dam No. 38; 503 miles below Pittsburgh, Pa. and 42 miles above Carrollton, Ky., May 4, 1939 to November 30, 1939	Average	22	n.d.	70	[2] 84	2.6	76,620
	Maximum		n.d.	81	[2] 94	3.4	201,000
	Minimum		n.d.	45	[2] 54	1.7	930
f. Red River, Station No. 4—Highway Bridge at Oslo, Minn., 1950-51	Summer Av.	4	2,490	64	83	0.8	24,000
	Winter Av.	3	1,930	32	78	2.1	17,000
g. Kansas River, at bridge east of North Topeka, Kan., May 20, 1948 to December 13, 1948	Average	19	6,100	61	82	3.6	99,000
	Maximum		26,700	79	96	5.0	1,100,000
	Minimum		1,585	36	66	2.3	9,300

River, location where sampled, and date	([1])	Number of samples	Discharge in cfs	Temperature (°F)	Dissolved oxygen (per cent saturation)	Five-day biochemical oxygen demand (ppm)	Coliform organisms (most probable number per 100 ml)
h. Walla Walla River, Wash., above confluence with Columbia River, August 15, 1951 to May 12, 1952	Average	17	n.d.	48	[2][3] 78	4.8	3,333
	Maximum		n.d.	70	[2] 83	7.1	11,000
	Minimum		n.d.	32	[2] 65	0.8	430
i. Colorado River, 4 miles below Parker Dam, May 1951 to December 1952	Average	20	20,600	68	[2] 87	n.d.	1,600
	Maximum		23,000	81	[2] 102	n.d.	23,000
	Minimum		13,500	48	[2] 73	n.d.	3

n.d. = no data.

[1] Although the organic character of water at any one point of examination is highly individualized, data have been selected to suggest conditions within the larger area where they are situated. Averages are given in all cases; they have been supplemented where possible with the maximum and minimum reported values. These figures constitute the reported maximum and minimum measures for each reported item, but collectively they do not represent the condition of the water at any one time.

[2] Computed from data provided in published report expressing the dissolved oxygen content in ppm, the presence of chlorides being ignored and a barometric pressure of 760 mm being assumed.

[3] Based upon four summertime samples only.

Sources:

a. State of Maine, Department of Health and Welfare, Division of Sanitary Engineering, *Report on Water Pollution in the State of Maine, 1950* (Augusta, Me., 1950, mimeo.), p. 51.

b. Works Project Administration and Massachusetts Department of Health, *Report on Sources of Pollution—Housatonic River Valley* (Boston, Mass., 1946, mimeo.), p. 89.

c. North Carolina State Stream Sanitation Committee, State Board of Health, *Yadkin River Basin—Pollution Survey Report No. 1* (Raleigh, N. C., 1952, mimeo.).

d. State of Tennessee, Stream Pollution Study Board, *Stream Pollution in Tennessee, 1943-44* (Nashville, Tenn., undated).

e. Report of the U. S. Public Health Service, *Ohio River Pollution Control; Supplements to Part Two* (House Document No. 266, 78th Congress, 1st Session, 1944), p. 1351.

f. U. S. Public Health Service, Division of Water Pollution Control, *Missouri-Souris Development Area—Water Pollution Investigation, 1952* (Water Pollution Series No. 32, mimeo.), p. 245.

dissolve almost twice as much oxygen as at 86°F.[32] The resulting effect of temperature upon water quality is a complex one, as is the influence of the volume of stream flow; yet these are two important additional factors which must be taken into account in such assessments.

Recognizing the high degree of local variation which occurs in the organic content of water, a few summary comments may be made about geographical occurrence of organic materials in water which raises problems for human consumption.

1) Streams having very low flows because of climatic, vegetative, water-use, or geological conditions are likely to exhibit very high concentrations of coliform bacteria, high B.O.D., and low dissolved oxygen content when and where untreated wastes are introduced into them during low flow periods. Such streams are most apt to be found in arid and semi-arid parts of the country, although by no means limited to them. The Kansas River and Walla Walla River stations (table 4) are illustrative.

2) Many streams exhibit undesirable low dissolved oxygen concentrations at some time during the course of their flow. They show little geographical pattern.

3) A number of streams, although fewer, exhibit an undesirable B.O.D. at some time during the course of their annual flow. Again this is most apt to be experienced under extreme low flow conditions.

4) Environmental conditions favoring high coliform bacteria concentrations and other features of organic pollution are most apt to be found within or in the vicinity of densely settled areas. That is, intensive use for waste disposal will create the same conditions as described for low flows in (1) above, whatever the region.

The organic content of water of greater interest to most water users is that of the microorganisms harmful to people's health, but there are

Sources for Table 4 (cont.)

g. U. S. Public Health Service, Division of Water Pollution Control, *Kansas River Basin—Water Pollution Investigation, 1949* (mimeo.), pp. 257-8.

h. U. S. Public Health Service, Division of Water Pollution Control, *Water Quality Studies on the Columbia River*, Cincinnati, Ohio, 1954, pp. B-31-32.

i. Colorado River Basin Regional Water Pollution Control Board, *Appendix: Report on Water Pollution Study of the Colorado River Basin Region of California, 1953* (mimeo.), p. 3.

[32] American Public Health Association, American Water Works Association, and Federation of Sewage and Industrial Wastes Association, *Standard Methods for the Examination of Water, Sewage, and Industrial Wastes* (10th ed.; New York: American Public Health Association, 1955), p. 254.

other variations in organic content which exhibit both geographical variation and economic significance. While of little concern to domestic water consumers, concentrations of dissolved organic matter from vegetative sources significant to industrial use may be found wherever marshes, bogs, or swamps occur in numbers. Undrained glacial terrain (Wisconsin, Minnesota, New England) and poorly drained parts of the Atlantic coastal plain, the Gulf coastal plain, and the lower Mississippi valley offer such conditions. Lesser amounts of dissolved organic matter are contained in many other streams. A characteristic feature of such water is its brownish tint.

The capacity of water for quality change. As has been suggested in connection with the discussion of organic and microorganism content, one of the most remarkable properties of water is its capacity to change under use. Sediment content of water may change within a year, or even a week, if a watershed is denuded or a sediment-producing farming or grazing system introduced. The introduction of raw organic wastes or chemicals in quantity within a period of days may increase the biological oxygen demand to the point where the actual content of oxygen in the water is near zero. A single industrial district may increase the temperature of a substantial stream in some special situations by 50 to 100 degrees Fahrenheit. The Mahoning River at Youngstown, Ohio, was reported to have temperatures approaching 140 degrees Fahrenheit in the winter of 1949,[33] and 117 degrees Farenheit in July, 1941.[34]

The acidity or alkalinity of a stream may be altered for hundreds of miles downstream by industrial waste discharge, and the dissolved solids content may be changed on a regional scale by the evaporation and leaching processes associated with irrigation of arid land. This latter event is illustrated by analyses of water from two rivers, the Cache la Poudre and the Arkansas, in Colorado. Within a distance of about forty miles through an irrigated district, a change from 37 ppm dissolved solids to 1,571 was recorded in the water of the Cache la Poudre; the Arkansas within 100 miles showed a change from 148 ppm to 2,134 ppm.[35]

[33] Thomas, *op. cit.,* p. 219.

[34] W. P. Cross, M. E. Schroeder, and S. E. Norris, *Water Resources of the Mahoning River Basin* (Circular No. 177), U. S. Geological Survey, Washington, 1952, p. 32.

[35] Clarke, *op. cit.,* pp. 70-71. The Rio Grande shows similar changes, but over longer distance. At Otowi Bridge, New Mexico, dissolved solids content of the Rio Grande is about 180 ppm, while at Fort Quitman, Texas, more than 400 miles downstream, it is 1,770 ppm (Fireman and Hayward, *op. cit.,* p. 321).

Quality measurements of the water of the San Joaquin River and some of its lower tributaries after many uses and reuses for irrigation upstream also have shown high dissolved solids content.[36] Discharge of untreated sewage from towns or cities can introduce into streams and ground water over wide areas pathogenic bacteria possessing epidemic potentialities. Water's great versatility thus provides it with a capacity for harm which must be constantly considered as it is exploited. The geographical relations of water flow are such that one city's waste may have to be another community's supply of drinking water, and an irrigation district's gain may be an industry's or a city's loss. Some of the most serious of the artificially induced water quality changes have been those in the vicinity of oil fields. Sodium content increases of irrigation water from oil field brine leakages or discharge can make the water injurious to crops, or unusable.[37] This problem has arisen within a number of southwestern oil fields, where cultivation and oil pumping are in proximity.

Water properties also vary over time with changes in natural conditions. The dissolved solids content of surface water in particular may vary by season, by year, or by periods of years, according to the amount of precipitation received. The lower the precipitation, the higher the average dissolved solids content of the water which does flow. Thus, the total dissolved solids content of the Pecos River below Red Bluff Dam near Orla, Texas, averaged 2,535 ppm during the period of the two water years from 1940 to 1942, when exceptionally large river discharges were experienced, while during the three dry years from 1948 to 1952 this figure more than doubled, averaging 5,650 ppm.[38] Such changes may be experienced wherever drought periodically takes place. Their occurrence may also affect the quality of the ground water with which surface waters are in contact.

Likewise water quality may change with conditions of flow from one

[36] Joint Committee on Water Problems, California Legislature, *Tenth Partial Report: Drainage in the San Joaquin Valley* (Sacramento, Calif., 1957), pp. 15, 19, 22.

[37] See, for example, accounts of specific cases on the High Plains of Texas in *The Cross Section,* August 1957 and September 1957, pp. 2-3 in each issue. (Published by the High Plains Underground Water Conservation District No. 1, Lubbock, Tex.)

[38] Texas Board of Water Engineers, *Chemical Composition of Texas Surface Waters* (prepared in co-operation with the U. S. Department of the Interior, Geological Survey, and others), Austin, Texas, annually published, mimeo.

section of a stream to another. The aeration provided by rapids may eliminate biological impurities and restore oxygen content. On the other hand, oxygen content may be reduced by storage in major impoundments.

Summary of the Physical Nature of Water Supply

In the above description of the nature of water occurrence several significant features have emerged upon which succeeding analysis of technology and water development will be based. They are:

1) The only inexhaustible supply of water is to be found in the oceans, which are high in dissolved solids content.

2) The amount of moisture contained in the atmosphere at any given time is small relative to the total amount on earth. However, it is important because it recirculates rapidly through the hydrologic cycle and because it is the only natural source of soil moisture and runoff.

3) The most widespread source of water on land is soil moisture, which also is a small amount of the total supply of water available on earth. Within the United States soil moisture is highly deficient in several regions, and the optimum, or perhumid, soil conditions are rare.

4) The most readily available *concentrated* source of water is runoff. Because it is so closely related to the vagaries of atmospheric precipitation, the amount of runoff available in different drainage systems exhibits wide variations geographically and over time. Geological structure, surface configuration, atmospheric temperature, and possibly vegetation may amplify or moderate the direct relation of runoff to amount and timing of precipitation, but regional variation is wide. So is seasonal variation in amount of water within any given drainage system. Even the daily range of flow is typically great; most streams have minimum flows far below their potential continuous capacity to supply water under regulation. Finally, the runoff of the United States is characterized by flow within large to very large drainage basin systems.

5) Aquifers contain a vast supply of water underground, both fresh and saline. Much of this water is not related to the circulation of water through atmospheric precipitation, runoff, and evaporation, although most aquifers near the surface are so related. Water in the aquifers receiving small recharge is particularly important west of the 100th meridian, and must be considered as having the same exhaustibility as a

mineral deposit. Like runoff, aquifers with the capacity to support wells of substantial flow exhibit much variation geographically. Their distribution, furthermore, does not correspond to the regional variations in runoff.

6) Water properties vary widely according to location and source. Both suspended materials and dissolved solids alter the nature of water which flows on or is stored within the ground. Surface streams also vary in their bed load. Water is capable of carrying, and on occasion does carry, large quantities of material in suspension, both organic and inorganic. Its chemical composition is commonly affected by more than a dozen different elements which may enter water solution and at least forty-six elements have been found dissolved in water. Regionally, there are important chemical differences between the chloride-sulfate waters (arid and semi-arid areas) and the carbonate waters (humid regions), but a great variety of regional and local differences may be introduced by elements which enter water solution even in small amounts. Variations in dissolved gases (e.g. oxygen), and in temperature also exhibit a geographical distribution. In general, ground waters show more stable temperature characteristics than surface-water flows.

7) Water shows much capacity for change under use. Its capacities as a solvent and as a transporting medium make it susceptible to relatively rapid changes in dissolved solids content and in the materials it may carry in suspension. It therefore tends to be responsive chemically and physically to any disturbance of the environment in which it occurs, whether the disturbance is natural or artificial.

CHAPTER 3

The General Nature of Water Use
in the United States

Classification of Water Use—withdrawal uses; flow uses and on-site uses. Tolerance of the Several Uses for Different Types of Supply. Seasonality of Periodicity of Water Demand. Location of Demand—size and growth characteristics of the population; resources other than water, and their combinations; space characteristics of the land; popular attitudes toward natural amenities; political structure and political policy; location of water demand: the two major regions. Summary of Water Demand.

The characteristics of water use, no less than those of water occurrence, must be described as part of the background of examining technology in water development. The way in which water is needed and used helps to determine the application of and the opportunities for technology. There are many local disparities between demands for water services and water supply, and it is toward the lessening of these disparities that engineering and applied science have been directed.

Classification of Water Use

The uses of water may be generally classified into three major forms: the withdrawal uses, the flow uses, and the on-site uses.

Withdrawal use is one in which water is diverted from its natural channel or ground-water site, either temporarily or permanently. Withdrawal uses further may be distinguished as disappearing withdrawals and returnable withdrawals. A returnable withdrawal is water which

47

will be returned to a natural waterway or to the ground. Most withdrawal uses have a component of both disappearance and return flow. Domestic, municipal, irrigation, and manufacturing (or industrial) uses all are withdrawal uses. Disappearing and returnable withdrawals may be found within each of these purposes, so it is relatively difficult to discern disappearance precisely.[1]

Flow uses are those which in some way depend upon water in movement. Hydro power production, waste carrying (as of sewage or industrial wastes), some game fisheries, and some types of recreation (e.g. scenic enjoyment) are the principal examples. No water withdrawal is involved.

On-site uses are those which can be most advantageously pursued on water that has only a slow rate of flow. (Very little water has no flow, even in lakes.) Navigation, some forms of recreation, certain fisheries, and wildlife support are examples.[2]

WITHDRAWAL USES

The minimum life needs (requirements) of a human being for water are relatively small. An individual probably could get along comfortably on one gallon a day for drinking and cooking.[3] However, civilized life

[1] The terminology here suggested is somewhat different from that previously employed by a number of writers on water development. A consistent set of terms for all water uses is sought, with avoidance of double, or even triple, meanings. The word "use" here is confined to its general meaning of "employment," and applies to all ways in which water is employed. This is consistent with the etymology of the word. In previous professional terminology "use" has been applied with this general meaning and, in addition, with the specific meaning of "disappearance" when applied to withdrawals. The word "consume" has had similar dual meanings. Often the two words "use" and "consume" have been combined in another term. Thus, "consumptive use" has meant disappearance. ("Consumptive depletion" is an alternative term also in the literature of the field.) "Use" (or "depletion") has been distinguished from "requirement," meaning the amount *withdrawn* (but not necessarily disappearing) for irrigation, domestic, municipal, or industrial purposes. No change is here proposed in the meaning of the term "requirement" or in its application.

[2] The duplication of recreation and fisheries within the flow and on-site uses is only apparent. Different forms of use are involved in each instance, as between slack-water and white-water fishing.

[3] This amount was estimated to be the minimum necessary for "drinking" in a study undertaken by M. A. Pond, as reported by H. E. Jordan in "The Problems that Face Our Cities," *Water, The Yearbook of Agriculture, 1955,* U. S. Department of Agriculture, Washington, 1955, p. 651.

demands much more. The minimum for laundry, face and hand wash-ing, and water closet flushing is about 19 gallons per person per day more. If a three-minute shower bath is added, an additional 15 gallons are required. Actually, in 1950, residential use within the municipal water systems of the United States was estimated at 50 gallons, rather than the 35-gallon minimum suggested. Public uses, like fire control, street washing, etc., added another 10 gallons per person per day, and commercial and undifferentiated industrial uses, 20 gallons more. Sys-tem loss was 10 gallons. Thus the average urban or suburban dweller is part of a household, public-order, and trading system which now demands water at the rate of about 90 gallons per capita per day. Of this amount only about 10 per cent actually disappears. Ninety per cent is returned to watercourses, or to ground water.[4] Within rural areas use by homes with running water has been estimated at about 60 gallons per person per day, and in homes without running water, at 10 gallons.[5] Presumably much of the 60-gallon withdrawal is also returned to the ground.

On these bases, it is likely that about 10.5 billion gallons per day were withdrawn for municipal residential, commercial, and public uses in 1955. About 1.0 billion gallons more were used by rural households. Actual *disappearance* of water for all may have been on the order of about 1.15 billion gallons a day. Total disappearance for the year might reasonably be estimated at 1.25 million acre-feet (table 5).

TABLE 5: *Withdrawal of Water in the United States, "Domestic" Purposes, 1955*

	Million gallons per day
Municipal residential, commercial and public	10,500
Rural households, no running water	400
Rural households with running water, but not on a water system	600
Total use	11,500
Actual disappearance (10 per cent of above)	1,150

Data Sources: K. A. MacKichan, *Estimated Use of Water in the United States— 1955* (U. S. Geological Survey Circular No. 398), and H. E. Jordan, "The Problems that Face our Cities," in *Water,* U. S. Department of Agriculture Yearbook, 1955.

[4] *Ibid.*

[5] K. A. MacKichan, *Estimated Use of Water in the United States—1955* (Cir-cular No. 398), U. S. Geological Survey, Washington, 1957, p. 6.

The trend in water requirements for domestic purposes is a slowly rising one. Recently, increase in rate of withdrawal has amounted to about 1 gallon per person per day annually.[6] If this trend is followed, an increase in total disappearance of around 250,000 acre-feet per year might be expected for the nation by 1975. Total disappearance at that time for this purpose then would be about 1.5 million acre-feet per year. As will be seen, this amount is small in comparison to the total water supply within United States territory, and in comparison to disappearance associated with other uses.

Consumption by livestock in 1955 probably added another 1.5 billion gallons a day to the water disappearance for life consumptive purposes, or about 1.7 million acre-feet a year. While some of this was returned to the soil in the form of animal excreta, animal consumption for the purpose of this study is assumed to be water which did not return to watercourse or ground-water deposit.

The largest disappearance of water for life uses from concentrated supplies is for irrigation. Although the total use can be estimated only in rough manner, about 123 million acre-feet of water are thought to have been applied on 34 million acres of farm land in United States irrigation in 1955. Again roughly, about two-fifths of this probably was returned to ground water or stream flow after application to crop or pasture (table 6, footnote 7). Disappearance of water for this purpose therefore may have approximated 74 million acre-feet, through transpiration from crops or pasture, and from evaporation which took place in the process of transportation or field application of water. Within the western states, where 96 per cent of all irrigation water is presently withdrawn, "return flow" irrigation water, as already noted, is not the same quality of water which was originally applied to the field. It almost always is altered by the addition of dissolved solids, or salts, in varying degrees of concentration, and organic matter. Therefore it may be unfit for some uses without ameliorating treatment.

Any of the above uses is minute in comparison to the disappearance of water which is received, stored, and used from soil moisture, or which is taken from the ground-water table by crops and other plant growth. While any estimate must be rough and subject to qualification, it is likely that a billion acre-feet are used from these sources for the support of nonirrigated pasture and the growth of nonirrigated crops. The

[6] H. E. Jordan, "The Problems that Face Our Cities," *Water, The Yearbook of Agriculture, 1955,* p. 651.

TABLE 6. *Estimated Annual Water Disappearance in the United States as Compared with Total Receipts of Water, 1955*[1]

	Million acre-feet (approximately)
1. Precipitation received	4,750
2. Disappearance of dispersed supply through evapotranspiration (except irrigated crops)	3,380
a. Farm crop and pasture, nonirrigated .... [2] 1,100 ±	
b. Forests, browse, vegetation, etc. .... [3] 750 ±	
c. Evaporation plus transpiration from non-economic vegetation .... 1,530 ±	
3. Total concentrated supply	1,380
a. Runoff .... [4] 1,372.4	
b. "Mined" water added to current supply from ground[5] 6.1	
4. Total disappearance concentrated demand	90
a. Household and municipal consumption, disappearance .... [6] 1.3	
b. Livestock watering 1.7	
c. Irrigation .... [7] 74.0	
d. Industrial—from municipal supply .... [6] 0.8	
e. Industrial—individual supply .... [6] 12.3	
5. Unconsumed runoff	1,290

[1] Data on consumption are for 1955, other data are estimates generalized from available information. Figures have been rounded because of approximate nature of some components.

[2] Estimated as average disappearance of 12 inches annually through transpiration for 1,100 million acres.

[3] Estimated average disappearance through transpiration of 18 inches yearly for 300 million acres of forest and pasture, and 12 inches yearly for 300 million acres of forest and pasture.

[4] See table 11.

[5] Principally overdraft pumping in California, Texas, and other southwestern states. The total ground-water withdrawal has been estimated at 23 million acre-feet. *A Program to Strengthen the Scientific Foundation in Natural Resources* (House Document No. 706, 81st Congress, 2nd Session), November 1950.

[6] Disappearance estimated as 10 per cent of total withdrawal.

[7] Disappearance estimated as 60 per cent of total withdrawal for irrigation. This estimate, like that for household consumption, must be regarded as a rough approximation. It is the same as, or close to, the figure used by other students of water use. The range of irrigation efficiency probably is somewhere between 50 and 70 per cent, the individual estimate depending on the methods of computing efficiency. One variant, for instance, is the efficiency of sub-irrigation, where employed. See H. F. Blaney, "Climate as an Index of Irrigation," *Water*, U. S. Department of Agriculture Yearbook, 1955, p. 344; K. A. MacKichan, *Estimated Water Use in the United States 1955* (U. S. Geological Survey Circular No. 398, Washington, 1957); and A. M. Piper, "The Nation-Wide Water Situation," in *The Physical and Economic Foundation of Natural Resources, IV—Subsurface Facilities of Water Management and Patterns of Supply—Type Area Studies* (Interior and Insular Affairs Committee, House of Representatives, United States Congress), 1953, p. 6.

amount that is probably consumed by forests and other nonfarm vegetation is less than the farm-land use of water, although still tremendous when compared with the totals drawn from concentrated supplies (table 6). The disappearance of water through vegetation which has some economic use, direct or indirect, probably exceeds in amount that moving from the earth through evaporation.

The vast amount of water used by nonirrigated plant growth on farm land still may be much less than crop and pasture plants might use if they received the amounts of water optimum to their growth. As previously noted (page 16), only a very small percentage of the land surface of the United States has a perhumid climate. A high percentage of the farm land of the country suffers water-deficit periods of varying length and severity in a majority of seasons. The median annual deficiency in rainfall for twenty-one eastern states has been calculated at eight inches.[7] Conceivably, therefore, a major potential demand for water to supplement soil moisture supplies received by crop and pasture plants exists over most of the United States. This deficiency increasingly is being recognized in the expansion of irrigation in the thirty-one eastern states. While the bulk of irrigation is still in the seventeen western states (26.9 million acres, 1954), irrigated acreage increased 70 per cent (to 2.6 million acres) in the thirty-one eastern states between 1949 and 1954.

Under our existing state of food production technology, the giant users of water among the life-consumptive purposes potentially are still the most thirsty. Their ultimate demands upon available stores of water can be enormous. The large existing gap between water disappearing from the concentrated supplies and the unconsumed runoff is consequently not a clue to the future situation.

Few common industrial products are manufactured without the employment of water which weighs many times the finished material itself. Water is used as a solvent or diluent, as a medium of flotation, suspension, and heat exchange, and in other ways. Except air, it is the material of largest volume moving through factories. An individual illustration of the volume of use is the average automobile. The ton of steel in each vehicle required the use of 285 tons of water in its conversion from iron

[7] J. R. Davis, "Future of Irrigation in Humid Areas," *Journal, American Water Works Association,* 48 (1956), 989. Deficiency estimates depend on assumed patterns of crop and vegetative cover, for which moisture requirements have not been well determined. Such estimates therefore must be considered rough at best.

ore. About 115 tons more were used in processing other materials (plastics, textiles, metals other than iron, etc.) which also are part of the automobile. Thus each automobile means the withdrawal of about 400 tons of water in its manufacture.[8] Each gallon of gasoline consumed as the vehicle moves along the road represents the previous use of about 25 gallons of water in its manufacture. While the use for different products varies greatly, a number require as heavy or heavier use of water than steel and petroleum products, which are admittedly large water users.

A survey conducted by the National Association of Manufacturers in 1950 indicated that among large factories in the United States (water intake of more than 10 million gallons a day) 54 per cent of the water used was for cooling, 32 per cent for "processing," 9 per cent for boiler feed, 6 per cent for sanitation and cleaning, and 4 per cent for miscellaneous operations. Among smaller plants the distribution by use was somewhat different: cooling 23 per cent, processing 36 per cent, boiler feed 12 per cent, sanitation 26 per cent.[9]

These uses probably totalled approximately 131 million acre-feet of withdrawal in 1955, and undoubtedly have grown since. While there is much variation within the different processes, it is unlikely that more than 10 per cent of the factory water intake represents actual water disappearance. The net disappearance thus represents an amount on the order of 13.1 million acre-feet for all United States industry.

Industrial requirements for water may be one of the more rapidly growing demands upon concentrated supplies. The President's Materials Policy Commission estimated in 1952 that the industrial requirements for water intake in 1975 would be about two and three-fourths times those of 1950.[10] The major increase is forecast for the production of energy (thermal electricity, petroleum refining, steam power). Already the largest industrial water requirement is for these purposes, accounting for a little more than half of the total industrial water intake in 1950. Water use for energy production was expected by the Commission to rise to 65 per cent of the total intake in 1975, thus constituting the major share of an expected total annual intake of around 240 million

[8] P. Weir, "Public Water Supply in the Future," *Journal, American Water Works Association,* 48 (1956), 755.

[9] H. E. Jordan, "The Increasing Use of Water by Industry," *Water, The Yearbook of Agriculture, 1955,* p. 654.

[10] *Resources for Freedom,* Report of the President's Materials Policy Commission, Vol. 5 (Washington: U. S. Government Printing Office, 1952), p. 95.

acre-feet. Disappearance of water from a total use of such proportions may rise to as much as 25 million acre-feet, which still places industry far below present-day irrigation in its demands upon concentrated supplies. By 1975 it may be farther below actual irrigation consumption.

FLOW USES AND ON-SITE USES

The withdrawal uses are only part of the services given by water. Most of the remaining functions relate to concentrated water supplies, but they are important segments of our economic and social life. Water requirements for navigation, for energy production, support of fish and wildlife, water recreation, and waste carrying, all must be taken into account in water development.

Four of the above water uses occasion comparatively little water disappearance and very little change in quality. Water disappearance is involved only where water must be stored, and evaporation takes its

TABLE 7. *Estimated Annual Water Evaporation Losses, Eleven Western States*

(acre-feet)

State	Large lakes and reservoirs[1]	Principal rivers	Small lakes, reservoirs, and streams	Totals
Washington	683,000	487,000	80,000	1,250,000
Columbia River[2]	173,000	372,000	—	545,000
Oregon	503,000	101,000	100,000	704,000
Snake River[2]	—	106,000	—	106,000
Idaho	968,000	248,000	84,000	1,300,000
Montana	1,188,000	272,000	143,000	1,603,000
Wyoming	431,000	71,000	85,000	587,000
Colorado	208,000	83,000	120,000	411,000
Utah	390,000	161,000	63,000	614,000
Nevada	538,000	3,000	30,000	571,000
California	1,352,000	202,000	203,000	1,757,000
Colorado River[2]	1,135,000	230,000	—	1,365,000
Arizona	94,000	209,000	119,000	422,000
New Mexico	192,000	35,000	81,000	308,000
Eleven western states ..	7,855,000	2,580,000	1,108,000	11,543,000

[1] More than 500 acres.
[2] Portion of stream coinciding with state boundary.
Source: U. S. Geological Survey, Water Resources Division, news release (April 7, 1958).

toll from surface storage for navigation, hydro power production, or recreation. In the humid sections of the country this is at present of little immediate importance, but under arid or semi-arid conditions, where every drop of water counts, storage evaporation may be a factor to be reckoned with in planning. Flow and on-site uses can be consumptive to this extent, although few data exist on the quantitative importance of this disappearance. It has been estimated that 11.5 million acre-feet of water are evaporated annually from water surfaces in eleven western states (table 7). Comparatively little of the evaporation loss in this instance is chargeable to nonwithdrawal uses. Across the country a relatively small disappearance probably is assignable to these purposes.

Flow and on-site water uses are significant in water demand because they present certain aspects of conflict with consumptive uses, and even with each other. For instance, waste carrying may be so pursued that "pollution" makes water unfit for consumptive and certain other uses without remedial treatment. Domestic, industrial, stock water, recreational, fish and wildlife uses all are affected when a stream is pre-empted as a waste carrier. The ideal for navigation is maintenance of the water in the river or other navigation channel; hence any withdrawal use that drops the level of the river channel is in conflict with navigation. The ideal for hydroelectric power production is the use of every acre-foot of water through all of the drop that it can fall before reaching sea level. Therefore any withdrawal use that involves disappearance of the water before it has reached sea level is in conflict with maximum use of the water for hydroelectric power production. Almost any purpose that causes reservoir fluctuation is in conflict with recreational uses.

Water's great versatility in use therefore gives its demand some dimensions not encountered in the demand for most other commodities. Conflicting uses for the same water present important technical challenges to the development of multiple use and reuse of the same water. The potentiality of multiple use and reuse distinguishes water use from most other economic production functions.[11]

Tolerance of the Several Uses for Different Types of Supply

In distinguishing between salt water and fresh, some anticipation was made of the tolerance of different uses for specific types of water supply. Further distinctions here may be made between the quality of

---

[11] Suggested by Blair Bower, letter to the authors, 1957.

TABLE 8. *Tolerance of Specific Water Uses for Different Types of Water Supply*

Use	Content of dissolved solids			
	Low (50 ppm or lower)	Moderate (50-400 ppm)	Moderately high (400-2,500 ppm)	High (2,500 ppm+)
1. Domestic supply	X	X	X	No
2. Livestock	X	X	X	Some to 15,000 ppm
3. Irrigation cropping:				
ditch	X	X	X	} Some to {
sprinkler	X	X	X	} 3,500 ppm[2] {
4. Public municipal	X	X	X	X
5. Navigation	X	X	X	X
6. Industrial				
a. Process	X	Many limitations by specific industry		
b. Cooling	X	X	X	X
c. Boiler feed	Preferred	X	Treatment desirable	
d. Sanitary and service ...	X	X	X	X
7. Hydro power	X	X	X	X
8. Recreation	X	X	X	X
9. Waste disposal	X	X	X	X
10. Fish and wildlife	X	X	X	X

X denotes use tolerance.
[1] High content of suspended solids or colloids.
[2] M. Fireman and H. E. Hayward, "Irrigation Water and Saline and Alkaline Soils," *Water,* U. S. Department of Agriculture Yearbook, 1955, p. 321.

High turbidity[1]	High bacterial content	Temperature 32°F.-60°F.	60°+F.	Mode of occurrence Ground	Surface
No	No	X	X	X	X
X	Some limitation	X	X	X	X
Some limitation	X	Used but some disadvantage	X	X	X
Limited	X		X	X	X
Preferably not	X	X	X	X	X
X	X	X	X	—	X
No	Food, pharmaceuticals, and others limited	X	X	X	X
No	X	Preferred	X	X	X
No	X	X	Preferred	X	X
No	Some limitation	X	X	X	X
Preferably not	X	X	X	No	According to elevation and site
Preferably not	Preferably not	X	Preferred	No	X
X	X	X	X	Some	X
Preferably not	X	X	Some limitation by species (e.g. salmonoid fishes)	No	X

the water and the mode of its occurrence when use tolerance is examined. Quality is taken to include dissolved solids and suspended solids, including bacterial content as well as temperature, B.O.D., dissolved oxygen content, algal content, and presence of oil or other floating wastes. Mode of occurrence may be ground, surface, or soil moisture.

A pairing of the different forms of water demand with the several forms of supply is interesting in the relatively wide tolerance shown by uses as a whole for a very wide range of supply characteristics (table 8). The outstanding limitations are for domestic and industrial use. Domestic needs cannot be served by untreated supplies which have high dissolved solids or high bacterial content. A large number of industries are sensitive not only to the total dissolved solids content, but also to the particular elements contained in the dissolved solids. Paper, chemical, and textile processing are illustrative, where some processes are made impossible even by minute quantities of a single element, such as manganese, dissolved in the water supply. Some manufacturing, like food and pharmaceuticals, also demands water with low bacterial content.

Like manufactural processing, irrigation also has some special limitations on dissolved solids content. Three criteria generally are considered to be important: the total dissolved solids content; the presence of salts or ions which are toxic in low concentrations; and the presence of more common ions which are nontoxic in low concentrations, but which may be toxic or otherwise injurious in larger amounts.

The total dissolved solids content of irrigation water preferably should not be above 1,500 ppm.[12] However, some waters used as irrigation supplied in the United States contain as much as 3,500 ppm total dissolved solids.[13] Continued use of such waters probably has unfavorable effects on the structure of the soil to which they are applied, and probably indirectly upon the soil's fertility.

Boron is an example of an element toxic to crops in low concentrations. Toxicity for some crops may develop with concentrations as low as 1 ppm, and the most tolerant crops (e.g. alfalfa and sugar beets) cannot be grown with boron concentrations greater than 3 ppm in irrigation water.[14] A fraction of 1 ppm of lithium will be toxic to citrus

[12] R. D. Hoak, "Greater Reuse of Industrial Water Seen," *Chemical and Engineering News,* 33 (March 28, 1955), 1279.

[13] M. Fireman and H. E. Hayward, "Irrigation Water and Saline and Alkaline Soils," *Water, The Yearbook of Agriculture, 1955,* p. 321.

[14] *Ibid.,* pp. 323-24, and Hoak, *op. cit.*

trees. Selenium and fluorine are undesirable because they accumulate in plants and are toxic to animal life.

Sodium, chloride, bicarbonate, and sulfate ions may cause toxic reactions in plants if present in sufficient amounts. Sodium is one of the most harmful in that it may affect soil structure unfavorably, impeding the movement of water and air. A limiting ratio for sodium appears to be: $Na/Ca+K+Mg+Na=0.60$. Beyond this, adverse effects on soil may ensue quickly.

The following comparison of sea water with irrigation tolerances is interesting in this connection:

	Irrigation crop tolerance (ppm)	*Sea water content (ppm)*
Total dissolved solids	1,500 preferable 3,500 possible	35,000
$Na/Ca+K+Mg+Na$	.60	.84
Boron content	3	4

The gap to be closed in sea water desalinization for successful irrigation use therefore would appear large insofar as total dissolved solids content is concerned.

Another major point of tolerance relates to the use of water to produce energy. The value of water for hydro power depends entirely on its relation to surface flow and the position of the water above sea level. Ground water does not enter calculations for hydro power production, and storage of water underground has been faced with this disadvantage in multipurpose use. Still other uses—recreation, fish, and wildlife—cannot be paired with ground-water supply.

Other points of interest are: (*a*) Industrial cooling water (the largest industrial use) and irrigation (the largest over-all use of concentrated supply) are opposite in their temperature preferences. Warm water is preferable for irrigation and cold water for industrial cooling. For this reason ground waters are favored (other things being equal) for industrial cooling purposes, and surface waters for irrigation. (*b*) One of the sharpest quality limitations on water use is high sediment content. Only four of the ten listed major uses of water are tolerant of water with a high content of suspended solids.

Seasonality or Periodicity of Water Demand

A final aspect of water demand worth consideration is its timing. Most water uses are not of continuously similar extent or intensity. This is well illustrated in the case of agriculture.

Even under the most extreme drought situations, irrigation water is demanded only during the growing season, which varies with latitude and continentality.[15] In the lower Colorado River Valley or southern Florida this may be almost year-round; in northern Montana or central Rocky Mountain valleys it may be ninety days or less. Even within the growing season water is not required over the entire period of crop growth and maturation, but its use is periodic and may stop some time before the crop is harvested. Where irrigation is practiced on "humid" lands, the incidence of demand is even more limited in time. Water should be applied only when soil moisture content is depressed to the point where plant growth is affected. This may occur during only a relatively few days of each growing season, and not necessarily within the same weeks from year to year. While the demand thus may be both of short duration and erratic incidence, its satisfaction is usually critical to optimum crop output. Agricultural demand therefore exhibits a high degree of seasonality, or fluctuation in time, within this country. Only a few areas, mainly in southern California, southern Arizona, and southern Texas, have growing seasons long enough and drought conditions continuous enough so that they exhibit something approaching a year-round demand for irrigation water. Elsewhere the shorter growing seasons and naturally supplied soil moisture limit the demand for irrigation water to certain months, or even certain days, in the year.

Demand for other water services is on the whole more constant than demand for irrigation. While diurnal fluctuations are characteristic of several of the nonagricultural uses, seasonality is much less apt to be pronounced. Domestic water supplies are needed in winter no less than in summer, livestock must be watered the year round, and most industries seek a schedule of continuous production. The principal exception is in recreational demands, which peak in the same season as irrigation.

---

[15] "Continentality" is the degree to which marine or land mass influences determine the thermal characteristics of a climate. A "continental" climate has a greater range of temperature, both seasonally and diurnally, than a "marine" climate. Sharper and more rapid fluctuations also distinguish the continental from the marine.

A few industries, like food processing, show summer or autumn peaks in demand. On the other hand, demands for electric energy, which fall in part on hydro power production, tend to peak in late autumn or winter (with the exception of areas with large irrigation pumping projects, or cities with heavy air-conditioning loads). This has become especially pronounced in southeastern United States, where electrical space heating has become popular. Within many urban areas domestic water demand, like irrigation demand, may have a summer peak, reflecting lawn watering, air conditioning, swimming pool use, etc.

Waste carrying, navigation, and fishery and wildlife may be expected to have more continuous demand patterns in time. *Use,* as distinguished from *demand,* of course may show seasonal characteristics reflecting the unusability of available water resources at certain periods of time (e.g., frozen stream, lake, or canal, or near-dry channel). Periodically recurring unusability (as in a winter-frozen stream) tends to cause adjustment of demand (e.g. navigation) to availability.

Location of Demand

The locational characteristics of demand have important bearing upon the problems of matching water supply and demand. Viewed in its simplest form, a supply in its natural state sooner or later gives rise to matching demands where there is a social group to generate demand. Thus flowing water in desert or semi-arid country is an invitation to irrigation, a large stream of easy gradient is an invitation to navigation, a large stream on a fall line encourages a search for hydro uses, and so on. There are, however, forces operating in our economy and society which generate demands quite without reference to the existence of water supply. For example, people need certain amounts of water for life, and wherever people congregate permanently in large numbers demands for water may have little relation to its availability.

Interest here centers on the concentration or dispersal of demand, and on events or situations influencing those characteristics, and observable trends therein. Consideration is given mainly to the demands that will be satisfied from concentrated water supplies.

The differentiation of water demand among the several regions of the country has been influenced by five features of land or culture: (1) size and growth characteristics of the population; (2) the character of natural resources other than water and their peculiar combinations within the regions; (3) the space characteristics of the land that help to determine the location of service functions in the economy and the people dependent upon them; (4) popular attitudes toward climate and other outdoor amenities; and (5) political structure and political policy. Some understanding of each of these forces is necessary if trends and potentials in demand of water are to be evaluated.

SIZE AND GROWTH CHARACTERISTICS OF THE POPULATION

Demand for water has a basic relation to numbers of people, and their regional and local distribution. Demands for the use of land and other resources, and demands on water use which those resources generate in turn, depend upon the number of people who must be supported within the nation and upon their standard of living. Since 1900 the population of the United States has doubled. This in itself accounts for many of the water development problems the United States has encountered in that period. Within the next twenty years there possibly will be 52 million people more in the country.[16] The number could be even larger.[17]

As this growth proceeds, the national demand for water will be increased at least proportionately over the present. However, differing pressures will be placed on water supplies in the several regions of the country. Those parts of the country where settlement is relatively recent, like the Pacific Northwest, quite logically may expect further growth on the basis of their unused but potential resources. On the other hand, a population increment of this size is not likely to be absorbed within the few remaining undersettled parts of the nation. Nor is the increment apt to be distributed evenly over the country. Trends in the location of water demand are likely to reflect other forces than simple growth of the population. Additional demands for water may be expected on the basis of population growth alone, but the incidence of that demand may be uneven geographically. Where people are and will be is in part dependent on the location of resources other than water (e.g. Oklahoma City); on the location of efficient service functions in

[16] Based on four projections made by the U. S. Bureau of the Census (C. Taeuber, *Population Research and Trends,* May 10, 1955 [12 pp. processed], p. 10). Also based on Resources for the Future memorandum study by Neal Potter, February 1 and 26, 1957.
[17] See *U. S. News and World Report,* August 9, 1957, pp. 46-54, for a detailed estimate which places growth at about 60 million between 1957 and 1975.

the economy (Indianapolis); geographical residential preferences (Los Angeles); and upon political or administrative considerations (Washington, D. C.).

RESOURCES OTHER THAN WATER, AND THEIR COMBINATIONS

The cultivable land, minerals, and other basic resources are distributed over United States territory quite without regard to the presence of water. If the presence or absence of water for the moment is ignored, large amounts of land suitable for cultivation exist in both the West and the East. Accordingly, there is either latent or actual demand for water to apply to cultivable land in both the West and East, depending on the existence of interested entrepreneurs, the adequacy of the soil moisture supply, the relative productivity of the soils, and market relations. Where soil moisture supply has been adequate to support crop or pasture production, the demand for irrigation water has been largely latent until now. On the other hand, where soil moisture supply has been small, deficient, or totally lacking, the demand for irrigation water to apply to cultivable lands has been very active and intense. Whether the demand for water is potential or effective, it is important to note that the existence of a land resource under our current technology is a predisposing cause for water demand.

The same may be said for other natural resources. The existence of a mineral deposit of commercial grade also creates a demand for water with which to process it, and for energy to convert it into usable form. The energy may be supplied from hydro power, or water may be used for transport to bring together ore and coal for conversion. Likewise where there are forests of pulping species there are also demands for processing water in their conversion to paper or textiles.

Like that of land, the distribution of mineral and forest resources has no direct relation to the distribution of concentrated supplies of water. Again, however, where these resources present opportunity for production to meet man's needs, a demand for water or its services automatically is created.

It follows that where the combination of resources other than water is richest and most varied, the demands for water are most intense, assuming a maturity of development. Thus the region including the northeastern states and the Midwest north of the Ohio and east of the Mississippi had a notable combination of natural resources which predisposed it to the development of manufacturing for the nation. The occurrence within the region of iron, coal, salt, petroleum, and other minerals, once admirable forests, and fertile agricultural land destined it to be a center of American manufactural production and a densely populated region. The demands for water and most of its services consequently have been great: for industrial processing, for waste carrying, for navigation, for domestic use, and other purposes.

In some instances, the peculiarity of a regional resource combination may create special demands for water. This has occurred in the Pacific Northwest, where a regional lack of the mineral fuels has brought forth a keenly felt demand for hydro power production. In another day the same situation prevailed in New England, which lacked regional supplies of fuel other than wood, and sought out many small direct water power installations as energy supplies for its nascent factories. The Columbia Basin situation is of further interest in that a combination of navigational access, hydro power, and other resources may form a basis for further industrialization which will require water withdrawal.

A basic appraisal of the nature and intensity of regional water demand can be made only in the light of the peculiar natural resource combination known to exist within the region as it is exploitable under the prevailing technology. This is particularly important in the appraisal of demand trends, and potential demand.

SPACE CHARACTERISTICS OF THE LAND

The combination of resources peculiar to each region gives the distribution of water demand its basic form, which may be supplemented, altered, or intensified by regional space relations. The position of the northeastern states opposite the populous and advanced European peninsula, and with marine access to the resources of the world has intensified the development of the Great Lakes–Northeastern area mentioned above. Both the manufacturing industries and the service functions of this region have been magnified by its position. As a result the urbanization of the region has been a notable feature of American growth, and with it the water demands which come from urbanization and industrialization.

The Pacific Northwest and California, opposite the great population centers of the Far East, have a regional position similar to that of the Northeast. This also is true for the Gulf Coast states, opposite Latin America. The effects, however, have been of a lesser size in both instances.

Space relations affecting water demand are not limited to those of international scope. They are of local importance in each region as they shape the growth of settlements and urban areas servicing the region or, in some instances, the nation. Within every modern economy certain urban areas are destined to become huge (the so-called "primate" cities, like New York, Chicago, and Los Angeles); all are of national importance. Others will be large and of regional importance, like Denver, New Orleans, or Cleveland; and still others of lesser size, serving subregions, and so on.

The growth force of these areas as population centers, and consequently as centers of intensified water demand, is not directly related to the presence or absence of water. This is illustrated by the case of Los Angeles, which has carried through its phenomenal growth in spite of a sharply limited local supply.

POPULAR ATTITUDES TOWARD NATURAL AMENITIES

Within recent years the United States has witnessed on a relatively large scale the appearance of a new force helping to shape demands for water and its services. This is the response of the people of the country to certain natural amenities, and their mobility in satisfying their preferences.[18] Climate has been the most important among the amenities, illustrated by the movement to Florida and California. However, other environmental features also figure, like scenic attractions, outdoor recreational opportunities, sparsity of settlement, and even common "elbow room." This force again has little direct relation to water supply; indeed, it exists in part as an inverse relation, for the absence of precipitation is one of the principal amenities appreciated.[19]

While other factors share in the recent migration to the western states, some indication of the meaning of this force is provided in the 1940-50 populations of six far western states. Between 1940 and 1950 the population of the United States as a whole grew about 15 per cent. The slowest growing of six far western states was 25 per cent larger in 1950 than in 1940; the most rapid increase was over 53 per cent (table 9). Considering an increase in the proportion of aged in the population, likely national decreases in work hours, and the momentum gained by

[18] E. L. Ullman, "Amenities as a Factor in Regional Growth," *Geographical Review,* 44 (1954), 119-32.
[19] *Ibid.,* p. 129.

TABLE **9.** *Population Increase, Far Western States, 1930-55, and as Estimated to 1975*

State	Population 1930	1940	1950	1955	Percentage increase 1930-40	1940-50	1950-55	Estimated percentage increase 1956-75
		(in thousands)						
California ..	5,677	6,907	10,586	13,032	21.7	53.3	22.4	91.0
Oregon	954	1,090	1,521	1,669	14.2	39.6	10.8	66.8
Washington .	1,563	1,736	2,379	2,570	11.2	37.0	9.6	57.0
Nevada	91	110	160	225	21.1	45.2	47.1	96.4
Utah	508	550	689	781	8.4	25.2	15.7	44.5
Arizona	436	499	750	980	14.6	50.1	34.3	93.7

Sources: U. S. Bureau of the Census, *Statistical Abstract of the United States, 1957,* and earlier editions; 1956-75 percentage estimates by Economic Unit, *U. S. News and World Report,* following Bureau of the Census methods.

the amenity-favored areas, it seems safe to forecast that people of all ages will continue to migrate to these areas, perhaps in increasing proportions. [20] Only a serious economic depression of national scope would be likely to arrest this trend.

The amenities are influencing water demand in other ways. The relatively large expansion of settlement in suburban and semirural localities within the United States in the last twenty years has tended to disperse municipal and domestic demands locally, enabling satisfaction of those demands from more widely dispersed supplies. This change in settlement habits, which has had repercussions upon the location of industry as well, is a response to the attraction of cleaner atmosphere, more residential space, and greater opportunity for outdoor recreation.

POLITICAL STRUCTURE AND POLITICAL POLICY

Political considerations have helped to influence water demands through the territorial divisions of the states of the United States, and through needs for modern defense strategy.

The territorial boundaries of the forty-nine states have little corres-

[20] A doubling of California's population between 1957 and 1975 is considered a probability. (*U. S. News and World Report, op. cit.*)

pondence to natural water occurrence or water supply. While state boundaries have operated in part to restrict movement of water supplies beyond state boundaries, the existence of a state *per se* may affect demand for water development, or at least the timing of demand. Thus the states of the Upper Colorado River Basin came into possession of certain agreed-upon shares of Colorado River water with ratification of the Colorado River Compact. Being in possession of the water, the Upper Colorado states politically created a demand for federal construction of facilities now included in the Upper Colorado River Basin project. Federal construction of the project was sought because the immediate actual demand for Colorado River water and water services within these states did not promise adequate return on the investment within a reasonable time. The demand for development in the early 1950's arose because of the existence of the states as political entities, ambitious to establish beneficial use of allocated water within their boundaries before economic demands from other states pressed upon them. While this is not a factor shaping demands within broad regions, it is of some influence in forming local and state demands for water and water services at any given time.

Within the last fifteen years, and perhaps indefinitely in the future, defense needs must be considered in water demands. Modern defense operations have three aspects significant in shaping water demand: (1) the need for huge tracts of land for testing, training, and maneuvering exercises; (2) the value of isolation for these operations and certain types of defense experiment and manufacturing; and (3) the value of dispersal and decentralization of settlement, administrative activities, and economic activities in densely settled sections of the country.

The impact of the first two of these defense-related factors again is without regard for the incidence of plentiful water supply. As in the case of the amenity consideration, there is even an inverse relation, for isolation and large tracts of low-value land are to be found precisely in those regions where water is lacking. Many localities in the West have felt these effects. Albuquerque, New Mexico, is an example. [21] Between 1940 and 1955 this urban area grew from approximately 40,000 people to 200,000. Within the same period its water withdrawals from the Rio Grande Valley streams and ground water multiplied about seven times. Still other defense installations, like those in the Alamogordo vicinity,

---

[21] Other activities have contributed to Albuquerque's growth, but defense-related pursuits have been major factors.

continue to add new and relatively large water demands in this area where water is already scarce. The same story may be repeated for a number of other western localities.

On the other hand, pressures for dispersal and decentralization in the more densely settled sections of the country may have a somewhat more favorable effect on water demand, lessening the concentration somewhat as vital industrial establishments find suburban or rural locations. Dispersal of both business and government administrative facilities, like that being undertaken to some degree in the Washington, D. C., metropolitan area, may disperse water demands, spreading the impact of the demand over a wider geographical range of supply. [22]

LOCATION OF WATER DEMAND: THE TWO MAJOR REGIONS

As in the case of water supply, the United States may be divided roughly into two major water demand regions, again East and West. A characteristic pattern of demand for water is found in each of the regions, with different degrees of demand intensity for the several services given by water.

The Eastern Region. Generally described, the East is a region where the cultivable land has a soil moisture supply usable by crops, extensive forest lands, and a widely distributed and well-developed supply of mineral fuels. Relatively dense settlement is found, increasingly in urban agglomerations, but also with a relatively numerous and well-dispersed rural population. The dominant economic activities (from the point of view of population and water demand) are in the manufacturing based on the minerals, forest, and agricultural products of the region, or upon the accessibility to the resources of other lands (figures 10, 11, 12, and 13). Demand for concentrated supplies of irrigation water within the region has been largely latent. On the other hand, demand for naviga-

[22] Dispersal of residential, government administrative, and industrial buildings within suburban and rural areas is far more than a defense-stimulated phenomenon. However, defense-stimulated movement is one element in the changing location of these facilities; this part alone is here considered. Within a small region the effects of dispersal upon water supply will depend upon sources. Where a single stream is the major supply, effects may not be significant except to increase distribution problems. Where ground-water supplies are useful, dispersal of demand may be helpful in improving the supply situation. Locational changes within the range of a larger region, on the other hand, obviously can affect demand incidence.

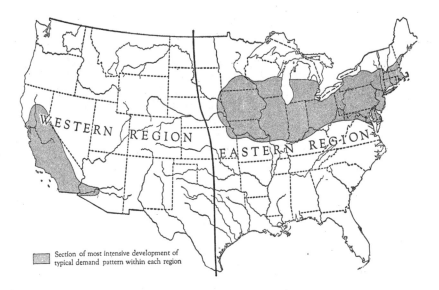

Figure 10. Regional Division of Demand Patterns for Water and Water Services.

Figure 11. Metropolitan Built-up Areas in the United States, 1950.

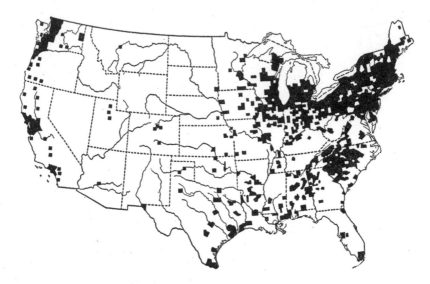

Figure 12. Manufacturing Counties in the United States, 1950, showing counties with more than 2,500 employed in manufacturing.

Figure 13. Mining Counties in the United States, 1950, showing counties with more than 1,000 employed in mining. Certain of the larger western counties here and in figure 12 are represented by a symbol, the size of a typical United States county.

tional services, industrial processing water, and waste carrying by water has been and is intense. Water demand is related to energy production more through thermal plant operation than by hydro development. Except perhaps in Florida, and to a degree also in Texas, the amenities and defense demands have meant little in shaping water demand. However, an undetermined but probably large demand for recreational uses of water exists.

Water demand in the East is characterized by concentration in some limited sites. A majority of the population of the region is urban, with a number of large urban areas. In 1950, 86 per cent of the 172 standard metropolitan areas, as defined by the United States Bureau of the Census, were located within the thirty-seven states of the East, Midwest, and Great Plains. Manufacturing is concentrated in these urban areas, or within favored valleys. In the absence of major irrigation water demands, the principal demands for water services therefore are concentrated within limited areas. Examples are the Ohio Valley, and the valleys of southern New England. The most advanced demand pattern is in the Great Lakes–Northeastern states area, which exhibits the above characteristics very clearly. Forty per cent of the United States anticipated population growth of 50-60 million between 1957 and 1975 is likely to take place within the Great Lakes–Northeastern area (shaded, figure 10). Thus within a relatively few years a total population of 110 million may be resident within the area, which comprises about 15 per cent of the total land area of the country. This large grouping of people will be overwhelmingly urban, so the typical problems of municipal-industrial supply may be expected to intensify throughout the Great Lakes–Northeastern section of the eastern region.

For the eastern region as a whole, future demand pattern is likely to be changed in an extension of the Great Lakes–Northeastern intensity of demand to some adjacent sections, and by the addition of further irrigation water demands, as more intensified cultivation becomes desirable within the region. The lower Mississippi Valley would seem a particularly favorable site for the development of irrigation agriculture on a large scale. Its extensive flat lands, the length of its growing season, and accessibility to transportation facilities predispose the valley to an untimely large irrigation water demand. Periodic short droughts create another condition which favors irrigation development not only in the valley, but also in other parts of the Southeast.

The West. Within the West a completely different demand pattern for water has evolved, and may continue. This is an area within which

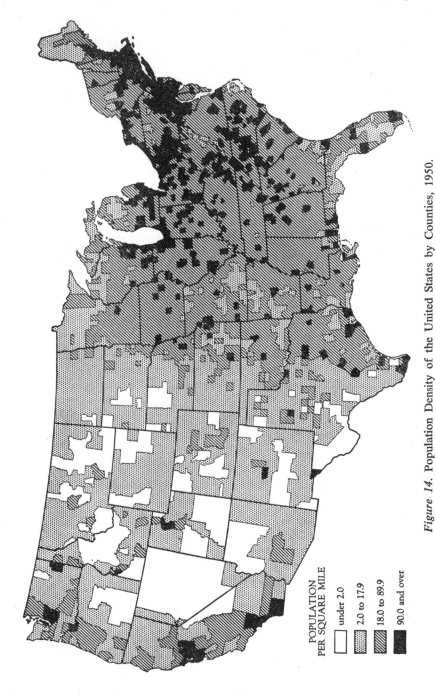

POPULATION
PER SQUARE MILE

☐ under 2.0
▨ 2.0 to 17.9
▨ 18.0 to 89.9
■ 90.0 and over

Figure 14. Population Density of the United States by Counties, 1950.

the abundant cultivable land resources at best are meagerly supplied with soil moisture. Consequently there is an almost insatiable demand for irrigation water in all but a few parts of the region.[23] On the whole it is a region of sparse settlement (figure 14), a condition which is considered an amenity by some people and is an asset in the location of some defense and defense-related activities. While the western region is a well-mineralized area, in the past it has been less well supplied with the conventional mineral fuels than the East. [24] Consequently demands for hydro energy have been more urgent than in the eastern region, even considering the sparser population. This demand is well illustrated in the Pacific Northwest. There the nation's largest hydro construction program still has failed to meet the growth in demand, which was brought about in part by the region's access to sea-borne raw materials.[25]

Within this region nearly all of the possible natural amenities (climate, scenic attractions, space, etc.) are forces in growth that may be expected to continue.

Where the economy of the region has developed as fully as in California, demands for water have reached dimensions found in no other part of the country. In addition to all of the demands found in the East there are the huge irrigational demands and the pressure for hydro energy production. Both the resource and the nonresource factors shaping demand for water are intensive. The likely continuing combination of amenity attraction, "open-space" and decentralized defense activities, industrial growth on the varied mineral and forest base, and local preferences for irrigated agricultural development suggest that water demand will continue to be a much more aggressive feature of the regional economy than it will be in the East. All indications suggest that the California–Southern Arizona area will be one of the most densely settled sections of the entire country, with a population of about 27 million at the end of the next two decades, or nearly twice the present number of residents. Such a concentration of people in an arid or semi-arid

[23] These exceptions are located mainly in the Pacific Northwest, in northern California, western Oregon, and western Washington State.

[24] The southern 20 per cent of this region (southern California to Texas), with substantial petroleum and natural gas deposits, and a few other more limited areas, do not conform to this statement. Recent petroleum and natural gas exploitation on the northern plains and extensive lignite and sub-bituminous coal reserves also indicate future fuel adequacy in other parts of the West.

[25] Demand in the Pacific Northwest eventually will require the construction of steam power facilities, probably within the next twenty years. (Frank L. Weaver.)

natural environment will produce one of the largest demands for extra-regional supply that has ever been experienced.

By far the largest withdrawals on a per capita basis in the country are within the western states. Ninety-six per cent of all irrigation water used in the United States is placed on land in the seventeen western states. These seventeen states have less than a fourth of the total population of the country, but they withdraw more than half of the water. [26]

A final aspect of western water demand is the wider dispersal of demand for development of concentrated water supplies than in the East. Whereas eastern demands are at their most intensive in industrial valleys

TABLE 10. *Irrigation Withdrawals as a Per Cent of Total Water Withdrawals—Western States, 1955*

(in thousands of acre-feet per year)

State	Withdrawals for irrigation	Total withdrawals	Irrigation withdrawals as a per cent of total withdrawals
Arizona	7,745	8,084	95.8
California	25,809	34,418	75.0
Colorado	7,065	7,938	89.0
Idaho	16,926	17,290	97.9
Montana	10,936	11,322	96.6
Nevada	2,149	2,290	93.8
New Mexico	2,818	3,005	93.8
Oregon	7,615	8,357	91.1
Utah	4,674	5,158	90.6
Washington	5,638	7,180	78.5
Wyoming	12,366	12,496	99.0
Total, eleven states	103,741	117,538	88.3
Kansas	829	2,505	33.1
Nebraska	2,858	3,736	76.5
North Dakota	136	457	29.8
Oklahoma	252	1,094	23.0
South Dakota	31	275	11.3
Texas	11,466	19,253	59.6
Total, seventeen states	119,313	144,858	82.3

Source: K. A. MacKichan, *Estimated Use of Water in the United States—1955* (U. S. Geological Survey, Circular No. 398, 1957).

[26] MacKichan, *op. cit.*

and in the vicinity of urban centers, demand for the use of concentrated supplies exists throughout the West in both rural and urban areas. As compared to the East, the smaller number of urban areas, the importance of irrigation, and the wide dispersal of mining enterprise account for this distributional difference.

The two regional types of water demand also show differences in the seasonality of use. The West, within which four-fifths of the withdrawn water is used for irrigation, thus far has shown a high concentration of water use in the crop-growing seasons (table 10). Within the eastern states irrigation is a potential rather than an effective use of water. As irrigation demand grows in the East a higher degree of seasonality will be introduced into eastern water demand. Conversely, as industrial and domestic demands grow in the West some increase in winter demand may be expected, although the pronounced seasonal peak may be expected to remain for withdrawal uses.

Summary of Water Demand

Several characteristics of water demand and use in the United States are significant in appraising the impact of technology on water development.

1) The most desired and most usable water has both a low dissolved solids and a low suspended solids content. Only a few uses tolerate much sediment in water supply. Almost all uses tolerate a moderate dissolved solids content, with special sensitivities for some elements. Within both agriculture and industry mere traces of some minerals are harmful or disqualifying for a water supply containing them.

2) By far the heaviest actual disappearance of water occurs through the use of soil moisture. Moreover, the potential additional agricultural demand for water from concentrated supplies is large because few soils in the United States have perhumid conditions.

3) Total disappearance from human use of concentrated water supplies is relatively small, probably not more than about 4 per cent of total runoff.

4) The major withdrawal use and the largest disappearance of water is for irrigation. Seventy-seven per cent of all water disappearance in economic use occurs in western United States.

5) Both flow uses and withdrawal uses may cause notable alterations of the quality of water, conflicting especially with withdrawal and on-site

uses of water. Waste carrying probably causes the most notable alterations of water quality, although irrigation causes many undesirable additions to dissolved solids content of water.

6) Demands for water may arise quite independently of the presence of water supply, both regionally and locally (table 11). Because
of this, water may have a strategic position in economic development
where it is locally scarce. The rate of development of all resources and
economic growth may depend on the manner in which the water demand
is met.

7) The major demands for water are seasonal, and even intermittent.

8) A continued future growth of water demand may be expected on
the basis of population increase alone. The incidence of the demand will

TABLE 11. *Estimated Population, Annual Withdrawal of Water,
and Runoff in the United States, by State, 1955*

	Estimated midyear population (thousands)	Per capita withdrawal of water (gallons per day)	Total withdrawal (1,000 acre-feet per year)	Estimated total runoff (1,000 acre-feet per year)	Withdrawal as a per cent of runoff
Western States	38,640	3,348	144,858	439,633	32.9
Arizona	980	7,359	8,084	4,250	190.1
California	13,032	2,356	34,418	71,937	47.8
Colorado	1,549	4,572	7,938	22,798	34.8
Idaho	609	25,328	17,290	37,877	45.6
Kansas	2,060	1,085	2,505	12,285	20.4
Montana	633	15,957	11,322	29,823	38.0
Nebraska	1,381	2,413	3,736	7,414	50.4
Nevada	225	9,080	2,290	3,537	64.7
New Mexico	795	3,372	3,005	3,893	77.2
North Dakota	642	636	457	1,886	24.2
Oklahoma	2,168	450	1,094	19,018	57.5
Oregon	1,669	4,467	8,357	67,238	12.4
South Dakota	677	362	275	2,875	9.6
Texas	8,563	2,006	19,253	49,909	38.6
Utah	781	5,892	5,158	8,152	63.6
Washington	2,570	2,493	7,180	76,375	9.4
Wyoming	306	36,431	12,496	20,366	61.4
Southeastern States .	33,502	870	32,617	478,065	6.8
Alabama	3,033	993	3,375	57,802	5.8
Arkansas	1,789	846	1,697	48,149	3.5
Florida	3,452	815	3,152	40,600	7.8
Georgia	3,621	608	2,467	47,101	5.2

	Estimated midyear population (thousands)	Per capita withdrawal of water (gallons per day)	Total withdrawal (1,000 acre-feet per year)	Estimated total runoff (1,000 acre-feet per year)	Withdrawal as a per cent of runoff
Kentucky	3,005	1,148	3,867	36,626	10.6
Louisiana	2,927	1,767	5,797	43,995	13.2
Mississippi	2,111	703	1,663	53,442	3.1
North Carolina	4,285	517	2,482	47,793	5.2
South Carolina	2,283	417	1,066	24,844	4.3
Tennessee	3,417	1,252	4,796	45,061	10.6
Virginia	3,579	562	2,255	32,652	6.9
Midwestern States ..	43,581	1,045	51,009	217,720	23.4
Illinois	9,361	1,055	11,073	30,079	36.8
Indiana	4,330	1,622	7,871	23,226	33.9
Iowa	2,692	669	2,019	13,809	14.6
Michigan	7,236	962	7,806	37,258	21.0
Minnesota	3,174	578	2,058	16,141	12.8
Missouri	4,128	549	2,541	40,877	6.2
Ohio	8,966	1,200	12,056	26,382	45.7
Wisconsin	3,694	1,371	5,675	29,948	19.0
Northeastern States .	47,703	765	40,851	236,990	17.2
Connecticut	2,241	897	2,254	5,877	38.4
Delaware	387	961	416	1,646	25.3
Maine	905	610	619	44,286	1.4
Maryland	2,669	726	2,171	9,025	24.1
Massachusetts	5,016	490	2,757	10,129	27.2
New Hampshire	557	460	287	12,405	2.3
New Jersey	5,420	798	4,851	8,359	58.0
New York	16,124	551	9,958	58,168	17.1
Pennsylvania	11,159	987	12,346	48,356	25.5
Rhode Island	845	471	446	1,424	31.3
Vermont	378	302	128	12,812	10.0
West Virginia	2,002	2,058	4,618	24,503	18.8
Total United States[1]	163,426	1,472	269,425	1,372,408	19.6

[1] Excludes District of Columbia.

Sources:
Column 1, U. S. Bureau of the Census, *Statistical Abstract of the United States, 1956*, p. 10.
Columns 2 and 3, computed from data available in K. A. MacKichan, *Estimated Use of Water in the United States, 1955* (U. S. Geological Survey, Circular No. 398, 1957).
Column 4, computed from data available in the offices of the U. S. Geological Survey.

vary regionally and locally, according to the presence or absence of other resources that require water in exploitation, the space relations that determine urban agglomeration, popular residential preferences, defense needs, and political policy. The growth of demand within the western region may be expected to continue in the California pattern, responding to amenity attraction, continued population mobility, and defense developments. While per capita demands in the western region may not increase significantly, the total demand is likely to grow substantially, as will multiple-purpose demands. Within the West the most notable contrast between water availability and withdrawals occurs in the north Pacific Coast region, where there presently are large unused surpluses.

Within the region of eastern demand pattern, on the other hand, per capita disappearance and per capita requirements are likely to grow appreciably, as irrigation is recognized as needed or feasible. The change in future total water demand therefore is likely to be most marked within the eastern region and existing multiple demands there are likely to be intensified. The areas of heaviest surplus supply as compared to withdrawal are in the southeastern states and the three northern New England states (table 11).

In both East and West continued shifts of population to urban residence and urban occupation, assuming a continuation of twentieth century trends in the United States, will bring increased concentrations of water demand for metropolitan and other urban areas. Intensification of water demand on those sites which are favored for manufacturing or service industries therefore is to be expected in the future.

CHAPTER 4

Matching Water Supply and Water Demand:

PROBLEMS AND OPPORTUNITIES PRESENTED TO TECHNICAL ACHIEVEMENT

Problems of Timing Discontinuities—seasonal; long-range. Problems of Quality Requirements. Problems of Quantity—water collection and transportation; conservational consumption of water; substitution of other resources for water; multiple use of water and water facilities.

There are three major groups of problems of matching water demand and supply in the United States: (1) those of timing discontinuities when fluctuations in the availability of water supply do not correspond to those in demand for water and water-derived services; (2) those of quality, where the natural supply does not meet quality requirements of one or more important uses; and (3) those of quantity, where there is a consistent inequality between demand and available supply in a region or locality. Technology has made contributions toward partial solutions within each of these major problem areas in the past, and promises further achievement in the future. An understanding of the opportunities and challenges facing the application of technology to water development may well begin with a description of each of these problems and the approaches that have been used toward their solution.

Problems of Timing Discontinuities

Discontinuities in timing between the incidence of demand and the availability of water supply have both seasonal and long-range aspects.

Seasonal discontinuity refers to the fluctuations of supply and demand which occur within a year, and which recur in the same pattern in succeeding years, although usually differing in amount. Long-range discontinuities may reflect changes in the trend of demand, and the longer "cyclical" fluctuations in the precipitation of moisture from the atmosphere. Although the latter are most pronounced in arid and semi-arid climates, they occur in every climate.

SEASONAL DISCONTINUITIES

Many demands for water and water services are on a continuous, day-in day-out basis, allowing for some diurnal fluctuation. Such are withdrawals for domestic, municipal, industrial, and livestock use. Waste carrying, and many fishery uses are also in this category, although the former may fluctuate from season to season, and anadromous fish may present seasonal demands. Most irrigation has an extended seasonal demand, or a demand during those times when receipt of water from precipitation is inadequate for expectable crop growth. Depending on local temperature conditions, recreation and navigation demand may be either continuous or seasonal. Water supply, on the other hand, is notably fluctuating over the seasons. This especially is true for the concentrated surface-water supplies, which in the past have offered the most economical sources of water and water services to meet concentrated water demands. The time at which peak demands occur may be the trough for supply. While this problem of discontinuity can be met in part by regulating demand (adjusting demand to supply availability), methods to improve supply have been more favored.

Since prehistoric times the problem of seasonal discontinuity has been met plainly by artificial surface storage, or by the use of ground water. Artificial underground storage is another device, although not widely used even yet. Costs, of course, are associated with the storage of water in any form and with the recovery of ground water. In addition, in some regional situations, large volumes of water must be captured if full availability and full use of natural supply is to be achieved. It follows that any technique which contributes to reducing the cost of storage or ground-water recovery makes inroads in this problem area. Any technique which enables more complete capture of the natural supply likewise is applicable.

For surface water, such contributions have been made through the invention of mechanical water lifts, the invention and development of earth-moving equipment, through improvement of cement making and concrete use, through hydraulic turbine design, and techniques of designing dams. Contributions also have been made through attention to a number of special problems associated with the construction of dams and the operation of reservoirs. Malaria control, evaporation control, minimizing ice damage, strengthening geologically hazardous foundations, passing valuable fish runs upstream, and passing fingerlings downstream, are cases in point.

For ground water, contributions toward the same end have been made through improved pump design and well-drilling techniques, and through the development of techniques for underground water storage.

LONG-RANGE DISCONTINUITIES

The long-range discontinuities have been much more difficult to treat than the seasonal discontinuities. The basic problems are those of forecasting any economic phenomenon on a long-range basis, and forecasting important recurrent but probably noncyclical fluctuations in moisture precipitation from the atmosphere. Here we are concerned with movements of regional and national scope which are likely to be effective on large scale. Any technique which helps to improve our knowledge of the availability of water through time, or which adds to our capacity to forecast demand for water and water services more accurately, will lessen this discontinuity. If the incidence of demand can be more accurately foreseen, and the probable availability of water more accurately charted in time, development of needed facilities to match supply and demand is more likely to be undertaken.

Appraisal of the long-range demand for water and water-derived services basically must depend on the analysis of a complex of demographic, economic, political, and technical factors. Some of the former factors already have been mentioned in describing the localization of water demand (chapter 3). Always important, these are sometimes so difficult to forecast as to be almost imponderable. But technology, too, can influence long-range water demands in striking ways. The demand for withdrawal uses has been visibly affected by the appearance and development of the rayon, petro-chemical, and other chemical industries, by new concepts of sanitation, by air conditioning, by light-weight metal piping, and other technical developments. The demand for water-derived services in the flow and on-site use of water has been particularly affected; this is well illustrated in the demands for electric power and recreational use which stem directly or indirectly from technologic change. The widespread use of electric motors as prime movers, the introduction of mass-produced home appliances, new metallurgical processes, and chemical processes demanding electric energy have had profound effects upon the demand for electricity, and accordingly for the development of hydro power facilities. Demands also have been affected by water-conserving devices, like systems for manufacturing plant reuse. Yet the over-all effects of these changes, in their combined impact upon the demand for water services, have not always been clearly understood.

In addition to the more permanent long-range changes in demand, there are fluctuations of a quasi-cyclical nature dependent upon economic and social conditions. Fluctuations in demand which follow business cycles exert a real effect on the demand of water no less than upon other production inputs. Supply must be so constituted as to accommodate the periods of high economic activity as well as those of low activity.

The nature of water supply over the long run also is not clearly understood. Adjustments of both water supply facilities and of water demand, to meet the vagaries in the longer-range natural availability of water, have been handicapped by imperfect knowledge of long-range precipitation fluctuations and the forces influencing them. They have also been handicapped by fragmentary knowledge of ground-water movement, by an inadequate knowledge of the hydrology of streams, and by a lack of quantitative understanding of the hydrologic effects accompanying physical alteration of watersheds under grazing, cultivation, deforestation, and other economic uses of land. Lack of a definitive understanding has helped to nourish controversies, like that concerning the merits of upstream works as compared to main-stem river regulation.

Technology is of interest in two further respects in approaching the long-range discontinuity problem: (a) in devising tools for the accurate appraisal of water availability; and (b) in mechanized agriculture and forestry, which have hydrologic effects. Devices like radio-recording hydrographs, the instruments of geophysical exploration, radiosonde balloons and missiles have all improved our capacity to appraise the long-range availability of water. On the other hand, the plow that broke the plains, and its successors, introduced potent new factors which still have to be taken into account in hydrologic study.

Two key problems in this area at the present time would appear to be long-range fluctuations in the receipt of moisture from the atmosphere; and appraisal of the dimensions, content, and dynamics of aquifers underground. In arriving at a better understanding of them, we may expect scientific and technical contributions from as diverse sources as knowledge of events on the surface of the sun and of the behavior of hydrogen isotopes as tracers.

Problems of Quality Requirements

In view of the sensitivity of a number of uses, including irrigation, to sediment and dissolved solids content, and the rare occurrence of "pure" water in natural supply, techniques of adjusting water quality and the quality requirements of demand are extremely important. Technical effort within this area has had many beneficial results.

The quality problem has been met in three ways: (*a*) quality improvement by treatment of supply; (*b*) development of "salt"-tolerant uses, pairing of tolerant uses with low-quality supply; and (*c*) restriction of use to prevent quality deterioration.

Depending on the specific situation and the nature of both supply and demand, the following supply treatments have been used successfully for quality improvement: dilution, as in mixing water of high dissolved solids content with water of low mineral content; chemical alteration, as in water "softening," or chlorination; distillation; filtering and precipitation of sediment.

The pairing of tolerant uses and low-quality supply has long been practiced in agriculture, where salt-tolerant crops and salt-tolerant animals (e.g. sheep) have been paired in use with water supplies of relatively high dissolved solids content. Within industry, especially in recent years, the use of brackish or salt water for cooling and cleansing is another type of illustration.

Restriction of use to prevent quality deterioration has been applied particularly to the disposal of industrial and municipal wastes. Most water-transported wastes can be treated to reduce their capacity to impair water quality. Since costs are involved, any technique which reduces the expenses of treatment contributes in this direction, whether by making wastes into economically usable products, by avoiding flowage disposal, by treatment process simplification, or by other means.

Considerable quality deterioration of water occurs where irrigation is practiced in arid or semi-arid climates; thus any technique which reduces evaporation of water, whether in the application to crops, in transit, or in control of seepage, will also contribute to quality improvement in this situation.

The importance of quality-improvement or quality-adjustment techniques in American water development is enhanced by the size of most American drainage basins. While water-quality problems are always met locally, any upstream deterioration of water quality is of regional concern within a basin. In the extreme situations found throughout much of the West (high initial dissolved solids content and intensive demands) cumulative quality deteriorations may eventually cause some waters to become unusable without treatment in the lower sections of the basins. Quality improvement by treatment and preventive restrictions on use thus should be considered together, and on a regional basis, within the susceptible drainage basins.

Problems of Quantity

There is a geographical aspect to the matching of water demand and supply, just as there is a time and a quality-adjustment aspect. Demand for water or water-derived services arises in many instances without reference to the local, or even the regional, characteristics of supply, and water supplies quantitatively are very unevenly distributed over United States territory. Techniques of balancing these geographical inequalities therefore become a very important part of water development technology.

In describing this problem it is useful to distinguish between demand that is really dispersed and demand that is really concentrated. The technical challenges presented by both may be quite different for the several water development purposes. Dispersed small-scale demands for withdrawal, like farm household or livestock watering, generally have been met easily by exploiting local shallow-lying ground water or collecting from small surface supplies like farm ponds. Except under extreme conditions, the technical challenge has been small. On the other hand, dispersed demand for electricity, illustrated in the same farm units, was never met successfully until improved techniques of transmission line construction warranted dispersed distribution. However, the principal problems of geographical inequality are those attending concentrated demands.

The classic answer to these problems has been to transport water from concentrated supplies to localities of concentrated demand. The transmission of electric energy from hydro installations may be regarded as a variant of water transportation.

Other methods of treating the problem have been water conservation, substitution of other resources for water, multiple use of water, and conscious dispersal of demand.

WATER COLLECTION AND TRANSPORTATION

Uneven water distribution would be no problem if demand for water coincided geographically with location of supply. In the case of soil moisture use, a dispersed demand, location of supply does in fact determine the location of use. However, concentrated demands often have different sites from concentrated supplies. A number of cities are located on or are adjacent to streams or lakes. Some, like Chicago, Cleveland, St. Louis, or New Orleans, have grown because of the presence of a stream or lake. But there are others, like Los Angeles, San Francisco, Oklahoma City, Boston, Indianapolis, Atlanta, or San Antonio, whose present size has little relation to fresh or near-fresh water sources. When the correlation of demand for supplementing soil moisture—or irrigation—is examined, the geographical distance separating demand and supply is even more notable, as in arid regions. The distance separating every major irrigation project in western United States is measured at least in tens of miles, and sometimes there are also substantial topographic barriers. The irrigation of the Columbia "basin" of central Washington State, the transmontane diversion involved in the Colorado-Big Thompson project of Colorado State, and the irrigation of the Imperial Valley of California are examples of the separation of lands and water sources in arid regions. The situation reaches an extreme in California, where water demands of southern California and the San Joaquin Valley have been separated from the water sources which serve them by as much as 450 miles. Even more distant supplies are in prospect for the mounting population of this state.

From the earliest days of organized water supply in the ancient world, the principal solutions to problems of concentrated demand have been through engineering projects capable of transporting water from areas of concentrated supply. Three principal functions have been characteristic of these works: collection, transportation, and distribution to consumers. In some cases collection and storage have been combined, thus treating the time distribution and the geographical problem together; in other instances, like the ancient foggaras,[1] or Indian tanks,[1] collection works have been necessary to assemble diffused natural supplies.

Construction of the first collection and distribution works is lost in prehistory. We have evidence of a well-developed irrigation and water distribution system about forty-two centuries ago (Indus Valley, Pakistan),[2] and a dam is thought to have been constructed in Egypt about 5,000 years ago.[3] One system constructed more than 2,000 years ago is still in use (Min River Delta, China). Aqueducts date at least from the time of Solomon (*circa* 950 B.C.). Mesopotamia, Persia, and several Arab cultures developed to a high degree the art of collecting water from concentrated natural supply and transporting it to the areas of concentrated demand within their territory.

While engineering techniques and available construction materials have improved greatly within the last century, the approach to problems of supplying concentrated water demand remains basically the same as it has been since the construction of the first large-scale irrigation works: collection and transportation of water from areas of surplus supply in such a manner as to give reasonable assurance of sustained-yield supply to meet the municipal or other demand during its season.

Several technological developments have been useful in this area. Much of the technique of collecting and lifting ground water is applied to the transportation problem. Contributions have been made in the techniques of ground-water collection (like the foggaras, tube wells, and fracturing consolidated aquifers), well drilling, and in pumping design. Of equal interest have been techniques of tunnel construction and equipment therefor, cement and concrete technology, equipment for canal construction, ditching machinery, and canal lining techniques.

---

[1] See Glossary, p. 661.

[2] B. Frank, *Water, The Yearbook of Agriculture, 1955*, U. S. Department of Agriculture, Washington, p. 1. See also Stuart Piggott, *Some Ancient Cities of India* (London: Oxford University Press, 1945); *idem, Prehistoric India to 1000 B. C.* (Harmondsworth, England: Penguin Books, 1950).

[3] Frank, *op. cit.;* R. S. Mehta, "India was a Pioneer 5,000 Years Ago," *World Health Organization Newsletter,* VII, 6 (June 1954); "The Oldest Dam in the World," *Public Works,* Vol. 83, No. 12 (December 1952), p. 117.

The transportation of a water-derived service—the transmission and economical distribution of electricity—is an aspect of this whole area which should not be overlooked.

CONSERVATIONAL CONSUMPTION OF WATER

Transportation of water or water services essentially is directed toward fulfilling demand through increasing supply in the locality of demand. Conservational consumption is designed to help satisfy demand through reducing use. This may be brought about by water reuse, wastage elimination or restricted withdrawal, equipment design for water saving, crop plant breeding, and water-conserving irrigation techniques.

Conservational consumption may be effective either in reducing the actual disappearance of water, or in reducing the need for withdrawal facilities. An example which involves relatively little reduction in actual disappearance of water occurs in various techniques of reuse—in the recycling of water in industry for cooling or other purposes, in its reclamation from municipal waste discharge, and in closed circulation for air conditioning.

Savings also are possible through more efficient transportation of water for irrigation. Most irrigation canals in the United States are unlined, a condition which accounts for the loss of about 32.5 million acre-feet of water in conveyance each year.[4] Techniques for economical lining of some canals could make the actual irrigation withdrawals more effective, although little water disappearance would be avoided. "Loss" in conveyance may actually become aquifer recharge, and can be beneficial.

An example of water saving for flow use is the device developed for minimizing diversions for fishery use at dams having hydro generators. Known as a "skimmer," it has no effect on disappearance, but it does assist conservation for turbine use.

On the other hand, crop plant breeding for drought tolerance and for lowered water consumption may actually reduce water disappearance. So would irrigation equipment and application practices designed to reduce the amount of deep percolation and evaporation which takes place in irrigation.

[4] K. A. MacKichan, *Estimated Use of Water in the United States* (Circular No. 398), U. S. Geological Survey, Washington, 1957, p. 7.

SUBSTITUTION OF OTHER RESOURCES FOR WATER

Some of the services provided by water also may be given through use of other resources. Inland transportation of heavy goods, for instance, does not depend on water alone; in fact it more commonly is carried on by rail, highway, or, for petroleum, by pipelines. Electricity may be produced by the use of fuels in thermal generating processes, and increasingly is being produced in this manner. Accordingly, a degree of water conservation can be attained by substituting other sources of electricity generation for hydro. Such conservation is only relative, however, being localized in time and place. For example, supplies of irrigation water available during the growing season may be augmented if hydro generation located between the irrigable lands and the storage sites for irrigation water is foregone during nongrowing seasons, or is used only for necessary pumping purposes.[5] Water thus may be "conserved" for use on the irrigable lands. Thermal generation would be substituted for hydro to meet the energy demands, and *local* conservation of water thus would be achieved. Looking at a river basin as a whole, however, the actual disappearance of water for electricity generation will be greater when thermal generation is employed, because of the evaporation of cooling water. The continued water inputs required when resource substitutions are made in the interest of water conservation may leave the net absolute savings small, even though such savings can be important locally.

At the same time, substitution of the product of other resources for some water services may ease or avoid local and regional problems of matching supply and demand. The principal services that can be provided conveniently from the product of resources other than water are electricity, transport, recreation, and waste disposal. The principal resources that are affected when these substitutions are employed, either consciously or unconsciously, are the mineral fuels and the minerals used in construction or manufacturing. (See checklist, table 12.)

For both water conservation and substitution of other resources, pricing and other economic considerations are important determinants. The cost of obtaining water for a given purpose, or the cost of the water-derived service, in the end determines the feasibility of applying

---

[5] If water supply is limited, conflict will occur between the hydro thus located and irrigation use at all times except those when delivery of irrigation water is needed.

TABLE 12. *Some Water Uses and Their Known Substitutes*

Water use	Substitute	Other resources required for substitute			
		Land	Fuel minerals	Other minerals	Forest
Hydro power	Thermal generation		X	X	
Recreation	Scenic, travel, indoor	X	X	X	X
Navigation	Highway, rail, pipe-line	X	X	X	
Waste carrying	Settling, soil filter-ing, combustion, by-product manu-facture	X	X	X	

conservation or substitution techniques. Thus public measures that underwrite part or all of the cost of water development may discourage the use of conservation or substitution by camouflaging true costs.

MULTIPLE USE OF WATER AND WATER FACILITIES

It is only a short step from consideration of substitution to multiple water use. Multiple use is nothing more than making the same water or the same facility serve effectively all the existing forms of demand for withdrawal, flow use, or on-site use. It is particularly within the last half-century that multiple use has emerged in this country and in other industrial nations. It has been dependent upon technical advances, and is far from having run its course in contributing to the problem of matching water supply and demand. Although not everywhere prerequisite, a key physical facility for most multiple use is the storage reservoir, as it is used for surface water. A key technique is the design of reservoir rule curves for water detention and release. The many facilities essential to multiple use include electrical generators and transmission equipment, penstocks, pipelines, filtering and waste-removal plants, canals, and locks. Efficient multiple use must be considered both from the point of view of supply and of demand.

Several significant distinctions may be made as to the nature of multiple water use. There is multiple use of water itself and multiple use of supply or storage facilities. There is multiple use on or from a given site, and multiple use at successive sites in the path of a given stream flow. There are competitive and complementary uses. Competitive uses may be physically exclusive, as illustrated by the impossibility of using the same water at a given location for both irrigation and municipal supply.[6] They may also be physically exclusive in requiring specific assignment of storage space in a given storage facility during a given period, as when flood-prevention storage and conservational storage for other purposes must come at the same season. Or competitive uses may conflict where one use renders water unfit for subsequent supply for other purposes. Thus waste carrying may make a stream unfit for municipal supply, industrial supply, recreation, or fish and wildlife use downstream from the point of waste discharge. Complementary uses generally have different incidence either in place or time, one from the other. For example, use of water for generating power at a hydro site in the growing season may be complementary with the use of the same water for irrigation immediately downstream from the hydro plant. Or storage of water for both these uses in the same reservoir space is complementary. Likewise, the winter use of reservoir space for flood prevention where floods occur during wintertime is complementary to use of the same space for storage and withdrawal during other seasons for irrigation, hydro power, recreation, or other purposes.[7]

The primary objective in multiple-use management of either stream or ground water is to eliminate or minimize competition between different demands for a given water supply, and to promote complementary use. While local water use may be limited to one or two purposes, without exception in the United States regional demands for water are multiple demands covering all or nearly all purposes served by water. On a regional basis, a major part of the concentrated supplies of water also must serve these multiple demands. Water use considered on a regional basis therefore inevitably is multiple-purpose use. The technique of effective multiple use accordingly is basic to efficient water development and management.

Competition among uses appears in six different ways.

1) Constant withdrawal uses may conflict with seasonally or diurnally peaking withdrawal uses. Year-round withdrawal for hydro power generation or for industrial cooling or processing water is not compatible with the high seasonal demands for agricultural use in most of

[6] Downstream reuse of course is possible, assuming no prohibitive deterioration of quality.

[7] The authors are indebted to Maynard Hufschmidt for helpful suggestions on the nature of multiple use.

the areas where irrigation is practiced. The constant withdrawal demands of waste carrying likewise are not compatible with the diurnally peaking demands of some industries, or the diurnally and periodically peaking demands of hydro power production.

2) There can be conflicting simultaneous incidence of demand. Irrigation again offers an important example. Irrigation water demand may appear at the same season as the peak demands for navigation use, air-conditioning use, recreational use, and fish and wildlife support—the warm season of the year.

3) Seasonal uses of one type may conflict with other uses at other seasons. For example, heavy summer releases for navigation along the Missouri River can conflict with needs for heavy hydro power generation during the winter months in the service area of the Missouri River plants. Likewise, storage of water for irrigation during the non-growing seasons may reduce or eliminate hydro power production during the season of storage accumulation.

4) Major flow regulation of any kind may be incompatible with some uses. Major storage, diversion, and detention facilities for all purposes can displace entirely recreational and fish and wildlife activities which depend on water flow. The now well-known controversies over power diversion and recreational benefits at Niagara Falls, the Echo Park dam, and game fish *vs.* flow regulation on the Rogue River in Oregon all illustrate this incompatibility. However, in other cases flow regulation may benefit the same purposes. Such is thought to be the result of regulation on the Upper Feather River, California.[8]

5) Some uses cause quality deterioration, which bars subsequent use of the water for some other purposes. Waste carrying offers the best example of this conflict, which is now found on many streams in the United States, and on nearly all reaches of a few (figure 15 and table 13).

6) On-site uses of water (recreation, fish and wildlife, navigation) may cause water disappearance through evaporation of stored water. Water availability for other purposes thereby is reduced.

Flood prevention recently has been a major purpose in the construction of water storage facilities, even though it is not a *use* of water. Ideally, flood prevention as a major purpose in management within the United States should disappear as more intensive use of water places higher and higher value on the conservation storage of peak stream

[8] Blair Bower, letter to the authors, 1957.

TABLE 13. *Organic Pollution—Selected Drainage Basins in the United States*

Drainage basin	Area (mi²)	Number of pollution discharge points reporting		Reported pollution (PE/mi²)		
		Munici-pal	In-dustrial	Munici-pal	In-dustrial	Total
1. Penobscot	8,570	51	15	8.8	281.6	290.4
2. Kennebec	5,870	59	41	16.6	263.2	279.8
3. Androscoggin	3,450	40	22	3.4	731.4	734.8
4. Merrimack	5,010	91	56	563.3	159.6	722.9
5. Connecticut	11,265	236	259	66.6	154.0	220.6
6. Hudson	12,650	206	169	81.4	112.1	193.5
7. Tennessee	40,910	185	87	26.4	34.4	60.8
8. St. Johns	7,612	41	27	35.8	36.7	72.5
9. Red River of the North	34,265	90	83	3.8	2.2	6.0
10. Minnesota	16,920	20	10	4.2	7.9	12.1
11. Lower portion of Upper Mississippi .	23,000	51	77	58.0	22.0	80.0
12. Mississippi-Wapsipinicon	9,530	43	10	83.0	44.0	127.0
13. Mississippi-Salt ...	9,530	16	—	5.1	—	5.1
14. Kansas	60,000	163	15	13.0	12.0	25.0
15. Trinity	21,000	110	84	17.4	25.4	42.8
16. Yellowstone	70,400	48	21	1.4	17.1	18.5
17. Spokane	6,640	18	22	32.6	27.8	60.4
18. Yakima	6,000	20	11	13.0	20.0	33.0
19. San Joaquin	16,850	23	23	17.6	30.3	47.9

PE = Population equivalents.

Source: New England—New York Interagency Committee, *The Resources of the New England—New York Region,* Part I; and Federal Security Agency, U. S. Public Health Service, *Water Pollution Series.* (See sources for figure 15, p. 674.)

flows. Until this happens, however, maintaining empty reservoir space to meet flood detention needs, as is necessary on nearly all important American rivers, will continue to be an essential part of river regulation. Ultimately, this storage space can be converted to serve other purposes when the needs appear, and the water of floods can be stored to meet the needs. Thus flood damage prevention still will be served, but beneficial use of the water will be the objective of development.

The major areas of competition in multiple-purpose use are those attending irrigation use of water, waste carrying, and flood prevention

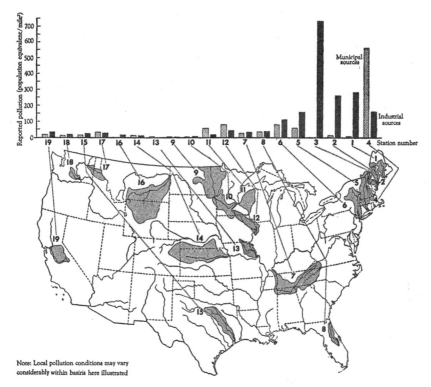

Figure 15. Organic Pollution—Selected Drainage Basins in the United States. (See table 13.)

(table 14). Besides the technical developments associated with constructing and managing storage reservoirs for multiple purposes, several other technical contributions have been made in the direction of minimizing conflict among water purposes. In this class one may include all devices or techniques which reduce or eliminate the capacity of industrial wastes to cause water quality deterioration; processes for economic treatment of municipal and residential wastes; nonstorage techniques of flood damage prevention; and domestic water filtering and bactericidal treatment.

On the other hand, any technical change tending to increase demand for water or water-derived services may also increase the competition. This has been particularly apparent in the rapid development of the chemical and pulping industries with their attendant problems of waste disposal. Problems of quantitative increase of a given water service

TABLE 14. *Competitive Water Uses by Purpose*[1]

	WITHDRAWAL	INDUSTRIAL PROCESS	MUNICIPAL AND OTHER INDUSTRIAL SUPPLIES [2]	DOMESTIC SUPPLY	LIVESTOCK	IRRIGATION	FLOW USE	HYDRO POWER	INDUSTRIAL WASTE	MUNICIPAL AND RESIDENTIAL WASTE	IRRIGATION RETURN FLOW	RECREATION, FISH AND WILDLIFE	ON-SITE USE	NAVIGATION [3]	RECREATION	FISH AND WILDLIFE	FLOOD PREVENTION
WITHDRAWAL																	
INDUSTRIAL PROCESS						X		X	X		X						X
MUNICIPAL AND OTHER INDUSTRIAL SUPPLIES [2]								X	X	X					X		X
DOMESTIC SUPPLY								X	X	X					X	X	X
LIVESTOCK								X							X		
IRRIGATION		X						X	X	X		X		X	X		
FLOW USE																	
HYDRO POWER	X	X				X						X		X	X	X	X
INDUSTRIAL WASTE	X	X	X	X		X						X			X	X	X
MUNICIPAL AND RESIDENTIAL WASTE			X	X		X						X			X	X	X
IRRIGATION RETURN FLOW	X					X										X	
RECREATION, FISH AND WILDLIFE						X		X	X	X				X			
ON-SITE USE																	
NAVIGATION [3]						X		X			X						
RECREATION			X	X	X	X		X	X	X							X
FISH AND WILDLIFE				X				X	X	X	X						X
FLOOD PREVENTION	X	X	X					X	X	X					X	X	

X *indicates competitive demand,* or conflicting effect on water quality or availability.

[1] This table is not to be construed as indicating that the conflicts noted inevitably occur where two or more competitive types of demand appear. Instead the table indicates that such conflicts have been observed in the past, or that unplanned use may result in conflict, or that compromise among two or more competitive purposes may be necessary under planned use.

[2] "Other industrial supplies" are used mainly for cooling.

[3] Navigation has been classified as a flow use in the past. However, it is not dependent on the flow characteristics of a stream. Flow may be either an aid or a handicap to powered vessels, but is not essential to their propulsion.

are raised, and also questions of balance among the multiple uses served by a regional supply.

An important objective in water management in recent years has been a reduction of competition and an increase of complementarity among multiple uses. On an ideally managed stream or other water supply the possible uses are brought to a state of complementarity. However, adjustments of each use often are necessary to achieve this state, and maximum use for a given purpose may have to be foregone in order to permit partial use of the water for other purposes. The objective of management should be the adjustment of the possible uses to a condition of complementarity which results in the production of maximum economic and social benefits at the investment and operational costs incurred.

PART II

TECHNIQUES AND TECHNICAL EVENTS
SIGNIFICANT IN THE
PROGRESS OF WATER DEVELOPMENT

CHAPTER 5

Techniques and Technical Events Affecting
Water Development and Its Administration

INTRODUCTION

Techniques and technical events have a specific meaning for administrative organization if an efficient matching of supply and demand is
sought. Some are of limited interest and automatic application, some
may be ignored (at least for a time), and others are profoundly dis-
turbing whether they are reflected in administrative organizational
adjustments or not. Furthermore, different types of administrative
interest or involvement are found in several technical events. Some
give rise to problems of public monitoring and regulation, others to
problems of territorial jurisdiction and co-ordination, some to the
public organization of research effort, others are clearly the province
of privately administered business, and so on.

Some of the technical changes that have influenced the course of
water development in the United States have been consciously arrived
at, with solution to a water problem a foremost objective. Other and
probably more numerous technical influences are of a windfall nature,
the product or by-product of other needs and objectives than water
development. And it is possible that these, as a group, have had the
greater influence on the course of water development. However, even
indirect technical influences require a degree of conscious effort in their
application to particular situations.

The technical events that have influenced the recent course of water
development in this country are illustrated in this section of our study
by twenty-two brief case descriptions. The windfall nature of some,
like the invention of earth-moving equipment, the rise of the chemical

industry, and invention and manufacture of household electric appliances, will readily be apparent. It follows that their appearance and impingement on water needs and facilities have been more or less random and out of the course of conscious organization for water development. Since the windfall events and the consciously developed techniques are incident on the same water development, a classification which takes account of both is the only valid one. Such a unified classification is helpful to administrative perspective because, without it, important technical influences of random incidence may be obscured until they assume commanding or even crisis proportion.

The impingement of technical events on the problem of matching water supply and water demand (also a problem for administrative organization) may be viewed generally somewhat as follows. Take a region of the dimensions of any moderate-sized river basin. Assume initially no significant long-range alteration of climate. Assume further an equilibrium between effective demand and water supply, recognizing the size of population and its consumption habits; the nature, location, and productivity of other resources; and the external markets for the region's products. Assume also that water has been treated more or less as a free good, and that development has been undertaken entirely under private entrepreneurship.

The situation of equilibrium may be altered by any of the following: a growth of population, a change in its standard of living, or other social change; a demand for further resource exploitation caused by demand from other markets, or other extra-regional demands; or technologic changes bringing new water uses, new demands for water services, and new geographical concentrations of demand for water.

These new demands may be met by: (1) exploration for, discovery, and exploitation of new supplies; (2) improved production from known supplies; (3) transportation from localities of surplus to those of deficit; (4) interrelated management of a group of individual supplies to meet a variety of demands; and (5) reduction of water demand.

Technology may enter all of these means of matching supply and demand. It can alter, even radically, the prospect of matching demand with supply, not only because of possible effects on demand, but also because of supply improvements. In some instances the application of technology to a problem of supply may represent a specific development for a particular purpose; in other instances it may consist in the use of devices already in existence and originally developed for other purposes.

As will be described later (chapters 17-22), administrative organization may and does cover the same ground. For some measures it is
even the indispensable accompaniment of technology.

Accordingly, one may classify technical events into those which affect
or influence the size and nature of water demand, and those which are
applied to alteration or improvement of supply. Those affecting water
demand may be grouped as: (*a*) increasing demand—these are often
fortuitous and unexpected; (*b*) decreasing demand—they usually are
consciously applied and often specifically devised. The techniques
affecting water supply are: (*a*) those promoting exploration and the
discovery and exploitation of new supplies; (*b*) those improving the
productive efficiency of supplies already in use. This may occur through
extending the use of a given unit of supply, or through economies of
scale achieved in interrelating several or more units of production to
one another and to the demands they may meet.

These considerations form the basis for the classification suggested
below, in which all techniques affecting water development are included
within five groups.

1. Technical Events Increasing Water Demand

One influential group of techniques may increase demand for water
and water-derived services. Many of the influences within this group
are windfalls, and it is in organizational response to them that some
of the most challenging problems of organization occur. Four case
descriptions have been chosen for discussion of this type of influence
in chapter 6. They are the relation of refrigerated transportation and
plant breeding to western irrigation, the effect of the chemical manufacturing industries upon the need for water as a waste-carrying
medium, water application techniques in irrigation, and the relation
of certain appliances to the residential demand for electricity. Numerous
other examples might have been added, like the rapid growth of electrical energy demands because of the development of industrial
processes requiring high electric inputs (e.g. light metals), and the
development of rural electrical distribution systems suited to low-density
service. Each of these technical changes has helped to create a demand
for water resource exploitation beyond that to be expected by the
growth of population in the United States and the accompanying growth
in use of other natural resources in the absence of the change.

2 and 3. Techniques Decreasing Demand or Extending the Services Afforded by a Given Unit of Supply

Some technical changes have a "horizontal" reach. Another important group has a "vertical" orientation. The techniques within the latter may be directed toward making one gallon of water serve where two otherwise would be required. Alternatively, they may permit the use of water for services that would be missing in the absence of these techniques; in other words, they may allow greater water use for existing services. Most of them are consciously generated as a part of water development, and in their development the matching of supply and demand is met on the basis of a discrete physical unit, whether that unit is the growing crop plant, the water-consuming residence or factory, or a water-regulating dam.

TECHNIQUES DECREASING WATER DEMAND

These techniques have a wide range of character and applicability, from agriculture to industry. For instance, any material or device which reduces loss of irrigation water in transit also will extend the use covered by a given supply of water. Any industrial or agronomic improvement which increases output for given inputs of water will have the same effect. Crops that are more drought resistant are also crops that conserve water, and devices that increase thermal efficiency of steam electric plants also reduce water consumption per kilowatt-hour of output.

Four techniques that have helped to decrease water demands in specific situations have been selected for description in chapter 7. Two are from agriculture. They are techniques of pairing tolerant use with low-quality supply, as shown in the use of saline water in western agriculture, and in the development and use of drought-resistant crops. Some parallel industrial experiences occur in the use of salt water for processing. The last two cases are industrial, dealing with methods of treating wood pulping wastes, and with spray disposal of food processing wastes.

EXTENDING THE SERVICES AFFORDED BY A
GIVEN UNIT OF SUPPLY

Among the techniques falling under this head are several which might be termed "scheduling." They are techniques of multiple-purpose water regulation, underground water storage, and dilution of low-quality supply. All of the scheduling techniques have the objective of "stretching" the available water supply in a manner permitting fulfillment of all effective demand for each of the services that can be provided by water.

In addition to the scheduling techniques, other technical works may be directed toward the same end. They are the water-conserving and water-treatment devices. Methods developed for the chemical or filtration treatment of water belong in this category. They make suitable for domestic or other high-quality use water which otherwise would be allowed to remain or flow on unused, or which would afford a service with an injury risk attached.

· This subject will be treated in chapter 8 with the following case descriptions: for the scheduling techniques—multiple-purpose reservoir operation, and underground storage; for the water-conserving and water-treatment devices—water-conserving land use, and "water-softening" equipment.

4. *Technical Improvements Promoting the Scale Economies*

Because of the size of American river basins and the extent of United States territory, the technical improvements that have promoted scale economies are among the more important from an economic point of view. In this class we might place all of the materials and design techniques that have increased the size of dams and the capacity of storage reservoirs, the equipment and techniques enabling the construction of facilities for long-distance transportation of water in large volume, and the equipment that makes possible the long-distance, high-voltage transmission of electricity. These techniques and forms of equipment have permitted capitalization upon the assets of diverse resources in water-scarce areas through construction of inter-basin diversions and water-exchange systems. They have permitted the merging of large geographical areas of diverse load and generating characteristics within a single transmission and distribution system. They have permitted the joint management of all regulation works within a single drainage basin as a unit, and even the joint management of the waters of several basins. They have made possible the storage of extreme flows for beneficial use over several seasons.

The scale economies are the techniques having a "horizontal" dimension, as contrasted with those concerned with a discrete unit of supply. Principally they have to do with the interrelated operation of several or numerous units of supply, but they also are concerned with scale interconnection of units of supply and units of demand. These qualities are perhaps best illustrated in the organization of a large-scale, geographically extended electric power generating and distribution system, with generating units of diverse characteristics and loads of equally diverse nature.

Six case descriptions are discussed for this group in chapter 9: the development of earth-moving equipment, dam design and construction, techniques of canal and tunnel construction, electrical transmission equipment and techniques, integrated basin water control,[1] and dispersal of demand.

5. Techniques Extending the Physical Range of Water Recovery

Much of the technical effort which in the past has extended the physical range of water recovery has concerned the exploitation of ground water. Deep-drilling techniques, geophysical prospecting, and study of ground-water storage and movement all have contributed to our "proven reserves" and usable supply of water. However, every step taken in the direction of more complete knowledge of the long-range availability of surface waters also extends the physical range of recovery, because it makes planning for exploitation feasible when exploitation economically is warranted. The most significant actions in this area are likely to come in the future. Weather modification techniques, desalinization, use of ground-water tracers, study of hydrometeorology, management of natural vegetation, and other technical and scientific areas now being investigated make the general field of extending the physical range of recovery a fascinating one.

The most extended discussion of these techniques appears in Part III of this study (chapters 11-16), dealing with future development and organization. However, some examples worthy of note will be found

---

[1] Integrated basin water control is to be distinguished from multiple-purpose use of single facilities. In effect, it is multiple-facility management for multiple purposes, the facilities being planned, constructed, and operated in co-ordinated manner.

in the recent history of water exploitation and use. The examples relevant to past development (chapter 10) are mainly from the field of ground-water use. To illustrate application of techniques in the use of surface water, malaria control on reservoirs is included as a case description.

CHAPTER 6

Technical Events Increasing Demand for Water

FOUR CASES

(1) Transportation, Plant Breeding, and Western Irrigation—the development of refrigerated transport; California iceberg lettuce; russet potatoes. (2) The Growth of the Chemical Industry—the problem of chemical wastes. (3) Water Application Techniques—lightweight piping and sprinkler attachments; plastic piping. (4) Residential Demand for Electricity, Associated Operating Problems of Electricity Systems, and Hydro Generation—the nature of changes in residential demand for electricity; significance of increase in residential electric use.

Among the many technical events that have influenced the rate of water exploitation, four have been chosen for case description: (1) refrigerated transport, and plant breeding; (2) growth of the chemical industry; (3) sprinkler irrigation; and (4) the residential demand for electricity.

These examples are intended to represent several functions, economic pursuits, research and technical disciplines, and regional applications. Agriculture, industry, and domestic consumption are represented. Among the water development purposes affected are irrigation, electric energy production, industrial water supply, and waste carrying. The disciplines represented are applied botanical research and agronomy, chemistry and chemical engineering, mechanical engineering, soil science, and the technique of transportation. Developments characteristic of the two major sections of the country, western arid-land irrigation and eastern industrial processes,[1] are included.

All of the events here described are influences increasing the rate of water exploitation. This is in key with the period through which the nation recently has moved, and each event should be considered within its own economic history context. It is not suggested that technical events alone are responsible for the course followed by demand pressures upon the rate of water exploitation. As suggested in chapter 3, growth characteristics of the population as a whole, political structure and political policy, popular attitudes toward the natural amenities, the use of resources other than water, growth of per capita productivity and income, and other factors also help to shape demand, perhaps overwhelmingly. But technical changes may intensify demands that have their roots in the social, political, or economic environment; and they may turn potential demand into effective demand. The cases here related to irrigation might well be considered with this view in mind. For instance, the expanding national market for agricultural products would have meant little to eastern irrigation if technical improvement had not brought irrigation equipment suited to the physical conditions and economics of production on eastern farms.

1. Transportation, Plant Breeding, and Western Irrigation

Until very recently the economic structure based on western irrigated agriculture was highly dependent on markets outside the region, and mainly in the eastern United States. It follows that the water requirements of such agriculture were indirectly related to extra-regional markets. The rapid growth of California's population and industry and the more modest growth of the Pacific Northwest have changed somewhat the market orientation of western irrigation districts. However, the condition of the eastern market and the receptivity of the eastern consumer to their products are still subjects of serious concern to a large number of the operators of western irrigated farm lands, particularly the producers of specialized horticultural crops.

Because of the great distance separating western farmer and eastern consumer, the connection between farmer and market was not achieved

[1] Developments related to industrial processes also concern the humid lands of the Pacific Northwest.

automatically. Specific technical events had an interesting place in bringing the eastern United States and other distant markets within the sales range of the western farmer. Some time-honored methods of food preservation, like fruit drying and, later, canning, were effective devices. And within the last two decades quick frozen foods in moisture-proof packages have vastly increased the market potential. However, effective counteraction of distance has required the marketing of as wide a range of products as possible, including fresh fruits and vegetables out of northeastern season.

THE DEVELOPMENT OF REFRIGERATED TRANSPORT

Among the significant effects of transportation improvement have been the transformation of agriculture in some regions and the creation of a national market for areas of specialized agricultural production. This has been especially evident in many localities in the West, where agriculture was made possible by irrigation. With the introduction of reliable methods of refrigerated transport, the potential of this development was greatly increased.

The impact of refrigerated transportation upon irrigated farming in the western states can be traced from 1890, the year that Philip Armour ordered a thousand new refrigerated cars in order to handle fresh fruits and vegetables as well as dressed meats. Possessing few customers for the new transportation service he inaugurated, Armour was forced to send representatives into the developing agricultural areas of the West to demonstrate the practicality of refrigerated cars and to promote production for his transit facilities. Traffic developed rapidly in the following decades; for example, the first carload lot of deciduous fruit was dispatched from California to New York in 1889, but by the beginning of the century 4,000 cars were being shipped each year, and by 1924, 65,000 carloads moved to New York from the Pacific Coast states.[2]

The design of refrigerated cars developed slowly, and specifications for a standard insulated car, or "reefer" as it is termed, were not firmly established until 1918. This had about 10,000 pounds of ice capacity. Subsequent innovations steadily increased both the capacity and efficiency of these cars. Load capacity per car was about 26 tons at the beginning of the century, and was standardized at 35 tons in the cars

[2] W. P. Hedden, *How Great Cities are Fed* (Boston: D. C. Heath, 1929), p. 40.

built during the 1920's. It now is twice the latter figure in the newest 50-foot cars that are equipped with mechanical refrigerating systems.[3]

At the same time that refrigerated rolling stock design was being standardized, progress was being made in providing complementary facilities like transit icing stations. Travel time from the West Coast to the Atlantic Seaboard was initially in excess of one week even for the most rapid service; this for most perishables necessitated reicing the cars at least once every twenty-four hours. Manufactured ice became an important contribution to this development.

Another factor in extending the shippers' geographic range was the determination of handling and shipping techniques to preserve the freshness and consumer appeal of each perishable crop, and to minimize spoilage. Such knowledge came not only from hard-gained experience but also from vigorous experimentation. The proper amount of initial icing and of reicing, the quantity of salt to be added to the bunkers, and the best method of packing and loading the produce were problems which had to be determined independently for each type of produce. New measures of body icing were introduced at this time. The importance of removing the field heat of the produce as rapidly as possible had long been recognized. Under the customary practice, however, it took from twenty-four to thirty-six hours to bring the produce down to the desired temperature for shipment, and only comparatively recently have effective precooling measures been devised to reduce this interval. These include the vacuum cooling techniques developed in the western states, and the process of cooling with chilled water, a measure extensively practiced in many areas of the Southeast.

A major development in refrigerated transport occurring since World War II has been the introduction of cars equipped with diesel-electric refrigerating systems. As of 1956, there were 2,000 such railroad cars in operation. Although capable of maintaining ladings at any temperature between zero and 70°F., they have been developed especially to meet the demand of the frozen food industry for zero temperatures in transit. With such shipments, important economies in transportation are achieved. One carload of frozen citrus concentrate, for example, is the equivalent of thirteen carload lots of unprocessed fruit. Another significant trend has been the increasing volume of produce moved by refrigerated truck. This method of transit has been relied upon primarily

[3] American Railway Car Institute, *Railroad Car Facts, 1954*, New York, 1955, p. 16.

for shorter hauls, but during the last decade the proportion of long-distance shipments made by refrigerated trucking has been increasing rapidly.

With each of the innovations and technical developments in this field, the final condition of the transported produce has been improved and new agricultural commodities have been added to those previously handled. Thus, the impact of this development upon irrigated agriculture in the West has been progressively multiplied. The changes that have occurred in the national diet of the United States can in large part also be traced to it. While the use of cereals, pulses, and starches has steadily declined, the per capita consumption of fresh fruits and vegetables, increasingly made available on a year-round basis in all areas of the country, has risen spectacularly in the last half-century.

The increase in the irrigated acreage of the western states and the number of refrigerated cars in service in the United States show a chronologic co-variance (table 15). This is worthy of note, even though other factors also influenced the growth of both irrigated acreage and refrigerated transport. Without refrigerated rail cars, western irrigation would have been much more limited to forage crop production and to the fruit and vegetable markets available through drying and canning processes. Although the number of refrigerated cars has declined slightly in the last two decades, other considerations must be taken into account in assessing the influence of this development upon western agriculture. Thus, allowance must be made for the sharp rise in frozen food shipments in recent years, the more rapid transit service, and the increase in average storage capacity of refrigerated cars. It then becomes clear that the volume of perishable produce handled by rail shipment has continued to increase. At the same time, the rapid growth of trucking in the refrigerated transport field should be noted. Currently, for example, some 195,000 refrigerated truck and trailer units are estimated to be in operation.[4]

One of the most significant accompaniments of these developments has been the emergence of many highly specialized centers of agricultural production in localized areas suited to given crops. Many of these centers, which produce for the national or for a broad regional market, are located in the western states, most often in areas that are irrigated. Today virtually all of the apricots, almonds, dates, figs, prunes, and olives that are domestically grown come from the eleven states west of

[4] Aluminum Company of America, *Refrigerated Trucks and Trailers,* Pittsburgh, Pa., 1956.

TABLE 15. *Increase in Refrigerated Railroad Cars and in Irrigated Acreage in the United States, 1890–1955*

Year	Number of irrigated acres[1]	Year	Number of refrigerated railroad cars[2]
1890	[a] 3,716,000	1890	[3][b] 2,440
1900	[a] 7,744,000	1900	[3][b] 14,480
1910	[a] 14,433,000	1910	[3][b] 30,918
1920	[a] 19,192,000	1920	[4][c] 90,978
1930	[a] 19,548,000	1928	[d] 148,450
		1933	[d] 154,876
1940	[a] 21,004,000	1940	[e] 144,666
1949	[f] 25,269,000	1950	[e] 127,210
1954	[f] 28,466,000	1955	[g] 124,253

[1] All land irrigated in 17 western states, plus Arkansas and Louisiana; for 1954, preliminary figure for irrigated cropland harvested plus irrigated pasture.
[2] Includes all cars of Class I railroads, railroad-owned and controlled private refrigerator car lines, and private lines, unless otherwise noted.
[3] Refrigerator cars owned by Class I railways only.
[4] All refrigerated cars except those owned by meat packers.

Sources:
[a] U. S. Bureau of the Census, *Historical Statistics of the United States 1789-1945,* 1949, p. 122.
[b] U. S. Interstate Commerce Commission, *Statistics of Railways in the United States, Annual Report,* Washington, 1891, 1901, 1911.
[c] W. P. Hedden, *How Great Cities Are Fed* (Boston: D. C. Heath, 1929), p. 40.
[d] U. S. Interstate Commerce Commission, *Use of Privately Owned Refrigerator Cars,* Suspension Docket INS No. 3887, Washington, 1934.
[e] American Railway Car Institute, *Railroad Car Facts, 1954,* New York, 1955, pp. 16, 17.
[f] U. S. Bureau of the Census, *Statistical Abstract 1956,* p. 605.
[g] American Railway Car Institute, *op. cit.,* p. 17; and Association of American Railroads, *Annual Report of the Car Service Division, 1955,* Washington, 1955, p. 21.

the Rocky Mountains, while 95 per cent of the grapes, 85 per cent of the sweet cherries, 75 per cent of the avocados, pears, and cantaloupes, 65 per cent of the asparagus, 50 per cent of the peaches, and more than 50 per cent of the commercial truck crops are produced in this region. The truck crops include 90 per cent of the lettuce produced commercially.[5]

[5] W. R. Nelson, "Development of Natural Resources," *Hydrodynamics in Modern Technology* (Cambridge, Mass.: Massachusetts Institute of Technology, 1951), p. 45.

CALIFORNIA ICEBERG LETTUCE

The story of lettuce production in California illustrates the impact of refrigerated transport, and of plant breeding, upon irrigated agriculture. Today about two-fifths of the nation's total lettuce crop for commercial shipment is from the Salinas-Watsonville district,[6] a narrow and comparatively small valley area formed by the Salinas and Pajaro rivers in Monterey County of central California. It has been observed that probably "nowhere else in the United States does so small an agricultural area produce so large a percentage of a widely consumed commercial crop."[7] The ability of this area to command such a large proportion of the nation's lettuce market is, of course, essentially a product of developments in the field of refrigerated transport. However, important innovations were first required in the type of produce grown and in the method of cultivating and handling it before this potential could be realized. The increasing availability of artificial fertilizer and the development of specialized equipment such as field levelers, multibed listers, and planters were also factors which made possible the mechanized type of production characteristic of horticulture in this area today. However, the basis for the expansion of lettuce production in California in large part depended upon successive varietal improvements, adjustments which facilitated shipment of this produce, increased the quality, and promoted market acceptability in distant consumption centers.

Prior to the First World War lettuce was grown in California only to supply local needs. However, year-round production for eastern shipment expanded with great rapidity shortly thereafter. In 1911 only 700 acres were devoted to this crop, and in 1916, when shipments were first beginning, this figure had increased to about 2,000 acres. Eight years later, more than 24,500 acres were planted in lettuce.[8] Initially, a substantial share of the lettuce crop was produced in Los Angeles County, but from the mid-twenties on almost the total output came to be concentrated in two different growing areas, the Salinas district and the Imperial Valley (table 16). Lettuce production was divided between these two areas on a seasonal basis. In the Imperial Valley lettuce is

[6] Hereinafter called the Salinas district.

[7] P. F. Griffin and C. L. White, "The Lettuce Industry of the Salinas Valley," *Scientific Monthly,* 81 (August 1955), 77.

[8] S. S. Rogers, *Lettuce Growing in California* (Circular No. 160 [Berkeley: University of California Agricultural Experiment Station, 1917]).

Plate 1. About two-fifths of the United States commercial lettuce crop is grown in the Salinas-Watsonville district, California.

TABLE 16. *California Lettuce Shipments, 1920–1955*

Year	Shipments of lettuce from California[1] (carloads)			Total rail shipments from all states (carloads)	California total as a per cent of U. S. total	Truck shipments in carload equivalents
	Imperial Valley	Central District[2]	Total for state			
1920	(2,440)	n.d.	7,358	13,788	45	n.d.
1925	10,308	8,448	21,618	37,306	57	n.d.
1930	12,942	22,199	37,473	55,636	67	n.d.
1935	5,794	24,674	31,502	46,999	67	[3] 1,035
1940	6,755	27,214	34,116	49,898	68	8,190
1945	12,543	33,757	47,651	68,471	70	9,155
1950	12,726	39,309	53,030	75,241	71	12,171
1955	12,967	38,818	55,316	79,070	70	9,132

n.d. = no data.

[1] Figures for 1935 and later include romaine.
[2] Principally the Salinas-Watsonville district.
[3] For Imperial Valley only.

Source: U. S. Department of Agriculture, Agricultural Marketing Service, *Fresh Fruit and Vegetable Shipments by Commodities, States, and Months.* Summaries for the years reported in the table.

harvested and shipped from mid-December to mid-March. During the remaining months of the year the high soil and air temperatures of this area do not favor lettuce production. In the Salinas district, where cool and often foggy conditions prevail during the summer, lettuce can be successfully cultivated any time of the year except from late December to the latter part of March. Although three crops can be grown per year, the general practice is to produce two. The peak of the harvest of the first crop occurs in May, but heavy shipments extend from March through November. Production in the Salinas and Imperial Valley districts thus is scheduled so that harvesting in one area or the other continues without great fluctuation or competition throughout the year.

As the number of contract shipments made from these two areas from 1920 to the present rapidly increased, California's share of the nation's commercial lettuce production steadily expanded until recent years, when it has leveled off at about seven-tenths of the total. Behind this statistical record two important developments can be noted: first, the dominance which the iceberg type of head lettuce has secured, and second, various measures which were instituted to meet the transport requirements of this commodity.

Lettuce varieties are generally classified in accordance with the four following main types: crisp-heading, butter-heading, loose leaf or non-heading, and cos or romaine. Cos lettuce possesses oblong, headlike clusters of spoon-shaped leaves; it stands erect and forms a compact but not hard head. Although they are securing increasing consumer favor cos varieties are of secondary importance in the total commercial lettuce crop. There is only a limited demand for loose-leaf, butter-heading (Boston), bunching or semi-heading (Bibb) varieties. It was production of the crisp-heading type of lettuce which made possible the rapid growth of the industry in California. Today most of the commercial crop is of this type, having become known in the produce trade as "iceberg" lettuce.

During the years following the First World War, four main varieties of crisp-heading lettuce were being grown in California.[9] The most popular of these were used as foundation stock, and the several strains of Imperial lettuce which today dominate the market were developed.[10] They are remarkable vegetables. They mature in the form of a slightly oblate or ellipsoid "head" with tightly layered leaves, which will not wilt for many days under mild refrigeration. There are relatively few outside green "wrapper" leaves. The leaves do not shell off or break easily, yet they have a delicate, crisp texture and a flavor which makes them very palatable. These qualities permit much easier harvesting and packing, with considerably less wastage than for other varieties of lettuce. Sanitary condition of the marketed product also is easier to maintain, a useful quality for a vegetable eaten raw.

Lettuce is sensitive to relatively slight differences in environmental conditions. Consequently, much effort has been directed toward breeding strains adapted to special growing conditions. Head size also has been increased. The effect of these improvements has been to encourage further both grower and consumer acceptance of different crisp-heading varieties of lettuce of the iceberg type.

At the same time, important steps were being taken to reduce transit losses and to maintain the quality of the produce to insure its market appeal. Following the First World War much attention was given to establishing and enforcing standards of quality; these were

[9] H. A. Jones and A. A. Tavernetti, *The Head Lettuce Industry of California* (Circular No. 60 [Berkeley: California Agricultural Extension Service, 1932]), p. 8.

[10] U. S. Department of Agriculture, *The Imperial Strains of Lettuce* (Circular No. 596), Washington, 1941, p. 3.

formulated in grading specifications which were put into effect through various inspection systems operating at the leading commercial shipping points. An optimum shipment temperature of between 32° and 35°F. was established at an early date for lettuce shipments, and at the same time the utility of package icing came to be recognized.[11]

After World War II new packing and precooling techniques were introduced. After collection, the field-packed lettuce is now passed through large vacuum tubes in which the air is extracted and the exterior moisture of the lettuce is evaporated, causing it to drop in temperature rapidly. Not only are important cost savings on crates, ice, and labor effected by this process, but it also eliminates water soaking as well as ice bruising and discoloration, and it ensures a fresher, greener, and crisper product. In the newest production system put into operation in Arizona, handling has been reduced to only two operations, loading the packed lettuce from the field to trucks, and transferring the load from truck to rail car. When 640 cartons of lettuce have been loaded, the rail car is admitted to a giant vacuum chamber in which the entire lading is cooled.[12]

Today, more than three-fourths of the California lettuce crop is shipped to markets east of the Mississippi River, and shipments to the North Atlantic states alone account for almost one-third of the total.[13] The importance of western lettuce is even more marked when the output of rapidly developing production centers around Yuma and Phoenix, in Arizona, is considered. Today the West not only meets the "off-season" demand, but increasingly supplies the nation's lettuce market even during the season when local produce of the Boston and other types is available east of the Mississippi.

Lettuce is, of course, a crop of minor importance when agriculture in the western states as a whole is considered, and the demands it places upon water are unquestionably less than for most of the commodities grown on irrigated lands. Thus, in 1950, 32,700 acres in California were devoted to the winter lettuce crop, 31,000 acres to the spring crop, 28,400 acres to the summer crop, and 37,300 acres to the

---

[11] U. S. Department of Agriculture, Bureau of Plant Industry, *Investigation on the Body Icing of Vegetables,* Washington, 1927 (mimeo).

[12] "Giant Chambers Cool Lettuce," *Refrigerating Engineering,* 63 (June 1955), 70.

[13] Griffin and White, *op. cit.,* p. 83; and W. O. Jones, *The Salinas Valley: Its Agricultural Development, 1920-40* (Palo Alto, Calif.: Stanford University Press, 1947).

fall crop.[14] More than two-thirds of this total was in the Salinas district where, it is stated, "a crop of lettuce needs not more than four acre-inches of water to reach maturity."[15] Considerably more water than this probably is applied.

In 1945 a study was made of the amounts of irrigation water applied to some 2,032 acres of lettuce crops in the Salinas district.[16] Data given in this study show that a total of 4,414 acre-feet of water was applied to the spring, summer and fall crops together during that year, or about 2.2 feet per acre. The cropped acreage covered in the study comprises only about 4 per cent of the total lettuce acreage of the Salinas district. Because many factors influence local irrigation water consumption, the consumption of the study-area cannot be extrapolated to the entire district. However, it does suggest a high total water consumption for lettuce growing.

Despite its minor place in the expansion of western irrigated agriculture as a whole, the growth of lettuce production typifies the response of many phases of that agriculture to technologic changes.

RUSSET POTATOES

A further illustration of the way in which demand for irrigation water has increased is the growth of potato-producing areas in the Far West, especially in the state of Idaho. In 1954, Idaho grew more than 40 million bushels of potatoes and produced more than 11 per cent of the total United States crop. This production was exceeded only by Maine. At present, Idaho is growing twice as many potatoes as it produced on a yearly average between 1928 and 1932, and recent harvests have been running ten times greater than in 1910.[17] On the average between 1945 and 1949, the amount of land in Idaho devoted to potatoes was 160,000 acres—four times the acreage of 1920.[18]

[14] U. S. Department of Agriculture, *Agricultural Statistics, 1952*, Washington.

[15] Griffin and White, *op. cit.,* p. 80.

[16] Harry F. Blaney and Paul A. Ewing, *Irrigation Practices and Consumptive Use of Water in Salinas Valley, California*, U. S. Department of Agriculture, Soil Conservation Service, Washington, 1946.

[17] U. S. Department of Agriculture, *Agricultural Statistics, 1955*, Washington, p. 233; *1936*, p. 157; and *The Yearbook of Agriculture, 1910*, p. 556.

[18] Idaho Potato and Onion Shippers Association, 22nd Annual Convention, Section III, *Potato and Onion Statistics* (Sun Valley, Idaho, 1950), p. 42; and U. S. Department of Agriculture, *The Yearbook of Agriculture, 1920*, p. 618.

While some potatoes in Idaho are grown without irrigation, most of the crop is grown on irrigated lands. Potatoes occupy about 7 per cent of the total irrigated acreage in the state.[19] The main production centers, such as the Idaho Falls district and the Boise area, are in the Snake River Valley.

The main factor contributing to the growth of potato production in Idaho is the suitability of soil and prevailing growing conditions to the cultivation of varieties able to command greater favor in the nation's retail markets than most of those grown nearer to areas of dense settlement.[20] However, consumer preference did not come about without a struggle. As the volume of potato shipments increased in this country, the market was at an early date parceled out among different production centers, each of which organized in its sales activity around one of three types of potato crop: the early, the intermediate, or the late crop. Idaho potato raisers, primarily engaged in the production of a "late" crop in a northern location, were able to force their way into the nation's commercial market largely because theirs was a product possessing distinctive characteristics. Idaho producers at present have another small advantage in that their potato yields somewhat exceed the national average. However this advantage is more than offset by the state's remoteness from major consumption centers and by the production costs of irrigated farming.

The great market acceptability and the premium price of Idaho potatoes in most markets can be attributed to the superior table quality of the "russet" type first produced in Idaho. In recent years, 95 per cent of the Idaho crop has been of the Russet Burbank variety,[21] and its growing acceptance in other areas is attested by the increasing proportion of the national total of certified potato seed production which is of this variety. These long, cylindrical potatoes, also known as Idaho Russets, Golden Russets, and Netted Gems, are very mealy when grown under favorable conditions and are therefore especially desirable for baking. They now are popularly known as "Idaho bakers." They also are well suited to the manufacture of chips and other products. Being

[19] U. S. Department of Agriculture, *Potato Growing in the Western States* (Farmers' Bulletin No. 2034), Washington, 1951, p. 24.

[20] The maintenance of federal government price supports for potatoes after World War II undoubtedly contributed to the size of western potato acreage for a period. This account, however, is concerned with those aspects of the case in which a technical consideration is prominent.

[21] Idaho Potato and Onion Shippers Association, *op. cit.*, p. 53.

reasonably resistant to rot and to bruising, they can be stored and shipped with comparative ease.

The origin of this variety is not known.[22] Whatever it was, much work has subsequently been done in state and federal agricultural experiment stations and by private concerns to improve it. The Russet Burbank variety has proved particularly susceptible to *verticillium* wilt or "early dying," and considerable attention is being directed to the development of a resistant variety which possesses all of its desirable characteristics. Great promise is thought to be shown by a recently released variety known as Early Gems.[23] The result of crossing the Russet Burbank variety with a blight-resistant seedling, it is in most respects comparable to the former, though it matures considerably earlier.

Idaho producers have in recent years encountered increasing competition from other potato producing areas. Outside of Idaho, successful cultivation of the russet type of potato was for long confined to a limited area of the western Rocky Mountains and the Pacific Coast. However, this regional advantage is disappearing, because the Russet Burbank variety can be adapted to other areas by altering cultivation practices.[24] The familiar term "Idaho baker" can be an asset to a grower, wherever located.

Greater emphasis is therefore being placed upon quality control measures to upgrade the output in Idaho, about two-thirds of which have been graded as U. S. No. 1 in recent years. To reduce bruising in shipment and to minimize handling costs, a portion of the crop is being packaged in 100-pound crates and in either kraft or polyethylene 10-pound bags.[25] Studies have indicated that much of the injury to the potato occurs as a result of careless handling practices between the wholesale warehouse and the consumer, and attempts have been made to initiate corrective measures.[26] Individual wrapping in tissue paper is

[22] C. F. Clark, *Descriptions of and Key to American Potato Varieties* (Circular No. 741), U. S. Department of Agriculture, Washington, 1951.

[23] F. J. Stevenson *et al.,* "Early Gem: A New Early, Russet-Skin, Scab Resistant Variety of Potato Adapted to the Early Potato Producing Section of Idaho," *American Potato Journal,* 32 (1955), 79-85.

[24] G. H. Rieman, D. C. Cooper, and M. Rominsky, "The Russet Burbank Variety," *American Potato Journal,* 30 (1953), 98-103.

[25] *Modern Packaging,* 28 (1954), 142-46.

[26] W. C. Sparks, "Injury Studies on Idaho Grown Russet Burbank Potatoes," *American Potato Journal,* 27 (1950), 287-303.

resorted to for choice grades. The use of fan-cooling and air-condition-ing equipment in storage plants is also helping to reduce losses and, in some instances, to double the storage life of the crop.

Recently, chemicals have been developed which inhibit sprouting and prevent rotting, and the possibility of employing atomic radiation for the same purpose is being investigated. One such chemical, maleic hydrazide, is claimed to act as a growth regulator and is applied in the field two weeks prior to harvest; a second product, 3-C1-IPC (3-chloro-isopropyl-N-phenyl carbonate), is used after the crop has been col-lected. All such measures, by facilitating storage, can be expected to stabilize the marketing potential of the area and they can therefore be expected to influence Idaho's future position as a potato producing region and its future need for irrigation.

2. The Growth of the Chemical Industry

In contrast to the West's prevalent use of water for agricultural pur-poses, an outstanding feature of water problems in eastern United States has been the demands for water by manufacturing industry. In its many uses as a heat exchange medium, solvent, flotation medium, culture medium, waste carrier, and even as a raw material, water has an unusually broad industrial application. From the industries with large demands for water and water-derived services (e.g., steel), chemical manufacturing has been selected for case description. It is most clearly the product of recent advances in technology and in the course of its growth it has been particularly closely related to water use.

The chemical industry dates only from the First World War as a major industrial force in the American economy. In its modern form, its development has depended heavily upon the increasing availability of electricity, needed in large blocks in the electrolytic production of caustic soda and chlorine, phosphoric acid, calcium carbide, calcium cyanamid, and numerous other basic materials. But in the early years of this century there emerged many decisive new methods for obtaining raw materials, such as the Frasch process of sulfur recovery (1903) and the Haber process of directly synthesizing ammonia from nitrogen and hydrogen (1910), which, augmented by the beginning of the coal tar industry and the introduction of rayon and plastics, constituted the foundation of this now huge and vital industry.

The physical output of the chemical industry has expanded rapidly in this century (table 17). For example, production of sulfuric acid, one of the most important and cheapest of the acids with hundreds of uses in virtually every branch of the industry, increased more than tenfold from 1937 to 1955, while the production of phosphoric, nitric and hydrochloric acid was four times as great in 1955 as in 1947.[27]

TABLE 17. *Index of Production of Chemicals and Allied Products in the United States, 1899–1955*

(1935-39 average = 100)

1899	17
1909	28
1919	47
1925	63
1930	87
1935	89
1940	130
1945	284
1950	264
1955	365

Sources: U. S. Bureau of the Census, *Historical Statistics of the United States, 1780-1945,* 1949, p. 180; and *Statistical Abstract of the United States, 1956,* p. 792.

Output of the important alkalies (lime, soda ash, caustic soda, and ammonia) and of the major salts, including alum (aluminum sulfate), sodium silicate, sodium sulfate (salt cake), and sodium phosphate, has risen almost as rapidly.

However, the character of this development is even more dramatic than is suggested by the production data for these basic inorganic chemicals, for such data fail to indicate the importance of today's new uses. A better measure of the rapid expansion of chemical manufacture is the value added by manufacture. In 1947 the value of shipments of chemicals and allied products, less the cost of materials, energy, and

[27] In 1937, 1,657 thousand short tons of sulfuric acid (100 per cent H_2SO_4) were produced (U. S. Department of Commerce, Bureau of Foreign and Domestic Commerce, *1940 Supplement: Survey of Current Business,* 1940), while a preliminary compilation indicated 16,758 thousand short tons, including sulfuric acid of oleum grades, for 1955 (U. S. Department of Commerce, Bureau of the Census, *Statistical Abstract of the United States,* 1956, p. 823).

labor, was $5,345 million; in 1954 the corresponding figure was $9,150 million, an increase of more than 70 per cent.[28] Even when a correction is applied for change in price index the increase is about 65 per cent. Yet during this same period the volume of chemical production was only half again as large,[29] and the number of persons employed in the industry had risen by only a little more than one-sixth.[30]

The chemical industry to an important extent is still a pioneering industry, probably more dependent upon scientific research than any other major industrial group. As a result, the development and expansion of new products and new uses for old ones today is the most characteristic attribute of its growth. The manufacture of plastics, resins, and plasticizers in the period of only a few decades has come to command an annual revenue of half a billion dollars, and in 1954 more than 179,000 tons of pesticides were produced.[31] The market for chemicals has greatly expanded with the increasing number of automobiles in operation; the chemical industry annually processes 160,000 tons of anti-knock compounds and 170,000 tons of additives to lubricants and greases, as well as 860,000 tons of chemicals, chiefly methanol and ethylene glycol, used in anti-freeze solutions.[32] Most recent is the production of synthetic rubber and other elastomers. An entire new line of synthetic fibers, including nylon, Orlon, Dacron, Acrilan and Dynel, has been added to the old-line fibers like acetate and viscose rayon. The rise of synthetic detergents in the postwar years has been equally dramatic. Doubtless, in the future the manufacture of special chemical fuels for rockets and the processing of nuclear fuels will have great importance.

The paper and pulp industry is sometimes grouped statistically with chemical manufactures under "chemical and allied products," or "process industries." Although displaying a somewhat less rapid growth than chemicals generally, paper and pulp manufacture has also expanded in the last few decades (table 18). Uses for pulp products have steadily multiplied and per capita paper consumption has risen

---

[28] U. S. Department of Commerce, Bureau of the Census, *Statistical Abstract of the United States,* 1956, p. 794.

[29] U. S. Federal Reserve System, Board of Governors, *Federal Reserve Bulletin,* 42 (1956), 386.

[30] U. S. Department of Commerce, Bureau of the Census, *op. cit.,* p. 794.

[31] *Idem,* p. 822.

[32] J. M. Weiss, "Chemical Industry Growth," *Chemical and Engineering News,* 33 (1955), 2363.

TABLE 18. *Increase in Pulp and Paper Production in the United States, 1899–1954*

Year	Wood pulp production (1,000 tons)	Paper and paper-board production (1,000 tons)
1899	1,180	2,168
1909	2,496	4,121
1920	3,822	7,185
1930	4,630	10,169
1940	8,960	14,484
1950	14,844	24,372
1954	18,336	26,652

Sources:
    1899–1940—U. S. Bureau of the Census, *Historical Statistics of the United States, 1789-1945*, p. 126.
    1950 and 1954—U. S. Department of Commerce, Office of Business Economics, *Business Statistics, 1955: A Supplement to the Survey of Current Business*, pp. 173, 175.

sharply. Paper and paperboard production, it has been estimated, totalled 31.5 million tons in 1956.[33]

The chemical industry, excluding all the "allied products," is the third largest industrial water user,[34] diverting about 8 billion gallons a day. The pulp and paper industry uses somewhat more than half this amount and is followed closely by petroleum refining.[35] More than one-fourth of the water withdrawn by the chemical industry is used in the production of industrial inorganic chemicals. In the last two decades the water requirements of such manufacturers have increased more than fourfold.

Almost three-fourths of the water withdrawn by the chemical industry is taken from surface supplies. A small amount of it is brackish, and in a few instances treated sewage effluent is being employed. Ground-water withdrawals constitute almost a quarter of the total, and may increase where larger ground-water withdrawals become technically

[33] "Paper Producers Race for New Markets," *Chemical Week*, 79 (September 15, 1956), 30.
[34] First, steam electric power generation; second, iron and steel.
[35] U. S. Bureau of the Census, *U. S. Census of Manufacturers: 1954*, Washington, 1957. U. S. Federal Power Commission, *Water Requirements of Utility Steam Electric Generating Plants, 1954*, Washington, 1957.

feasible. One of the largest wells recently completed, for example, is capable of delivering 10 million gallons a day.[36] Three-fourths of the total intake of the chemical industry is recirculated. As the demand rises further and the costs of new water supplies increase, it is clear that a larger proportion will be reused (see chapter 16).[37]

Because the chemical industry includes a diversity of basically different processes and products, and because its organization is so complex,[38] it is not easy to summarize types of water use. Water is the solvent in which many chemical reactions take place; it is usually required in such basic processes as hydrolysis, electrolysis, fermentation, and extraction. It serves as a raw material in the production of numerous chemicals such as caustic soda, sulfuric acid, or hydrogen peroxide; it is also a raw material for organic solvents and pharmaceuticals. Because most chemical reactions of industrial importance liberate heat which must be removed, water has a further important role in this industry as a cooling agent. In addition, the heavy energy requirements of the chemical industry as a whole place a substantial demand upon water for boiler feed and for condensing purposes in steam electric power plants. Most of the water required by the industry is used for cooling, but tremendous volumes are also employed in various washing operations; the latter, for example, is the largest use of water in pulping. Finally, one of the most important uses of water in chemical industry is in the disposal of wastes.

Several generalized long-range conclusions on water use by the chemical industry are possible. Requirements per unit of output have followed changes in type of processing. A typical history starts with a batch process; this later is replaced by continuous flow operations. Subsequently, as quality control has assumed greater importance, further process design improvement has tended to reduce water need. The demand placed upon water for waste-disposal purposes, although rising rapidly in total, has lessened per unit of product wherever recirculation and reuse practices have been employed and effective recovery and treatment techniques have been instituted. Economic and social condi-

---

[36] E. B. Brien, "Shell Water Well to Produce 10,000,000 Gallons per Day," *Petroleum Engineering*, 26 (March 1954), C21-27.

[37] *1953 Annual Survey of Manufactures*, and R. D. Hoak, "Greater Reuse of Industrial Water Seen," *Chemical and Engineering News*, 33 (1955), 1278-82.

[38] It might be noted, for example, that each of the six largest chemical companies, du Pont, Union Carbide, Allied Chemical, American Cyanamid, Dow, and Monsanto, produces from 500 to 1,000 or more different chemical products.

124   *Technology in American Water Development*

tions, and consequently legal measures, increasingly have favored the
use of waste treatment, and thus have promoted water recovery and
reuse particularly within this industry.

Information is available regarding the water requirements for a
variety of different chemical products (table 19). Among all manufactures few consume more water than those producing acetate rayon

TABLE 19. *Amount of Water Required to Produce One Ton of
Selected Chemical Commodities*

Commodity	Gallons per ton	Specific product	Source
Acetic acid, from pyroligneous acid	100,000		a
Alumina	7,000		b
Aluminum	32,000		b
Ammoniated superphosphate	27–30		a
Ammonium sulfate	200,000		c
Ammonia, synthetic	31,000–94,000	liquid NH$_3$	a, c, d
Ascorbic acid	1,760,000		d
Butadiene	320,000		e
Carbon dioxide	23,000		c
Caustic soda (lime-soda process) .	18,000–21,000		a, c
Charcoal and wood chemicals ...	65,000	crude calcium acetate	a
Cottonseed oil hydrogenation	1,200		d
Hydrogen	660,000		c
Magnesium hydroxide (from sea water and dolomite)	500		d
Nitric acid, synthetic	14,000–50,000		d
Phosphoric acid (strong-acid process)	7,500	35% P$_2$O$_5$ acid	a, c
Portland cement	750		a, d, e
Soda ash (Solvay process)	15,000–18,000	58% soda ash	a, c
Sulfathiazole	1,000,000		c
Viscose rayon yarn	180,000–200,000		a, c
Wool, scouring and bleaching	40,000	goods	e

Sources:
   a. *Chemical and Metallurgical Engineering Flow Sheets* (4th ed.; New York: McGraw-Hill, 1944).
   b. Howard L. Conklin, *Water Requirements of the Aluminum Industry* (Water Supply Paper No. 1330-C), U. S. Geological Survey, 1956.
   c. G. E. Symons, "Treatment of Industrial Wastes," *Water and Sewage,* 82 (November 1944), 44.
   d. "Estimating Requirements for Process Steam and Process Water," *Chemical Engineering,* 58 (April 1951), 110-11.
   e. *Journal of the American Water Works Association,* 37 (September 1945), 4.

yarn. It has been reported that from 250,000 to 400,000 gallons are required for each ton of acetate yarn,[39] while about half as much is used to make viscose rayon.[40] About 100,000 gallons (62.5 tons) of water are used in producing a ton of gunpowder, and to obtain a ton of lactose twice this amount is required.[41]

Not only are withdrawal requirements of importance, but water disappearance during use is also of significance. In the production of a barrel (42 gallons) of oil by the Fischer-Tropsch process for the hydrogenation of coal, between 550 and 650 gallons of water are needed, four-fifths of which disappears.[42] The average amount of water used in making a ton of carbon black by twenty-two plants employing the furnace method (collectively responsible in 1953 for more than two-thirds of the output of this industry) was found in a recent survey to be 6,680 gallons, nearly all of which disappears.[43]

The localization of chemical industry water demands may be illustrated by the requirements of a few individual plants. Several of the largest heavy chemical producers each withdrew more than 200 million gallons of water a day from a single locality, and are accordingly located within easy access of such supplies. The largest ammonium sulfate fertilizer works require almost as much. The new paper and pulp mill of the Bowaters Southern Paper Corporation at Calhoun, Tennessee, was equipped with a capacity for handling 25 million gallons a day.[44] The plant of the Ethyl Corporation at Baton Rouge, Louisiana, is another industrial installation with heavy requirements, using in excess of 125 million gallons of water a day.[45]

[39] *Water in Industry* (New York: National Association of Manufacturers and The Conservation Foundation, 1950), Appendix C.

[40] "The Water Problem," *Chemical Engineering,* 56 (July 1949), 119.

[41] H. E. Degler, "Increase of Water Usage Makes Selection of Water Cooling Equipment for Reuse of Water Vital," *Heating and Ventilation,* 48 (April 1951), 130; and "Industrial Wastes," *Water and Sewage Works,* 102 (1955), R-314.

[42] A. N. Sayre, *Water Resources of the United States,* U. S. Geological Survey, Washington, 1949 (mimeo), quoted by Harold E. Thomas, *The Conservation of Ground Water* (New York: McGraw-Hill, 1951), p. 214.

[43] H. L. Conklin, *Water Requirements of the Carbon Black Industry* (Water Supply Paper No. 1330-B), U. S. Geological Survey, Washington, 1956, pp. 88, 94-95.

[44] Roscoe C. Martin, *From Forest to Front Page* (University: University of Alabama Press for Inter-University Case Program, 1956), p. 46.

[45] M. A. Lea, "Ethyl Corporation's Baton Rouge Plant Water System," *Proceedings of the Fifth Annual Water Symposium,* Engineering Experiment Station Bulletin No. 55 (Baton Rouge: Louisiana State University, 1956), p. 13.

Substantially more than half of the output of the industry, measured in terms of the total value of product, was accredited in 1947 to only five states: New Jersey, New York, Pennsylvania, Michigan, and Texas.[46] However, one cannot assume that the demand for water is proportionately distributed. At present, about one-third of the industry is located in the South, but this proportion is expected to rise sharply in the years ahead because of water availability and the location of petroleum production. One estimate, for example, suggests that more than half of the nation's chemicals will be produced in the South by 1975.[47] However, it is probable that a disproportionately large share of the "wet process" plants, which generally require large volumes of water and most often produce extensive liquid wastes, will be included in this total. Three-fourths of the petrochemicals produced in the United States now come from the 600-mile crescent which extends from Baton Rouge, Louisiana, to Brownsville, Texas. It is estimated that about one-fourth of all the chemicals produced during a year are derived from natural gas and petroleum,[48] the raw materials of which are abundant in the Gulf Southwest.

Much of the water used by the chemical industry is employed for cooling purposes and, unless recirculated, the only significant qualitative change it undergoes between intake and discharge is an increase in temperature. Other things being equal, the water subsequently is rapidly restored to its original temperature through the process of evaporation. However, materials involved in industrial production are often incorporated in process water and are carried away in the outflow unless recovered. Finally, certain unwanted solids and solutions are intentionally disposed of by being discharged into available water courses.

THE PROBLEM OF CHEMICAL WASTES

The wastes of plants manufacturing chemicals and chemical products vary in character. The discharge from some plants is ligneous or

[46] M. Queeny and R. M. Laurence, "The Chemical Industry," chap. XIV in J. G. Glover, W. C. Cornell, and G. R. Collins (eds.), *The Development of American Industries: Their Economic Significance* (3rd ed.; New York: Prentice-Hall, 1951), p. 444.

[47] A. P. Black, "The Water Resources of the South," Engineering Experiment Station Bulletin No. 55, *op. cit.* p. 54.

[48] W. D. Seyfried, "Developments in the Southern Petrochemical Industry," *Proceedings of the 129th Meeting, American Chemical Society,* Dallas, Texas, 1956.

resinous. In other cases it is salty or soapy or contaminated with acids. Still others produce an effluent containing quantities of oily or greasy floating matter, or animal and vegetable material in suspension. Some wastes are merely offensive to the eye and nose, but in sufficient concentration they may be poisonous, corrosive, or otherwise harmful.

Chemical wastes may contain one, and often several of the following materials: (1) excessive acid or alkali; (2) deleterious amounts of organic matter; (3) pathogenic organisms or substances toxic to human, aquatic, or other life; (4) damaging loads of suspended solids, of dissolved solids, or of undesirable floating matter; (5) hazardous radioactive particles; or (6) substances responsible for objectionable odors and tastes. Wastes of the first type can be dealt with quite readily by neutralization. Very frequently discharges containing strong acid solutions are treated with lime, while alkali solutions may be neutralized by the addition of sulfuric acid. The mixing of wastes from a number of sources often may serve the same purpose.

Wastes from the pulp and paper industry, from tanneries, breweries, distilleries, and from the production of acetone, vitamins, and antibiotic drugs, to name but a few, tend to be high in organic content. Wastes of an organic character are often serious because of their action in reducing the dissolved oxygen content of water. However, the effluent discharged by the chemical industry is objectionable in most instances because of some particular toxic or corrosive property. Often the presence of a very minute amount of some substance in water causes concern. For example, it has been suggested that metallic or alkali cyanides and soluble copper compounds in concentrations as low as 0.2 to 0.3 ppm are lethal to fish,[49] while various phenolic compounds equally diluted, are often found to make water unpotable. However, there are no universally applicable standards regarding the use of water for the disposal of wastes. A discharge which constitutes a reasonable use of a stream under one set of conditions may become an abuse under another. Furthermore, each branch of the industry presents its own distinctive pollution problem.

While there is no simple measure of the polluting characteristics of the wastes discharged by the chemical industry, or even of their total volume, information from several different fields may help to suggest the dimensions of the problem. For example, published data show that

[49] M. M. Ellis, *Detection and Measurement of Stream Pollution* (Bureau of Fisheries Bulletin No. 22) U. S. Department of Commerce, Washington, 1937, pp. 365-437.

the plants of the Dow Chemical Company at Midland, Michigan, have discharged 200 million gallons of waste waters per day.[50] A somewhat larger total volume of wastes has been indicated in the operation of the two plants of the Union Carbide Corporation at South Charleston and Institute, West Virginia. It is estimated that this discharge contains organic wastes having an oxygen demand as high as 30,000 pounds per day.[51] At the Frankford Works of the Allied Chemical and Dye Corporation at Philadelphia, wastes totaling a million gallons of soda sludges and 115,000 gallons of sulfuric sludges were discharged annually in the manufacture of a coumarone resin, until a new process for its manufacture was introduced. Likewise more than 18 million gallons of wastes each year, containing over 5 million pounds of chemicals, were discharged in the course of producing various synthetic phenolic compounds[52] until recent process changes greatly reduced these totals.

Chemical wastes are also produced by other industries. It has been estimated, for example, that iron and steel plants annually dispose of 500 to 800 million gallons of waste pickle liquor, composed of water containing about 5 per cent sulfuric acid and 15 per cent ferrous sulfate.[53] Prior to the widespread sealing of abandoned coal mines in the 1930's, the annual acid pollution load from mines in the bituminous fields east of the Mississippi River was estimated to have been 2.7 million tons of sulfuric acid each year. Since almost half of this total originated from active mines, this source of pollution remains important.[54]

[50] I. F. Harlow and T. J. Powers, "Pollution Control at a Large Chemical Works," *Industrial and Engineering Chemistry,* 39 (1947), 572.

[51] G. F. Jenkins, "Process Wastes from Chemicals Manufacture," *Sewage and Industrial Wastes,* 27 (1955), 717-18.

[52] "1955 Industrial Wastes Forum," *Sewage and Industrial Wastes,* 28 (1955), 667; also letter, H. E. McNutt, Technical Supervisor, Barrett Division, Allied Chemical and Dye Corporation, Philadelphia, January 8, 1958.

[53] R. Norris Shreve, *The Chemical Process Industries* (New York: McGraw-Hill, 1945), p. 61. A figure of 600 million gallons is given by R. D. Hoak in "Waste Pickle Liquor," *Industrial and Engineering Chemistry,* 39 (1947), p. 614.

[54] U. S. National Resources Committee, 3rd Report of the Special Advisory Committee on Water Pollution, *Water Pollution in the United States* (House Document No. 155, 76th Congress, 1st Session, 1939), p. 8. For a more recent discussion, see J. R. Hoffert, "Acid Mine Drainage," *Industrial and Engineering Chemistry,* 39 (1947) 642-46; and W. W. Hodge, "Waste Disposal Problems in the Coal Mining Industry," in W. Rudolphs (ed.), *Industrial Wastes: Their Treatment and Disposal* (New York: Reinhold, 1953), pp. 312-418.

Various regional estimates have been made of the volume and character of wastes from the chemical and the pulping industries as a part of total wastes. In the Ohio River Basin (200,000 square miles) it has been estimated that the chemical industry is responsible for almost one-fourth of the total industrial wastes.[55] Somewhat more than a fifth of this originates with the pulp and paper industry. The wastes considered in these estimates were evaluated in terms of their organic character expressed as the biochemical oxygen demand (B.O.D.) which they create. In a comparable survey of wastes in the city of Chicago, in contrast to the Ohio study, more than one-third of the pollution load was attributed to the packinghouse industry, reflecting a different industrial pattern.[56]

The volume of effluent discharged per unit of product, its biological oxygen demand, and its suspended solids load vary widely among the chemical and other water waste discharging industries (table 20). In many cases there also is a wide spread between the highest and lowest figures for each product. A 1952 general estimate suggested that 22.5 pounds of B.O.D. and 7.4 pounds of suspended solids are discharged in the wastes of the chemical industry per employee per day.[57]

As the volume of these wastes has expanded with the growth of the chemical industry, and as the resulting problems have multiplied, effluent treatment has assumed increasing importance. The chemical industry was reported in 1939 as spending more than $28 million, and the pulp and paper industry $129 million, for treating wastes.[58] No comparable figures of more recent date are available, largely because waste-treatment processes in many instances have been merged with waste-recovery practices. However, it has been suggested that the chemical industry is currently investing about $50 million a year in new water pollution control facilities.[59] Today thousands of industrial

[55] Report of the U. S. Public Health Service, *Ohio River Pollution Control*, Part II (House Document No. 266, 78th Congress, 1st Session, 1944), p. 167. Tanning, brewing and distilling, and textile manufacture were not considered as parts of the chemical industry.

[56] F. W. Mohlman, "Waste Disposal Problems," *Chemical and Metallurgical Engineering*, 50 (1943), 128.

[57] Edmund B. Besselievre, *Industrial Waste Treatment* (New York: McGraw-Hill, 1952), p. 84.

[58] U. S. National Resources Committee, 3rd Report of the Special Advisory Committee on Water Pollution, *op. cit.*, p. 54.

[59] "Preserving Water Quality," *Chemical and Engineering News*, 34 (1956), 5371.

TABLE 20. *Waste Discharges Involved in the Production or Processing of Certain Commodities*

Industry	Unit of input or output	Waste water (gals. per unit)	B.O.D. (ppm)	Suspended solids (ppm)
Brewing	1 bbl. beer	470	1,200	650
Alcohol (from sugarcane molasses)	100 gals. 100 proof alcohol	[a] 12,840	[b] 15,000–50,000	n.d.
Grain distilling (combined wastes)	1 bu. grain mashed	600	230	360
Meat packing and slaughtering	100 hog units kill	710	3,100	1,580
Milk creamery and condensery	1,000 lbs. raw milk	260	2,550	1,410
Tanning (vegetable)	100 lbs. raw hides	800	1,200	2,400
Groundwood pulp .	1 ton dry pulp	[c] 12,000	645	n.d.
Soda pulp	1 ton dry pulp	[d] 58,000	110	1,720
Sulfate (kraft) pulp	1 ton dry pulp	[c] 64,000	123	n.d.
Sulfite pulp	1 ton dry pulp	[c] 48,000	443	n.d.
Paper mill	1 ton paper	40,000	19	452
Paperboard	1 ton paperboard	14,000	121	660
Strawboard	1 ton strawboard	26,000	965	1,790
Rayon manufacture	1 cord wood distilled	680,000	30	n.d.
Cotton textile (processing and dyeing)	1 ton goods	79,000	n.d.	n.d.
Oil refinery	1 bbl. (42 gal.) oil	[a] 770	0.2	0.5
Synthetic rubber (GR-S)	1 ton rubber	[e] 6,550	[e] 45	[e] 124
Butadiene, including distilling	1 ton product	[f] 300,000	[g] 2,550	[g] 27.6

n.d. = no data.

Sources:

    Unless otherwise noted, Report of the U. S. Public Health Service, *Ohio River Pollution Control* (Supplements to Part II, House Document No. 266, 78th Congress, 1st Session, 1955), p. 33.

    [a] "The Water Problem," *Chemical Engineering,* 56 (July 1949), 120.

  [b] E. B. Besselievre, *Industrial Waste Treatment* (New York: McGraw-Hill, 1952), p. 179.

  [c] O. D. Mussey, *Water Requirements of the Pulp and Paper Industry* (Water Supply Paper No. 1330-A), U. S. Geological Survey, 1955, p. 33. The figures available in this publication refer to water requirements, not water discharged; however, the amount of water that disappears in use is probably very small, as noted in this reference on pp. 34-35.

  [d] Technical Association of the Pulp and Paper Industry, Monograph No. 1, *Industrial Water for Pulp, Paper, and Paperboard Manufacture* (New York, 1942).

  [e] G. M. Hebbard, S. T. Powell, and R. E. Rostenbach, "Industrial Wastes: Rubber Industry," *Industrial and Engineering Chemistry*, 39 (1947), 593.

  [f] H. E. Degler, "Increased Water Usage Makes Selection of Water Cooling Equipment for Reuse of Water Vital," *Heating and Ventilation*, 48 (April 1951), 130-34.

  [g] G. A. Rohlich, "Engineering for Industrial Wastes," *Selected Papers from the Engineering Institute on Industrial Waste Problems* (Madison, Wis., 1952), p. 6 (mimeo).

plants treat their wastes; recently, when reviewing the increasing number of treatment plants that have been built since World War II, the United States Public Health Service characterized more than a thousand of the existing installations as possessing "adequate capacity" and an even greater number as "satisfactorily operated."[60]

While manufacturing industry in general is increasingly instituting measures to cope with the problem, its dimensions are rapidly expanding. It has been estimated that between 1950 and 1955 self-supplied industrial withdrawals of water increased 43 per cent, from 77 billion gallons to 110 billion gallons a day.[61] The total volume of wastes discharged by the chemical industry probably rose even more sharply during this period. New products, such as synthetic detergents, insecticides like DDT, herbicides like 2,4-D, and many others, are contaminating the waters with the wastes associated with their manufacture. Some of these and still other contaminants are difficult to remove from waste waters, and numerous technical problems concerning them remain to be solved. Thus, while waste treatment practices have improved and their application has spread, the industry's vitality has created a stream of new waste disposal needs.

  [60] U. S. Federal Security Agency, *Water Pollution in the United States* (Public Health Service Publication No. 64), Washington, 1951.

  [61] K. A. MacKichan, *Estimated Use of Water in the United States, 1955* (Circular No. 398), U. S. Geological Survey, Washington, 1957.

3. Water Application Techniques

The growth of the chemical industry, and particularly of the pulping segment of chemical manufacturing, has affected the use of water in some regions of the West as well as in the East. Nevertheless, these industrial changes, as well as many others characteristic of the growth of manufacturing in the United States, notably have been located within the eastern states. The same may be said for some changes in irrigation techniques that have appeared within recent years. They center principally on the application of water to ungraded or unterraced land, and on the use of low-cost, lightweight piping to facilitate this. The techniques are applicable wherever irrigation can be employed, including areas in the West, but their impact is most striking in the development of water demand in the eastern states.

Pump improvements also have had an effect upon the growth of irrigation water demand. Since ground water is the most readily usable source for much eastern irrigation, pumps are becoming particularly important in this area. Their discussion as a case description, however, is reserved for chapter 10.

LIGHTWEIGHT PIPING AND SPRINKLER ATTACHMENTS

At the close of World War II the nation had a large aluminum production capability for which outlets were sought. This, around 1946, led to the introduction of lightweight aluminum piping, which, together with the development of simple types of couplings, made it practical to lay mobile pipe lines and distribution systems for water and other commodities. This type of system is relatively cheap and can be moved from field to field with a minimum of labor. It is usable over the surface of an entire farm and does not seriously interfere with cultivation and harvesting. The development of lightweight aluminum piping greatly widened the economic feasibility of sprinkler irrigation.[62]

Before this development nearly all irrigation had been confined to relatively level land which economically could be graded for flooding.

[62] John R. Davis, "Future of Irrigation in Humid Areas," *Journal, American Water Works Association*, 48 (1956), 982-90.

Sprinkler irrigation was undertaken only through fixed overhead systems for horticultural crops, as for example, the overhead systems used for truck crops in the Boston suburban area.[63]

There are limitations to level-land or terraced irrigation surface. As an example, a large field of deep soil with slopes less than 1 per cent (1 foot in 100) can be field leveled or broad-bench leveled; slopes of 1 to 3 percent can be leveled in narrow benches. The average leveling job moves 360 cubic yards of soil per acre at the cost of 15-18 cents per yard ($55-$65 per acre).[64] When the volume to be moved exceeds 800-1,000 yards per acre it is usually uneconomic to level the land.[65] The depth of the topsoil also has an effect since it is not advisable to remove more than one-half of its depth.[66]

In the western states millions of acres of land are level, or can be leveled at reasonable cost. Similar land is, of course, found within the East, notably in the Lower Mississippi Valley, on the Atlantic and Gulf coastal plains, and in the midwestern states; however, a large amount of eastern cultivation takes place on sloping land surfaces. A technique adapted to several types of land surface was a prerequisite to any large-scale irrigation in the East. While fixed steel overhead piping was physically adapted to the surface conditions, it was impractically expensive except for truck crops with favorable market relations. Adaptation to mechanized cultivation also was difficult.

Once an economically practical system of irrigation was available, compatible with mechanized cultivation, climatic conditions in nearly all eastern states favored the extension of irrigation. Plant water requirements for optimum growth and natural soil moisture availability frequently are not synchronized as crop growth progresses, and for most crops, even in the humid East, there is a normal deficiency of 4-15 inches of rain during the dry season, despite an average annual rainfall of 30-40 inches.[67] A crop that gets no rainfall for the entire month of July may be just as much a failure as one that has had no rain all year long. These facts were well illustrated to eastern farmers by the

[63] E. A. Ackerman, "Sequent Occupance of a Boston Suburban Community," *Economic Geography,* 17 (1941), 61-74.

[64] H. N. Smith, U. S. Soil Conservation Service, letter to Louis Koenig, 1956.

[65] J. C. Marr, *Grading Land for Surface Irrigation,* Circular No. 438 (Berkeley: Division of Agricultural Sciences, University of California, 1954).

[66] Smith, *op. cit.*

[67] Davis, *op. cit.;* C. W. Thornthwaite and J. R. Mather, *The Water Balance,* Laboratory of Climatology Publications, VIII, 1 (Centerton, N. J., 1955).

droughts of 1953-54.[68] Even aside from outright failures, recent agricultural research has shown that greatly increased yields are obtained by maintaining an optimum moisture content in the soil at all times.[69]

In the decade 1939-49, when lightweight sprinkler systems became available, irrigated acreage in the seventeen western states increased by 41 per cent while that in the twenty-eight eastern states increased by 283 per cent.[70] Most of the eastern increase was due to sprinkler irrigation, which in 1949 accounted for 72 per cent of the total eastern irrigated acreage. The comparable figure for the West was 2 per cent.[71] By 1954, irrigated acreage in the East had increased another 287 per cent to almost 600,000 acres,[72] while that in the seventeen western states had grown by only 8.5 per cent. It is probable that the large increase in eastern irrigated acreage was due mainly to sprinkler irrigation, and that sprinkler irrigation today constitutes probably 90 per cent of all irrigation in the twenty-eight eastern states.

Nationwide, it is estimated that sprinkler-irrigated acreage will increase from 1.3 million acres in 1955 to 4.2 million in 1975.[73] Some 2.3 million acres of this total will be in the East, if West and East maintain the same relative positions in sprinkler irrigation. In view of the relative recent increases, this assumption may be conservative. Since sprinkler irrigation results in less water withdrawal and disappearance than ditch irrigation, its use is likely to expand also in the West.

Assuming a 7.2-inch average annual water deficiency, 2.3 million acres will call for use of 1,350,000 acre-feet of water a year in the twenty-eight eastern states by 1975. A major part of the increase in irrigation water use will probably be supplied from ground water. Most eastern farms are small, individual enterprises with little or no access

[68] "Eastern Water Shortage and Drought Problem—A Symposium," *Journal, American Water Works Association,* 47 (1955), 203-209.

[69] M. M. Tharp, and C. W. Crickman, "Supplemental Irrigation in Humid Regions," *Water, The Yearbook of Agriculture, 1955* (U. S. Department of Agriculture), pp. 252-58.

[70] Davis, *op. cit.* The twenty-eight eastern states are those that lie mostly or wholly east of the 93rd meridian, excepting Louisiana, Arkansas, and Florida. The seventeen western states are those west of this meridian.

[71] U. S. Department of Commerce, Bureau of the Census, *Census of Agriculture, 1950,* Vol. III, *Irrigation of Agricultural Lands,* table 3.

[72] Davis, *op. cit.;* and H. T. Critchlow, "Water Rights in Humid Areas," *Proceedings of the American Society of Civil Engineers,* Vol. 80, No. 398, 1954.

[73] Walter L. Picton, *Summary of Information on Water Use in the United States* (Business Service Bulletin No. BSB-136), U. S. Department of Commerce, Washington, January 1956.

Plate 2. Sprinkler irrigation of pasture, near Clear Lake, California.

Plate 3. Irrigation pump and engine, Grant County, Nebraska. A 6-inch sprinkler system is shown in operation in the background.

to surface-water supplies of sufficient size to irrigate their acreage during low-water season. And few important storage projects that will affect the available supply of many farms are contemplated, or are likely to appear within the period here mentioned. Small watershed projects, like those envisioned by the Soil Conservation Service, may create some storage of water for local use, but, on the whole, farm irrigation is most likely to be from wells, because the major storage facility available to the farmer is the natural ground-water reservoir. According to a 1954 survey, 36 per cent of eastern irrigated acreage in that year was supplied by ground water.[74] The figure may rise to 70 per cent by 1975,[75] in which case ground-water withdrawal for irrigation will be about 950,000 acre-feet a year. This is about ten times the volume of ground water withdrawn for irrigation in 1950.

Such an increase would be important in eastern agriculture, but its impact on total eastern ground-water use would be small. The 95,000 acre-feet of ground water used for irrigation in 1950 was only 1.3 per cent of all ground water used in the twenty-eight eastern states. Projecting this to 1975, and taking as a basis the nation's total ground water use, which it has been estimated will increase somewhat more than twofold between 1950 and 1975,[76] the 950,000 acre-feet of ground water used for sprinkler irrigation in the East in 1975 will amount to about 8 per cent of the East's total ground-water use. Even if a 3 million-acre increase in sprinkler acreage were to take place in the twenty-eight states, this would draw upon only 12 per cent of the total expected ground-water use there, and 2 per cent of that in the nation. Thus, it appears that a continued increase in eastern sprinkler irrigation will be less significant in expanding demand for ground-water withdrawal than will eastern industrial and municipal requirements.

PLASTIC PIPING

Within the last few years there has been substantial development in the use of plastics for piping natural gas, water, and other fluids.[77] Plastic piping is being improved continually and may influence not only

[74] Critchlow, *op. cit.*

[75] John R. Davis, Purdue University, letter to Louis Koenig, 1956.

[76] Picton, *op. cit.*

[77] General information on this topic may be found in a series of articles, "Plastics as Materials of Construction," *Industrial and Engineering Chemicals,* 47, No. 7 (1955), 1292-367. Also "Plastic Pipe—over the Hump," *Chemical and Engineering News,* 33, No. 30 (July 25, 1935), 3062-64.

the use of ground water, but also the availability of surface supplies.

Introduced in the oil fields to move waste salt water, plastic pipe immediately proved superior to steel in corrosion resistance, ease of installation, and over-all economy. Applications are now common in industrial plants for movement of salt water and many chemicals not satisfactorily handled in metal. In plants using saline water for cooling, plastic piping has been substituted for steel with substantial savings. Very recently, it has come into use in rural water supply systems and in household water service piping. When plastic piping was first introduced its pressure limits were comparatively low, but now it is being used as a substitute for steel or copper in service lines at pressures up to 100 pounds per square inch. Underground lawn sprinkler systems employing plastic piping are becoming more common than the metal systems they replace. Advantages are ease of installation, simple connections, light weight, transportation convenience, semi-flexibility, and resistance to damage by freezing and corrosion.

In 1950, only 5 million pounds of plastics were used for pipe in the United States. By 1954 this amount had grown to 30 million pounds, and it was estimated that 37 million pounds of plastics were used in pipe manufacture in 1955.[78]

Polyethylene and polyvinyl chloride are the principal materials being used in plastic piping. These polymers are highly stable and resistant to attack by nearly all natural materials in the ground and in water supplies. With suitable wall thicknesses, they have sufficient strength for use in high-pressure municipal distribution systems. Although there has been only limited application in sizes above 4 inches, the use of plastic piping of larger diameter in water laterals and mains in municipal systems will become significant as techniques and materials continue to improve. Plastic pipe now can be obtained in sizes up to 12 inches in diameter.

The principal advantage of plastic piping in large installations is the cheapness with which it can be laid. Although plastic materials are five to ten times as expensive by weight as iron or steel, their lower density brings the cost of fabricated pipe into the same general range as iron and steel pipe. For example 4-inch plastic pipe with pressure specification up to 75 pounds and suitable for potable water supply may cost from $1.00 to $2.00 per linear foot[79] compared with about

---

[78] *Chemical and Engineering News, loc. cit.*

[79] Republic Steel Corporation, "Consumer Price Lists on Plastic Pipe and Fittings," March 1956, September 1956.

$1.50 for cast iron.[80] Typical installation expenses are 40 cents per linear foot of plastic pipe and 65 cents per foot of iron pipe.[81] Not only can a pipeline be laid with considerable saving in cost, but the time requirement also has been remarkably reduced. In one case, a 12-mile plastic pipeline was installed by three men in 96 work hours.[82]

An interesting future possibility for plastic pipe in long water-transmission lines is its production on the site by using portable extrusion equipment fed with the granulated plastic. With this technique, shipping comparatively bulky materials can be avoided; continuous lengths of joint-free pipe can be made and installed in an integrated mechanical operation of ditch digging, pipe production and laying, and covering. Should the technique prove economically feasible, transportation of water piped over longer distances might help to balance exploitation of both ground- and surface-water supplies. Crop and pasture irrigation in the twenty-eight eastern states could thereby be given further impetus.

4. Residential Demand for Electricity, Associated Operating Problems of Electricity Systems, and Hydro Generation

The case descriptions presented thus far have concerned principally the withdrawal uses of water. Mention has been made, also, of one flow use, that of waste carrying. Another flow use closely related to demand for water development has been electric energy generation. Electric energy generation may be undertaken to meet industrial, residential, farm, municipal and other public demands. The most notable increases in the demand for electricity have come from industrial sources. The growth of electro-metallurgical manufacturing industries, like aluminum, has had a large share in the total growth of electric energy demands within the last twenty years. The vast electric energy needs of atomic energy installations have been even more striking. Preceding both these developments, industry in general contributed to the growth in the use of electricity as its superior qualities for supplying energy to prime movers came to be appreciated.

[80] R. S. Means, *Building Construction Cost Data* (15th ed.; Duxbury, Mass., 1956).

[81] Rocky Mountain Engineering Company, personal communication to George O. G. Löf, April 1957.

[82] *Chemical and Engineering News, loc. cit.*

The incidence of specific industrial demands, however, is likely to be localized, and characteristics of demand vary from one industry to another. Residential use of electricity, however, is of interest and application in almost every corner of the land. While demands vary somewhat from one region to another, potentially they have a high degree of uniformity. A case description of electricity demand has therefore been chosen from the field of residential use. The situation of the TVA power service area has been selected because electricity demands are considered advanced in that area, and are illustrative of the national trend in this phase of electricity use. Effects upon seasonal and daily loads are singled out for specific attention because of the future significance of hydro capacity in meeting peak demands for electricity. Although once-removed from water use, residential electrical demand is closely related to water consumption through both the hydroelectric requirements and the need for water in the condensers of steam-electric power plants.

THE NATURE OF CHANGES IN RESIDENTIAL DEMAND FOR ELECTRICITY

The demand for electricity in the United States has been significantly affected, especially during the last two decades, by the rapid increase in the number and variety of electrical appliances used within the home. Electricity long was used by domestic consumers primarily for illumination. Today the importance of this use, despite a severalfold increase, has been eclipsed by other household electricity demands. Few technical developments are more striking than the phenomenal increase in the numbers of modern electrical household appliances.

The extent of the changes that have taken place is suggested by statistical data on the appliances now common in American homes (table 21). Many of the appliances have attracted a mass market only in the years since World War II, but the use of earlier appliances has greatly increased also.

Statistical data also reveal the great increase in residential sales of electricity (table 22). The amount of electricity used by residential consumers has mounted at a substantially more rapid rate than have total sales of electricity during the last ten years. This is in large part attributable to the sharp rise in the use of electric appliances within the home, the most significant in recent years being the increasing provision of electric cooking stoves, television sets, and air conditioning units. Today that ideal of many spokesmen for the electric utilities, the "all-electric home," is approaching reality.

Not only has the total demand for electricity been significantly influenced by these developments, but the daily and seasonal incidence of that demand has also been affected. Each residential use for electricity possesses its own distinctive load characteristics. For example, there is an obvious diurnal variation in the demand for electricity for illumination, and a definite, though less marked, seasonal fluctuation. Use of some major electrical appliances is largely confined to the daylight hours; the time during which others are employed is limited to

TABLE 21. *Extent of Use of Major Residential Electrical Appliances in the United States and in the TVA Area, 1955*

Appliance	Per cent of homes served by electricity	
	United States[1]	TVA service area[2]
Refrigerator	92	94
Electric iron	90	n.d.
Washing machine	81	n.d.
Television	74	42
Vacuum cleaner	62	n.d.
Food mixer	34	n.d.
Electric range	27	66
Portable electric heater	25	n.d.
Water heater	12	44
Food freezer	12	20
Attic or window fan	n.d.	35
Room air conditioner	4	10
Electric heating (sole source)	n.d.	11
Electric heating (auxiliary)	n.d.	9

n.d. = no data.

[1] Based on 44,787,000 domestic and farm electric customers.

[2] Data based on information from 13 distributors of TVA power, representative of all distributors in terms of average amount of electricity used per year and in other respects. Data on the TVA area are included because more extensive use is being made of electrical appliances in the Tennessee Valley region than is average for the nation. Trends in use are thought to be indicated by the TVA area data.

Sources:
William R. New and William A. Bell, Jr., "Serving the All-Electric Home," *Electrical World*, March 19, 1956, pp. 124-29.
U. S. Bureau of the Census, *Statistical Abstract of the United States, 1955*, p. 846.

TABLE 22. *United States Sale of Electric Power to Residential and Rural Consumers as a Part of Total Electric Sales, 1920–55*

Year	Residential consumption (million kw-h)	(per cent of total)	Rural consumption[1] (million kw-h)	(per cent of total)	Total sales (million kw-h)
1920	3,190	[2] 9.8	n.d.	n.d.	32,536
1925	6,020	[2] 11.9	n.d.	n.d.	50,461
1930	11,018	14.7	[2] 1,498	2.0	74,906
1935	13,978	18.0	[2] 1,242	1.6	77,596
1940	23,318	19.7	1,991	1.7	118,643
1945	34,184	17.7	3,669	1.9	193,558
1950	67,030	23.9	7,400	2.6	280,539
1955	120,524	25.1	10,751	2.2	480,921

n.d. = no data.

[1] Sales on district rural lines.
[2] Computed from available data.

Sources:
   Electrical World, 147 (January 28, 1957), 132.
   Edison Electric Institute, *Electric Industry in the United States—Statistical Bulletin for the Year 1955* (No. 23), New York, 1956.
   U. S. Bureau of the Census, *Historical Statistics of the United States, 1789-1945,* 1949, p. 159.

some part of the day or the year. In the case of still others, use may be more evenly distributed over a 24-hour period. As electric heating and air conditioning have assumed increasing importance, the load peaks of the affected systems have become far more pronounced, and as a consequence operating problems have arisen. When the range between the peak load and the average load increases, the total cost of providing the additional unit of electricity required for peak operation rises quite sharply. At the same time, the total load factor decreases, with a resulting loss in operating efficiency. Thus recent developments that have increased the use of electricity within the home have been important in two respects: they have greatly augmented the over-all demand for electricity, and they have in several instances significantly altered the daily and seasonal peaks of the demand.

The experience of the TVA power service area with electrical heating. There are a number of distinctive features attending the increased residential use of electricity within the area served by the Tennessee Valley Authority. There, both the increase in per capita income and the extent of new home building have exceeded the national average. The area also has low residential electric rates, each kilowatt-hour of electricity being about half as costly as the national average. Natural gas has not been available in many communities, and petroleum offers no special economic advantage. In addition, the diurnal and seasonal temperature ranges of the southeastern climate make electrical heating more attractive than in northern regions where temperature variation and heating demands are greater.

In part because of these factors, the average annual number of kilowatt-hours consumed per home in this area is about twice as great as the national average.[83] Chattanooga was the first large city in the United States in which average consumption per customer exceeded 10,000 kilowatt-hours per year. Because a substantial part of this load is accounted for by electrical heating, this development in the TVA area will be briefly reviewed.

In 1956 the 150 distribution systems retailing TVA power were serving some 155,000 consumers who used electricity as the sole energy source for domestic heating purposes (table 23).[84] This means that one of every eight homes in the Tennessee Valley region was then electrically heated. The use of electricity for residential heating received its first real trial in this area in 1934 when electric heating systems were installed in 250 of the 350 homes erected in the TVA construction town of Norris. Although subsequently electric heating steadily grew

TABLE 23. *Number of Electrically Heated Homes in the Tennessee Valley Region, 1934–56*

Period	Number
1934	250
1947-48, heating season	7,600
1950, August	35,000
1955, January	131,500
1956, January	155,000

Source: Various releases, TVA Division of Power Utilization.

[83] Tennessee Valley Authority, *Annual Report of the Tennessee Valley Authority, 1956,* Washington, 1957, p. 37.

[84] Tennessee Valley Authority, Division of Power Utilization, *Estimated Saturation of Residential Electric Space Heating—Distributors of TVA Power—Fiscal Year 1956,* Chattanooga, 1956 (mimeo), p. 1.

in popularity, by 1946 less than 1 per cent of the houses in the Tennessee Valley were so heated.[85] However, with the spurt in residential construction which occurred after World War II, the number of homes using electricity for complete house heating greatly multiplied. In many areas, 75 to 90 per cent of the new homes built during the last decade have been equipped with electrical heating systems and numerous older residential structures have been converted to electricity.

A number of supporting factors, of interest on a national as well as a regional basis, have encouraged this development. Basements and chimneys, which represent up to 20 per cent of the total construction cost of small houses, are not required where electric heating systems are installed. Elimination of the need for fuel storage space makes additional savings possible. Ease of temperature control, freedom from the dirt which accompanies most coal- and oil-fired heating systems, elimination of the need for prepurchase of these fuels, as well as safety and limited servicing and maintenance requirements, have all contributed to the popularity of electrical heating.

One of the most popular types of installation now is the radiant heat panel. Some homes have been equipped to allow the circulation of electrically heated water in the floor or ceiling. A more recent method of radiant panel heating involves the use of low-resistance heating cable which is embedded in the plastered wall or ceiling area. In other structures, use is made of rigid panels with electrically conductive sheets sandwiched between them, the surface temperatures of the radiant panels reaching 115°F. when necessary. Auxiliary portable electric heaters, which are numbered in the hundreds of thousands in the TVA area also contribute substantial electrical demand.

The postwar increase in the use of electricity for residential heating is exerting a significant influence on the load characteristics of the TVA system. More than one-fifth of the total residential demand can be attributed to this use. In 1956 residential consumers accounted for the purchase of about half of the total electricity retailed by the distributors, which in turn constituted about one-eighth of the total power generated by the TVA.[86] However, there is a decided seasonal character to the residential demand for electricity which the heating load creates. The heating season begins in October and terminates in May, with the

[85] William R. New, *Load Characteristics of the Residential Consumer Using Electricity for House Heating* (Chattanooga: TVA Division of Power Utilization, Special Studies Section, 1955), mimeo, p. 1.

[86] TVA, *Annual Report, 1956, idem,* pp. 26-27, 34.

peak electric heating load generally occurring at the beginning of February. During January, the month of greatest total electricity consumption for home heating in this area, residential consumers possessing electric heating systems consume almost nine times as much energy as those utilizing some other source of energy for heat.[87] At the same time, the use of electricity for heating homes has a strong load-building influence in terms of the demand for other electrical appliances; thus, most of the electrically heated homes also use electricity for cooking, water heating, refrigeration, small appliances and improved lighting. The average consumption of electricity in such homes in the TVA area during the summertime, it has been found, is more than twice as great as that of homes heated by other means.[88]

It should be recognized that the conditions in the TVA service area are peculiarly well suited to the application of electric space heating, and that such extensive development elsewhere in the United States may not be achieved. The low-cost power and moderate winter weather prevailing in this region are ideal circumstances for electric heating, not generally encountered in central and northern states. It is probable, however, that this type of heating will find increasing application in the Pacific Northwest and in other sections where cost differentials encourage it.

Air conditioning as a nationwide example. The marked seasonality of the demand for electric heating in the TVA area is being compensated in part by rapidly growing electricity requirements for summer air conditioning. However, the situation in other regions of the country is often found to be quite different. For example, in the service areas of many southern electric systems and several large municipal systems in the central and northeastern states, it is air conditioning, and not residential heating that augments loads most. Depending upon the prevailing climatic conditions, the situation varies in each area of the country. Accordingly, problems of compensating for the consequent uneven distribution of the demand for electricity also vary.

The rise in the demand for electricity for residential air conditioning has been dramatic in many areas during the last few years. In 1947 only 43,000 room air conditioners were sold, but by 1953 sales totaled

[87] TVA, *Estimated Saturation of Residential Electric Space Heating,* previously cited, p. 5.
[88] William R. New and William A. Bell, Jr., "Serving the All-Electric Home," *Electrical World,* March 19, 1956, p. 125.

more than half a million units,[89] and by the following year, this number
had increased to 1,230,000 units.[90] It has been estimated that future
annual sales of room air conditioners will probably approach 2 million
units.[91] Central residential air conditioning systems, it is anticipated,
also will assume increasing importance. In 1955 a quarter of a million
new homes were so equipped and it was expected that the number of
fully air-conditioned homes put into operation in 1958 would be about
three times as great.[92] According to another estimate, by 1965 the
annual sale of central air conditioning systems will have risen to a
million units.[93] Evaporative coolers, which use water directly and
electricity only for a fan, are decreasingly important in air conditioning
because their performance characteristics are less desirable than those
of the refrigeration type.

The rapid spread of air conditioning has already had a discernible
impact upon the total load of various power systems. First, it has
caused the daily peak in many areas to be moved from mid-afternoon
into the evening. It has also compensated for the winter peak demand
and, in some areas, has established a summer load substantially in
excess of that of the winter months. The occurrence of peak loads in the
summertime is undesirable because the generating potential for both
hydroelectric and steam-electric power is usually lowest in this season.
Furthermore, transmission facilities have lower capacity in warm
weather. The character as well as the volume of this demand has
accordingly been the cause of considerable concern.

Many persons intimately acquainted with the situation view the
development and further application of the heat pump as the most
promising solution to this problem in some sections of the country.[94]
Only a limited number of heat pumps, which can be operated both as
a heating and a cooling system, are in use in the United States, primarily
because the initial cost of such installations has been too high to attract

[89] Philip Sporn, "All-Electric Home Seen at Hand," *Electrical World News,*
143 (April 25, 1955), 16.

[90] C. A. Tatum, "Air Conditioning—Big Business," *Electrical World,* 143 (June
27, 1955), 98.

[91] J. R. Hertzler, "Air Conditioning Peak Seen in 1963," *Electrical World,* 142
(November 1, 1954), 81.

[92] "Charges for Residential Air Conditioning," *Journal, American Water Works
Association,* 47 (November 1955), 1089.

[93] "What's Ahead for Air Conditioning?" *Electrical World,* 144 (September 26,
1955), 55.

[94] See, for example, Sporn, *op. cit.*

a mass market. However, substantial cost reductions are beginning to be made, and recent records show sales of 12,000 to 13,000 heat pumps in a year.[95] One estimate suggests that at least 900,000 heat pumps will be in American homes by 1964.[96] Because a typical residential heat pump consumes from 10,000 to 16,000 kilowatt-hours of electricity a year, this suggests that at least 9 billion kilowatt-hours will be required to operate the systems. This load represents three times as much electricity as the estimated use for residential heating and air conditioning in 1954. However, it is probable that within the next decade heat pumps will not require a major fraction of the total residential electricity used for heating and cooling purposes.

It has been estimated that by 1963-64, 55 billion kilowatt-hours of electricity will be required to operate all of the various forms of electric heating and air conditioning which will then be in use.[97] If these projections prove correct, residential consumption in 1963-64 will be increased 40 per cent over present residential consumption by the increment added in air conditioning and space heating alone.

SIGNIFICANCE OF INCREASE IN RESIDENTIAL ELECTRIC USE

The foregoing discussion of increased domestic electric demand is only indirectly related to demand for water, but the effect of the one on the other is highly significant. This may be illustrated by historical comparison of the portion of total water withdrawals for thermal power for which residential electrical demand was responsible. In 1955 residential electricity consumption was about one-fourth of total sales (table 22 and chapter 12). Excluding 18 per cent generated by hydro, water withdrawn for thermal generation was 60 billion gallons per day (p. 312), of which therefore about 12 billion gallons could be allocated to residential electrical demand. In contrast, only 1 billion gallons were required to satisfy daily residential electrical needs in 1920. It is also of interest to observe that the current water demand for residential electrical generation exceeds that for water actually domestically used (table 5, chapter 3) by several billion gallons per day.

[95] "Electric Space Heating: What Utilities Are Doing about It," *Electrical World*, 147 (January 21, 1957), 72-75.

[96] H. M. Brundage, "Heat Pump Market is Opening Up." *Electrical World*, 142 (November 15, 1954), 120-22.

[97] Brundage, *op. cit.*, p. 120; and Hertzler, *op. cit.*, p. 81.

The use of water either in hydroelectric generation or in the condensers of steam-electric power plants is nearly in direct proportion to the total amount of power produced. This is in turn governed by the power requirements that are now becoming so substantially affected by air conditioning and heating loads in many localities. Furthermore, the character of the load, particularly as related to peak demands, affects the operation of a generating system containing both hydro and thermal power plants. As explained more completely in chapter 12, the trend is toward the use of large steam-electric generators to supply base load requirements and the use of hydro for carrying the peaks. This trend, coupled with the peaking characteristics of heavy air-conditioning or electric-heating loads, will affect the design and operation of reservoirs and hydro generation plants. Where applicable, it also will affect the increased use of water in these facilities as well as in thermal power plants. Increased domestic electric use, therefore, has a marked influence on the rate of exploitation of water and the incidence of demand.

CHAPTER 7

Techniques Capable of Decreasing
Demand for Water—

FOUR CASES

Two Western Agricultural Problems Affecting Demand: (1) The Use of Saline Water in Western Agriculture and Cultivators' Adaptations to Saline Soil Conditions—water classification; salt-tolerant crops; the salt balance; livestock adaptations; geographical distribution of saline conditions; salinity problems and water development. (2) Drought-resistant Crops. *Two Waste-Disposal Problems and Techniques for Their Solution:* (3) Spray Disposal of Food Processing Wastes; (4) Developments in the Disposal of Wood Pulping Wastes—waste treatment used in the sulfate pulping process, waste treatment used in the sulfite pulping process; effects of waste treatment methods on sulfate and sulfite pulping processes.

Where requirements for water exceed the supply available from existing facilities under existing practices, conscious adjustment of either supply or demand may be possible so as to bring supply and requirements into balance. In such cases technology can help in achieving this. In almost every instance the techniques are consciously applied where supply deficiencies are evident, as they are in arid lands, or where a single use may preclude most subsequent uses of the same water, as in waste carrying. For example, techniques for reduction of waste-carrying requirements can make available for other beneficial use water which would not otherwise have been available.

As illustrations, eight cases have been selected in this and the succeeding chapter, four to describe the techniques that can affect demand, and four that may be applied from the viewpoint of supply. They cover a range of different scientific and technical skills, impact upon different regions of the country, ground- and surface-water use, and industrial, agricultural, and residential application. Of the four cases discussed in this chapter for the capacity to decrease demand for water, two, dealing with spray disposal of food wastes and disposal of wood pulping waste, are of nationwide application. Two others concern agriculture and the western states.

The agricultural sciences and hydraulic engineering have been the disciplines most concerned with these techniques in the past, although chemical engineering and microbiology have made more recent contributions.

TWO WESTERN AGRICULTURAL PROBLEMS
AFFECTING DEMAND

The cases concerned with saline water use and with drought-resistant crops are peculiar to western agriculture. (See also water-conserving and drought-resistant land use, chapter 8.) Their description here reflects the fact that these techniques come into use most prominently where water scarcity is experienced.

1. The Use of Saline Water in Western Agriculture and Cultivators' Adaptations to Saline Soil Conditions

One set of techniques exists because most water needs are intolerant of highly mineralized or organically contaminated supplies. Ideally, the objective of these techniques is to pair tolerant uses, insofar as the products concerned are required or accepted in the economy, with supplies of limited applicability. The use of saline irrigation water, saline stock water, and saline soil moisture supplies in western agriculture offer examples.

In the West adaptation to or treatment of soils containing injurious concentrations of salts is not a new problem, particularly in many of the irrigated areas.[1] However, the use of saline and alkali soils has become a problem of increasing severity in recent years. Attention to it has increased correspondingly. Especially within the last few decades, the salinity problem in western agriculture has been viewed dynamically in terms of adjusting land-use practices so as to prevent the detrimental consequences of progressive soil salt accumulation which irrigation may cause in many localities.

The salinity problem exists wherever high soil concentrations of all soluble salts or exchangeable sodium, or both, are present. Where soil productivity is impaired by the presence of soluble salts—consisting mostly of various proportions of the cations sodium, calcium, and magnesium, and the anions chloride and sulfate—saline soil is said to exist. Alkali soils, according to generally accepted usage, are those in which productivity is depressed by the presence of exchangeable sodium. Most often, however, such soils contain an excess of both soluble salts and exchangeable sodium, and these are technically referred to as saline-alkali soils.[2]

Salts that accumulate in the soil have their origin in the rock of the earth's crust. These salts, made soluble and released through the process of weathering, are carried away by water. In dry areas where the flow of water is limited they are rapidly precipitated and deposited. They then tend to accumulate in the surface soil because plant roots take in water but absorb very little salt from the soil solution, and because water (but not salt) is lost in the course of surface evaporation and leaf transpiration. The resulting accumulation of salts is thus generally greatest in hot and arid or semi-arid regions.

The process of soil salinization and alkalinization may be intensified by the application of water for irrigation purposes and by the existence of high ground-water tables. The greater the load of salts present in irrigation water, the greater the salinity hazard. Each acre-foot of irrigation water used in the Imperial Valley of California contains a ton

[1] The occurrence of saline and alkali soils was repeatedly reported in nineteenth century surveys of the West, and the possible use of such lands was explored by pioneers in this field like F. S. Harris and E. W. Hilgard in the early years of this century.

[2] While these terms are not yet universally used, they are the definitions employed by the United States Salinity Laboratory and are likely to gain increasing acceptance. See L. A. Richards (ed.), *Diagnosis and Improvement of Saline and Alkali Soils* (Agriculture Handbook No. 60), U. S. Department of Agriculture, Washington, 1954, p. 1. The terms are more technically defined on pp. 4-6 of this publication.

of salts.[3] They must be removed from the soil surface if the productivity of the soil is not to be impaired. Where the ground-water table is near the surface, salt accumulation may be due to the solvent action of ground water rising by capillary action, then disappearing by evaporation and plant use. A similar process of salt concentration takes place in impermeable soils, which may become "waterlogged" after continued irrigation.

Excessive salt concentration in the soil is harmful to plant growth because it reduces the availability of water for absorption by roots. As salts accumulate, the soil solution progressively increases in osmotic pressure, which adversely affects the rate of water uptake and therefore of plant growth. When the soil solution becomes more concentrated than that of the plant cells, water generally cannot pass as freely from the soil to the plant. Under very severe salinity conditions, vegetation is thus deprived of water. Sodium salts also disperse the colloidal material of the soil and the resulting clogging checks or prevents drainage, areation, and aerobic bacterial action. Visible symptoms of salt injury to plants may occur if the salt concentration of the soil substrate is very high. When excessive quantities of soluble salts exist in the soil, chlorosis and leaf tip burn are frequently observed, while sap flow may be prevented because the external tissues of the plant become burned.[4]

Not only is crop yield adversely affected by salinity, but quality also is impaired. Where the soil moisture available to a plant is restricted by salt, plant behavior is modified in the same manner as when drought causes water deficits in tissues. Some forage plants may absorb so much salt that they become unpalatable to livestock other than sheep, which may tolerate some plants. The sugar content of sugar beets is lowered when they are grown in highly saline soils, and the amount of ash which they contain is increased. Cereals grown under such conditions tend to produce small and shriveled kernels. Similarly, the bolls of cotton are generally fewer and smaller, and the length and fineness of their fibers are decreased. The texture, flavor, and storage quality of different types of fruit are often unfavorably influenced where these crops are produced on saline soils.[5]

[3] U. S. Department of Agriculture, Agricultural Research Service, *Control of Salinity in the Imperial Valley, California 1955* (ARS-41-4, Washington), p. 1.

[4] H. B. Roe, *Moisture Requirements in Agriculture* (New York: McGraw-Hill, 1950).

[5] Ivan E. Houk, *Irrigation Engineering,* Vol. 1 (New York: John Wiley, 1951), p. 481.

Plate 4. Barren area in flax field caused by alkali formation over high water table, Imperial Valley, California.

The presence of excessive concentrations of salts in the soil thus is a significant agricultural problem because of its adverse effects upon plant growth.

WATER CLASSIFICATION

An important step in identifying and controlling salinity is classification of the salinity and sodium hazard in irrigation water. Salinity reflects the total concentration of soluble salts. It is measured by the capacity of the ionized inorganic salts in the water solution to conduct an electrical current and is expressed in terms of the specific conductance of the water. According to this classification, water of low salinity (specific conductance less than 250 micromhos per centimeter at 25° C., or about 150-175 ppm dissolved salts) can be used for irrigating most crops on most soils with little likelihood that excessive soil salinity will develop. Medium salinity water (250 to 750 micromhos per centimeter) can be used without special practices where a moderately salt-tolerant crop is grown. High salinity water (750 to 2,250 micromhos per centimeter) cannot be employed on soils of poor drainage, while water of very high salinity (more than 2,250 micromhos per centimeter) is not suitable for irrigation under ordinary conditions. It can be used only with very salt-tolerant crops and then only if special practices are followed, including a high degree of leaching.[6]

The sodium hazard of irrigation water similarly has been classified. This is done by measuring not only the total concentration of salts in the water but also the sodium-absorption-ratio. The latter expresses the relative activity of sodium ions in exchange reactions with the soil. It is possible to use low sodium water (in this classification) for irrigation with little danger of the development of harmful levels of exchangeable sodium. Medium sodium water presents an appreciable sodium hazard in the case of certain fine-textured or permeable soils high in organic content. High sodium water may produce harmful levels of exchangeable sodium in most soils and requires special soil management such as the provision of adequate drainage, extensive leaching, and the addition of organic matter. Very high sodium water is generally unsatisfactory for irrigation unless special measures, such as amendment of the

[6] B. Irelan, "Chemical Quality Standards for Irrigation Waters," *The Cross Section*, 3 (December 1956), 1, 4; U. S. Department of Agriculture Handbook No. 60, previously cited, pp. 79-81.

soil with gypsum, are practiced.[7] In practice, total salinity and sodium absorption ratio must both be appraised, because the limits of sodium hazard are affected by total salinity. Tolerance for sodium in water of low salinity is roughly twice as great as in waters of very high salinity. By using both the sodium and salinity measurements it is possible to obtain a fairly comprehensive guide to the practices which must be followed if crop productivity is to be maintained.

SALT-TOLERANT CROPS

It has long been recognized that some crops are more salt tolerant than others.[8] A second important line of endeavor within this area has been an examination of the resistance of different crops to the injurious effects of salinity.[9] Sugar beets, after they are well established, barley, Bermuda grass, asparagus, and dates are examples of crops that are classified as possessing a high salt tolerance. Others, such as alfalfa, wheat, onions, lettuce, and potatoes, have been placed in the medium salt tolerance category, while beans, strawberries, and most clovers, to identify but a few, are characterized as very sensitive to salt injury.

As information on the salt tolerance of different crops has accumulated, and as knowledge of the salinity and sodium hazards of irrigation waters has increased, it has become possible to adjust agricultural practice to minimize the risk of crop injury. Crops may be selected on the basis of the extent of the hazard. In some areas difficulty has been avoided by pairing salt-tolerant crops with irrigation water possessing high concentrations of soluble salts and exchangeable sodium. In addition to this step, however, a number of important methods have been devised to reduce the accumulation of salts in soil under irrigation and thus maintain the productivity of the soil.

[7] *Idem.*

[8] T. H. Kearney and L. L. Harter, *The Comparative Tolerance of Various Plants for the Salts Common in Alkali Soils* (Bulletin No. 113), U. S. Department of Agriculture, Bureau of Plant Industry, Washington, 1907; and T. H. Kearney and C. S. Scofield, *The Choice of Crops for Saline Land* (Circular No. 404), U. S. Department of Agriculture, Washington, 1936.

[9] The most definitive classification of the salt tolerance of different crops is that compiled by the United States Salinity Laboratory. See U. S. Department of Agriculture Handbook No. 60, cited previously.

THE SALT BALANCE

Introduction of the "salt balance" concept two decades ago was an important step in studying salt accumulation. The salt balance concerns the ratio between the quantity of dissolved salts in the irrigation water delivered to an area and the quantity removed from the area by drainage. If irrigated areas are to be kept in continuous production, it is necessary to maintain a favorable salt balance, in which the outflow of salts equals or exceeds the input.[10] The most important means of obtaining this result is artificial leaching, whereby more water is applied than a crop uses. The excess water passes through the surface horizon and removes the salts. The leaching necessary to maintain a favorable salt balance can be carried out in several ways, provided soil drainage is satisfactory. Water can be applied in excess of the amount required for plant growth with each irrigation or in very heavy irrigations at intervals. The latter practice is called winter flooding or "pre-emergence irrigation."[11]

Another important way of coping with the salinity problem is frequent irrigation. By keeping the concentration of salts in the soil solution dilute, the harmful effects of salinity on plant growth in many cases can be decreased. Where the rate of salinization is intensified to a serious extent by the impermeability of the soil and by high ground-water tables, measures can be undertaken to improve natural drainage and to eliminate waterlogging.

Even though leaching is done with the same water that causes a soil salinity problem, the treatment is effective if adequate quantities are used. The capacity of the water to dissolve salts accumulated in the soil by evaporation is ordinarily sufficient to effect nearly complete removal. The leaching requirements, or amount of irrigation water that must pass through the root zone of the soil to avoid salt accumulation, can be computed easily. The amount varies with the original salt load of the water used, as well as with other factors.

Leaching involves the removal of excessive amounts of salts, but other measures modify the harmful effects of salts. Some soil amendments

[10] C. S. Scofield "The Salinity of Irrigation Water," *Annual Report of The Smithsonian Institution, 1935,* Washington, 1936, pp. 275-87. *Idem,* "Salt Balance in Irrigated Areas," *Journal of Agricultural Research,* 61 (1940), 17-30.
[11] L. V. Wilcox, *Classification and Use of Irrigation Waters* (Circular No. 969), U. S. Department of Agriculture, Washington, 1955.

serve this purpose. If gypsum (hydrated calcium sulfate) is applied to a black alkali soil (high sodium content), for example, it reacts to form the less harmful and more soluble Glauber's salt (hydrated sodium sulfate) and lime, a desirable soil constituent.[12] Other materials that can be added advantageously to saline soils include sulfuric acid, aluminum sulfate, ammonium sulfate (also as fertilizer) and ferrous sulfate.

A favorable salt balance may also be obtained if crops are effectively rotated. Salts will accumulate in the surface soil during the growth of a lightly irrigated forage or field crop. If it is followed by a heavily irrigated row crop, salinity will be reduced. Mulching and maintenance of a vegetative covering are beneficial insofar as they serve to reduce evaporation and capillary movement toward the surface. Deep and regular tillage helps to prevent the formation of surface encrustations of salts and, by keeping the soil well mixed, increases the effectiveness of downward leaching.

As knowledge of this field improves, other prescriptions for surmounting the hazards of salinity are being devised. Recently it has been found that a new shape of the double row beds used for furrow-irrigated crops will cause seeds to germinate in topsoils initially so high in salinity that successful germination hitherto was inhibited. Most of the salts are carried upward toward the center of the new type of beds, leaving the shoulders relatively free of salts. Seed can be planted safely on the slope below the zone of salt accumulation on such beds. The salts are carried away from the soil around the seed instead of accumulating in that vicinity.[13]

This practice has spread with great rapidity, especially in the cultivation of sugar beets. Once established, sugar beets are very salt tolerant, but they are susceptible to injury during germination and as seedlings.[14] In 1954 almost a million acres in the United States were devoted to the cultivation of sugar beets. More than four-fifths of this total was located in the western states.[15] California alone produced about one-third of the total tonnage of sugar beets, while Colorado and Idaho were each responsible for more than one-tenth of the country's output. Nearly all

[12] Black alkali soil is a dark-colored soil composed of organic matter and highly alkaline (sodium) inorganic materials.

[13] U. S. Department of Agriculture, ARS-41-4, previously cited, pp. 14-15.

[14] F. S. Harris, *The Sugar Beet in America* (New York: Macmillan, 1919), p. 66.

[15] U. S. Department of Agriculture, *Agricultural Statistics, 1955* (Washington, 1956), table 106, p. 71.

Plate 5. Leaching of soil to remove excessive salts, Imperial Valley, California.

of this acreage is in irrigated areas where the salinity problem is encountered, except that part of the crop produced in the fog-belt coastal areas of California.[16]

Almost half of the total alfalfa acreage of the country is located in the seventeen western states and, because a substantial share of the western plantings of this crop is irrigated, the West accounts for an even larger proportion of total production.[17] The salinity problem exists in many of the irrigated areas where alfalfa is grown. Although difficulty is often encountered in establishing new stands of alfalfa where salinity conditions are present, the plant is able to withstand the accumulation of a considerable amount of salts over the period of several growing seasons. However, there appear to be substantial differences between different varieties and strains. It seems likely that further selection and development of more tolerant strains and types may make alfalfa still more useful in areas of saline soil.

LIVESTOCK ADAPTATIONS

Livestock grazing within extensive areas of the seventeen western states represents another aspect of adaptation to conditions of soil salinity and alkalinity. Just as certain crops exhibit a greater tolerance than others to saline water supplies, likewise it has been observed that sheep are more tolerant of high salt intakes than are other forms of livestock. For example, no adverse effects were observed when a test ration fed to fattened sheep included sodium chloride to the extent of about 10 per cent of the total diet.[18]

This tolerance has been important in the land use of the West because the soils of many of the most important sheep ranges, covering a total area of many thousands of square miles, possess heavy accumulations of salts. The sheep ranges are particularly widespread in the Intermontane region, between the Rockies to the east and the Cascade and Sierra Nevada mountains to the west. Rangeland over parts of the more level terrain within this area is composed dominantly of shrub types which grow under semi-arid climatic conditions on soils surcharged with salts. Sheep have demonstrated their ability to thrive on forage which may

[16] J. L. Haddock, "The Irrigation of Sugar Beets," *Water, The Yearbook of Agriculture, 1955,* U. S. Department of Agriculture, Washington, pp. 400-405.

[17] *Ibid.,* p. 264.

[18] J. H. Meyer and W. C. Weir, "The Tolerance of Sheep to High Intakes of Sodium Chloride," *Journal of Animal Science,* 14 (1955), 412-18.

include an important admixture of shrubs highly tolerant of saline conditions, including saltbush, (*Atriplex* sp.), shadscale (*Atriplex confertifolia*), alkali sacaton (*Sporobolus airoides*), and winterfat (*Eurotia lanata*), or the less desirable rabbitbrush (*Chrysothamnus* sp.).

These salt-tolerant vegetation associations are widely distributed in the Great Basin, as well as over the lower and drier areas of Utah southeast of the Wasatch Mountains, and to a lesser extent in the Snake River plains of southern Idaho and the lower areas of southwestern Wyoming.[19] They are interspersed with other types of range dominated usually by sagebrush (*Artemisia tridentata*), growing on less saline soils, and are accordingly used by graziers at the same time as the sagebrush plains. The salt-desert shrub may be used for a few weeks or a few months during the fall, winter, or spring, when weather severity precludes use of rangelands at higher elevations. To some extent these ranges are used primarily for spring and fall grazing by sheep bound to and from the higher summer pastures in the mountain areas. Saline soils are most widespread at the lowest elevations within the Intermontane area, where evaporation of water which has seeped down from higher areas is pronounced. The saltbush-shadscale association therefore is most useful as a winter range. As such, its role in the livestock economy of the West is important, since it is seasonally complementary to the summer use of the sheep-fattening meadows within the higher mountain areas.

Today the frontier of the western livestock industry lies in the application of management practices which will restore and maintain ranges at their maximum production of both forage and livestock. An important aspect of this is the proper handling of soils where the salinity problem is known to exist. This requires prevention of the spread of undesirable and nuisance plants, notably Russian thistle (*Salsola Kali* var. *tenuifolia*). In general, a short but extensive vegetative covering is more likely to minimize the further accumulation of salts in the soil than are taller but sparser types of vegetation.

GEOGRAPHICAL DISTRIBUTION OF SALINE CONDITIONS

The natural occurrence of saline and alkali soils in Western United States is highly variable (figure 16). Severe salt accumulations, such

---

[19] Some salt-tolerant shrub vegetation also occurs farther south, particularly along the Rio Grande and Pecos rivers and in southern and western Arizona below 3,000 feet. Much of this scrub, however, is unpalatable to sheep.

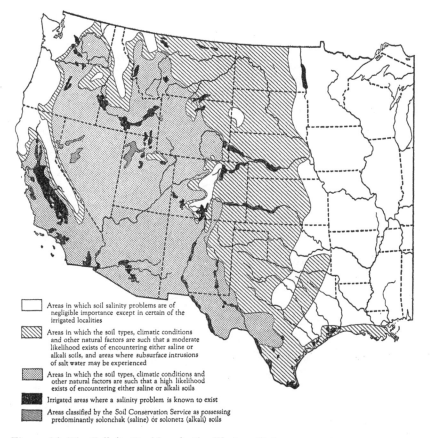

Areas in which soil salinity problems are of negligible importance except in certain of the irrigated localities

Areas in which the soil types, climatic conditions and other natural factors are such that a moderate likelihood exists of encountering either saline or alkali soils, and areas where subsurface intrusions of salt water may be experienced

Areas in which the soil types, climatic conditions and other natural factors are such that a high likelihood exists of encountering either saline or alkali soils

Irrigated areas where a salinity problem is known to exist

Areas classified by the Soil Conservation Service as possessing predominantly solonchak (saline) or solonetz (alkali) soils

Figure 16. The Salinity Problem in the Western States.

as spots of black alkali or white surface encrustations, are generally of limited extent and are frequently widely separated. Furthermore, the problem often may be highly localized, even to spots within a single field. Injurious salt accumulation serious enough to inhibit crop germination may exist in patches in an otherwise productive field. Where the salinity problem exists in an irrigated area, scattered incidence of harmful conditions is characteristic. Severity of the salt problem is determined by the salt load of the water used for irrigation, the amount of water applied, the effectiveness of management practices, and the character of the soil and the adequacy of the prevailing drainage. Largely as a result of trial and error, variable soil conditions have been adapted to

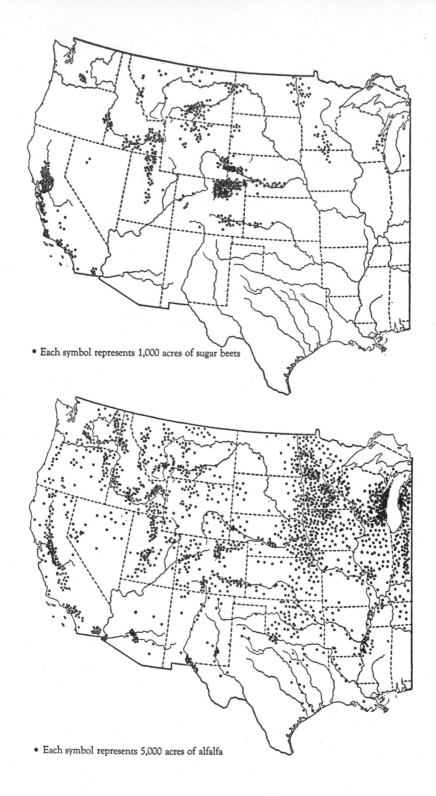

• Each symbol represents 1,000 acres of sugar beets

• Each symbol represents 5,000 acres of alfalfa

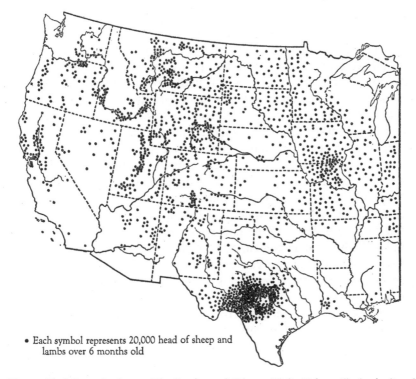

• Each symbol represents 20,000 head of sheep and
  lambs over 6 months old

Figure 17, left and above. Distribution of Three "Salt Tolerant" Agricultural
Activities in Western United States.

uses best suited to them. Hence there is a good correlation between
saline and alkaline soil distribution and the distribution of sheep hus-
bandry, alfalfa acreage, and sugar beet acreage within the same area
(figures 16 and 17).

SALINITY PROBLEMS AND WATER DEVELOPMENT

Where salinity problems are encountered in western agriculture they
are continuing problems. The irrigated areas with a tendency toward excessive salinity for healthy crop production must have a favorable salt
balance continuously maintained if they are to remain in permanent use.
The hazard can be so high that in the course of a single growing season
a hitherto productive soil may acquire an injurious load of salts. The

productivity of many irrigated areas already has been impaired seriously by alkalinization and salinization of the soils. The problems associated with accumulation of soluble salts and exchangeable sodium are increasing in the western states. This is true partly because the salinity hazard tends to increase as the period a soil is under irrigation lengthens. Moreover, as the demands upon water in the West have mounted, there has been a marked trend in irrigated agriculture toward the use of drainage water and the return flow from other irrigated areas. The increased salt content of such water already is intensifying the problem.[20]

In addition to its unfavorable effect upon crop productivity, the salinity problem adds to agricultural production costs by increasing the quantity of water required per unit of product. In many irrigated areas more than half of the water which is spread over the land is used for the purpose of leaching the salts from the soil.[21] If the water supply is initially high in saline content and the soil possesses a heavy salt concentration, the salt balance can be maintained only when a very large proportion of the water withdrawn is devoted to this purpose.

In many of the irrigated areas of the western states more water is applied than is required directly by the growing crops. As greater knowledge of the water requirements of different crops is obtained, and if adjustments are possible in the systems of allocating irrigation water, water use per unit of product generally should decrease. Studies have shown that yields of numerous crops can be improved by reducing the amount of water applied by most growers. In areas where the salinity problem is seriously aggravated by poor drainage and by high groundwater tables, such a trend would lessen the salinity hazard. Where drainage conditions are favorable, however, water demands may be expected to continue high, for soil leaching needs are not likely to lessen. The competitive position of irrigated areas suffering salt accumulation handicaps (and consequent fertility losses) therefore may be in doubt. Questions may arise as to the economics of irrigation water use in areas of salt hazard where other potential water uses are not served.

There is a growing awareness of the problem, and it seems likely that in the future the cultivation of salt-tolerant crops will be paired with

[20] H. E. Hayward and C. H. Wadleigh, "Plant Growth on Saline and Alkali Soils," *Advances in Agronomy*, Vol. 1, A. G. Norman, ed. (New York: Academic Press, 1949).

[21] It may be noted that the process of leaching also removes the plant nutrient elements from the soils so treated, thereby lessening one of the natural advantages possessed by dry region soils.

soils that are highly susceptible to the hazards of soluble salts and exchangeable sodium. This may occur more extensively and with far greater conscious intent than at present. While emphasis in the past has been given to classifying different crops in terms of their tolerance to salinity conditions, a major line of endeavor in the years ahead will be to locate and to develop salt-tolerant varieties and strains within different plant species. Much remains to be done in obtaining the best distribution of crops of different salt tolerances and water requirements to farming areas with the various salinity hazards. And as saline soils become more widely usable, water demands (as for leaching) may be decreased as related to the total product obtained.

2. *Drought-resistant Crops*

Drought is an even larger problem than saline soils in the western states. Indeed, the two problems are closely associated, and saline conditions in a sense are but one facet of drought. Lack of adequate precipitation, however, does have other facets. The most obvious agricultural countermeasure to drought is irrigation, but the amount of land surface that can be irrigated is a minor percentage of the total in the West. Even under the humid land conditions of the eastern states, potentially irrigable acreage is not a major part of the total. Consequently, extending the use of soil moisture by introducing plants adapted to natural moisture conditions is of wide geographic significance. "Adaptable" plants are those capable of enduring the vicissitudes of natural soil moisture supply over the limit of their life, and also of being economically productive. Two characteristics are sought: the capacity to endure drought without serious tissue injury, and the capacity to make rapid and advantageous use of moisture when it comes.

These plants are of greatest importance in the regions of the United States where drought has been most felt, particularly the Great Plains and other semi-arid sections where enough soil moisture is periodically present to encourage crop cultivation.

As illustrated in the discussion of soil salinity, a number of factors limit plant growth, but the availability of soil moisture is among the most important. Water is a key substance in the metabolism of all forms of vegetation. In the arid and semi-arid regions of the United States the inadequacy of natural moisture is the chief restriction on the possibility of agricultural development. However, periods of deficiencies in soil moisture are experienced at some time during the growing season in virtually all parts of the country. When one of these periods is sufficiently long to injure plant growth seriously, it is known as a drought. This term commonly is used to refer to a protracted period without appreciable rainfall. However, the agricultural problem which drought presents is generally viewed in terms of the prevailing climatic conditions of a locality. A subnormal amount of precipitation experienced in one area of the country would be an abundance in another.[22] While the concept of drought can be readily understood, it proves difficult to define in terms of plant growth. For the agronomist, definition of drought centers upon a consideration of soil conditions. Agronomically it can mean (1) a chronic deficiency of soil moisture; and (2) occasional periods when the available moisture supply is less than the average. In the following remarks, the term "drought" will include both meanings.

There are wide variations not only in the water requirements of different types of plants but also in their drought-resistant characteristics. The physiological and genetic basis for the drought-resistant capacity of such plants is not fully understood. Some species, like various ephemerals, are drought escaping. They are annuals which can complete their life cycle in a comparatively short period of time before the soil moisture becomes exhausted. Thus they avoid rather than withstand drought. Agriculture in many areas has been adjusted to the use of short-season field crops of this type. There is also an important group of drought-evading plants which are adapted to a limited supply of moisture because of small size, restricted growth, wide-ranging root systems, or low water requirement.

In still another category are drought-enduring plants. The various forms of cacti, which can provide for the retention of considerable amounts of water and can reduce water demand, belong to this group. For the most part, they are able to endure drought by passing into a dormant condition. Another distinct group is that of plants with cells capable of withstanding substantial desiccation without irreparable injury. In many cases these plants are able to survive dry conditions because of their previous development of extensive and deep root systems when moisture was available. However, they have no superior ability to grow in a dry soil and generally no superior efficiency in the use of

[22] Ivan R. Tannehill, *Drought; Its Causes and Effects* (Princeton, N. J.: Princeton University Press, 1947), pp. 15 ff.

water.[23] If the environmental conditions of drought-enduring plants become more favorable in terms of moisture availability, they will then grow more satisfactorily and produce more in a given period of time.

Although the vegetation of the arid regions often possesses economic value, the return per unit of area is in most cases so little that no attempt is made to harvest or otherwise use it. It has long been an aim of plant breeders to develop the usefulness of desert plants, from Luther Burbank's attempts to develop a thornless cactus to those more recently concerned with possible industrial raw materials from such plants. However, the very characteristics which tend to adapt these plants to desert conditions—their form, composition, relatively slow growth, and limited extent—also greatly restrict the potentialities of their use.[24]

The true arid regions of the country are separated from areas with more adequate moisture supplies by transitional zones, except where sharp mountain divides exist. These zones of transition, often natural grasslands, are very extensive, and appear to afford one of the most promising possibilities of increasing agricultural productivity. The Bureau of Plant Industry of the Department of Agriculture in 1951 estimated that average production for all of the grasslands of the nation could be tripled at least.[25] Although increased grassland productivity will be sought in many different ways, such as improved range management and greater use of fertilizers, one of the most important is the development and introduction of plants better adapted to the prevailing moisture conditions. In line with this concern, attention today increasingly is being directed toward the improvement of the drought-resistant characteristics of grasses and legumes growing under conditions of persistently reappearing soil moisture deficiencies.

There have been two main approaches in the effort to develop and propagate drought-resistant plants of agricultural importance: (1) the introduction and propagation of strains, varieties, and species into an

[23] This classification of favorable adaptations that plants can make to drought conditions first was advanced many years ago. However, it still is a useful way to summarize the different mechanisms of plant response to such deficiencies. See T. H. Kearney and H. L. Shantz, "The Water Economy of Dry Land Crops," *The Yearbook of Agriculture, 1911*, U. S. Department of Agriculture, Washington, pp. 351-61. The same classification was restated in H. L. Shantz, "Drought Resistance and Soil Moisture," *Ecology*, 8 (1927), 145-57.

[24] W. G. Whalley, "Arid Lands and Plant Research," *The Scientific Monthly*, 75 (1952), 230.

[25] Tom Dale and V. G. Carter, *Topsoil and Civilization* (Norman: University of Oklahoma, 1955), pp. 242-43.

area where previously they did not exist; and (2) programs of plant breeding. The two overlap to an important extent. Both have the objective of making available plants with superior drought-resistance characteristics through a process of selection. The importance of the first type of activity is suggested by a recent estimate that more than 65,000 different plants have been introduced into the United States from various parts of the world for this purpose.[26] Some introductions have been inadvertent, and some have proved undesirable, creating annoying problems. Yet through this approach many valuable species, varieties, and strains have been added to the store of existing vegetation in the United States.

Among the most important are various members of the genus *Sorghum*. Kafir (*Sorghum vulgare caffrorum*) and milo (*Sorghum vulgare* form), the two most widely grown classes of grain sorghums, have been used principally for livestock feed in areas too hot and too dry for successful corn (maize) production. These plants are able to withstand the effects of hot dry weather better than other grain crops. They can remain dormant during dry spells and resume growth when moisture becomes available, unless the period of moisture deficiency is prolonged.[27] In the last twenty-five years the acreage devoted to grain sorghums has increased fourfold,[28] and in many areas of the South and Southwest it has become a leading crop. A number of grasses belonging to the sorghum family are also of at least equal importance; Johnson grass (*Sorghum halepense*), brought to the United States more than a century ago, has spread steadily westward from the southern Atlantic area where it was first used. In many areas it was initially considered an objectionable weed, but it has gained popularity because of its ability to provide good dry-weather pasture during the summer.[29] Sudan grass (*Sorghum vulgare* var. *sudanense*) is another such species; first located near Khartoum in the eastern Sudan of Africa in 1909, it is one of the first examples of a plant type that was intentionally sought in another

[26] B. T. Dickson, "The Challenge of Arid Lands Research and Development for the Benefit of Mankind," *The Future of Arid Lands*, ed. by Gilbert F. White (Washington: American Association for the Advancement of Science, 1956), p. 60.

[27] J. H. Martin, "Grain Sorghums Highly Drought Resistant," *The Yearbook of Agriculture 1933*, U. S. Department of Agriculture, Washington.

[28] U. S. Department of Agriculture, *Agricultural Statistics*, Washington, 1956.

[29] G. W. Burton, "The Adaptability and Breeding of Suitable Grasses for the Southeastern States," *Advances in Agronomy*, ed. by A. G. Norman (New York: Academic Press, 1951), pp. 197-241; and H. W. Staten, *Grasses and Grassland Farming* (New York: Devin-Adair, 1952), pp. 198-201.

part of the world for introduction into the United States. It has been valuable particularly because of its capacity to produce crops of forage and hay while enduring long periods of drought in the moisture-deficient regions of the Great Plains. It has been used effectively in other areas as well.[30]

Crested wheatgrass (*Agropyron cristatum*), introduced from Russian Turkestan, has become one of the most important drought-resistant forage plants of the Great Plains region. The United States Department of Agriculture experimented with this grass before 1900, after which the grass achieved some popularity; but it was primarily after the drought of the 1930's that crested wheatgrass became well established in this country. A bunch type of grass with deep fibrous roots, it was employed extensively in attempting to effect the recovery of many of the seriously depleted areas of the western range. It is well adapted to areas that experience moisture deficiencies in the summer and autumn, due to its root system and its ability to lie dormant when the supply of soil moisture is exhausted. After entering dormancy, it remains in this state until a considerable amount of precipitation falls and its growth is then renewed. Russian wild-rye (*Elymus juncens*), brought to this country a number of years ago, also is rapidly emerging as economically important to this region.

In the southern Great Plains area much attention has been centered on a number of lovegrasses (genus *Eragrostis*), which were imported to this country from Africa. Few grasses appear to be more drought resistant than weeping lovegrass (*Eragrostis curvula*), and in recent years the extent of its North American range has rapidly increased.[31] Various subterranean clovers (*Trifolium subterraneum*) also are coming into use. They were first introduced into California two decades ago from Australia.[32] These plants, native to Europe, Asia, and Africa, are characterized as subterranean because they bury their seed beneath the soil surface. Once established, a stand of this legume is accordingly very hardy. It is well adapted to a climate of winter rainfall and summer drought and has been grown successfully where annual precipitation is

---

[30] *Ibid.*

[31] M. M. Hoover *et al.*, "The Main Grasses for Farm and Home," *Grass, the Yearbook of Agriculture, 1948*, U. S. Department of Agriculture, Washington, 1948; and S. G. Archer and C. E. Bunch, *The American Grass Book: A Manual of Pasture and Range Practices* (Norman: University of Oklahoma, 1953).

[32] E. A. Hollowell, "Clovers that Make a Crop," *Grass, The Yearbook of Agriculture, 1948*, previously cited.

as low as 15 inches a year.[33] Another more recent clover introduction, *Trifolium hirtum,* was imported from Turkey toward the end of World War II and has proved of great value because it will grow on phosphorous-deficient soils.[34]

Although the West comprises a great diversity of climatic regions and soil types, it is impossible to find a single area in which new introductions of a drought-resistant character have not either already become important or indicated much promise.

In the last few decades the search throughout the world for promising plants to introduce into the United States has become far more systematized than in the past. The expedition conducted by the Soil Conservation Service to Turkey, Turkestan, and China more than twenty years ago is an example.[35] Turkestan bluestem (*Adropogon ischaemum*), a hardy and drought-resistant grass, has become established in the southern Great Plains as a consequence of this work. During 1950 the Soil Conservation Service harvested over a half million pounds of seed of this plant in order to extend its present range.[36] This latter type of activity is widespread today. For example, in the recent work of the Texas Agricultural Experiment Station, several thousand plants of more than one hundred species have been collected from Mexico and the southwestern states, and now are undergoing plot tests.[37] To a great extent, the effort to develop drought-resistant crops in this state is an attempt to combat the effects of prolonged overgrazing and a succession of dry years. More than 25 million acres of range land in Texas were considered to be in poor condition in 1957, some of it nearly denuded of soil.

Comparatively little plant breeding has been carried out specifically to develop vegetation able to endure drought conditions.[38] In large part this is due to the complex nature of plant reactions to drought. Many

[33] *Subterranean Clover: Widely Adapted Legume for Non-Irrigated Pasture* (San Francisco: C. M. Volkman and Co., 1950).

[34] R. M. Love, *op. cit.,* pp. 353-54.

[35] C. R. Enlow, "Promising Introduced Species," *American Soil Survey Association Bulletin No. 17* (State College: Pennsylvania State College, 1936), pp. 1180-20.

[36] Staten, *op. cit.,* p. 264.

[37] "Grasses to Withstand Drouth," *Texas Agricultural Progress,* 2 (January 1956), 12-13.

[38] T. Ashton, *Technique of Breeding for Drought Resistance in Crops,* Technical Communication No. 14 (Cambridge, England: Commonwealth Bureau of Plant Breeding and Genetics, 1948).

features are responsible for the physiological capacity of some plants to endure desiccation, and it is probable that they are not the same for all families of plants. Drought resistance is significantly related to the ability of a plant to resist high temperatures. Association also has been found between drought resistance and frost resistance, both in the mode of plant behavior and in the responsible mechanisms.[39] Drought-resistant plants are, in general, characterized by higher osmotic pressures, a fact which is presumably related to their ability to resist evaporation but which may also be significant in other respects.[40] Increasing attention today is being given to the relation between the drought-resistant capacity of a plant and its tolerance to saline and alkaline soils, for the effects of salinity are in many respects comparable to those of drought. Most research workers in this field consider the capacity of increased protoplasmic hydration of some plants to be the basis for their drought-resistant characteristics. Others, however, question this contention. It has also been suggested that the drought resistance of certain grasses depends upon their ability to accumulate food reserves.[41]

Because so little is known of the physiological characteristics of drought resistance, the main emphasis in plant breeding work has been to develop vegetative types that possess greater adaptability to prevailing conditions, and that indicate more promise, than the currently available types. Thus the development of drought resistance in plants constitutes but one of the objectives of an effective breeding program. A new drought-resistant plant is obviously of little value unless it can be established with relative ease and possesses, for example, yield and acceptance by livestock superior to those characteristics of the vegetation it replaces. Once established, such plants must have the capacity to compete successfully with other vegetation. Greater drought tolerance can therefore not be sought at the expense of other desirable characteristics.[42]

[39] G. W. Scarth, "Dehydration Injury and Resistance," *Plant Physiology,* 16 (1941), 171-79. J. Levitt, "Frost, Drought and Heat Resistance," *Annual Review of Plant Physiology,* Vol. II, ed. by D. I. Arnon and L. Machlis (Stanford, Calif. Annual Reviews, 1951) pp. 245-68; and J. Levitt, *The Hardiness of Plants* (New York: Academic Press, 1956).

[40] W. G. Whalley, "Arid Lands and Plant Research," *The Scientific Monthly,* 75 (1952), 228-33.

[41] O. Julander, "Drought Resistance in Range and Pasture Grasses," *Plant Physiology,* 20 (1945), 573-99.

[42] A. A. Hanson and H. L. Carnahan, *Breeding Perennial Forage Grasses* (Technical Bulletin No. 1145), U. S. Department of Agriculture, Washington, 1956, p. 3.

The general procedure followed in a grass breeding program is to select, by field inspection of the range, desirable plants, and move them to the breeding nursery where they can be studied side by side and compared. Superior plants can then be isolated, further selected, and the seed character improved.[43] Drought-resistant grasses can be combined with those of high nutritional value or other outstanding characteristics, and if the genotypes within a species possess high combining ability, hybrid vigor can be obtained.[44] All of the established techniques of breeding are currently being employed, such as inbreeding, vegetative propagation, and the use of radiation and chemical mutagens to increase the mutation rate. However, the breeding of grasses proves in many cases to be very difficult, and it is always a slow and laborious process.[45]

The practical consequences of this work are illustrated by the development of an intermediate wheatgrass, the product of crossing two imported types of wheatgrass. Improvement of some of the native grasses like buffalograss (*Buchloe dactyloides*) and the various gramas, *Bouteloua,* are further illustrations. The severe drought of 1933-34 in the central and southern Great Plains killed a large percentage of these native grasses, previously thought resistant to moisture deficiencies, once established.[46] Several strains and hybrids of buffalograss and of sideoats grama (*Bouteloua curtipendula*) have been obtained which appear to possess considerable promise of being more drought resistant than their predecessors.

Most species of grasses vary greatly in plant type and growth characteristics. Until recently, the bulk of the breeding work directed toward the development of drought-resistant plants has been concentrated upon the comparison of species and the isolation of the most promising for propagation. It seems likely that future emphasis increasingly will center upon analysis within species. Many workers in this field feel that the chances of obtaining important new drought-resistant plants through introduction from other parts of the world are progressively becoming more remote. However, a fertile area for experiment appears to lie in efforts with the presently available species.

[43] Staten, *op. cit.,* p. 274.

[44] Whalley, *op. cit.,* p. 232.

[45] Breeding procedures for forage grasses have recently been comprehensively reviewed by Hanson, *op. cit.,* pp. 40 ff.

[46] D. A. Savage, *Drought Survival of Native Grass Species in the Central and Southern Great Plains, 1935* (Technical Bulletin No. 549), U. S. Department of Agriculture, Washington, 1937.

As each advance toward the ideal of an economically useful drought-resistant flora is made, better use of the water that is available only as soil moisture becomes possible. Plant breeding efforts that result in a superior drought-resistant variety therefore are just as surely a contribution toward efficient use of available water within the West as a concrete dam or an irrigation canal. The achievement of a range flora with decreased water requirements at critical periods can be as important a stabilizing factor in agricultural production as water storage.[47]

TWO WASTE-DISPOSAL PROBLEMS
AND TECHNIQUES FOR THEIR SOLUTION

Wastes discharged in volume into any stream often have the unfortunate characteristic of precluding many other water uses. In other words, waste-carrying may have an unusually high demand for water in relation to the service afforded by the economic product obtained in the process. Consequently, a number of the techniques designed to reduce demand for water and water-derived services have centered on the problem of industrial and municipal waste disposal. The history of municipal sewage treatment techniques might well have been included here as a widely applicable example. However, industrial wastes have presented some of the more complex and difficult problems of disposal, and undoubtedly will continue to be important sources of water demand for waste-carrying purposes. Therefore, two techniques from very different types of industry have been selected for description: spray waste disposal in the food processing industry, and the development of techniques for disposal of wood pulping wastes.

3. *Spray Disposal of Food Processing Wastes*

Few waste disposal problems have proved more difficult in the United States than that of the food processing industry. In the course of canning, freezing, and dehydrating fresh fruits and vegetables, large

---

[47] Crop selection and plant breeding are not the only means of reducing crop requirements for soil moisture. Others worth mention are adjustment of the rate of seeding, and thinning of seeded crops or range plants.

quantities of water are used for washing, blanching, and grading the produce, for moving it hydraulically within the processing plant, and for keeping equipment sanitary. For example, Seabrook Farms, a large integrated farming-freezing operation in southern New Jersey, uses from 10 to 12 million gallons of water a day at its seasonal peak.[48] This water becomes heavily contaminated with suspended and dissolved organic matter. It therefore cannot be discharged into existing streams and waterways without causing serious pollution.

The volume and characteristics of the waste waters of this industry vary greatly with the particular product processed. The indicated biochemical oxygen demand of the liquid wastes from asparagus canning, for example, is not more than 100 ppm, but the comparable values for squash waste run as high as 4,000 to 11,000 ppm, and with such concentrated wastes as pea vine ensilage juice, a B.O.D. of 75,000 ppm is encountered.[49] Since the strength of domestic sewage effluent in terms of B.O.D. is usually near 200 ppm, it can readily be recognized that disposal of the wastes of the food processing industry may give rise to acute local problems.

Because of the serious polluting hazard of the liquid wastes characteristic of this industry, treatment is required in most states and attempts are made to meet the prevailing requirements in numerous ways. However, there are several reasons why the industry has not practiced treatment more extensively. First, the operation of most food processing plants is highly seasonal in character. Large capital investments in waste treatment facilities consequently often have appeared prohibitive. The fluctuating character of the waste is a second consideration. Not only do the objectionable features of the wastes change as different products are processed, but there are also substantial daily fluctuations in their strength and volume. In addition, the processing of certain products is accompanied by special problems. Examples are the high color level of the waste water which has been used in processing red beets; the waste odor from sauerkraut processing; the heavy load of suspended solids associated with all root crops; the highly acidic character of citrus fruit wastes; and the alkaline nature of wastes arising from the treatment of beets, carrots, peaches, potatoes, and other

[48] C. W. Thornthwaite, "Agricultural Climatology at Seabrook Farms," *Weatherwise,* 4 (April 1951), 29.

[49] R. A. Canham, "Current Trends in Handling Canning Wastes," *Proceedings of the Fifth Annual Water Symposium,* Engineering Experiment Station Bulletin No. 55 (Baton Rouge: Louisiana State University, 1956), p. 68.

products when lye is employed to facilitate peeling. When extremes in the hydrogen ion concentration of the water are encountered, neutralization by the addition of chemicals may be required before the waste will respond to treatment.

Waste disposal methods for the food processing industry must be comparatively inexpensive. They must be capable of accommodating fluctuating volumes of discharge which change in organic matter load and other characteristics; and they must not cause other objectionable effects in the course of eliminating the pollution hazard. A disposal practice of possible extensive applicability which meets all these requirements has been developed only in the last decade. This is spray disposal. But a variety of other methods have been, and still are, used by the industry although each of them has definite limitations. Among these methods are mechanical screening, chemical precipitation, and lagoon impoundment.

Mechanical screening has always been considered an indispensable preliminary step in achieving safe disposal of food processing wastes, regardless of subsequent additional treatment. Effective screening generally removes the larger solids which will float or settle out, but only the largest of the suspended solids are eliminated. Use of rotary or vibrating screening units has become standard practice without, however, effecting a significant reduction in the biochemical oxygen demand even when the finest practicable mesh is employed.

Chemical precipitation, which generally follows screening, has been an extensively employed method of treatment, although today it is infrequently used. Even in the most efficient operation, only about half of the B.O.D. of such wastes as those from pea, corn, and tomato processing can be removed by this method, and only about a quarter or less from fruit wastes.[50] Other treatment practices, like biological filtration and aeration, have also been employed for the same purposes by some food processors, but usually these methods are impractical because of high investment costs and operational difficulties.

Where the effluent is not directly released to streams or other available waterways, the most extensively employed disposal practice has been discharge into impounding lagoons. Lagooning involves the reten-

---

[50] R. A. Canham, "Some Problems Encountered in Spray Irrigation of Canning Plant Waste," *Proceedings of the Tenth Industrial Waste Conference,* Extension Series No. 89 (Lafayette, Ind.: Purdue University Engineering Extension Department, 1955), pp. 120-34.

tion of liquid waste, usually after screening, in one or more open ponds or reservoirs. Usually sufficient organic decomposition of the residues occurs during the normal period of plant idleness. After decomposition has taken place the waste water can be discharged into a stream without danger of pollution. The attractiveness of lagooning wastes is diminished by the offensive odors produced in the course of bacterial stabilization, even when odor-reducing compounds are added.

Still other methods of waste disposal have been employed by the food processing industry to avoid discharge into streams. They include water spreading on broad absorption beds, over gravel deposits, or through ridge and furrow irrigation systems. While these practices have proved satisfactory in some cases, there are few localities where the soil continuously infiltrates water rapidly enough to avoid odors and other unsatisfactory conditions.[51]

The spray irrigation method of waste disposal developed during the last decade appears very promising as an inexpensive process which will neither pollute streams nor create offensive odors. First practiced in 1947 in eastern Pennsylvania, it was rapidly adopted by many food processors. By 1951 more than 24 plants in the canning, dairy, and frozen food field were using some form of spray irrigation for the disposal of processing wastes. In 1955 a survey of 365 food canning plants in only six midwest states showed that 73 were using some type of spray irrigation waste disposal system.[52] There are now hundreds of users of this system.[53]

The spray disposal technique can be traced to the interest of the integrated farming-packing operator in employing used processing water for field irrigation. It was only a short step from humid land irrigation (stimulated recently by the appearance of portable lightweight piping) to sprinkling liquid wastes on the soil surface, thereby using them advantageously to promote plant growth. Usually, however, irrigation water requirements are not coincident with liquid waste availability. Waste water therefore could be used for field irrigation only during a part of the year. When the effluent could not be used for irrigating growing crops, the spraying equipment was operated experimentally over idle land. It was then found that the soil possessed a tremendous

[51] Canham, *op. cit.*, p. 77.
[52] R. A. Canham, unpublished data.
[53] N. H. Sanborn, "Disposal of Food Processing Wastes by Spray Irrigation," *Sewage and Industrial Wastes,* 25 (1953), 1036.

capacity to absorb water under certain conditions without becoming fully saturated.[54]

In the course of experimentation with this method, the results obtained by Thornthwaite at Seabrook Farms are especially impressive. A phenomenally high permeability of woodland soil was demonstrated when, after a ten-day test during which more than 150 inches of waste water were discharged by sprinkler into a scrub oak woodland area, it was seen that the soil was still absorbing water as fast as it fell.[55] Since 1950, all of the wastes at Seabrook Farms have been disposed of by spraying over a single wooded tract less than 60 acres in extent. The area receives over 200 inches of water a month during the processing season. There now is evidence that trees may be damaged or killed when water is applied at this very high rate, but even lesser rates of absorption still would be remarkably useful.[56] The optimum flora to be maintained on spray disposal plots of this kind still needs determination.

In most localities where wastes are disposed of by spraying, the waste water irrigates cover crops. The extent of the soil's vegetative covering is probably the most important factor determining the operating efficiency of this technique. If the water strikes the bare ground, the soil aggregates at the surface are fractured and broken into finer particles which then seal the surface. The upper layer thereupon becomes susceptible to erosion.[57] A dense cover crop prevents soil compaction and erosion by providing needed storage space for water, by increasing the infiltration capacity of the soil, by providing a soil binder through the root system, and by intensifying the rate of the process of evapotranspiration. An effective vegetative cover is also essential for the purification of the waste water, which occurs in the layer of humus or decaying organic material that develops at the soil surface. It is this layer which removes the suspended and dissolved solids from the water

[54] The essential condition for rapid water absorption appears to be prior existence of a grass, shrub growth, or tree cover for a sufficiently long period to allow complete development of a root mass and the associated microbiota, and elimination of the effects of compaction by cultivation. Porous substrata are required for adequate dissipation of water absorbed in the overlying soil.

[55] Thornthwaite, *op. cit.,* p. 30; and John R. Mather, "The Disposal of Industrial Effluent by Woods Irrigation," *Proceedings of the Eighth Industrial Waste Conference,* Engineering Bulletin No. 83 (Lafayette, Ind.: Purdue University Engineering Extension Department, 1953), pp. 439-54.

[56] R. A. Canham, letter to authors, 1957.

[57] R. A. Canham, "Some Problems Encountered in Spray Irrigation of Canning Plant Waste," previously cited, p. 82.

as it flows over, and just below, the surface of the ground. The organic matter in the waste water is absorbed onto the leaves and stems of the vegetative growth and onto the leaf litter covering the soil surface, and is then decomposed by bacterial activity. If the system is properly designed and operated, no serious clogging of the soil occurs. In fact, the waste material, if nontoxic to the vegetative cover, is rapidly broken down, adding to the existing layer of humus. The growth of the vegetation is thus accelerated and the efficiency of the disposal system increased, making it possible to increase the rate of waste water application.

Although the principal purpose of spray disposal of food processing wastes is *quality* improvement in surface streams, there is an indirect quantitative effect in reducing the amount of water required for food processing and waste carrying. If wastes are discharged into surface streams, large water volumes are required so that the concentration of organic matter is acceptably low. Conversely, spray disposal dictates use of the least practicable amount of water in processing and waste carrying. In an entire basin, return flow of ground water to the streams results in increased availability of high quality water for other uses, rather than decreased availability because of pollution.

The use of spray irrigation for the disposal of food processing wastes in many ways is the opposite of the application of supplemental water for agricultural production. In the case of spray waste disposal irrigation, effort is directed toward spreading the maximum amount of water upon a given land area without causing damage to the soil or the crop, while in agricultural irrigation the aim is to supply as little water as is necessary to insure the optimum conditions of soil moisture for plant growth.[58]

No one manner of employing this technique is applicable to all operators. Each food processing plant faces its own problems in terms of the character and volume of its wastes, the size of the available disposal area, the initial permeability of the soil, and other factors. While generalizations therefore are difficult, where disposal within wooded areas is not practical the most desirable cover crop may be a dense grass or combination of grasses. Alfalfa also has been employed for this purpose, where the water table is well below the root zone. The choice of a cover crop also depends in many cases upon the possible return to be derived from the land as, for example, through the production of

[58] R. A. Canham, "Current Trends in Handling Canning Wastes," previously cited, p. 78.

a hay crop. To some operators, the favorable effect that spray waste disposal has in raising the ground-water level may be important and accordingly influence its practice.

In addition to canneries, dairies, and frozen food plants, spray waste disposal has already been used by a number of meat packing and poultry dressing plants.[59] It has also found application in meeting the disposal problem of several tanneries.[60] In the future it is probable that it will be employed in still other industrial operations where a high load of organic wastes is encountered. Where appropriate disposal areas are not available in the immediate vicinity of the plant, it might still be economic to transport the wastes a considerable distance to locations where they could be applied. Although the method cannot be prescribed as a cure-all to all waste disposal problems of this character, where local conditions are favorable it appears to be a possible solution. Further investigations and experiments are necessary to ascertain more clearly the limitations of this method, such as the maximum amount of suspended or toxic matter that can be safely handled, the danger of soil clogging, and the optimum vegetal cover.

4. Developments in the Disposal of Wood Pulping Wastes

Perhaps one-fifth of the total pollution load in the effluent discharged by industry into American streams originates in groundwood and chemical pulping mills.

The pollution-causing discharge involved in the production of groundwood pulp is today a comparatively minor problem, since the wood is only mechanically treated and most of its constituents can be used in the final product. Wastes consist of residual quantities of groundwood particles, either so small as to escape retention in screening operations, or so coarse that they are undesirable and are sorted out for rejection. Improved filters have reduced losses of the first type, while reprocessing

[59] J. W. Bell, "Spray Irrigation for Poultry and Canning Wastes," *Public Works*, 86 (September 1955), 112; L. C. Lane, "Disposal of Liquid and Solid Wastes by Means of Spray Irrigation in the Canning and Dairy Industries," and F. J. McKee, "Spray Irrigation of Dairy Wastes," *Proceedings of the Tenth Industrial Waste Conference*, Engineering Bulletin No. 89 (Lafayette, Ind.: Purdue University Engineering Extension Department, 1955), pp. 508-13 and pp. 514-18.

[60] J. F. Eick, "Tannery Waste Disposal by Spray Irrigation," *Industrial Wastes*, 1 (1956), 276-77.

techniques have been developed which facilitate further refining of the larger particles.[61] In 1930 groundwood pulping accounted for more than a third of the total pulp output in the United States.[62] Today, however, only one-eighth of the wood pulp produced is made in this way. Several chemical pulping processes account for the remaining seven-eighths. It is for the chemical processes that the most significant developments in the disposal of wastes have occurred.

The chemical processes depend on the use of various chemicals to digest wood chips under heat and pressure. This treatment dissolves and removes the lignin from the cellulose fibers. The two most widely employed chemical methods of pulping are the sulfite and the sulfate processes. The sulfite process, which is more than ninety years old, in almost all cases is based on a calcium bisulfite cooking liquor prepared by treating a lime slurry with sulfur dioxide gas. In the newer sulfate or kraft process, the pulping agent is caustic soda and sodium sulfide. The sulfate process was first limited to the pulping of long-fibered coniferous woods like southeastern pine species, but it subsequently has been adapted to the use of other species and to the production of an increasing number of pulp grades.

The soda pulping process which has been used for many decades also bears mention. In operation it is very similar to kraft pulping. The main difference is that caustic soda is the principal constituent of the solution employed in digesting the wood. The effluents produced by these two processes are of similar strength and composition, except that the sulfate (kraft) pulping wastes possess a substantial sulfur content.

The extent of the waste disposal problem of the chemical pulping industry is indicated by the large volume of the material that must be eliminated. There are first the chemical residues of the various washes, bleaching agents, and spent digesting liquors. In addition, large quantities of lignin, wood sugar, and other materials from the wood enter the cooking liquors in the digestive process. Since almost one-third of untreated wood consists of lignin, the potential volume of these wastes can readily be appreciated. It was estimated a few years ago that the sulfite pulping industry in North America annually discharged 7 million

[61] A. D. Hamilton and Douglas Jones, "Pulp and Paper Mills Progress in the Use of Water Resources and Pollution Control," *Pulp and Paper Magazine of Canada,* 56 (September 1955), 100.

[62] U. S. Department of Commerce, Bureau of Foreign and Domestic Commerce, *Supplement to the Survey of Current Business 1940,* Washington, 1940, p. 142.

tons of lignin into surface streams as wastes.[63] Even more damaging, however, is the effect of that part of the organic materials consisting of wood sugars, of which half a million tons are believed to be discharged by sulfite mills in the United States each year.[64] Such sugars, in heavy concentrations, can cause a serious pollution problem because they undergo a microbiological oxidation by natural organisms. These organisms in turn consume the oxygen available in the water.

There are a number of other materials in the mill effluent which deplete the oxygen supply of the waters into which they are discharged, as well as various residues of kraft pulping such as resin acids and mercaptans, which even in very low concentrations are toxic to fish. Chlorine, which is discharged from the bleaching operation, is lethal in very weak solutions for many types of aquatic life.[65] The wood pulping industry not only discharges a large volume of effluent per ton of product (see table 20), but this effluent is of a type that can cause serious problems when it enters natural streams untreated.

The character of these wastes and the high ratio of wastes to final product long have made the disposal problem of major importance in the industry. Several decades ago, mills usually discharged their effluent directly into streams and rivers without treatment. Since then, the industry's concern with pollution abatement has become widespread, not only to comply with enacted legislation and to maintain favorable relations with the public, but also to recover as salable by-products some of the materials going to waste. At the same time, the rapid expansion of the industry has required the development of alternative, unobjectionable methods of waste disposal, because many streams into which the mill effluent had been discharged were unable to carry a further pollution load.

One of the great advantages of the kraft and soda processes for wood pulping, as compared with sulfite pulping, is the practicality of recovering the chemicals employed in the cooking liquor, of using the separated lignin as a fuel, and, in some cases, by-product production.

[63] Egon Glesinger, *The Coming Age of Wood* (New York: Simon and Schuster, 1949), p. 177.

[64] O. D. Mussey, *Water Requirements of the Pulp and Paper Industry* (Water Supply Paper No. 1330-A), U. S. Geological Survey, Washington, 1955, p. 21.

[65] B. A. Westfall, *Stream Pollution Hazards of Wood Pulp Mill Effluents* (Fishery Leaflet No. 174) U. S. Department of the Interior, Fish and Wildlife Service, Washington, 1946 (mimeo.), p. 7. "Free chlorine is highly toxic to fish, 1 ppm or less being lethal for many warm water fish, and much lower concentrations are detrimental to various food organisms."

WASTE TREATMENT USED IN THE SULFATE PULPING PROCESS

In the sulfate (kraft) pulping process, the system of waste treatment inherent in the method of handling spent cooking liquors makes possible recovery of more than 95 per cent of the chemicals and the wood substances present in them.[66] The demands placed upon water for the disposal of these very objectionable wastes are accordingly reduced. However, the plant effluent may contain relatively high concentrations of waste materials derived from the other pulping operations and from the recovery process. Some highly odorous compounds are particularly troublesome. Various improvements in recovery practice, in engineering, and quality control measures have reduced the requirements of water per unit of product, thus further decreasing the quantities of these wastes. Accordingly their character has been made much less objectionable in recent years.

Even considering the many process improvements made, the typical biochemical oxygen demand of sulfate pulping effluent is undesirably high. The presence of suspended or floating solids and toxic materials as well as substances causing objectionable odors indicates the desirability, and often the necessity, of waste treatment. Three types of treatment practice are important: controlled discharge of the mill effluent, submerged dispersal, and reduction of its biochemical oxygen demand by mechanical or biological means. In the first method, the effluent is impounded in large storage lagoons or retention basins from which it can be discharged into receiving waters as conditions warrant. The mill effluent often varies considerably in strength in the course of a day's operation, and one of the advantages derived from the use of storage lagoons is that the fluctuating pollutional load of the wastes can be equalized. More important is the fact that the wastes can be retained during seasonal periods of low flows in the receiving waters. In some cases, sufficient storage capacity is provided to hold the entire waste for a portion of a year. Discharge can then be scheduled only when stream flows are at their peak so that serious pollution can be avoided by adequate dilution.

Other methods of treating sulfate mill effluent are directed toward B.O.D. reduction. In almost all cases an attempt is made to reduce the biochemical oxygen demand of the liquid wastes mechanically by re-

[66] H. W. Gehm, "Pulp, Paper and Paperboard," *Industrial Wastes, Their Disposal and Treatment,* ed. by W. Rudolfs (New York: Reinhold, 1953), p. 195.

Plate 6. Discharge of effluent from paper manufacture, Potomac River, Luke, Maryland.

moving suspended and floating solids by filtration. These filters frequently are augmented with settling basins and clarifiers. Before discharging mill wastes to natural waterways, their B.O.D. may be lowered further by various biological oxidation techniques, the most important being retention in oxidation ponds, natural purification, and accelerated aeration. Whereas the impoundage basins or lagoons used for controlled discharge are often of considerable depth, efficiently operating oxidation ponds are shallow. They have proved effective in many areas of the South where the effluent can be maintained in an environment conducive to biological stabilization on a year-round basis; B.O.D. reductions in excess of 25 per cent are obtained.[67]

Where the process of natural purification is relied upon in biological treatment, the waste effluent is passed down a stream or canal, in which a culture of appropriate biological organisms hastening oxidation has been developed. Under favorable conditions, significant B.O.D. reductions can be secured in this way.

The accelerated aeration method of effluent oxidation is carried out in an aeration tank by a biological sludge or floc. Recently a large sulfate mill has begun to treat its effluent, in excess of 16 million gallons a day, by the activated sludge process. Because of the very high B.O.D. reductions obtained, this constitutes one of the most significant recent developments in the handling of pulping wastes.[68]

WASTE TREATMENT USED IN THE SULFITE PULPING PROCESS

Today the major pollution problem of the pulping industry centers upon the disposal of the calcium-bisulfite waste liquors involved in sulfite pulping. For each ton of pulp produced, about a ton of soluble materials is discharged in the spent liquor and the subsequent washes. In contrast to the sulfate and soda processes, there is no simple way to recover the pulping chemicals or to evaporate the liquor to the point where this organic matter can be burned. As a result, the average biochemical oxygen demand caused by the wood sugars and the free

[67] E. W. Luce, "Pollution Abatement Measures in the Kraft Pulp and Paper Industry," *Proceedings of the Fifth Annual Water Symposium,* Engineering Experiment Station Bulletin No. 55 (Baton Rouge: Louisiana State University, 1956), p. 28.

[68] "Bio-oxidation of Process Wastes Begins Operation at Covington, Virginia," *Chemical Engineering,* 62 (August 1955), 112-14.

sulfur dioxide, fatty acids, acetone, methanol, and other cellulose residues in typical effluents from sulfite mills is from ten to fifteen times
higher than that from sulfate (kraft) mills. Nearly all of the total
B.O.D. of the sulfite pulping effluent originates in the spent cooking
liquor, from 10 to 12 per cent of which consist of dissolved solids.[69]

Because of the difficulties and expenses of evaporating sulfite pulping
liquor, considerable effort has been expended in developing methods
for recovering high-value by-products from the residues. Limited success has been achieved. In the Howard process [70] developed for the
Marathon Paper Mills at Rothschild, Wisconsin, the spent digester
wastes are treated with lime to produce a solid lignin compound which
may be used either as a boiler fuel or as a lignin raw material.[71] In the
Paulson process, a multiple-effect evaporator is used to obtain an effluent sirup which can be burned in liquid form.

Although a substantial share of the market for vanilla flavoring is
being taken by synthetic vanillin made from sulfite waste liquor, this
high-value by-product is of relatively minor importance in waste disposal. United States sulfite pulp mills produce enough waste-liquor
lignin in two days' operation to meet the nation's demand for vanilla
for a full year.[72] Other limited uses of by-product lignin include the
preparation of tanning agents, dye bases, road binders, and fertilizers
and soil conditioners. It also can be employed as an additive in resinous
plastics and in special-purpose concrete. Experimental work to increase
the usefulness of lignin has involved development of a process of hydrogenation which indicates considerable promise.

In addition to these efforts to secure a high-value use for the lignin
constituents of sulfite wastes, a number of others concern the use of
material in the mill effluent. In some countries alcohol is produced
from recovered wood sugars; in the United States the sulfite wastes of
one mill in the state of Washington are being used for this purpose.
From the point of view of stream pollution, however, the process effluent is not materially improved by alcohol by-product production. The
spent cooking liquors of two sulfite mills in Wisconsin are being used as
a culture medium in the production of *torula* yeast which is marketed

[69] Gehm, *op. cit.*, p. 203.

[70] This recovery technique is summarized in: Guy C. Howard, "Sulfite Liquor
Developments," *Chemical and Metallurgical Engineering*, 46 (1939), 618-19.

[71] C. Placek, B. J. Buhmann, and R. T. Galganski, "Chemicals from Wood,"
Industrial and Engineering Chemistry, 50 No. 4 (1958), 570-76.

[72] Glesinger, *op. cit.*, p. 178.

for human food and animal feeds. Substantial reductions in the B.O.D. of the effluent discharged are being obtained.[73]

Because use of the sulfite wastes as fuel has generally not been economical, and because development of large markets for by-products is slow, much effort has been directed toward the development of cheap, practical methods of treating the final mill effluent. Lagooning has long been practiced, and various soil filtration techniques have gained considerable importance in recent years. Numerous other measures, such as the use of trickling filters and foam phase aerators, have been intensively explored, but they appear to be prohibitive in cost despite the impressive reductions in the B.O.D. which they make possible. Artificial stream aeration has shown considerable promise during experiments and may have applicability under certain conditions. This is the addition of a dissolved oxygen to a stream by compressed air diffusion. Because of the high concentration of wastes in sulfite pulping effluent, however, the treatment problem is difficult and it is not likely that an effective technique will be developed which is not expensive.

Another method of handling sulfite wastes ensures their rapid dilution by the receiving waters. Instead of discharge at one single point, the effluent is distributed into the receiving stream through a long submerged perforated pipe or some other device that facilitates its dispersal and dilution. Ample stream flow is an accompanying requirement for use of this disposal system.

EFFECTS OF WASTE TREATMENT METHODS ON SULFATE AND SULFITE PULPING PROCESSES

Satisfactory waste disposal thus has not been as successfully achieved for sulfite pulping as for the sulfate process. The simplicity and comparative economy of the recovery practices employed in sulfate pulping have undoubtedly promoted the industry's increasing reliance on this process. Not only is an important saving effected in the use of chemicals, but the recovered lignosulfonate is an excellent fuel. About half a ton of coal or its equivalent is needed in the manufacture of a ton of chemical pulp;[74] by burning the lignin, a substantial part of this need is met.

A major factor in the growth of sulfate pulp manufacture also has

[73] G. C. Inskeep, *et al.*, "Food Yeast from Sulfite Liquor," *Industrial and Engineering Chemistry*, 43 (1951), 1702-11.

[74] Glesinger, *op. cit.*, p. 176.

been the improvement in both quality and usefulness of the product. The introduction of bleached kraft has been especially important in this regard, making possible its use in place of sulfite pulp for a number of purposes. These factors, along with the lesser waste-disposal problem, account for the dominance which this method of pulping has achieved. In 1920 the kraft process accounted for only 3 per cent of the wood pulp output, while today more than half is made in this way.[75] In the last three decades almost the entire growth of the industry has been in sulfate pulping. The output of sulfite pulp, although not decreasing, is today less than one-third as important as it was.[76] The rapid expansion of sulfate pulping is suggested by the fact that the capacity of such mills increased fivefold between 1936 and 1953.[77]

However, a significant development of the postwar years is altering the waste-disposal situation and will unquestionably affect this general trend. For more than twenty-five years interest has been directed toward the possibility of replacing calcium in sulfite cooking liquor with more soluble bases which could be recovered practicably. In the Pacific Northwest an ammonia base cooking liquor was substituted for the conventional calcium base first by the Crown-Zellerbach Corporation at Lebanon, Oregon, and by 1957 substantial use of it had developed. Weyerhaeuser Timber Company has pioneered with a magnesia-base process at Longview, Washington. At least two other Pacific Coast mills have adopted this more soluble alkali. These modifications in the pulping process are important for waste disposal because they result in considerable reduction in stream pollution as well as providing substantial heat economies. Othmer has noted that "Semichemical pulp rose from almost nothing 25 years ago to about 1,600,000 tons last year. Growth is expected to continue at an even faster pace. Some 40 mills are operated in the United States and Canada by some 30 companies; daily capacity is over 5000 tons."[78]

The disposal of mill wastes so as to prevent serious pollution in the receiving waters continues to constitute a major problem for the pulping industry, but much progress has been achieved in the last decade or so.

[75] U. S. Department of Commerce, Bureau of Foreign and Domestic Commerce, *op. cit.*, p. 142; and U. S. Department of Commerce, Office of Business Economics, *Business Statistics, 1955,* Washington, 1955, pp. 173-74.

[76] *Idem.*

[77] Stanford Research Institute, *America's Demand for Wood, 1929-1975* (Report to the Weyerhaeuser Timber Company, Tacoma, Wash., 1954).

[78] D. F. Othmer, "Chemical Recovery from Pulping Liquors," *Industrial and Engineering Chemistry,* 50, No. 3 (March 1958), 60A.

Intensive research is being conducted with many new lines of approach.[79]

An indication of the extent of the accomplishment throughout the nation can be gained by comparing the rate of expansion in the production of kraft pulp, which increased fourfold between 1937 and 1952, with the change in the total pollutional load attributable to this product, which, it is estimated, underwent a slight decrease during this period. Thus the pollutional load per unit of product has been reduced by at least 75 per cent.[80]

Despite this accomplishment, severe waste disposal problems still exist. This is particularly true where there are numerous plants on the same stream, as on the Fox River in Wisconsin.[81] Costs are an added expense in manufacture. They are more of a financial burden in some areas than others. At the same time, the economic feasibility of recovery practices and of the use of by-products has materially increased, thereby partly supporting the development of these technical improvements.

Since a number of promising disposal methods recently have been developed, it is likely that further progress may be achieved merely by wider application of successful techniques. It seems certain that there will be no single waste-disposal practice meeting the needs of the entire industry, but the problem rather will continue to require a variety of solutions. As success is achieved, the water requirements for waste disposal correspondingly will be reduced.

[79] "The Utilization of Spent Sulfite Liquor: A Bibliography of the Literature Published during 1954," *TAPPI,* 38 (1955), 165A-80A.

[80] Luce, *op. cit.,* p. 30.

[81] "Pulp, Paper, and Pollution," *Industrial and Engineering Chemistry,* 50, No. 4 (April 1958), 33A.

CHAPTER 8

Extending the Services Afforded by
a Given Unit of Supply —

FOUR CASES

Two Scheduling Techniques: (1) Multiple-purpose Reservoir Operation. (2) Underground Storage of Water—water spreading; recharge by wells and trenches. *Other Water-conserving and -meliorating Techniques:* (3) Water-conserving and Drought-adapted Land Use—effect of water-conserving practices on runoff, summary. (4) Water-softening Practices and Equipment.

A second group of techniques is devoted to extending the services afforded by a given unit of supply. The approach in this instance is one of altering the supply, rather than demand, and the objective is to adjust the nature of the supply to the limitations of the consumer, whether these are of a seasonal, quality, quantity, or other nature. The four case descriptions selected to illustrate the range of these techniques come under the headings of "scheduling techniques" (multiple-purpose reservoir operation and underground water storage) and "other water-conserving and -meliorating techniques" (water-conserving land use and water-softening practices).

TWO SCHEDULING TECHNIQUES

The essence of the scheduling techniques is physical arrangement. The objective is physical manipulation of either supply units, demand units, or both, in such a way that available water can be made to serve as many units of demand as possible.[1] Probably the most ancient scheduling technique is that which stores water in a surface reservoir during a rainy season, for use on crops during a drier growing season. The central idea embodied in this attempt to synchronize availability of supply and incidence of demand has been much elaborated in modern times. Two illustrations of relatively recent techniques are given below.

1. Multiple-purpose Reservoir Operation

Within the United States multiple-purpose operation of water facilities is about eighty years old.[2] While the general concept is applicable to many phases of water management, multiple-purpose use has come to be most closely associated with multiple-purpose reservoir operation. Particularly since the 1920's, the construction and management of multiple-purpose reservoirs for conservation storage, flow regulation, diversion, and power production have become an increasingly common part of water development. Within recent years few major storage facilities have been built which were not planned for multiple-purpose operation.

Multiple-purpose reservoir operation is derived from a relatively simple concept. Essentially it is the synchronization of water detention and release with two or more different types of demand which may be incident upon a single source or single flow of water. In terms used earlier (chapter 4), the objective is one of arranging supply so as to reduce competition among uses and increase complementarity among them. The principal water development purposes around which multiple-purpose operation has been planned in this country are navigation, irrigation, flood prevention, and electric power production. Hitherto secondary purposes which are likely to receive more attention in the future are domestic and industrial water supply,[3] waste carrying, and recreation.

---

[1] This is equivalent to saying maximum economic return.

[2] See chap. 18, p. 508.

[3] Domestic and industrial use have been important purposes in the past, but supplied principally from single-purpose storage of surface supply, from unregulated stream flow, or from ground water.

While the idea is simple, the technique is complex. The high degree of intra-seasonal and seasonal flow fluctuation which characterizes most streams in the United States (chapter 4) presents a complex pattern of supply. Some demands also have complex patterns of incidence, both diurnally and seasonally. Their nature also has been noted (chapters 3 and 4). The seasonal nature of irrigation demands; diurnal, seasonal, and locality fluctuation in electric power demand; the seasonality of northern navigation; the seasonal and chance arrival of flood prevention needs; and the simultaneous peak incidence of several demands— these are among the problems which must be analyzed and met from the side of demand. The greater the number of purposes, the more complex becomes the problem of scheduling the detention and release of water from the reservoir. Whatever the number of purposes served, efficient multiple purpose operation is essentially minimizing competition and maximizing complementarity among the incident demands for water and water-derived services.

Each use of a reservoir competes with other uses to some extent. For example, the water needed to lock a barge tow past a dam cannot be used for the generation of electricity. Water released for irrigation, power production, channel maintenance, or other low-flow regulation may eliminate on-site uses of water in a reservoir. Probably the most important conflict attends reservoir space requirements for flood prevention. In 1951 nearly a million acre-feet of water were spilled on the TVA reservoir system to obtain flood-control storage space. If the spilled water could have been retained for use in generating power, an estimated additional 560 million kilowatt-hours of electricity might have been generated.[4]

Compromise among the competing demands for reservoir space is achieved in two ways: reservation of definite space within a reservoir for a specific use; and continuous alteration of the use of all active storage space in the reservoir according to the demands and hydrologic conditions of the day, week, month, or season. The first has been called the "layer" method of operation, and is particularly applicable where large fluctuations in flow may occur in any season. Its employment is illustrated in the operation of the Norfolk Dam (Arkansas) of the United States Army Corps of Engineers, and partially in the Watauga,

---

[4] Tennessee Valley Authority, *TVA—Two Decades of Progress*, Washington, 1953, pp. 33-34. The Division of Water Control Planning, TVA, reports that no water was spilled for this purpose in the years 1952-56, and only a small amount in 1957.

South Holston, and Norris projects of the TVA.[5] Operational problems are relatively simple, the overriding rule being one of restoring detention space within the reservoir as soon as practicable after use of the space for flood prevention.

The second method, continuous management of the storage space in adaptation to current requirements and current hydrologic events, more properly is the scheduling technique, aiming to balance two or more demands with the vagaries of water supply. It is particularly applicable where the daily or seasonal incidence of demand for one or more purposes, and the need for storage, shows pronounced seasonal characteristics. Seasonal demands for irrigation water in the western states, the seasonal incidence of high stream flows (and consequent flood hazard) throughout the country, and the seasonal demands for water use in outdoor recreation, are illustrative of the conditions favoring employment of continuous management of reservoir storage space. It is in this technique that true complementarity is most likely to be achieved. For instance, conservational storage of water for withdrawal use can in part be serving the requirements of flood control in a given basin. As water demands become greater, and the economic feasibility of additional conservation storage increases, near complete provision for such storage may result in a situation in which most flood control is incidental to conservational regulation (but no less efficiently served). True complementarity thus will have been achieved.

While the operational requirements of almost every individual reservoir differ, the technique may be understood from a description of the operation of Norris Reservoir, a part of the TVA water regulation system. Norris does not have the most intricate or varied scheduling operation because irrigation is not a major water use along the Tennessee and because industrial water supply, municipal water supply, and waste-carrying demands have not yet become intensive enough to be scheduling problems on the Tennessee system. Some other facilities (e.g. on the Delaware and Sacramento rivers) are operated on schedules to serve as many as six different purposes. However, a description of the Norris operation presents the basic outline of the technique more clearly than would description of a more complex schedule.

Norris Dam, which created the Norris Reservoir, was the first major

[5] Tennessee Valley Authority, *Engineering Data—TVA Water Control Projects and Other Major Hydro Developments in the Tennessee and Cumberland Valleys* (Technical Monograph No. 55, Vol. 1), Knoxville, 1954, p. 3-2. Also Division of Water Control Planning, memorandum to the authors, January 1958.

regulation work of the TVA system. It is located on the Clinch River about 80 miles above its confluence with the Tennessee at Kingston, Tennessee, and controls a drainage basin of 2,912 square miles. The total volume of the reservoir is 2,567,000 acre-feet, and the useful controlled storage capacity (active storage) is 2,281,000 acre-feet. The backwater extends 72 river miles from the dam on the Clinch River, and 56 miles on the tributary Powell River. Two hydraulic turbines and two generators are installed at the dam, the installed generating capacity being 100,800 kilowatts.[6]

Operation of Norris Reservoir is carried out within the statutory limitations prescribed for the TVA system in its basic legislation, with due recognition of the climatic characteristic of the part of the drainage basin controlled by the reservoir, and with regard for the place of the power-generating facilities at the dam as part of the TVA power system.

The tributary basin regulated by Norris, like much of the eastern part of the Tennessee Valley, experiences moderately heavy precipitation, with a peak during the winter months and again in July. The total precipitation on the drainage basin above the dam averaged 46.4 inches in the period 1937-56. The rainiest winter month, March, has more than about twice as much rainfall (4.8 inches) as the driest, October (2.2 inches). While the concentration of rainfall during the wettest half of the year is distinct, it is not pronounced. Heaviest precipitation actually occurs during July, but July also is a month of high evaporation. About 55 per cent of the year's total falls during the wet period. However, the incidence of relatively heavy rains at times when the deciduous vegetation of the region is dormant, when transpiration is negligible and when the soil surface is saturated or frozen, causes pronounced runoff peaks during the period December-March. Surface runoff from a prolonged storm during the wet season may be as much as 75 per cent of the total amount of water falling in the storm.[7] The estimated maximum known flood at Norris was 128,000 cubic feet per second. By contrast, the minimum daily flow at the site was 200 cfs. The average calculated natural flow is 4,100 cfs.[8] These natural conditions obviously make the regulation of the Clinch and the Powell rivers of considerable economic value.

[6] *Ibid.*, pp. 23-3 to 23-4.

[7] This remark applies to the Tennessee Basin as a whole. Source: C. E. Blee, "Multiple-purpose Reservoir Operation of the Tennessee River System," *Civil Engineering*, July 1945, p. 263.

[8] TVA Technical Monograph No. 55, previously cited, p. 23-3.

Plate 7. Norris Dam of the Tennessee Valley Authority, Clinch River, Tennessee.

The statutory prescription for the management of Norris, along with other TVA reservoirs, was framed to take account of the flooding hazard created by the climatic conditions within the Clinch and Powell River watersheds. It also recognized the benefit of flow regulation to summer navigation. Thus Section 9a of the TVA Act states:

> The Board is hereby directed in the operation of any dam or reservoir in its possession and control to regulate the stream flow primarily for the purposes of promoting navigation and controlling floods. So far as may be consistent with such purposes, the Board is authorized to provide and operate facilities for the generation of electric energy at any such dam for the use of the Corporation and for the use of the United States or any agency thereof, and the Board is further authorized, whenever an opportunity is afforded, to provide and operate facilities for the generation of electric energy in order to avoid the waste of water power, to transmit and market such power.[9]

Within this physical environment and legislative framework the multiple-purpose management of Norris Dam has been designed and carried out. The manner of its operation is roughly as follows:

The first step was the establishment of a "guide-curve" or a "rule-curve" which serves as a basis for taking account of seasonal needs in the daily operation of the reservoir. The guide curve for Norris is shown on figure 18. One of the principal uses of this curve is to establish the levels above which the reservoir should not be filled during a given period except to regulate a flood. Like other curves of this type, it was developed from a study of valley-wide objectives, taking into account an analysis of the historic floods, the hypothetical floods, and other hydrologic data concerning the watersheds controlled by the reservoir.

The guide curve is only one of several tools for planning the actual releases of water from the reservoir. Another essential component of the equipment for such planning is a network of precipitation-gauging and stream-gauging stations within the watershed above the dam. There are thirty-one precipitation- and eleven stream-gauging stations above Norris Dam, of which ten precipitation and four stream gauges send daily reports to Knoxville. From this network the inflow of water into the reservoir is computed in the Knoxville offices of TVA, along with similar computation for other reservoirs in the TVA system.

Another useful tool in planning reservoir operation is the daily weather forecast. For the TVA system, a special forecast is prepared

[9] Act of May 18, 1933, Sec. 9a, 48 Stat. 58.

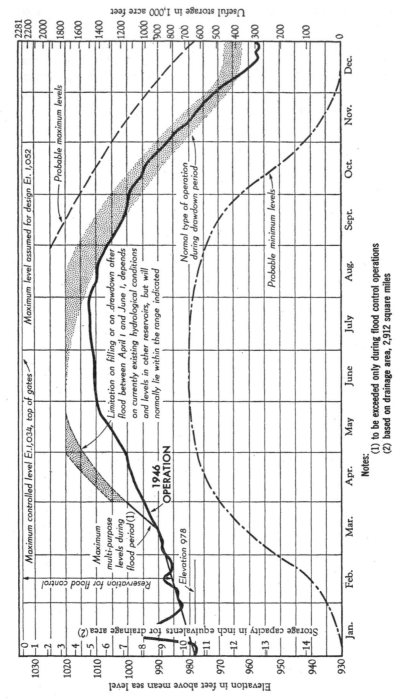

Figure 18. Guide Curve for the Operation of Norris Reservoir, TVA.

Notes:
(1) to be exceeded only during flood control operations
(2) based on drainage area, 2,912 square miles

Useful storage in 1,000 acre feet

Maximum level assumed for design El. 1,052

Maximum controlled level El.1,034, top of gates

Probable maximum levels

Limitation on filling or on drawdown after flood between April 1 and June 1, depends on currently existing hydrological conditions and levels in other reservoirs, but will normally lie within the range indicated

Normal type of operation during drawdown period

Probable minimum levels

1946 OPERATION

Maximum multi-purpose levels during flood period(1)

Reservation for flood control

Elevation 978

Storage capacity in inch equivalents for drainage area (2)

Elevation in feet above mean sea level

daily by the United States Weather Bureau, and twice daily during the flood season, covering geographical subdivisions of the basin regulated by the TVA facilities. In them forecasts are made of rainfall conditions two to five days in advance, thus alerting water control planners to unusual conditions, particularly anticipated floods. Long-range forecasts have not yet reached the stage of accuracy which permit the employment of synoptic meteorology for the anticipation of droughts or other seasonal conditions of equal interest in the planning of reservoir operation. However, when they do become accurate, long-range forecasts also will be added to the water control planner's kit.

The seasonal multiple-purpose operation of Norris Reservoir has been described by TVA as follows (refer to figure 18):

On January 1, the start of the flood season, the reservoir is at least as low as elevation 978, or 56 feet below the top of gates. It is allowed to fill slowly until March 15 as the probability of the occurrence of a succession of large floods decreases. From March 15 to April 1 it is allowed to fill at a more rapid rate. After April 1, most of the critical flood danger is past and, if water is available, the reservoir is filled to elevation 1020, which is 14 feet below top of gates, by May 1. During the summer season the top 14 feet of space in the reservoir is reserved for the control of runoff from summer floods. In the flood season this additional storage space is valuable for regulating large floods which require more space than is afforded below elevation 1020. During the dry season in the late summer and fall, the water which was stored in the spring is then released through the turbines to generate electric power both at Norris and at the mainstream dams and to sustain flows from the Tennessee River system into the Ohio and Mississippi Rivers. If, as the end of the year approaches, this release has not been sufficient to draw the reservoir to the January 1 elevation of 978, gate discharge is started to insure having the necessary flood control storage space available at that date.

.   .   .   .   .

The regulation of floods and the release of water in periods of low flow materially benefit navigation by stabilizing the (main stream) reservoir levels. Also, the maintaining of a relatively high discharge from the Tennessee into the Ohio during the low-water periods is of definite assistance to navigation on the lower Ohio and Mississippi Rivers.[10]

[10] TVA Technical Monograph No. 55, previously cited, p. 3-4. The difference between the requirements for flood storage space during the summer season, on one hand, and during the remainder of the year, on the other, bears additional explanation (first paragraph above). Winter and spring floods generally have

The above operation serves the "statutory purposes" prescribed in the TVA Act. In practice, the day-to-day operation may have many additional details. These probably are most complex during the flood season, when constant watch must be kept over anticipated runoff conditions, and actual runoff. Daily analyses of hydrologic conditions are the basis for instructions issued by the Chief Water Control Planning Engineer of TVA, and transmitted through the Superintendent of Power Operations of TVA to the Norris power house crew. At all times when the reservoir is near capacity determination as to whether a given volume of water is to be saved for release through the turbines, or "spilled," must be made in the light of the TVA power system load and the anticipated production of other generating plants in the system. Operation of the reservoir, like that of most other TVA impoundments, also is carried out with some understanding for the needs of recreation. When practicable, an attempt is made to maintain suitable water levels for boating, swimming, and fishing. Account is taken of the needs of fish propagation, by maintaining as stable water levels as possible during the spawning periods. Special discharges may be made for sanitation and water supply purposes during seasons of low water. TVA water releases also may take account of such needs as refloating stranded vessels, aiding farmers in fording or ferrying to cultivated islands, and other special needs.[11]

The gains from scheduling the use of storage space in a reservoir, and regulating water releases in a manner like that used at Norris, are obvious. Particularly where seasonal demands are met, and do not overlap seriously, the use of a single facility or set of facilities must correspondingly reduce investment as compared to single-purpose installations for water regulation. Norris is part of the TVA integrated system, and an appraisal of results from its multiple-purpose operation alone therefore is somewhat hypothetical. The benefits of multiple-purpose operation on the system of which Norris is part, however, have been studied carefully and adequately summarized in the past. If the three primary or statutory purposes alone are considered, the cost of provid-

large volumes as well as high peaks. Summer floods may have high peaks but less volume, and need less storage space for control than the winter and spring type. Conditions of vegetative growth on the watershed, differing infiltration capacities of the soil at different seasons, evapotranspiration differences, duration and type of precipitation contribute to the lesser volume of summer floods.

[11] N. W. Bowden, *General Design and Operation of Multiple Purpose Reservoirs* (Knoxville: Tennessee Valley Authority, 1946), p. 8.

ing similar services of flood control, navigation, and electric power production in the Tennessee Valley would have been almost half again as much as the actual cost of the sixteen-dam system which was studied for the sake of comparing allocation methods, and reported upon by the Federal Power Commission in 1949. The actual cost of this sixteen-dam system, including Norris and fifteen other dams, was $562,774,000. The minimum cost of the three alternative single-purpose systems affording the same benefits in the valley would have been $828,164,000,[12] or 45 per cent more. While no attempt should be made to apply the same ratio of savings to the Norris installation alone, these figures give some indication of the economic results attainable by application of the multiple-purpose management technique.

If the alternative justifiable expenditure measure is applied to Norris alone, the principal benefits to be reckoned would have been flood control and power production. The cost of the Norris Dam, power house, and reservoir was $30,508,000. To have provided the same flood control and power benefits in single-purpose installations might have cost approximately $51 million, or 66 per cent more than the actual installation, as operated in multiple-purpose fashion.[13]

Such measures of alternative justifiable expenditure on a single-purpose basis do not give the full story of the value of multiple-purpose operation. Good dam and reservoir sites are limited within any watershed—so limited that it usually would be impossible to construct a set of single-purpose systems which would provide equal benefits with a well-planned and efficiently operated multiple-purpose system. In the Federal Power Commission calculations referred to above, for instance, Norris Reservoir and Norris Dam site were assumed to be used in both the single-purpose flood control system and the single-purpose power system hypothesized. In reality, of course, another site or sites would have had to be used for one purpose or the other if two or more single-purpose facilities were the basis of water regulation on the Powell

---

[12] President's Water Resources Policy Commission, *Ten Rivers in America's Future,* Vol. 2 of Report (Washington: U. S. Government Printing Office, 1952), pp. 739-43.

[13] Basis for calculation: 1,635,000 acre-feet of flood control storage costing $19 million, and 120,000 kilowatts installed hydro capacity costing $31.6 million, the first as part of a single-use flood control system, and the second as part of a single-use electric power generating system. See Tennessee Valley Authority, *Report on Allocation of Costs as of June 30, 1953, pursuant to Section 14 of the Tennessee Valley Authority Act,* Knoxville, 1953, pp. 18, 21.

and Clinch rivers. The cost of such a set of control facilities undoubtedly would have been very high.

2. *Underground Storage of Water*[14]

Where natural surface-water storage sites are not available, or where evaporation rates are high in water-scarce areas, underground storage may be a very desirable answer to seasonal conservation storage of runoff for withdrawal uses. It also may be a desirable measure where permeable geological formations which are naturally sealed offer a convenient natural underground distribution system for water.[15] Co-ordination of surface flow with ground-water storage and use thus may reduce disappearance of water in evaporation and flood flows, and thereby extend the economic use of a given supply. It also may be useful in arresting salt-water incursions into fresh-water aquifers. The technique concerned is the artificial replenishment of aquifers.

Basically the objective here is the same as for surface storage: the capture and release of available water in a manner which best enables the matching of demand upon an aquifer, which for this discussion may be considered a kind of reservoir.

The hydrologic term for the addition of water to an aquifer is "recharge." Recharge can be natural or artificial. Natural recharge consists of those natural processes which increase the amount of water in storage, particularly where an originally stored supply has been depleted through natural or artificial means. "Artificial recharge" here does not include artificial aids to natural recharge, such as land treatment to increase the infiltration of rainfall. It does include those operations where water on the surface is physically moved or directed by works of man to locations where it would not have moved naturally.

Two methods of artificial recharge will be described briefly: water spreading, and recharge by wells and trenches. A third method is by sprinkling or spraying, which has been described in chapter 7 as a technique of disposing of cannery waste, but which basically is a water recharge method, its success depending on the infiltration of large volumes of water into the soil and down to aquifer formations.

[14] Materials supplied by Louis Koenig, under contract with Resources for the Future, were used in the compilation of this case description.

[15] Such as valley fill in many parts of the country, alluvial cones at the bases of western mountains, western bolsons, etc.

WATER SPREADING

Water spreading is the diversion of surface waters out of their normal channels so as to increase the area of wetted contact between water and land surface. Spreading may be accomplished by basins, by furrows, or by flooding. Flooding includes the use of small dams across wide flat-bottom stream beds over which the water passes as over a weir in a thin sheet both above and below the dam.

Water spreading has been known on this continent since prehistoric times, as in the flood-water irrigation of the Hopi Indians in the Southwest. In its modern form, spreading has been experimented with over several decades. However, only recently have detailed studies been made of the actual mechanism of water travel below ground in the localities where spreading is feasible. Water spreading was first developed as a major activity in California between 1910 and the 1930's.[16] Eighty-seven recharge projects were reported in California in 1956, almost three times as many as the other forty-seven states. Practically all of these were spreading projects.[17] Spreading was first applied on coarse alluvial fans where the infiltration rate was as much as 8-10 feet per day (or 8-10 acre-feet per acre of horizontal area). Costs at that time, including operating, maintenance, and capital costs, were variously reported as less than the equivalent of $1.50 (1954 dollars) per acre-foot replenished.[18] Later cost figures for the same general type of operation are $8.00-$9.00 per acre-foot (1954 dollars).[19]

The high percolation rates mentioned above were achieved in coarse gravelly materials in southern California close to the mountains. As it became desirable to use spreading more generally it had to be undertaken on fine sandy loams, like those in the Central Valley. On this type of material it was found that infiltration rates were only about 0.5 foot per day. For the past five or six years experiments have been under way to develop methods of increasing infiltration rates on these

[16] A. T. Mitchelson and Dean C. Muckel, *Spreading Water for Storage Underground* (Technical Bulletin No. 578), U. S. Department of Agriculture, Washington, December 1937.

[17] American Water Works Association Task Group 2440 R—Artificial Ground Water Recharge, "Artificial Ground Water Recharge," *Journal of American Water Works Association*, 48 (May 1956), 493-98.

[18] Mitchelson and Muckel, *op. cit.*

[19] F. B. Laverty, "Ground Water Recharge," *Journal of American Water Works Association*, 47 (1952), 677-81.

fine, sandy, agricultural soils. Improved construction methods and operating procedures have shown some promise. For example, use of lightweight machinery to avoid soil compaction has been helpful. But greatest success has been achieved with surface treatments. Incorporating cotton gin trash in the upper six inches of soil, and establishing stands of Bermuda grass on the basins or flooding areas, can increase the water infiltration rate from 1 foot to 6-8 feet per day. Use of chemical soil conditioners is reported to increase infiltration 5 and 6 feet per day. These improvements are immediate but their effect deteriorates after a short time; hence they are likely to be expensive.[20] Moreover, surface treatments cannot have significant effect unless the subsurface structure is sufficiently porous and permeable to permit the rapid infiltration attained at the surface.

Other difficulties for spreading arise where there is an impervious rock stratum overlying an aquifer. On such a surface, a lens of saturation will grow until it reaches the surface. When that happens the infiltration is controlled not only by the downward percolation but by the slow lateral movement of water. Accordingly, infiltration rates are slow. Where the impervious stratum is shallow-lying it may be penetrated by surface treatments such as subsurface ripping.[21]

A special form of spreading is used when increased infiltration is desired on a stream whose bottom has been made impervious by fine sediment. In that case, river water is diverted into basins especially constructed with permeable bottoms of sand or gravel, and a much greater infiltration rate is thus achieved. Such specially constructed pits with high heads of water (up to 10 feet) achieve infiltration rates up to 60 and 75 feet per day.[22] The same effects are achieved in places by using unlined irrigation canals.[23] In addition, efficient recharge may be possible through field over-irrigation.[24]

---

[20] Dean C. Muckel and Leonard Schiff, "Replenishing Ground Water by Spreading," *Water, The Yearbook of Agriculture, 1955,* U. S. Department of Agriculture, Washington, 1955, pp. 302-10; Dean C. Muckel, "Research in Water Spreading," *Proceedings, American Society of Civil Engineers,* Vol. 77, No. 11 (1951).

[21] Muckel and Schiff, *op. cit.*

[22] Max Suter, "High Rate of Recharge of Ground Water by Infiltration," *Journal, American Water Works Association,* 48 (April 1956), 355-60.

[23] E.g., San Joaquin Valley, California (Stephen C. Smith, personal communication, 1958).

[24] W. A. Hall, R. M. Hagen, and J. D. Axtell, "Recharging Ground Water by Irrigation," *Agricultural Engineering,* 38 (1957), 98-100.

A fourth variant of spreading is the practice of holding back stream flood waters in excess of the amount of influent seepage from the stream bed, then releasing them at a later time for seepage in the stream bed. This practice does not eliminate the need for surface storage but can greatly reduce the size of the storage required. Well-known examples are in the Santa Clara Valley, California, and at Queen Creek, Arizona.[25]

Queen Creek is an intermittent mountain stream and typical desert "wash," debouching on an alluvial fan which it has constructed on the adjacent valley floor. Close to the mountain the alluvium is coarse gravel, grading into sand and finally into silt in the middle of the valley floor. Queen Creek flows only during and a short time after the time when rains occur in its drainage basin. The alluvial fan of Queen Creek can infiltrate all of the water from a small storm, none of the water running beyond the sand area and onto the silt. In heavy storms, however, the capacity of the gravel and sand portion of the stream bed for infiltration is exceeded by the stream flow, and the excess water flows down to the silt areas where in times past it formed a playa and disappeared by evaporation. When the area was settled the silt alluvium was farmed and ground-water pumping for irrigation begun, with resultant decrease in ground-water levels. The wastage of excess flood water then became economically important, as was damage by the flood waters to the now-farmed silt lands. Studies showed that out of an average flow of 30,000 acre-feet a year, about 15,000 passed beyond the infiltration area onto the silt lands. However, these excessive flows occurred on only a relatively few days out of the year. Hydrological studies showed that the infiltration capacity of the stream bed for relatively silt-free water was about 1,000 acre-feet per day. A small reservoir to be constructed in the mountain canyon will store, for a few days, flood water in excess of this amount—in other words, it will detain water that would ordinarily run beyond the natural infiltration area. When a flood subsides the stored water then can be released so as to infiltrate completely.[26] This scheme of underground storage is expected to eliminate higher capacity surface storage with its attendant evaporation losses.

In similar physiographic situations underground storage of this type

[25] H. M. Babcock and E. M. Cushing, "Recharge to Ground Water from Floods in a Typical Desert Wash, Pinal County, Arizona," *Transactions, American Geophysical Union*, 1942, pp. 49-56.

[26] Babcock and Cushing, *op. cit.* The reservoir is the Whitlow Ranch Reservoir to be completed in 1960.

will undoubtedly find greater usefulness in the western valleys as water needs and population density increases, but many complicating factors require a separate evaluation for each case. It seems certain that during the next twenty years recharge by spreading will become a significant ground-water practice as well as an important part of surface-water management. Artificial recharge is now being practiced in twenty-four states, and at least thirty-nine of the forty-nine states have potential recharge areas not yet developed.[27] One of the largest scale projects of this type yet proposed is for the diversion of flood flows from the Snake River in southern Idaho into the lava aquifer which is drained by the Thousand Springs.[28]

Artificial recharge as now practiced uses surplus and flood waters of local streams. This practice may be extended by using imported water for recharge. It has already been shown in California, where salt-water encroachment on aquifers is a problem, that it is cheaper to use imported water for recharge (i.e. to store it underground) than it is to construct surface storage. Use of imported or local surface waters in this way may allow the use of local ground water that otherwise would deteriorate in quality through salt-water encroachment. Raising the water table by underground storage also lowers pumping costs. Averaged for the nation as a whole, these costs are about 4-5 cents per acre-foot per foot of lift. Power costs comprise 50 to 60 per cent of the total cost of pumping.[29]

Ground-water recharge is an ideal way to reclaim water from sewage, and studies are already under way, again in California, to develop methods for doing this in strict compliance with health and sanitation requirements. Sewage reclamation by spreading is quite likely to become the common method of effluent disposal in areas where there are usable aquifers.

[27] American Water Works Association Task Group 2440 R, *op. cit.* Also John J. Baffa, chairman, American Water Works Association Task Group 2440 R, personal communication to Louis Koenig, 1956.

[28] Roy Bessey, letter to the authors, 1958.

[29] Arizona Underground Water Commission, *The Underground Water Resources of Arizona,* Phoenix, 1952; Texas Agricultural Experiment Station, *Cost of Water for Irrigation on the High Plains* (Bulletin 745), College Station, Texas, 1952; B. Laverty and H. A. van der Goot, "Development of a Fresh Water Barrier in Southern California for the Prevention of Sea Water Intrusion," *Journal, American Water Works Association,* 47 (September 1955), pp. 886-908; I. D. Wood, *Pumping for Irrigation* (Soil Conservation Service publication SCS-TP-89), U. S. Department of Agriculture, Washington, 1950; C. Rohwer, "Wells and Pumps for Irrigated Lands," *Water, The Yearbook of Agriculture, 1955,* p. 293.

RECHARGE BY WELLS AND TRENCHES

Where there are subsurface impervious strata out of reach of surface soil treatment it becomes necessary to place the water below the impervious stratum. If it is not deep, trenches, pits, well bores, or shafts can be dug, through which the water may be introduced into the aquifer.[30] These excavations may be filled with gravel or other material to provide a permeable medium for flow.

These methods have not been used much in the past because they are expensive. Injection through well bores, for example, may cost $15.00-$25.00 per acre-foot on the average.[31] Except where land is valuable, this method therefore cannot compete with the cost of recharge by spreading. However, where impervious strata extend over large areas it will be necessary to take this higher expense, if recharge is to be employed.

Such a condition occurs in the Grand Prairie region of Arkansas, where an extensive impervious clay bed makes the overlying soil ideally suited to rice culture but at the same time prevents natural recharge of the pumped aquifer.[32] In this area a careful and comprehensive experiment on recharge by injection through well bores is being carried on. The key to successful practice of the technique is avoidance of sealing effects on the part of the aquifer entered by the injection well or trench. Great care must be taken to see that the water being injected does not damage the aquifer formation.[33] This can occur in three ways. First, suspended material in the water physically can seal the formation. To avoid this, settling and clarification of water before injection are necessary. Second, the dissolved mineral content of the injection water may be incompatible with that in the formation water and may cause mineral precipitation, also with sealing effects. Third, the injected water may be incompatible with the water normally con-

---

[30] Leonard Schiff, "The Status of Water Spreading for Ground Water Replenishment," *Transactions, American Geophysical Union,* 36 (December 1955), 1009-20; Muckel and Schiff, *op. cit.*

[31] Laverty and van der Goot, *op. cit.;* Laverty, *op. cit.*

[32] Eldon Dennis, U. S. Geological Survey, personal communication to Louis Koenig, October 18, 1956; R. T. Sniegocki, *First Open File Progress Report on Studies of Artificial Recharge in the Grand Prairie Region, Arkansas,* U. S. Geological Survey, Washington, April 15, 1955.

[33] *Ibid.*

tained in the aquifer so as to cause an undesirable colloidal effect, like swelling of the clays or other silicate materials in the aquifer. Swelling of these materials can seal the aquifer formation also. The necessary water treatment to avoid these dangers is expensive. The Arkansas experimentation will be carried on for several years, and observations on these problems will be carefully watched for their general application to other areas.

Another illustration of well injection was provided at El Paso, Texas. El Paso uses surface water from the Rio Grande, treated in a standard filter and softening plant; and also uses ground water from a nearby well field. The well field was threatened by salt-water intrusion because pumpage was 25 per cent in excess of recharge. During winter the El Paso demands on both well field and Rio Grande water were low, and requirements of Rio Grande water for irrigation were also low. This made the well field potentially rechargeable with Rio Grande water at this season. Successful experiments were run in which it was found that the well field would take recharge water at the same rate that it produced water by pumping. When the volume pumped out equalled the volume injected, about 70 per cent of the injected water had been recovered from the field. When the pumpage equalled approximately twice the injection volume, about 95 per cent of the injected water had been recovered. Salt-water encroachment was stopped in the vicinity of the experimental well and presumably would have been stopped throughout the aquifer if the practice had been expanded. Technically the experiment was an outstanding success.[34]

However, El Paso was not practicing recharge in 1957. An important reason was the geographical diffusion of benefits from recharge. The benefits of recharging the entire field for storage of winter surface waters would accrue not only to the city of El Paso but also to many surrounding industries independently withdrawing ground water from the same formation. The city was not successful in obtaining financial assistance from the industries benefited. Secondly, recharge involved a duplication of costs. If river water was used for distribution directly, only the treatment and purchase cost was incurred. When ground water was used, only the pumping cost was incurred. However, when ground water from recharge was used, this water was subject to treat-

[34] Raymond W. Sundstrom and J. W. Hood, "Results of Artificial Recharge of the Ground-Water Reservoir at El Paso, Texas," *Texas Board of Water Engineers, Bulletin 5206,* July 1952.

ment and purchase cost as well as pumping costs.[35] In view of the one-sided incidence of costs, and the much lower cost run-of-river flow in winter, the need for storage at El Paso had not yet reached a point where the use of well injection recharge was considered economically justifiable.

Some of the more successful experiments have taken place on the High Plains of Texas. One recharge well there is reported to have repaid the cost of drilling and casing within a month and a half of the time of well completion.[36] Data in a proposal for an extensive program of recharge well construction on the High Plains have indicated that costs might be no more than $1.00-$2.00 per acre-foot of water recovered.[37]

As water costs rise in line with increasing demands, the conservational contributions of well injection may become feasible in some local situations where recharge is a physically and economically practical means of capturing water which otherwise would be lost to beneficial use. However, recharge of aquifers by pit, trench, and well methods probably has limitations in the extent and purpose of application. While entirely feasible for industrial and municipal use, the volume of water required for the typical irrigation demand is so large that physical difficulties arise for the recharging operation.

OTHER WATER-CONSERVING AND -MELIORATING TECHNIQUES

Further and relatively numerous techniques or technical improvements have ends which are similar to the scheduling techniques. That is, they prevent natural water disappearance, promote an increase of desired products or services from water, or extend the range of uses possible with a given supply. Two illustrations are used: water-conserving land use as it assists in preventing natural disappearance

[35] Haskell Street, El Paso Water and Sewage Department, personal communication to Louis Koenig, October 17, 1956.

[36] "Another Recharge Well Does the Job," *The Cross Section* (Lubbock, Texas), July 1957, pp. 2-3.

[37] High Plains Underground Water Conservation District, *Proposal for Ground Water Recharge—High Plains of Texas, Lubbock, Tex.,* 1957.

of water; and water-softening practices. The first involves not only agronomy but also plant physiology, hydrology, soil science, agricultural engineering, microclimatology, and forestry. The last is a product of chemical engineering.

3. Water-conserving and Drought-adapted Land Use

The sections of the United States where the adaptation of drought-resistant plants has had widest application (chapter 7) also have shown decided interest in the kinds of land use that will conserve needed water and are adapted to drought.

The popular conception of natural conditions limiting agriculture in the western states has undergone much revision in the last seventy years. At one time the territory in the intermountain West and in the Southwest was viewed by many as "the Great American Desert." Others, eager to promote the settlement of the West, contended that virtually all the western lands were amenable to the same type of farming as that practiced in eastern United States.[38] As more information about these areas accumulated, still another view emerged: the amount of annual precipitation best indicated the suitability of different areas in the western states for dry farming. In the first decades of this century prevailing thought held that only those areas receiving more than 18 inches of precipitation a year were suitable to dry-land farming.[39]

These often conflicting popular views and opinions gradually have yielded to a growing body of results from scientific investigations. In many areas of the West dry-land farming has been successfully carried on where considerably less than 18 inches of annual precipitation falls.

---

[38] The initial contest between these two views has recently been summarized by Wallace Stegner: *Beyond the Hundredth Meridian: John Wesley Powell and the Second Opening of the West* (Boston: Houghton Mifflin, 1954).

[39] This characterization of the tendency of prevailing thought should be clearly distinguished from that of a number of pioneer figures who fully recognized the importance of scientific investigations in this field. In the 1890's for example, a Division of Agricultural Soils was created within the U. S. Weather Bureau to study the moisture and temperature conditions within the soils. See Milton Whitney, "Announcement of the Creation of the Division of Agricultural Soils" (Weather Bureau, Division of Agricultural Soils, Circular No. 1), U. S. Department of Agriculture, Washington, 1894, p. 1.

Conversely, the occurrence of a greater amount of rainfall provides no guarantee that the moisture requirements of cropping agriculture will be adequately met in a given area. As more information has been obtained, understanding in this field has deepened and a new conceptual framework has emerged. The factor upon which attention centers today is the amount of moisture available in the soil for plant use during the growing season. This depends only in part upon the amount and incidence of precipitation. Soil moisture constitutes that part of precipitation which remains within the root zone of the soil (chapter 2). Some soil moisture is absorbed in plant tissue through the process of photosynthesis, but most of it is transpired. The supply of moisture is also depleted by evaporation occurring within the upper layer of the soil.

Different types of vegetation possess their own distinctive water requirements and individual plants use water according to their root length, depth, and distribution. When the natural vegetative covering which existed in the Plains area and in the mountain West was first removed and the soil brought under cultivation, the pre-existing inter-relations between evaporation and transpiration were greatly altered. The effects of the disturbance and elimination of this growth upon soil erosion are well known. There is less understanding of the effects of the exploitation of these soil resources upon transpiration and evaporation.

It is doubtful that the total water yield of these areas has been significantly altered in historical time by the activities of man, but it is clear that the disposition of the moisture retained within the soils of a region has often undergone marked changes. Where the cultivation of row crops has involved elimination of the protection afforded by natural vegetation, the amount of soil moisture lost by evaporation has unquestionably increased. It frequently has been noted that overgrazing of arid and semi-arid ranges has started a trend in which drought conditions become intensified. The sparser the forage plants and other types of vegetation, the greater the evaporation from the soil and the less the amount of moisture available to them for their life and growth.[40]

---

[40] See, for example, O. Julander, "Drought Resistance in Range and Pasture Grasses," *Plant Physiology,* 20 (1945) 573-99. It may be noted that soil moisture content does not necessarily change; the important change is in ratio between the amount of moisture which is available for productive plant growth (and transpiration) on one hand, and that which disappears as evaporation, on the

The difficulty of measuring accurately plant water intake and soil moisture disposition and depletion unquestionably retarded improved analytical understanding in this field in the past. The emergence of new and more accurate measurement techniques and the development of indirect ways of estimating these factors is rapidly changing the situation. Improvement in both the quality and the amount of soil moisture information has stimulated efforts to conserve soil moisture for optimum crop growth. A number of water conservation practices have been developed, and are collectively exerting a profound effect in many semi-arid areas toward increasing the productivity of agriculture.

A number of means have been devised to retain and conserve soil moisture. They have been practiced under widely varying environmental conditions. Among them it is possible to identify five independent approaches toward these ends.

The first involves terracing, pitting, listing, strip cropping, contour cultivation, and other physical changes in the land surface which may minimize runoff and retain precipitation where it falls. These mechanical means of retarding surface-water flow also create conditions which allow time for rain and snow melt to infiltrate the soil. While much of moisture retention capacity depends upon the soil type, intensity of rainfall and other climatic conditions, the soil may be held at field capacities or well above the wilting point, where these measures are applied, for longer periods of time than otherwise.

A second group of practices designed to conserve soil moisture for plant growth is aimed at reducing surface evaporation. The provision of mulches and other protective coverings for row crops and fruit trees, and several techniques of stubble or "trash" farming are included in this group. The latter have gained popularity in recent years in the drier areas. A wide variety of materials has been used in mulching, but costs of all crop-residue mulching generally limit this practice to high-valued crops. In stubble farming, organic trash and crop residues are left on the soil surface to act as a mulch, and modified tillage instruments are employed which make cultivation possible with the minimum disturbance of the surface layer. In the past it was frequently contended that the surface soil could be managed as a natural mulch

other. At the same time total disappearance of soil moisture may be somewhat less in the evaporation from a bare soil than total disappearance through evaporation and transpiration when vegetation is present.

Plate 8. Small flood-water detention reservoir, Whitehead subwatershed, Nebraska, installed under the Pilot Watershed Program, U. S. Soil Conservation Service. Conservation land treatment is well illustrated on the area draining into the pond.

by repeated cultivation. However, it is now recognized that there are a number of reasons why moisture loss cannot be controlled by this means.[41]

A third group of practices is directed toward the elimination of weeds and other undesirable plant growth which use moisture that otherwise might be available for crop use. Cultivation formerly was the main means of controlling the competitive demands of weeds. Today herbicides are a powerful instrument for this purpose.[42] Where herbicides are employed, one of the leading arguments favoring cultivation is removed. Another key measure in controlling weeds is crop rotation.

A fourth type of measure concerns the manner of tillage. It has long been realized that the improvement of tilth and soil structure promotes an increase in the volume of water retained in the soil. Plowing, cultivation, and other management practices greatly influence the granulation and stability of the soil as well as other characteristics, such as its degree of aeration. These in turn influence its ability to hold moisture and to allow the infiltration of water. Because soil types differ so much in their physical properties and because of the great variations in the environmental conditions affecting them, there is no one best practice. Tillage in general is desirable, but if practiced too frequently it can destroy the structure of soil colloids in some soils, making them less pervious to air and water. Tillage practices consequently must be adjusted to several factors. In some cases subsoiling and deep plowing may prove beneficial. Where soil compaction constitutes a serious problem special techniques of claypan shattering may be effectively practiced. In some regions soil fertility is not the controlling influence in crop production, and yields are governed almost entirely by the amount of water available for plant growth.[43] However, in other areas

[41] The efficacy of this measure in terms of retention of soil moisture was challenged more than two decades ago in: F. J. Viehmeyer and A. H. Hendrickson, "Some Plant and Soil-Moisture Relations," *American Soil Survey Association Bulletin No. 15,* Houma, La., 1934, pp. 76-80. For a more recent review of this issue, *cf.* T. L. Lyon, H. O. Buckman, and N. C. Brady, *The Nature and Properties of Soils* (5th ed.; New York: Macmillan, 1952), p. 232.

[42] A. S. Crofts and W. A. Harvey, "Weed Control," *Advances in Agronomy,* Vol. I, ed. by A. G. Norman (New York: Academic Press, 1949), pp. 289-320.

[43] Herbicides are also important in phreatophyte control, which is being sought along stream courses and irrigation canals. See T. W. Robinson, "Phreatophytes and their Relation to Water in the United States," *Transactions of the American Geophysical Union,* 33 (February 1952), 57-60. A more recent summary is presented by H. C. Fletcher and H. B. Elmendorf, "Phreatophytes—A Serious Problem in the West," *Water, The Yearbook of Agriculture, 1955,* pp. 423-29.

soil fertility treatment can reduce the amount of water required for the per-unit output of a crop by as much as one-fourth.[44] The addition of amendments to the soil may under a variety of circumstances also favorably alter the ratio of available to unavailable water for plant growth. The development of soil conditioners, like "Krilium," has added a completely new but as yet economically unproven type of amendment.

Still other tillage practices have been devised to alter the conditions of soil moisture supply. Summer fallowing is probably the most widely used of the remaining tillage measures. In it, uncropped soil is used as a reservoir to catch as much of one season's rainfall as possible and carry it over for use during the following year.[45] Possibly one-half or even more of one year's precipitation may be carried over to the next.[46] Summer fallowing has been employed particularly on the wheat lands of northern Texas, western Kansas, eastern Montana and other parts of the Great Plains, and in the Palouse Hills of eastern Washington.

A fifth set of practices is composed of special vegetative coverings which may alter prevailing microclimatic relationships. In the United States the most extensive activity of this type has been the planting of shelterbelts and windbreaks on the Great Plains. It was estimated in 1948 that more than 120,000 miles of such plantings had been made in the United States since the middle of the last century.[47] However, the Soil Conservation Service estimated at the end of World War II that over 2 million acres of land were in need of such protection.[48] The primary purpose of these shelterbelts is generally to protect crops against drying winds. Within their effective range, which varies from ten to thirty times their height, they are thought to aid in preventing the firing of crops. They also serve to catch and store snow, thus

[44] V. C. Jamison, "How to Store More Moisture in the Soil," *Crops and Soils,* 8 (February 1956), 20-21.

[45] H. H. Finnell, *Soil Moisture and Wheat Yields on the High Plains* (Leaflet No. 247), U. S. Department of Agriculture, Washington, 1948, pp. 5-7.

[46] T. L. Lyon, *et al., op. cit.,* p. 234.

[47] J. H. Stockeler and R. A. Williams, "Windbreaks and Shelterbelts," *Trees, Yearbook of Agriculture, 1949,* U. S. Department of Agriculture, Washington, p. 192.

[48] U. S. Department of Agriculture, Soil Conservation Service, *Soil and Water Conservation Needs, Estimates for the United States by States,* Washington, 1945 (mimeo.), quoted by R. K. Frevert, G. O. Schwab, T. W. Edminster, and K. K. Barnes, *Soil and Water Conservation Engineering* (New York: John Wiley, 1955), p. 131.

increasing the moisture supply in the spring on a given site. The shelter-belts themselves unquestionably consume more moisture than would the cultivated crops which they directly displace, but under most circumstances they are thought to be a favorable influence in conserving soil moisture, as well as giving soil erosion control benefits.[49]

The type and management of vegetative covering maintained in an area is also of interest in nonfarming land use. Where maximum water yields are sought, as in the montane West, vegetative management is of interest both in the regulation of runoff and in the total water yields. Both are of direct interest to semi-arid and desert areas where water of mountain origin can be used for irrigation of crops. There is evidence that a well-maintained forest cover, including a complete litter or ground cover and underlying uncompacted soil, reduces the flash runoff following rainfall and thereby helps to regulate stream flow. This is probably the case only where the infiltrated water can reach a ground-water table. Where the forest soil is shallow and such moisture is rapidly discharged as interflow, less benefit can be expected.[50] Forest management also can cause differences in total water yield. It has been demonstrated that where soils are deep, clear cutting in small patches or strips, as well as heavy forest thinning in regions of plentiful snowfall and rainfall, can increase water yields by comparison to continuous cover. Such practices reduce transpiration and to some extent the quantity of snow or rain which is intercepted by the forest cover and re-evaporated. The covering given by the remaining forest also protects the snow pack on the ground from evaporation as compared to full clearing. Flow regulation also is experienced under such conditions because the snow will melt more slowly than in unprotected areas.[51] The character of the vegetative covering also probably influences temperature appre-

---

[49] N. P. Woodruff, *Shelterbelt and Surface Barrier Effects on Wind Velocities, Evaporation, House Heating, and Snowdrifting,* Kansas Agricultural Experiment Station Technical Bulletin No. 77 (Manhattan, Kan.: Kansas State College, 1954). However, in some investigations little evidence has been discovered of a significant shelterbelt influence upon soil moisture. *Cf.,* for example, W. J. Staple and J. J. Lehane, "Influence of Field Shelterbelts on Wind Velocity, Evaporation, Soil Moisture and Crop Yield," *Canadian Journal of Agricultural Science,* 35 (1955), 440-53.

[50] R. K. Linsley, M. A. Kohler, and J. L. H. Paulhus, *Applied Hydrology* (New York: McGraw-Hill, 1949), pp. 628-29.

[51] M. D. Hoover, "Effect of Removal of Forest Vegetation on Water Yields," *Transactions of the American Geophysical Union,* 23 (1944), pp. 969-77.

ciably, which in turn may affect the depletion of soil moisture.[52] The water-conserving properties of forest cover have been given their lowest recent rating in the so-called "Barr Report" of the Arizona Watershed Program, in which proposals have been made to maximize water yield on the Salt River Valley watershed by an extensive reduction in forest and other vegetative cover in the upper part of the watershed.[53]

EFFECT OF WATER-CONSERVING PRACTICES
ON RUNOFF—SUMMARY

The Barr Report on Arizona has illustrated in acute form diverging, if not competing, interests of water users in different sections of one watershed. Such competition is very plain where runoff from upland or montane slopes has been established as a source of irrigation water for adjacent lowlands. It is equally plain that there are some possibilities of varying the amount and time incidence of water flow by the manner of vegetative management within a catchment area. The problem is one of discovering what combination of land and water will yield the maximum long-range economic return within a given basin. Further, it is a question as to where the water supply available within a given basin is to be used. There were no conclusive answers to this question in 1958.

Most of the practices discussed in this section, however, have concerned lands that have not yet been established as water sources for downstream use in the manner of the western mountain slopes. Yet the same question of upstream-downstream competing interests has arisen about them.

Downstream interests frequently have shown anxiety over the possibility that water-conserving practices on upstream range and cropping lands may reduce stream flow. There is little conclusive evidence on these points.[54] By reducing surface runoff and making it possible for a larger proportion of the rainfall to infiltrate the ground, water-conserving practices tend to retard and delay the water yielded to the

[52] D. H. Miller, "The Influence of Open Pine Forest on Daytime Temperature in the Sierra Nevada," *The Geographical Review,* 46 (1956), 209-18.

[53] Arizona Watershed Program, *Recovering Rainfall,* Tucson, 1956, Part I.

[54] For recent evidence of the magnitude of possible changes in the peak rate of ground-water flow in one locality, see: J. M. Rosa, "Forest Snowmelt and Spring Floods," *Journal of Forestry,* 54 (1956), 231-35.

surface drainage system, once the soil acquires as much moisture as it can hold. This has the effect of regulating the fluctuations in stream flow, reducing the precipitous increases in the volume of water discharged immediately after a heavy rainfall, and raising the water volume at low water stages in the watershed area. Moisture conservation and management practices thus may have little effect upon the total volume of water draining from a given watershed, but they probably alter the timing and incidence of this discharge. The usefulness of a stream is affected, among other things, by the volume of minimum flow. Fluctuations above low water may be wasted unless the water is retained in reservoirs or otherwise impounded. Any redistribution of the total discharge, through changed land use which increases low stream stages, heightens the economic value of that stream flow, and may reduce the need for impoundment.[55] Here again, however, the optimum combination of uses from an economic point of view should be the deciding element in the land-use and water-yield practices that are maintained. It is entirely possible that downstream impoundment of flash runoff may represent the most economic use of water in some special situations.

Water-conserving farming practices are directed not only toward the infiltration of water which otherwise immediately would become runoff, but also toward the reduction of loss through on-site evaporation and unproductive transpiration. Such are mulching, weed control, and shelterbelt planting. To the extent that additional soil moisture thus is directed toward productive use, a net water gain is achieved.

In sum: insofar as water conserving practices divert to productive use soil moisture which would otherwise disappear through evaporation or transpiration by unproductive vegetation, they add to the available

---

[55] Actual measurements of this effect are available in the following references: Leon Lassen, Howard W. Lull, and Bernard Frank, *Some Plant-Soil-Water Relations in Watershed Management* (Circular No. 910) U. S. Department of Agriculture, Washington, 1952, p. 8; E. A. Colman, *Vegetation and Watershed Management* (New York: Ronald Press, 1953); E. G. Dunford, "Relation of Grazing to Runoff and Erosion on Bunchgrass Ranges," *Rocky Mountain Forest and Range Experiment Station Note No. 7*, Fort Collins, Colo., 1949; M. D. Hoover, "Effect of Removal of Forest Vegetation upon Water Yields," *American Geophysical Union Transactions, Part VI* (1944) (Washington: National Academy of Sciences, 1945), pp. 969-72; H. G. Wilm, "The Relation of Different Kinds of Plant Cover to Water Yields in Semi-Arid Areas," *Proceedings of the Sixth International Grassland Congress*, 1953, pp.1046-50; L. D. Love, "Watershed Management Experiments in the Colorado Front Range," *Journal of Soil and Water Conservation*, 8 (1953), 213-18.

water supply. Where they alter the timing of runoff or the site of moisture use, as in the mountainous basins of the West, the benefits are to
be found in other ways than in total water yield. In situations where
downstream withdrawal is clearly reaching limits of available supplies,
the economic relation of water-conserving upstream land use and provision of downstream supply needs much further study before the
optimum combination can be established.

4. Water-softening Practices and Equipment

Water-conserving practices and adjustments of use to conditions of
supply also are found in water uses other than agriculture. Water
conservation particularly has been a subject of attention in industry.
However, another kind of practice here is chosen to illustrate technical
attention within industry and for domestic uses: water melioration. As
has already been noted, water quality may be a barrier to economic
use, either because of mineral or of organic content.

Today virtually all municipally supplied water, a substantial share
of that withdrawn by industry, and even some used for irrigation purposes is treated prior to use. Numerous practices have been developed
for bringing the quality of locally available water into conformity with
the particular requirements of each different use, almost regardless of
its initial or natural condition. Water can be mechanically processed
by filtration, sedimentation, or aeration, and it can also be chemically
treated in a great variety of ways. The two primary objectives of
chemical treatment are purification and softening. Purification involves
the removal of organic matter and micro-organisms from water, whereas
softening is the removal or reduction of undesired effects of dissolved
solids, especially those contributing to hardness.

The term "hardness" refers to dissolved calcium and magnesium salts
in water, generally present as bicarbonates, chlorides, sulfates, or
nitrates. These salts react with soaps to form insoluble precipitates.
For each grain per gallon of such hardness, one and a half pounds of
pure soap are destroyed per 1,000 gallons of water used. If the hardness of water (expressed as calcium carbonate ($CaCO_3$) equivalent
of dissolved calcium and magnesium salts) were 120 parts per million,
10.5 pounds of pure soap would be wasted for every 1,000 gallons of

water used.[56] The rise in the use of detergents, particularly in the home, is largely due to the fact that they do not react with calcium and magnesium ions nor lose their effectiveness in hard water.

Use of hard water increases costs and lowers efficiencies in other respects as well. When these salts come into contact with metal parts, they may form sludges and adherent scales, which among other things seriously reduce the carrying capacity of pipes. Because thermal conductivity is also lowered, hard water is especially undesirable for use in boilers and water heaters. Hardness is also troublesome in many industrial uses in which such mineral salts react unfavorably with materials coming into contact with the water, as in textile processing.

One manner of meeting the problem of hardness is simple dilution, with "soft" water. However, soft water supplies are not frequently encountered in regions or localities suffering the disadvantages of hard water.

Carbonate or temporary hardness, caused by the presence of bicarbonates of calcium and magnesium, is commonly distinguished from noncarbonate or permanent hardness caused by sulfates and chlorides of calcium and magnesium. Whereas carbonate hardness can usually be greatly reduced by heating and filtering, noncarbonate hardness can be removed only by using chemical agents. There are six principal chemical processes for reducing the hardness of water, the cold and the hot lime-soda water-softening processes, the zeolite (sodium cation-exchange) water-softening process, the hydrogen cation-exchange process, the ion-exchange demineralizing process, and distillation.

The use of slaked lime and soda ash was for many years the only method in commercial use. In its modern application, a distinction is generally made between the cold and the hot lime-soda processes. In the cold process, the raw water is first treated at ordinary temperatures with lime, which reacts with the bicarbonates to form insoluble carbonates that precipitate; and then with soda ash, which reacts on the noncarbonate hardness and precipitates the calcium as carbonate, leaving in the water an equivalent amount of sodium sulfate, chloride, or nitrate. The lime also converts the magnesium salts to the hydroxide which precipitates. Most of the precipitated calcium and magnesium salts

---

[56] One grain per gallon of hardness is equivalent to 17.1 ppm. R. N. Shreve, *The Chemical Process Industries* (2nd ed.; New York: McGraw-Hill, 1956), p. 45. A table of the soap destruction of water of various degrees of hardness is provided in: E. Nordell, *Water Treatment for Industrial and Other Uses* (New York: Reinhold, 1951), p. 52.

are then removed by settling; a coagulant such as alum is often subsequently employed, and as a final step the water is usually filtered. The cold process is employed chiefly for partial softening in municipal water supply systems and in manufacturing plants where calcium bicarbonate hardness is troublesome but complete removal is not required.

The hot lime-soda process, first introduced at the beginning of this century, differs in that it is carried out in a pressure settling tank at or near the boiling temperature of water. Because the reaction proceeds more rapidly than in the cold process, precipitation is facilitated and generally no coagulant is necessary. A higher degree of softening is effected by this process. The hot process is used almost entirely for conditioning boiler feed water.

The zeolite process is the simplest and unquestionably the most widely used. It depends upon ion exchange reactions, whereby an undesirable ion is replaced by one that can be tolerated or is innocuous. Zeolites are complex crystalline bodies composed of silica, alumina, and sodium which are capable of exchanging a part of their sodium for the calcium and magnesium dissolved in water. Although the base-exchanging properties of zeolites had long been recognized, it was not until 1905 that it was discovered that they could be employed to soften hard water.[57] Their use rapidly spread in subsequent years with the development of appropriate equipment. Today a zeolite water-softening installation usually consists essentially of a pressure-type filter possessing a granular zeolite bed. As water is passed through the granular zeolite in a pressure-filter chamber, the calcium and magnesium cations, or positively charged ions of the water, combine with the zeolite and remain in the filter, while the sodium cations of the zeolite pass into the water. When the surfaces of the zeolite become saturated with calcium and magnesium, the softener is regenerated by passing a solution of common salt through the zeolite bed. This removes the calcium and magnesium in the form of their soluble chlorides and restores the sodium content of the zeolite.

Sea water and natural brines may be used as the source of the sodium solution employed in backwashing the zeolite to regenerate it and, wherever they are available their use may effect substantial economies. Elsewhere, the equipment must be periodically recharged with sodium chloride crystals or backwashed with a sodium chloride solution, the

---

[57] The history of the development of zeolite water softeners is summarized by Nordell, *op. cit.*, chap. XIII.

amount required depending upon the hardness of the untreated water and the quantity softened since the previous regeneration. Removal of hardness by the zeolite process may be so complete that water softened by this process is frequently referred to as "zero water," to indicate that all but insignificant traces of carbonate hardness have been removed.

In the hydrogen cation exchange process, hydrogen ions, instead of sodium, are exchanged for the undesired calcium and magnesium ions. Several types of sulfonated coal products are most frequently provided in the exchanger. They react with the bicarbonates in hard water to form carbonic acid (H_2CO_3), which immediately decomposes into water and carbon dioxide. The carbon dioxide thus formed can be removed by aeration, and the hydrogen cation bed regenerated as frequently as is required with sulfuric acid. Since the water produced by this method is somewhat acidic, it is generally either neutralized with alkali additives or further processed by being passed through an anion exchange material in which the negatively charged or nonmetallic ions in the water are replaced.

Demineralization processes involving anion exchange are of two types, highly basic and weakly basic. The anion exchanger medium most widely used in demineralization is a granular, complex amine resin of a weakly basic nature which effectively removes such strongly ionized acids as sulfuric, hydrochloric, and nitric acid, and the anions from solutions containing the corresponding salts of those acids.[58] For the removal of silicate and carbonate ions, highly basic anion exchangers are required. Regeneration of the anion bed is accomplished by backwashing with an alkaline solution of soda ash, caustic soda, or, in some cases, ammonium hydroxide.

When the processes of cation and anion exchange are combined in sequence, the metallic cations are replaced by hydrogen and the anions by hydroxide, thereby forming only water as a final product. The dissolved solids are thus completely eliminated, in contrast to the substitutions of the zeolite process.

The oldest process for freeing volatile liquids from their nonvolatile impurities is distillation. With modern equipment the mineral content of water can be virtually eliminated by vaporizing the water and then condensing the vapor back into liquid form. Water that is commercially distilled typically contains between 3 and 21 ppm of total salts, although even higher purity may be achieved for special purposes. Distillation

[58] Nordell, *op. cit.*, p. 394; R. Norris Shreve, *The Chemical Process Industries* (New York: McGraw-Hill, 1945).

costs are high, in some instances in excess of $8.00 per 1,000 gallons. Even in a large, modern installation they are more than $1.00 per 1,000 gallons.[59] Distillation thus is prohibitively expensive except for special uses and in unusual environments. Modifications in the design of multiple-effect evaporators and the development of practical vapor-compressor evaporators may permit substantial cost reductions, however.[60]

Installations of water-softening processes, particularly the zeolite, have increased greatly in recent years (table 24). According to a

TABLE 24. *Summary of Water-Softening Installations in the United States, 1940 and 1954*

| Type and field | Number in operation | |
of application	1940	1954
Zeolite, household	300,000	1,750,000
Zeolite, industrial	35,000	50,000
Zeolite, municipal	170	300
Chemical precipitation, industrial[1]	3,000	4,500
Chemical precipitation, municipal[1]	364	700

[1] The chemical precipitation type includes both cold and hot lime-soda installations.

Source: Estimates made by E. Nordell, presented in the following reference: R. N. Shreve, *The Chemical Process Industries* (New York: McGraw-Hill). Data for 1940 can be found in the 1st ed., 1954, p. 47; data for 1954 appear in the 2nd ed., 1956, p. 47.

recent estimate, sales of ion exchange equipment and material in the United States exceeded $40 million in 1954; water conditioning, including demineralizing, silica removal, and alkalinity reduction as well as softening, accounts for all but a small fraction of this total. Along with the rapid increase in sales of ion-exchange water-softening equipment, the consumption of regenerative chemicals has risen sharply.

[59] Nordell, *op. cit.,* pp. 402, 403.
[60] For a review of developments in the field of distillation, see Cecil B. Ellis, *Fresh Water from the Ocean* (New York: Ronald Press, 1954), chap. VII; and Sheppard T. Powell, *Water Conditioning for Industry* (New York: McGraw-Hill, 1954), chap. IX. A more complete discussion of this process is also presented herein (chap. 13).

Annual use, according to this estimate, amounts to 200,000 tons of salt and 50,000 tons of sulfuric acid, and an equal amount of soda ash. The equivalent of 300 carloads of $CaCO_3$ is removed per day in softening water by ion exchange.[61]

Probably three-fourths of the ion exchange equipment currently being sold is for household use, for which the zeolite softener is the only practical type. Zeolite equipment is the simplest and most convenient to operate, in both household and industrial installation. It has a special advantage in its capacity to supply water of near-zero hardness without attention or adjustment until regeneration is required. This is true even when variations are experienced in the hardness of the raw water which is treated. The extent to which zeolite softeners are employed today is suggested by an enumeration of no less than 320 different types of users of this process.[62] Industry relies upon the zeolite process of water softening especially where high-grade washing water is required and where process water is needed in the preparation of commodities for human consumption. Users include canneries, meat packing plants, edible oil refineries, creameries, bakeries, ice manufacturing plants, as well as tanneries, many types of textile plants, and commercial laundries. Many highly specialized water users (e.g. photographic studios) also employ the process.

Ion exchangers of the hydrogen-substitution type also are employed in many industries. Some of the most important applications are in the fermentation industries and in the manufacture of glues and gelatine, insecticides, and refined sugar. In addition, they are used in recovery of some of the rare earths and in treatment or recovery of a number of industrial wastes.

The list of industrial processes that require water practically free of mineral salts is rapidly growing. Already such water is used in the manufacture of plastics, synthetic rubber, drugs and fine chemicals, ceramics, and various food products. It also is employed for such other varied uses as diesel locomotive cooling, in broadcasting stations, oil refineries, electric power plants, and electro-plating plants.[63] It is clear that demineralization by ion-exchange processes will be very effective and not prohibitively expensive where exceptionally pure water is required. There are a number of special-purpose fields of application,

[61] F. C. Nachod and J. Schubert (eds.), *Ion Exchange Technology* (New York: Academic Press, 1956), p. 3.
[62] Nordell, *op. cit.*, pp. 144-50.
[63] *Ibid.*, p. 404.

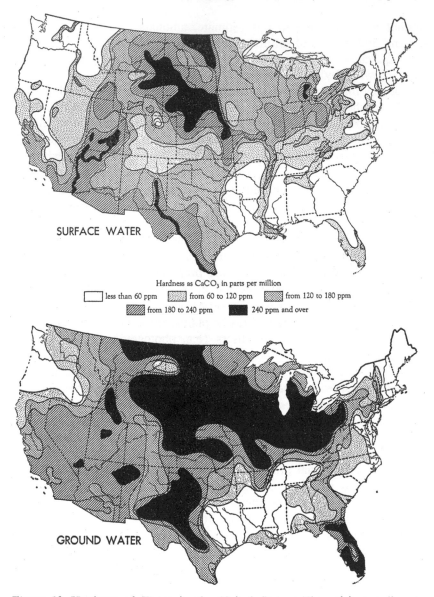

SURFACE WATER

Hardness as CaCO₃ in parts per million

less than 60 ppm — from 60 to 120 ppm — from 120 to 180 ppm
from 180 to 240 ppm — 240 ppm and over

GROUND WATER

Figure 19. Hardness of Water in the United States. The calcium carbonate content of water may vary considerably on a seasonal basis and from year to year, especially in areas of low precipitation; furthermore, the water at any one location may not be comparable with that of the surrounding area. The attempt here has been to delineate areas representative of average conditions on a highly generalized basis.

such as the provision of potable water supplies in remote places and the desalinization of sea water on ships, where distillation is likely to continue to be employed.[64] Should there be notable success with new distillation developments, application of this process will certainly show substantial increase.

The extent to which water-softening and -demineralizing equipment is used depends in large part upon the hardness of the existing supplies of water. Regional differences in the need for such treatment are suggested by the maps of water hardness in the United States (figure 19). The heaviest and most immediate demands for water-softening equipment have been in those areas where water, particularly ground water, is naturally hard, and where the density of population and manufacturing activity also are high, viz. the Midwest (figure 14). Demands for water-softening units, however, are yearly becoming more widespread, and in the future most sections of the country will have at least some dependence on them. Although these demands will be meliorated somewhat, particularly in the household, by the substitution of synthetic detergents for soap, substantial growth of water softening may be expected because: (1) industrial production is likely to require waters of increasingly high purity; (2) lower quality supplies may be brought into use as demands for water increase; (3) the household advantages of water softeners will be more widely understood; and (4) population density and urbanization will increase throughout the country. Water softening therefore may be considered an important aspect of water supply development in the future of the United States.

[64] For details regarding some recent installations for obtaining domestic water supplies from sea water, see "Equipment to Provide an Additional 2½ Million Gallons a Day of Water at Kuwait," *Chemical and Engineering News,* Fall, 1956; "New Multiple-effect Evaporator for Aruba," *Engineering News Record,* 157 (September 6, 1956), 72-76; and R.S.Y. Yoseph, "Demineralization of Brackish Water for Potable Supply at Matagorda Island," *Journal of the American Water Works Association,* 48 (1956), 579-84.

CHAPTER 9

Technical Improvements Promoting
the Scale Economies —

SIX CASES

Two Techniques Enlarging the Scale and Coverage of Single Water Facilities: (1) The Development of Earth-moving Equipment. (2) Concrete Dams—materials and techniques of using materials; dam design and construction; dam and reservoir site preparation; summary comment on concrete dams. *Interrelated Operation of Two or More Units of Supply in an Integrated System:* (3) An Example of Integrated System Water Control—benefits from integrated water regulation; changes in the status of water control within the TVA System. *The Interrelation of Systems of Supply with Systems of Demand on a Mutually Appropriate Scale:* (4) Developments in Canal Construction and Tunneling—canal construction; tunnel construction. (5) Electricity Transmission—techniques and equipment related to use of increased voltage; avoidance of natural hazards; line construction costs; present transmission distances and costs; distribution of dispersed load. (6) Dispersal of Demand for Water as an Influence on Quantitative Inequalities of Demand and Supply. *Summary Comment.*

Continuous improvements in the techniques that contribute to scale economies have had significant effects upon water supply. The techniques may be grouped as follows: (1) those that enable the construction of single facilities with wide effectiveness and coverage, like a large-scale dam and reservoir; (2) those that facilitate the interrelated operation of two or more units of supply into a co-ordinated, or integrated system; and (3) those that facilitate the interrelation of systems of supply with units of demand on an appropriately large geographical scale.

Of these three groups, the first is represented here by two cases, one describing the development of earth-moving equipment, and the other the construction of dams. The second group is illustrated by integrated river basin water control; and the third by cases of widely differing character—techniques of canal and tunnel construction, development of electrical transmission equipment, and dispersal of water demands.

These cases all concern engineering, with the exception of the last, in which techniques of regional planning are briefly treated. In them one finds some of the more important and, in outline, familiar events in the progress of methodical mechanical, electrical, and hydraulic engineering experiment and design. Probably, however, the basic significance of their application is appreciated fully only by specialists. Among all of the technical events that have raised problems for American administrative organization in water development, this set has posed some of the most complex and persistent questions.

TWO TECHNIQUES ENLARGING THE SCALE AND COVERAGE OF SINGLE WATER FACILITIES

The possibility of damming streams of almost any size, the physical feasibility of regulating watersheds including tens of thousands of square miles with a single master facility storing millions of acre-feet of water, and the possibility of regulating streams under a large variety of topographic and geological conditions have changed vastly the nature of water development within this century. The feasibility of constructing earth-fill dams has increased the range of topographic conditions in which large-scale conservation storage of water is practical, and it has helped to reduce the actual and potential costs of storage on many streams of the country. This has been an outgrowth of the development of modern earth-moving equipment and theoretical and applied soil mechanics, which together have effected some of the more spectacular changes wrought by man on the face of the earth.

Although the influence of earth-moving equipment and soil mechanics extends far beyond water development in our culture, these techniques have been major tools for the civil engineer concerned with dam construction. The improvement of concrete and the design inventions for large-scale reinforced concrete structures are of equally great significance to techniques of water storage and regulation. Progress in arch dam design, development of instrumentation for measuring internal stresses inside dams, and research in hydraulics have contributed greatly to the general advance of technology in this type of construction. Without all of these, multiple-purpose regulation probably would have been much less feasible, low-cost hydro power probably would not have appeared in large amounts, and complete control of some of the major streams in the country would have been an impossible goal. As it is, the United States is moving in the direction of such a final objective.

A land surface of continental scope, possessing predominantly continental climates, places a special premium on the construction of large-scale facilities for water regulation. The large and even great size of drainage basins in North America is conducive to streams of enormous volume. The continentality of the climates leads to a wide range between peak flows and low flows, both because of the influence of below freezing winter temperatures, and because of the path of moisture-producing storm systems. The consequent basic problem for hydraulic engineering in American water regulation comprises the use of design techniques and available materials in a manner which will create on any stream in the country, safely and at economically feasible costs, storage reservoirs capable of detaining, storing, or passing without damage to the structure peak flows often of enormous extent.

This basic problem of regulation is difficult enough in itself. It is made more difficult by the paucity of reservoir sites near the localities where demand is concentrated. The necessity of finding rock structures capable of supporting massive artificial barriers is one problem. The desirability of creating reservoirs with the smallest area of evaporating surface possible for a given volume of storage, the correction or avoidance of faulted and fractured foundation rock and porous (hence leaky) rock outcrops within otherwise good reservoir sites, and the avoidance of excessive flooding of land valuable for other purposes than water regulation, are additional problems.

Translated into engineering effort, these environmental conditions become problems of dam design, reservoir and dam site preparation, and materials development. In each of these three basic tasks the development of earth-moving equipment has been very important.

1. The Development of Earth-moving Equipment

The history of construction in the field of water engineering is as old as civilization itself, but man's present ability to dig and to fill, and thus to regulate the movement of water over the face of the earth, dwarfs all efforts of the past. Earth moving is of importance in many different types of water resource projects. It long has been necessary to swamp and marsh drainage, canal construction, irrigation land levelling, dredging, and embankment or levee construction. Earth and rock are frequently employed in the construction of large dams. Earth moving also usually is essential in providing access roads to water development projects, in land clearing for reservoirs, and in excavating for foundations of dams and auxiliary structures.

During the first years of this century, levees and embankments were still being built by relatively primitive methods. Earth was moved by wheelbarrow or by mule-drawn drags and scrapers, each having only a few cubic yards capacity in a day. Canals for irrigation and navigation were dug by means scarcely altered from those of past millenia. The development of railroad haulage greatly increased the potential of various filling operations. The most spectacular use of rail equipment was in 1906 and 1907 when thousands of freight cars of rock fill were dumped into the Colorado River to return its flow to the former and present channel after a period of flooding into the Salton Sea.[1]

Hydraulic construction methods also were a development which aided large-scale construction during the same period. After 1868 a number of hydraulic fill earth dams were built by a method employing water in all major operations. In this technique of construction, a dam is built of earth material which is hydraulically moved through flumes or pipes. When discharged, the larger particles carried by the water remain near the two faces of the dam where they are discharged, but the finer particles move toward the central area between the faces. The smallest particles are deposited in an intervening pool where they accumulate to form the core of the structure. Because the material becomes segregated after being discharged, the toe and faces of the final structure are more pervious than the core, permitting water to drain out from the interior of the dam. The principal disadvantage ac-

---

[1] R. A. Nadeau, *The Water Seekers* (New York: Doubleday, 1950), pp. 152-67.

companying this method is that its applicability is limited to sites where a large volume of water at high head is available and where suitable earth fill is ideally located with respect to the dam position. Moreover, there is some temporary risk of collapse until sufficient water has drained away from the core to permit consolidation of the solid particles.

Many earth dams have also been built by the process of semi-hydraulic fill, by which materials are moved to the dam site by a mode of transport other than water, but some of the fill is worked into its final position within the dam by the action of water.[2]

The huge earth-moving capacity at the disposal of the modern construction engineer is the product of greatly increased mechanical power which has been used in the design of specialized types of equipment. The first requirement was for more efficient mechanical digging tools, but progress in the development of such machines soon gave rise to the need for very mobile and versatile systems of traction to remove the excavated material and to deposit it where desired. A pioneering machine of the first type was built at the beginning of this century. It consisted of a fully revolving steam excavator weighing some 65 tons. It was mounted on skids and rollers and had an orange-peel bucket suspended from a wooden boom. It was soon equipped with a scraper or drag bucket, thereby becoming the prototype for all subsequent draglines.[3] Even in its original form, it constituted a great improvement over the dipper dredges then used in canal work.

Today draglines are built with buckets able to handle 10 to 20 cubic yards of material at a pass. A large modern dragline may weigh more than 1,000 tons and possess a boom up to 250 feet in length, being able to move up to 20,000 cubic yards of earth a day.[4] In some instances these machines have been made self-propelling by the addition of a walking mechanism which is activated by cams mounted on each end of large shafts that are connected to giant shoes straddling the machine. The comparatively low pressure of the machine on the ground is especially advantageous for work in difficult terrain where soft footings exist or in proximity to loose down slopes.[5]

[2] J. D. Justin, *Earth Dam Projects* (New York: John Wiley, 1932), p. 202.

[3] R. Finnie, *Earth-Moving: 50 Years on Heavy Engineering Projects* (Dallas and Los Angeles: Taylor and Taylor, 1945), p. 13.

[4] J. L. Allhands, *Tools of the Earth Mover: Yesterday and Today* (Huntsville, Tex.: Sam Houston College Press, 1951), p. 214.

[5] H. L. Nichols, *Moving the Earth: the Workbook of Excavation* (Greenwich, Conn.: North Castle Books, 1955), Sect. 13, p. 28.

Plate 9. Dragline moving earth from barrow pits to embankments for dam construction; U. S. Army Corps of Engineers Oahe project, Missouri River, South Dakota.

In spite of hydraulic and power equipment use, mules and scrapers still accounted for most of the earth moved in water development projects in the early 1920's. In that decade a profound change in earth-moving methods was initiated by the development of the gasoline crawler tractor by Holt and Best. It was later given diesel power and improved through a series of other innovations, making it the "caterpillar" of the present day. The fundamental accomplishment was establishing the means for a powerful but heavy machine to move rapidly over soft earth without severe sinking into the ground. As a "caterpillar," a pair of endless linked steel tracks are used as the load-distributing surface. With the development of techniques for making very large tires, load distribution on large wheels then became practical.

Although initially used solely for traction, the caterpillar tractor was soon converted into a self-contained work unit by adding a bull-dozing blade.[6] The bulldozer, either of the rubber-tired or crawler type, has subsequently demonstrated its ability to do a wider variety of jobs than any other earth-moving tool, in terms both of output and productive efficiency. At the same time it is but one unit in the fleet of equipment employed on present-day projects requiring movement. The operation of standardized power shovels, often so designed that they can be used either as a shovel, a dragline, a crane, or a backhoe,[7] is integrated with numerous belly-dump, side-dump, or end-dump trucks, while an increasing share of the earth being moved today is accounted for by carrying scrapers.

Carrying scrapers, also known as pans, are digging and grading machines which carry or drag their load behind a tractor. First developed by the Le Tourneau Company in the early 'twenties, there are today numerous self-powered scrapers in operation which can carry a load of 35 cubic yards of earth. The largest machines can haul this tonnage at speeds up to 30 miles per hour.[8] Two of these machines operating as a team can cut, load, haul for one and a half miles, and dump in its final position 3,000 tons of material in an eight-hour day. Although each construction project presents its own distinctive problems and accordingly differs in the type of equipment which most effectively can be employed, there is a growing reliance upon powered scrapers. Within the past decade, the pay load of many of these machines has

[6] Allhands, *op. cit.*

[7] A backhoe is comparable to a shovel, except that it digs toward the machine instead of away from it.

[8] Allhands, *op. cit.*, p. 127.

more than doubled, and giant new machines are constantly appearing.

Dam and levee construction by hydraulic fill methods has been supplanted by rolled fill embankments constructed by the aid of these machines. Aided by advances in techniques of soil sampling and soil density analysis, compacting instruments have assumed increasing importance in this type of construction. The sheepsfoot tamping roller is the tool most extensively employed. This spiked roller is dragged repeatedly over the same area until the feet no longer penetrate the surface. Depending upon the condition of the fill, from six to eleven trips are required to "walk the machine out of the ground."[9] Pneumatic-tired rollers, manually operated tampers, and other instruments are also frequently used to stabilize earth embankments. In addition, the crawler and rubber-tired prime movers help to compact the fill.

One of the more time-consuming and costly aspects of earth-moving projects has been land clearance, including removal of the surface layer of soil containing the roots of vegetative matter. This is now expedited by huge special-purpose rooters and rippers which break up terrain so compacted as to prevent the effective operation of bulldozers and scrapers.

Application of the continuous conveyor belt to earth-moving projects is another significant development. The "levee stacker," for example, is a machine mounted on rails and equipped with a portable boom which deposits fill conveyed by continuous belts to form a levee. Belt conveyance permits delivery of materials from a considerable distance.

Another machine which incorporates the conveyor belt in its design is the elevating grader. It is often employed for excavating earth materials on large projects where the conditions of the borrow area are suitable. The grader is equipped with vertical cutter blades which cut a slice of earth in the borrow area, mix it, and then place it onto a conveyor belt. The belt carries the material to a sufficient height to load large bottom-dump trucks. The excavating equipment is slowly moved by tractors and the trucks are loaded while in motion. At one Bureau of Reclamation dam construction project, 1,500 cubic yards of earth were moved by such a unit in one hour, resulting in the placement of more than 1 million cubic yards of rolled fill in one month.[10]

[9] *Earth Moving and Construction Data* (2nd ed.; Milwaukee: Allis Chalmers, 1949), p. 12.

[10] U. S. Department of the Interior, Bureau of Reclamation, *Dams and Control Works,* 3rd ed.; Washington, 1954, p. 193.

Still another method of using a conveyor belt was employed in the construction of the Anderson Ranch Dam at Boise, Idaho. There the clay required for the interior section of the dam was excavated by electric shovels and loaded into the hopper of a special "pendulum" type loader which deposited the material on a conveyor belt which carried it to the dam site. And in the construction of Shasta Dam in California, where more than 6 million cubic yards of concrete were poured, all of the aggregate required for the operation of the concrete mixing plant was transported a distance of more than ten miles by a continuous conveyor belt.

The most striking feature of developments in this field has been the rapid increase in the power and the work capacity of earth-moving equipment. Most of these tools can handle pay loads almost as great as their empty weight. Today, for example, there are available 50-ton rear dump trucks which can travel at 30 miles per hour; these huge vehicles are powered by two engines of 300 horsepower each.[11]

The huge work capacity of the recently developed earth-moving equipment is reflected in a number of rolled fill dams and reservoirs of unprecedented dimensions that have been completed in the last two decades. Where formerly only concrete dams could be considered, the much more massive earth-fill dams may now be competitive, even if power generation facilities are involved. Design and cost studies in several recent large projects show that where no power plant is required, earth-fill dams are generally cheaper than concrete if the site is suitable for earth-fill construction. Greater susceptibility to damage by severe floods or some other factor may make the earth dam inadvisable in a particular situation, however. A typical cost for cutting a cubic yard of earth, moving it several hundred yards, depositing, and compacting it in a large construction project is about 25 cents. By comparison, a typical cost of concrete placed in large projects is $16.00 per cubic yard.[12]

The use of earth-fill dams has been made more widely applicable by new seismic techniques for exploring and testing foundation rock, and by improved methods of foundation preparation, especially cement grouting. Better sealing of the earth fill to the foundation rock and the cut-off of percolation through the dam have been achieved by use of asphalt membranes, graded materials, and other means. Protection

[11] C. B. Colby, *Earth Movers* (New York: Coward-McCann, 1955), p. 54.
[12] L. G. Puls, personal communication, January 1957.

of fill material by asphalt liners has been developed with success.[13]

These various developments and methods of approach have resulted recently in earth-fill dam projects of tremendous scope. Green Mountain Dam, a part of the Colorado-Big Thompson project, is a rolled fill dam over 300 feet high which required more than 4.3 million cubic yards of embankment material. Davis Dam on the lower Colorado River, although less than half as high as Green Mountain, contains an equal amount of material. In the construction of the Anderson Ranch Dam, in Idaho, some 10 million cubic yards of material were used.[14] The Fort Peck Dam on the Missouri River in Montana is the largest earth dam ever built, containing some 128 million cubic yards of material, the equivalent of an area 1 mile square and more than 40 yards deep. The Garrison Dam, downstream on the Missouri River in North Dakota, is another project of more recent date which required the movement of some 75 million cubic yards of material. A new earth dam completed in 1958, on the Lewis River, Washington State, is the world's tallest at 450 feet.

Although less dramatic, the record of work accomplished in constructing the Buford Dam, which was recently completed on the Chattahoochee River near Atlanta, Georgia, is even more impressive. With the use of a comparatively small complement of earth-moving tools, consisting of twenty-two scrapers, twenty-five tractors, thirty-five belly-dump and eight end-dump trucks, and three shovels of 2½ yard capacity, fill was moved at an average rate of 2,000 cubic yards an hour.[15]

On the one hand, these developments have made possible various earth construction projects of enormous size. The Fort Peck dam can impound 14.9 million acre-feet of water, and Garrison Dam has a capacity of 18.1 million acre-feet. They are among the largest reservoirs made by man.[16] On the other hand, the new types of earth-moving equipment also are facilitating numerous small undertakings. Many agricultural areas are spotted with small ponds which have been con-

[13] T. F. Thompson, "Foundation Treatment for Earth Dams on Rock," *Proceedings, American Society of Civil Engineers,* 80, Separate 548 (November 1954).

[14] U. S. Department of the Interior, Bureau of Reclamation, *Dams and Control Works,* 3rd ed.; Washington, 1954.

[15] "Dirt Flies at Buford Dam," *Construction Methods and Equipment,* 37 (August 1955), 84, 91.

[16] N. O. Thomas and G. E. Harbeck, Jr., *Reservoirs in the United States* (Water Supply Paper No. 1360-A), U. S. Geological Survey, Washington, 1956.

Plate 10. Caterpillar Diesel tractor equipped with bulldozer. Construction of a cofferdam in the Clark Fork River, Noxon hydroelectric project, Montana.

Plate 11. Large earth carrier moving weathered shale; spillway construction, U. S. Army Corps of Engineers Oahe project, South Dakota.

Plate 12. An earth-fill dam under construction; U. S. Bureau of Reclamation, Palisades project, Idaho.

Plate 13. An aerial view of a completed earth-fill dam; U. S. Army Corps of Engineers Garrison Dam, Missouri River, North Dakota. Garrison reservoir serves flood control, irrigation, storage, hydroelectric generation, and navigation.

structed by bulldozers. Ten years ago it was estimated that there were some 360,000 acres of farm and ranch ponds in the United States;[17] the acreage probably is at least twice as much today. By 1957 the Soil Conservation Service alone had assisted ranchers and farmers in constructing about 842,000 ponds.[18]

Despite spectacular past achievements the use of earth-moving equipment still cannot solve some problems related to water development, such as the removal of inflowing sediment which reduces reservoir storage capacity. Reservoir service life usually is long, even centuries long, but where streams are heavily sediment laden and it is not hydraulically possible to pass the suspended solids through the structure, the life of water storage facilities of all types may be seriously curtailed. A high bed-load is particularly detrimental because of its certain deposition in a reservoir. As yet, only expensive means of removing accumulated deposits have been developed.[19] This and other problems still challenge the development and use of earth-moving techniques.

2. Concrete Dams—Materials, Design, and Site Preparation

Although one of the most revolutionary changes in dam design and construction was the development of the massive earth dam, nearly all of the best known hydraulic structures in the world are concrete dams. Concrete is indispensable in developing some of the best storage and power sites (e.g. rock gorges). Furthermore, concrete spillways, aprons, penstocks, and some other structures usually are parts of earth dams. The nature of materials available and their methods of handling, the site preparations for massive concrete structures, and the design of concrete dams are therefore an important part of America's capacity to store, regulate, and use on a large scale the flow of its streams.

MATERIALS AND TECHNIQUES OF USING MATERIALS

Concrete is a mixture of a cementing material and an indurated filler or "aggregate." Although some forms of concrete were known

---

[17] L. A. Walford, *Fishery Resources of the United States* (Washington: Public Affairs Press, 1947), p. 117.

[18] U. S. Soil Conservation Service, typed memorandum of June 30, 1957.

[19] E.g. settling basins upstream from hydraulic works.

as far back as the time of the Egyptian empires, and others were widely used by the Romans, its full versatility as a building material did not arrive until the introduction of reinforced concrete with Portland cement as the binding material. Portland cement, a vitrified mixture of lime and clay materials, was first introduced about 1845, but processing improvements which permitted the production of cement of uniform quality on a large scale did not appear until the powdered coal-fired rotary kiln became commonly used between 1877 and 1895. The resulting cement, of relatively uniform quality and of greatly superior strength to anything preceding it, was essential before concrete could be used in the masses required in a large dam.

The invention of a reinforced concrete (Monier, France, about 1868) which incorporates a network of small iron or steel rods was another important advance. Ordinary concrete has a high compression strength but relatively low resistance to shearing and tensile stresses. With the incorporation of steel rods, a material is produced which has most of the virtues of plain concrete in plasticity and handling, but which also has adequate resistance to the tensile and shearing stresses found in all large structures.

Since these basic inventions in modern concrete technique, Portland cement and concrete have undergone steady improvement in strength, durability, and dependability. Much of the development has been due to increased knowledge of the basic chemistry of the materials and the processes taking place in cement manufacture, their mixture with other materials to make concrete, the setting of concrete, and the aging of the final structure. As illustration, three types of improvement will be mentioned: the rapid hardening (high early-strength) cements, the low heat cements, and sulfate-resisting cements.

The rapid hardening cements are of importance in construction because of the need for speed to avoid excessive overhead where large commitments of equipment, labor, and the other appurtenances of construction are required. These cements are of two types, the aluminous cements, and the high early-strength Portland cements. The aluminous cements have a higher proportion of alumina (in relation to silica and lime) than Portland cements. They are said to develop the same compression strength within a single day that is found in ordinary Portland cement in twenty-eight days. They have particular importance in winter construction because of this rapid setting and because the high heat of hydration prevents freezing during the setting period. High early-strength Portland cements have the same qualities as the aluminous cements, and in the United States have been considered economically more desirable that the aluminous cements for most uses. They differ from standard Portland cements in having a higher tricalcium silicate content, and are more finely ground.

Low-heat cements have been developed to avoid as far as possible the cracks which may emerge in great masses of concrete during the hardening and curing period because of temperature changes resulting from heat liberated in the hydration reaction of the mass. This type was first used in the construction of Hoover Dam on the Colorado River, and subsequently it has been widely used in large concrete structures built by the United States federal government. The most widely employed has been a modified Portland cement, characterized by a low tricalcium aluminate and a high tetracalcium alumino ferrite content.

Of particular importance has been the development of cements which, when placed in concrete, resist the attack of certain soil chemicals, particularly sulfates. Sulfate-resistant cement therefore can provide longer life in concrete structures and hence increase over-all economy. The sulfate-resisting cements have been either a modified Portland, or the Portland-pozzolana cements. The modified Portland sulfate-resisting cement has low contents of tricalcium aluminate and tetracalcium alumino ferrite. In the second type, Portland cement is mixed or ground together with a pozzolana. The latter is a siliceous material, natural or artificial, which will react with lime in the presence of water at normally encountered atmospheric temperatures. A true pozzolana cement is a naturally occuring volcanic ash such as pumice, high in silica, with good cementing properties. This type of reaction with water is also exhibited by artificially produced pozzolanic materials such as fly ash, or calcined shales. By replacing a portion of the Portland cement in concrete mixtures with fly ash, considerable economies in construction can be achieved while actually improving the strength and durability of the concrete. Large amounts of fly ash have been used with excellent results in some recently built dams.[20] Hungry Horse Dam, Montana, is a well-known structure in which fly ash was used. It also has been used in some recent TVA structures.

Along with significant advances in the strength, curing qualities,

[20] E. M. Fucik, "Benefits to Guayabo Project Justify Cost of Adding Fly Ash to Concrete," *Electrical World News,* 142 (Oct. 11, 1954) 30-31. Also A. S. Pearson and T. R. Galloway, "Fly Ash Improves Concrete and Lowers its Cost," *Civil Engineering,* September 1953, pp. 542-45.

and chemical resistance of the cements, improvements also have been made in the selection and use of aggregate in concrete. Experience with a great amount of heavy concrete construction and the results of extensive laboratory studies show that certain types of rock aggregates react chemically with the alkali in cement. Loss of strength, fracturing and lowered moisture resistance then may develop in the finished concrete. Reactive aggregates are now identifiable in advance, and their inadvertent use in the construction of dams and other structures can be avoided.

Another important and comparatively recent development is air-entrained concrete. With the aid of chemical additives many small air bubbles can be trapped in wet concrete during mixing. In addition to making the wet mixture more plastic and thereby reducing needs for excess water, these bubbles remain in the concrete and impart several desirable properties. Moderate density reduction is accompanied by increased strength per unit of cement and decreased moisture permeability. Air-entrained concrete is now finding increased use in large concrete structures of many types.

Indirect but important assistance in obtaining improved materials has been provided by the development of testing machines capable of exerting compressive and tensile loads of a million pounds or more. With these new machines, much larger samples of concrete and other materials can be tested and more reliable results can be secured.

Among recent advances in placing steel reinforcements in concrete, one of the most significant is prestressed concrete. If reinforcing steel is placed in tension prior to casting of concrete around the steel, considerably higher strengths can be achieved with less material. Prestressed concrete has received considerable use in the construction of bridges, beams for reinforced concrete buildings, and other structures, but thus far there have been only a few applications in dam construction. In France, for example, a multiple-arch reservoir wall has the edges of the arches tied together by a flat concrete exterior slab. During construction, the reservoir was partly filled in order to stress the exposed steel tie rods. Concrete was then cast around the stressed steel, permitting a 40 per cent reduction in the quantity of concrete required.[21]

In the United States a recent application of prestressed concrete to design is in a bridge section of Dexter Dam (in the Willamette River

---

[21] "Unfinished Reservoir Filled with Water to Stress Hollow Arch Wall Construction," *Engineering News Record,* 152 (April 1954), 61-62.

Basin), Oregon, which was precast on the ground and placed in position between piers supporting large regulating gates.[22]

DAM DESIGN AND CONSTRUCTION

Technological change in the design of large dams is comparatively slow, chiefly because so few are built. Experimentation with such large-scale units must be with a wide margin of safety, because of the risk of great financial losses and loss of life in the event of failure. A radical design for a new pump can be tested and discarded if found unsatisfactory, but a large dam must serve its life of fifty years or more economically and safely.

It is generally agreed that the features and requirements of each site for a large new dam are so nearly unique that there is no single type of structure that can be employed universally, or even widely. Each dam must be specifically designed for its site.[23]

The basic design still used on many of the large structures in the United States, the gravity dam, was developed thousands of years ago. In the main it is a solid barrier of trapezoidal cross section anchored in the foundation rock of the stream bed and adjacent valley walls. While some changes have been made in the last hundred years toward reducing width-height ratio,[24] the basic design of the gravity structure is that of the first ancient masonry dams. The use of reinforced concrete, however, has increased enormously the size of structure that can be constructed on this design. Most of the famous dams of the United States are of this type, including Grand Coulee in Washington, and other Columbia River dams; Shasta in California; many of the principal dams of the Tennessee Valley Authority system; and Hoover Dam, Arizona-Nevada. Reinforced concrete construction techniques, as practiced in the United States, have seemed especially well adapted to gravity dams.

The single arch, or horizontal arch dam is the second most commonly used design in American large-scale dam construction. Most successfully used in gorges, this type is employed only where there are sound

[22] "New Techniques in Dam Building, Dexter Dam," *Electrical World*, 142 (July 12, 1954), 88.

[23] "Fifth International Congress on Large Dams, Paris," *Engineer*, 199 (June 10, 1955), 812-13 and (June 17, 1955) 851-53.

[24] The most important design change of the gravity dam was made in the nineteenth century, when it was discovered that width-height ratios could be reduced from 3-4:1 to somewhat less than 1:1.

rock abutments to take the thrust of the arch. In this design each horizontal section of the dam acts as an arch spanning the space between valley walls. Where it can be used the great virtue of the horizontal arch dam is its economy of materials. Under the most favorable conditions, as little as one-seventh of the reinforced concrete necessary for a modern gravity dam 400 feet high would be required for the horizontal arch structure. Although originally scheduled for use in Hoover Dam, a curved gravity type finally was built. The horizontal arch design is represented in this country mainly by secondary projects, like the Shoshone River Dam in Wyoming. Dams of the multiple arch, buttress, hollow gravity, and standard gravity types are also being constructed in the United States where conditions are favorable for those structures.[25]

The greatest innovations in dam design presently are being made in western Europe, where remarkably thin concrete arch structures are being used; substantial savings in cost of concrete and steel structural materials are possible. The Gage Dam (on the Gage River, a tributary of the upper Loire, in France), for example, is 125 feet high, 472 feet long, and is only 4 feet thick at the top and 8 feet thick just above the base. Other new and unconventional designs include an arch structure overhanging the downstream power house (Couesque Dam, on the Truyere River, a tributary of the Lot and the Garonne, in France), for which a 20 per cent saving in concrete is claimed in comparison even with the economical arch dam.[26] Another new dam has a "ski jump" spillway with the power house beneath, and a comparatively lightweight dam is anchored to the river bed by means of cables passing through the entire mass of the structure and held in tension to secure the dam to its foundation.[27] The new Grand Dixence Dam in Switzerland also has been constructed on an unconventional and economical design.

New construction in the United States has been more conservative, being largely based on the conventional arch and gravity types. Technical development has been more in the direction of improving the quality of materials such as Portland cement, concrete aggregates,

[25] "Fifth International Congress on Large Dams," *op. cit.;* also A. E. Komendant, "Economy and Safety of Different Types of Concrete Dams," *Proceedings, American Society of Civil Engineers,* 81, Separate 684 (May 1955).

[26] W. G. Bowman, "Exciting is the Word for French Hydro," *Engineering News Record,* 154 (June 30, 1953), 34-37.

[27] A. Coyne, "Water Control Structures: Dams," in *Proceedings of the United Nations Scientific Conference on the Conservation and Utilization of Resources,* Vol. IV, *Water Resources* (New York: United Nations, 1951), pp. 2224-43.

water-proofing agents, and joint sealing materials. The design and use of better equipment, such as concrete mixers, vibrators, earth-moving machinery; and the wider use and improvement in techniques such as concrete cooling and grouting, have led to construction economies. Most of these developments are resulting in "holding the line" on construction costs while at the same time improving the safety and service life of the structures.

The hesitancy of American dam designers and planners to adopt some of the innovations being used abroad appears based on a more conservative view in the matter of absolute structural safety. Moreover, the division of costs between materials and labor is usually quite different in the United States than it is abroad. This difference encourages American emphasis on economy in labor and greater investment in materials. Higher labor costs are associated with thin arches, unusual spillways, and other design innovation, primarily because of complicated concrete form construction. Their use under American conditions might not be an over-all economy. Finally, most European dams are designed as well as constructed under a competitive bidding system, whereas a government agency or an operating private corporation usually designs the structures in the United States. Hence there is a particular incentive in Europe for developing new types of structures permitting major materials savings.[28]

DAM AND RESERVOIR SITE PREPARATION

Major improvements in large-scale clearance of reservoir and dam sites have been made by the various new types of earth-moving and other heavy clearing equipment, as suggested above. Other new techniques worthy of mention are improved methods for locating and correcting weaknesses in foundation rock for heavy concrete structures.

In instances where excessive faulting has been suspected in the only near-feasible sites for certain dam construction projects, expensive precautions in foundation preparation have had to be taken. In other instances, construction has proceeded in the absence of complete soil mantle and rock formation data, corrective measures then having to be taken after leakage or other defect appeared.[29] Thus the feasibility of

[28] L. G. Puls, personal communication, January 1957.

[29] An example was the water leakage through a faulted formation in the bottom of the Granby, Colorado, reservoir. Extensive cement grouting was required after project completion.

entire projects may depend on this factor. Recent developments in underground structure determination appear to be applicable in the preliminary exploration of river beds and other structural sites. Geophysical exploration, although not infallible, can provide useful information on the severity of faulting in the underlying rock.[30] Increased knowledge of micro structure of minerals and the greater reliability of compressive strength measurements on core samples, have supplemented geophysical exploration and helped to promote significant improvements and increased confidence in foundation preparation.

A corollary development is the more effective handling of foundation problems where they are found to exist. Improvements in equipment and techniques for grouting a formation by injection of cement mortar into the rock structure beneath and surrounding a dam site have been effective in providing substantially an impervious base and sides for the dam structure.[31] Another application of this technique is in the sealing of rock cracks and faults which otherwise would permit excessive seepage of water from a reservoir bottom.

SUMMARY COMMENT ON CONCRETE DAMS

New techniques of producing materials and constructing large-scale water storage facilities have had a major impact on the feasibility of storing water and regulating the use of streams in the last hundred years. Many of them have come into most prominent use within the last fifty years. The work of centuries under previous technology can be telescoped into a relatively few years by our capacity to produce concrete of exacting specification, and our ability to move it into place speedily in large volumes. The most influential American technical changes in this respect have been in materials rather than design, but our capacity to meet large-scale construction problems has helped greatly to keep pace with our physical and economic need for water. However, the more modest progress in employed structural design suggests further technological opportunities to lower the cost of water storage on the remaining undeveloped sites through design innovation.

[30] H. Price, personal communication, January 1957.
[31] Thompson, *op. cit.*

INTERRELATED OPERATION OF TWO OR MORE
UNITS OF SUPPLY IN AN INTEGRATED SYSTEM

Large-scale operation of water facilities may be achieved not only with the construction of huge individual structures, but also through the technique of merging operation of two or more units of supply into an integrated system. In fact, the scale economies in water development and management are perhaps most clearly demonstrated in system operation of facilities within a river basin. But such operation need not be limited to a single river basin, provided geographical and engineering factors otherwise are favorable.

The principle of integrated operation appears to have received general acceptance within the United States. Its advantages have been most readily demonstrated for the single purpose of flood control and power generation. However, the term has been used to describe individual cases which differ widely in degree and scope. For purposes of a brief summary of the status of integrated operation here, three distinctions are suggested: integrated operation for a single purpose, fully integrated operation, and co-ordinated operation of facilities.

Integrated operation for a single purpose would imply operation of a system of complementary facilities under single over-all management (but not necessarily single ownership) in such a manner as to maximize the combined net benefits from the operation of the individual facilities. This has been done on a number of small and middle-sized streams in several parts of the country, including the Deerfield River, Massachusetts (New England Power Company and others), and the Feather and McCloud rivers, California (Pacific Gas and Electric Company). Water control and electricity generating plans are managed in close conjunction. Flood prevention on an integrated basis is also found on similar (but separate) streams, as for the Miami Conservancy District in Ohio.

Fully integrated operation would imply integrated operation as defined above, but as applied to all the beneficial purposes for which water or water-derived services are available and might be used within the area of service. There probably is no fully integrated operation in this country if literal interpretation of the term is taken. However, the TVA system comes close to an expression of the idea in its operation, which, incidentally, includes some privately owned reservoirs. For this reason, and because discussion of the TVA water control system here forms a logical sequel to the case description of multiple-purpose operation given in chapter 8, the TVA example of integrated operation is selected for description below. Among privately organized efforts, the Wisconsin Valley Improvement Association reports integrated operation for power, water supply, waste carrying, and recreation—the most important services afforded by that river, and required within its service area.

Co-ordinated operation would imply the presence of two or more management agencies with either full autonomy or a high degree of independence in managing groups of facilities or individual works. Co-ordinated management usually is directed toward achieving certain over-all benefits which may not be assured by individual operations of the separately managed facilities or groups of facilities. The over-all plan may be jointly agreed upon, or it may be determined by a single agency. The operation of flood control facilities on the Mississippi-Ohio-Tennessee river system under the over-all supervision of the United States Army Corps of Engineers is of this type. Thus far, the operation of facilities on the Columbia River system may be described best by this term, even though they are multiple purpose facilities. The Pacific Northwest Power Pool is now a well co-ordinated operation of major importance in the management of the Columbia River and neighboring northwestern streams. Co-ordination of reservoir operations and sharing of generating reserves under conditions of diverse system peaks and diverse types of stream flow have resulted in marked capacity gains estimated at least at a half-million kilowatts. The major control works and generating plants are federally owned and operated on the main stem of the Columbia; the primary transmission grid also is federally operated (Bonneville Power Administration). On the other hand, the major service areas either are privately managed or are under the administration of local public bodies. These latter agencies also operate a number of secondary and a few major generating plants. Co-ordinated scheduling has been made possible by close co-operation among the several public and private groups.

As present development on the Missouri River nears completion, co-ordinated operation also is certain to be established for those works. The Colorado River Storage Project, on which construction is just commencing, is another where co-ordinated operation, and possibly integrated operation, will be undertaken.

For the most part, the technique of integrated water regulation on a full multiple-purpose basis is in an elementary stage on the American streams, where it exists at all. Important opportunities lie ahead, as in the great Mississippi Basin, within which the ultimate possibilities of integrated multiple-purpose operation may only have been touched thus far. Some interesting technical opportunities also undoubtedly lie ahead. One deterrent to fully integrated operation of a large river system has been the laborious detail of analyzing the complex interrelations which exist. Experiments have been made with the application of high-speed computers to these problems, and undoubtedly this will receive increasing attention in the immediate future. Analysis with high-speed computer aid, for example, is being employed on the current Corps of Engineers' review of the "308" plans for the Columbia River development. It also has been used to assist operational planning on the Missouri River system.[32] The use of machine analysis and formal methods of programming have recently added tools to the equipment of the water control planner which cannot fail to increase his scope and precision.

3. An Example of Integrated System Water Control

The TVA water control system in 1956 was composed of thirty major dams and their associated facilities lying within the Tennessee River Basin (figure 20). Twenty-five of these are owned by TVA, and five, owned by the Aluminum Company of America, also are operated as part of the system by agreement between that company and TVA.[33] The combined gross storage capacity of the reservoirs behind these thirty dams is about 22.6 million acre-feet, of which about 14.9 million acre-feet are useful controlled storage. The remainder of the reservoir capacity is dead storage. A 650-mile continuous slack-water navigation channel from Knoxville, Tennessee, to the river mouth is part of the system. Hydro generating plants of slightly more than 3 million kilowatts installed capacity are a part of the water control system. The hydro plants also are part of the TVA electric power system, which included an additional 6.5 million kilowatts of thermal capacity in 1956.

[32] Harvard University Graduate School of Public Administration, "Water Resources Operations Studies Using Computing Machine Methods," memorandum February 17, 1957, p. 2.

[33] Tennessee Valley Authority, *Annual Report* (Washington: U. S. Government Printing Office, 1956), p. 16.

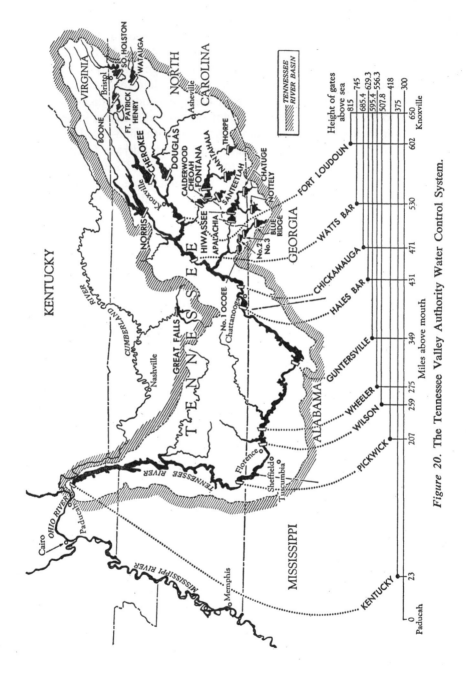

Figure 20. The Tennessee Valley Authority Water Control System.

The thirty reservoirs within the TVA system operate within the same physical environment and under the same statutory limitations described for Norris Reservoir (chapter 8, pages 187-94): that is, under an average annual rainfall of about 50 inches, with pronounced seasonal concentrations in runoff, over a basin of approximately 41,000 square miles; statutes direct operation for flood control and navigation, with incidental production of electric power.

Within these limitations water regulation is planned and undertaken in two phases: (1) seasonal or long-range control, and (2) day-to-day, or short-range regulation. Long-range control has been described by Blee as "that which is required to provide the proper reservation of storage space for flood control and to determine the general method of operating for power production to meet the forecast power load."[34] This definition, of course, applies to the Tennessee operations, and may apply to other purposes on other streams. Seasonal control obviously applies to irrigation, municipal and industrial water supply, navigation, recreation, and fishery releases where these purposes provide recognized benefits, no less than to power and flood control. Short-range operation, in the case of the Tennessee, is the continuous operational adjustment needed to meet the needs for routing heavy runoff through the system, or management of flow so as to meet the fluctuating power load. For other streams and other conditions, water supply, irrigation, navigation, recreation, etc., also may be included among the purposes to which the short-range as well as the seasonal operation applies.

Both long-range and short-range operations may be illustrated in the management of the system for flood control and for power production. The long-range operations have been described by Blee thus:

> . . . a chart or rule curve is prepared for each reservoir showing the limiting elevations above which the reservoir is not to be filled throughout the flood season [season shown in figure 21], except in controlling a flood. These limitations are established by the flood control requirements assigned to the particular reservoir as part of the designed requirements for the whole system . . . the objective is to regulate for substantial floods only. . . . The long-range control for hydroelectric power production is something of a complement to the flood-control operations, starting in where the latter leave off. As a guide to reservoir operations, a master guide curve is prepared each year, based on the predicted power load or range of loads for

[34] C. E. Blee, "Multiple-Purpose Reservoir Operation of Tennessee River System," *Civil Engineering*, 15 (1945), 220. Mr. Blee for many years was Chief Engineer of TVA. He retired in December 1956.

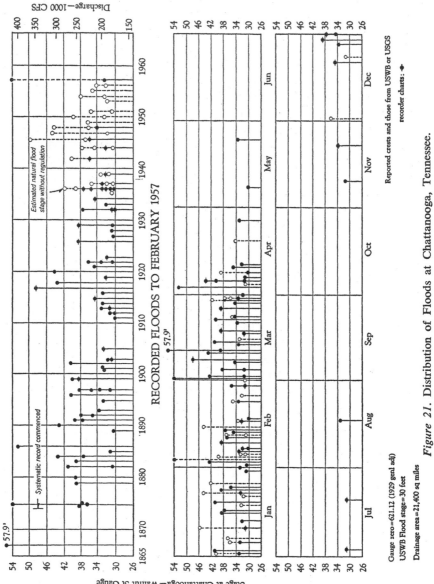

Figure 21. Distribution of Floods at Chattanooga, Tennessee.

the year. This curve is an envelope for a number of curves, each showing the amount of hydro energy which should have been in storage during certain critical dry years of record in order to carry the designated load. [figure 22] . . . In addition to the master guide curve, curves are prepared for each reservoir to indicate the desirable distribution of the storage among the reservoirs. The Tennessee River, with sites for storage dams toward the headwaters at relatively high elevations, is well adapted to an integrated operation whereby the power load is carried largely by the main river plants during the wet season, while the storage dams on the tributaries are shut down to permit filling. Then in the dry season the storage dams are drawn upon heavily, generating power at the plants located at each dam and utilizing the released water through all the downstream plants. Water from the Fontana Reservoir will ultimately be used through a total of 13 plants having a combined gross head of about 1,400 feet. The system had . . . to be planned and designed for integrated operation, with provision for the requisite installed capacity, with the necessary transmission capacity, and with the necessary transmission lines to accommodate this method of operation.[35]

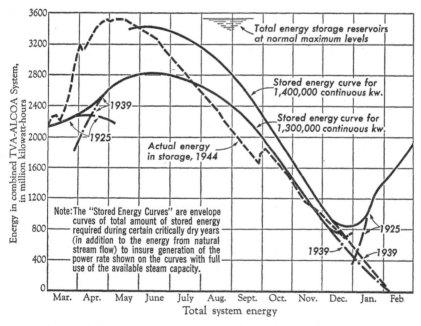

Figure 22. Multiple-purpose Reservoir Operation—TVA.

[35] *Ibid.*, pp. 220-22.

The day-to-day operations have been summarized by Bowden:

Handling the system of multiple purpose reservoirs in the Tennessee requires constant attention and at least daily scheduling of flows throughout the entire system for several days ahead. During floods, schedules must sometimes be revised 2 or 3 times daily to

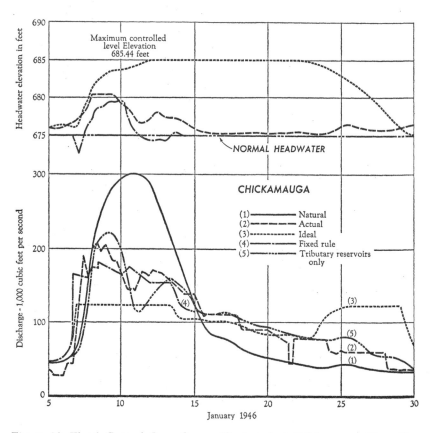

Figure 23. Flood Control Operation as Illustrated at Chickamauga Dam, Tennessee, 1946. (1) Natural—flow as it would have been in the absence of water control facilities. (2) Actual—water released through turbines or spilled on dates shown, taking account of fullest possible use of flood control storage space. (3) Ideal—theoretical regulated flow permitting maximum benefits from multiple-purpose use. (4) Fixed Rule—curve plotted in advance as general guide for water releases. Prediction of inflow is not required for fixed rule plotting. (5) Tributary Reservoirs—flow at Chickamauga with regulation achievable by facilities on tributaries upstream from Chickamauga, but not including Chickamauga reservoir.

take into account changes in rainfall and runoff as they occur and affect flows and stages, having due regard for weather predictions, which are available 3 times daily. This is accomplished through centralized control under the Chief Engineer and the issuance of instructions as often as necessary to the Department of Power Operations which is responsible for seeing that the required discharges are made at all of the dams. At the central office, hydrologic data are received, assembled, and interpreted; runoff is estimated; flows are routed; and operations are decided upon by engineers thoroughly familiar with the characteristics of the system. Frequent reports received by teletype and telephone from the Operating Department at Chattanooga furnish checks on operations at the dams and their effect on the reservoirs. This centralized control, where decisions in time of flood must be made and put into effect promptly if success is to be had, takes cognizance of the fact that no two storms or the floods they produce are alike and that each situation requires detailed treatment all its own, based, of course, on fundamental principles and procedures governing system operations.[36] [See figure 23.]

At the time when these descriptions of integrated operation were written, the principal objectives of integrated operation were for flood control and power production. The power production undertaken at that time was for the purpose of meeting the base load of the system, supplemental power being provided either thermally or by importation from other interconnected systems. Previous to 1944, when the reservoir behind Kentucky Dam commenced to fill, navigation also was a specific purpose in water control. Blee reported: "During the early stages of the development and, in fact, until the completion of Kentucky Dam, the operating requirements under conditions of open-channel navigation were quite exacting." [37] For the completed water control system, however, Blee has said that "very little in the way of special operating procedure at the reservoirs is required for the benefit of navigation." [38] While less demanding than formerly, navigation requirements still must be taken into account in water control planning for the Tennessee. The TVA Division of Water Control Planning states: ". . . there are no physical facilities that automatically ensure the maintenance of minimum pools for navigation. The pool levels must be maintained by careful estimating of inflows and regulation of discharges at all main river plants. Although the water supply needed to maintain navigation pools on the

---

[36] N. W. Bowden, *General Design and Operation of Multiple Purpose Reservoirs* (Knoxville: Tennessee Valley Authority, 1946), p. 17.

[37] Blee, *op. cit.,* p. 222.

[38] *Idem.*

Tennessee River is not great, the day-to-day operations for flood control and power continually reflect the navigation requirements." [39]

The benefits from integrated operation on the Tennessee system, as distinguished from planning and construction for integrated development, thus are traceable mainly in the direction of flood control and electric power production.

No detailed studies of the benefits of integrated operation, when separated from multiple-purpose operation, have been made for the Tennessee water control system, nor for any other co-ordinated multiple-purpose operation of river regulation facilities within a large area in the United States.[40] Studies of benefits in the past have merged the two techniques, which in fact have been closely allied in the TVA development and operations.

The benefits from flood control are clearly related to integrated operation, since it is unthinkable that an effective job of routing a major flood safely through the Tennessee Basin, and timely water detention for Ohio Valley and Mississippi Valley protection, could be accomplished without careful co-ordination of releases. The value of the system in providing flood protection was vividly demonstrated during the flood of January-February 1957. Without regulation this flood would have been the second largest at Chattanooga during ninety years of record, being surpassed only by the flood of 1867. It was the result of twenty days of rainfall within twenty-one consecutive days. During this period 12 inches of precipitation fell within the entire watershed above Chattanooga, and nearly as much elsewhere in the Valley. The TVA Division of Water Control Planning reports that: "The unregulated crest on February 3, 1957, would have been 54 feet at Chattanooga. The actual, regulated crest occurring on February 2 was 32.24 feet, about 2 feet

[39] TVA Division of Water Control Planning, memorandum to the authors, 1958.

[40] Studies of benefits from integrated operation of *single-purpose* facilities, mainly for hydroelectric power production, have been made in detail in this country and abroad, particularly since 1945. A system such as that of the Southern California Edison Company in the upper San Joaquin Valley, California, has a well-integrated system for power production, based on careful study of prospective benefits from several alternative patterns of operation for available facilities. (Blair Bower, note to the authors, July 1956.)

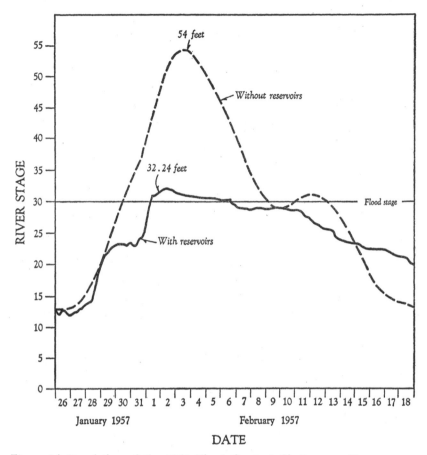

Figure 24. Regulation of the 1957 Flood Crest at Chattanooga, Tennessee.

above flood stage (figure 24). The unregulated flood would have caused damages estimated at \$66,000,000, and some 12,000 homes would have been flooded." [41] Less heavy, but still large damages were also averted by flood control operation of TVA reservoirs during 1946, 1947, and 1948 for floods which would have been the sixth, seventh, and eighth highest of record without the TVA water control system.

The total estimated damages prevented at Chattanooga by flood control operation in these three floods and in the 1957 flood were about

[41] TVA Division of Water Control Planning, memorandum to the authors, January 1958. See also TVA, *Annual Report,* 1957, pp. 7-20.

$103,000,000.[42] The Division of Water Control Planning reports further; "Total damages averted by TVA operations through 1957 at Chattanooga amount now to nearly $120,000,000 or 65 per cent of the entire system cost allocated to flood control. In addition, damages averted in the lower Ohio and Mississippi basins during this and other TVA flood control operations during fiscal year 1957 amounted to nearly $4,900,000. The total estimated damage averted at Chattanooga and in the lower Ohio and Mississippi basins, excluding substantial but not specifically estimated amounts for many other points in the Valley, thus amount to about $132,000,000 at the end of the 1957 fiscal year. This amount is equal to about 70 per cent of the total of $184,000,000 invested in the flood control features of the TVA system." [43]

On the Mississippi River, some twenty-four floods between 1936 and 1956, which would have exceeded a flood stage of 40 feet at Cairo, have been lowered by amounts varying from a few tenths of a foot to 2 feet.[44] It has been estimated that the present TVA system can reduce the most damaging floods on the lower Ohio and Mississippi rivers by as much as 3 feet, depending upon the type and size of the flood in the Tennessee River at the time, and the relation of flood peaks throughout the Ohio or Mississippi basins.

Some of the benefits attributed to the system presumably could be provided by a system of detention reservoirs and levees, but it undoubtedly would be at much greater cost than the present system. Cost estimates of a set of flood control facilities for unco-ordinated management have not been made, but they would be likely to exceed considerably those of the single-purpose flood control system which was designed for purpose of cost allocation estimates. This was a co-ordinated system of reservoirs costing about $261 million, as compared with the $182.4 million of direct and common costs attributable to flood control in the system existing in 1953 (now $184 million).[45] Furthermore, no dependable benefits on the larger rivers of which the Tennessee is tributary could be foreseen from an unco-ordinated system of detention basins

[42] *Idem;* and Edward J. Rutter, "Flood-Control Operation of Tennessee Valley Authority Reservoirs," *Proceedings, American Society of Civil Engineers,* 76, Separate No. 19 (New York, 1950), 32.

[43] TVA Division of Water Control Planning, *op. cit.*

[44] G. O. Wessenauer, "Success of the TVA Multipurpose River Development," *Civil Engineering,* 26 (1956) 439.

[45] Tennessee Valley Authority, *Report on Allocation of Costs as of June 30, 1953,* Knoxville, pp. 18, 23.

and levees. The reduction of flood crests on the Ohio and Mississippi thus might be taken as the minimum flood control benefits directly attributable to integrated operation of the Tennessee River system of reservoirs.

The benefits from integrated operation are most clearly seen in the effects upon hydro power production. An interesting basis for comparison was afforded on the Tennessee in two dams which had been built on the main stem of the river before the establishment of TVA. One had been constructed between 1905 and 1913 by the Tennessee Electric Power Company in Marion County, Tennessee, about thirty-nine miles downstream from Chattanooga. This is the Hales Bar Dam. The other had been built by the United States War Department at Muscle Shoals, Alabama, between 1918 and 1924. This is the present Wilson Dam. The Hales Bar Dam had an installed generating capacity with a rating of 51,100 kilowatts at the time it was purchased by TVA in 1939; the generating capacity of the Wilson Dam plant at the time of the establishment of TVA was 184,000 kilowatts.[46] After the construction of key upstream storage works like Norris, Douglas, Cherokee, Hiwassee, and Fontana dams, and the institution of integrated water regulation on the Tennessee, it became possible nearly to double the capacity of the Hales Bar plant and more than to double the Wilson generating capacity. The generating capacity of Wilson Dam since 1950 has been 436,000 kilowatts, and that of Hales Bar since 1952 has been 99,700 kilowatts.

The difference between regulated flow and integrated operation, on the one hand, and single plant regulation, on the other, is well illustrated in these figures, inasmuch as the original plants were not underbuilt as *single* facilities.

Any extension to the entire TVA main stem system of the capacity increases made possible in the two plants by integrated operation is, of course, purely theoretical. It is quite unlikely that a completely uncoordinated system of nine plants[47] would exist on the same river together, even under separate ownership of each. Furthermore, even uncoordinated operation would yield certain random, or windfall, regulatory

[46] Provision for an *ultimate* capacity of 450,000 kilowatts was made in Wilson Dam, since system regulation was envisioned as possible when the dam was built.

[47] At Kentucky, Pickwick Landing, Wilson, Wheeler, Guntersville, Hales Bar, Chickamauga, Watts Bar, Fort Loudon dams between the mouth of the river and Knoxville, Tennessee.

benefits which are difficult to determine. For the purpose of illustrating the benefits of integrated operation, however, it may be noted that present main stem generating capacity of the Tennessee is 1,654,100 kilowatts. A not unreasonable assumption is that the seven plants other than Hales Bar and Wilson, theoretically unco-ordinated, would warrant installations at 60 per cent of the present capacity. Nine unco-ordinated plants thus would have about 750,000 kilowatts less capacity than the present system, if the assumption is accepted. A final pair of assumptions is that the unco-ordinated plants might produce at a 75 per cent load factor, and net about 2 mills per kilowatt hour over costs. Under these assumptions the annual net additional income of the power system attributable to integrated water control operation would be on the order of $10 million. These estimates, of course, would vary according to the manner in which the added generating capacity is used. If instead of high load factor power, the generation were for peaking purposes or simply for a lower load factor, the estimate of net gain would differ.

CHANGES IN THE STATUS OF WATER CONTROL
WITHIN THE TVA SYSTEM

In recent years the position of hydro generation in the TVA electric power system has changed significantly. Since 1953 the base load of the system most of the time has been carried by thermal generation, while the hydro generators have become more and more valuable for meeting peak or emergency demands. Actually the operation of the system and individual plants within it must change as water supply and runoff conditions change. During high flow periods, as during the 1957 flood previously described, most of the main river plants are heavily loaded and take a position in the base of the system load. At other times, and especially during periods of low flow, the capacity of the hydro plants is used most effectively in the peak of the system load.[48] The shift away from base load position of the hydro thus is that which is compatible with the operation of a multiple-purpose water control system. Under these conditions it is obvious that the hydro will never be turned completely to peaking purposes, although it has become more important for this use within the system.

This pattern of operation has been encouraged by the lowering costs of generation in the large new thermal electric generating units, by

[48] TVA Division of Water Control Planning, *op. cit.*

Plate 14. Hiwassee Dam of the Tennessee Valley Authority, Hiwassee River, North Carolina. Hiwassee is the site of a large reversible pump-turbine installation.

previous development of all important hydro sites, and by the greater flexibility in operation of hydro as compared to the very large steam installations. Any technical change which tends to intensify load peaks may also favor the further development of higher cost hydro to meet those peaks.

The changed position of hydro may be reflected in three ways: (1) An altered pattern of operation for existing hydro installations, as has taken place in the TVA hydro plants, operating more frequently for capacity than formerly. (2) Construction of additional hydro units. Installation of larger hydro capacities at a given site may be warranted when operation can be for capacity, by comparison with the previous base load function of operating hydro stations. Sites considered hitherto uneconomic also may be valuable for development. (3) The development of new methods of managing water flows. Such methods are suggested in past experiments with temporary reversal of stream flow so as to improve capacities at time of greatest system demand. This has been done by using off-peak electricity for pumping water into upstream storage for later use.

The practice of using steam- or hydro-generated electricity during off-peak periods for pumping into upstream storage is not new in the United States, having been employed at the Rocky River (Lake Candlewood) development in western Connecticut for several decades. Water there is lifted from the Housatonic River during each spring to fill a reservoir on a tributary, from which it is released during the summer and autumn months.[49] The Buchanan project in Texas is a similar development, and a number of installations in Europe have made use of pumped storage.[50] However, this practice has received much increased interest since the development of the reverse turbine, like, for example, the TVA Hiwassee Dam installation, which was completed in 1956. This is the largest reversible pump turbine in the United States.[51] The Hiwassee pump turbine has been described by TVA:

[49] R. K. Linsley, M. A. Kohler, and J. L. H. Paulhus, *Applied Hydrology* (New York: McGraw-Hill, 1949), pp. 610-11.

[50] F. L. Weaver, "Prospect for Future Hydroelectric Development in the United States and What It is Likely to Cost," paper presented at Fifth Atomic Energy in Industry Conference, Philadelphia, March 14, 1957, p. 6.

[51] Another of prior construction is the Carter Lake installation on the Bureau of Reclamation Colorado-Big-Thompson project, Colorado. (See W. E. Blomgren, "Water Supply Procedures for Water Resource Projects in the United States," in International Commission on Irrigation and Drainage, *Annual Bulletin*, 1954, pp. 4-10.)

> Operated as a generator, the unit has a capacity of 59,500 kilo-watts. . . . It is operated as a conventional generator. . . . In reverse operation, however, the generator operates as a motor of 102,000 horsepower, turning the turbine which then acts as a pump. As a pump, the unit takes water from the reservoir of Appalachia Dam . . . and pumps it back up into Hiwassee Reservoir. In a test operation, the unit pumped 4,000 cubic feet per second against a head of 238 feet. . . . This operation helps to maintain the head of water for power generation at the project and provides additional water to be run through the Hiwassee turbines.[52]

The turbine is used for generating electricity only during periods of peak load. When capacity within the system sufficiently exceeds the load, surplus electricity is transmitted to the generator connected to the reversible turbine, and the turbine is operated as a pump.

The changing load and generating characteristics of electricity systems suggest that contrary to some popular impressions, this flow use of water may not yet have reached its greatest importance. Hydro generation still appears to be a water development purpose which must be considered in planning and constructing facilities for water use, and in their later management.

The recent changes in operation within the TVA system further suggest that integrated operation, while carefully designed for TVA water control, may not yet have reached its final stages or most detailed application. The successful operation of the Hiwassee reversible turbine indicates that the load growth of the region presents still further opportunities for scale economies with integrated water control.

THE INTERRELATION OF SYSTEMS OF SUPPLY
WITH SYSTEMS OF DEMAND
ON A MUTUALLY APPROPRIATE SCALE

The organization of a group of facilities into an integrated system of supply which takes advantage of, or makes possible, the scale economies in development and operation must be balanced by a distribution system for water or water-derived services which matches the scale of the supply system. It is of no practical use to have 2 million acre-feet of cheap conservation storage on a given site unless the water can be

[52] TVA, *Annual Report,* 1956, p. 11.

transported to the sites of current effective or potential demand for irrigation water, industrial process water, or municipal service. The creation of a hydro plant of a half million or more kilowatts installed capacity likewise is of little use unless the generated electricity can be delivered economically to the sites of effective demand or potential demand. The scale of distribution system organization for water and water-derived services therefore is of equal significance with the techniques that have promoted large-scale supply units, or systems of supply.

Two illustrations of techniques promoting large-scale distribution are given. One describes the two most important sets of techniques promoting the large-scale physical movement of water itself: specifically, canal construction and tunneling methods. The second is a description of electricity transmission over long distances.

These illustrations deal with what are essentially techniques of bringing supply within the reach of demand by transporting the water itself, or its products or services. Another approach deserves brief mention, although it has had limited application in the development of water resources in this country. This is the dispersal of large-scale demands in such a manner that the use of dispersed supplies can be promoted.

4. Developments in Canal Construction and Tunneling

Recent progress in tunnel and canal construction has exerted a profound influence upon the development and use of water resources. It has facilitated further additions to the irrigated acreage in the western states, and also municipal and industrial development, in two ways: first, by substantially reducing costs in many instances; and second, by greatly increasing the efficiency of facilities provided.

CANAL CONSTRUCTION

Irrigation canals must of necessity be constructed over a great variety of different terrain and the resulting engineering problems are extremely diverse. However, special-purpose earth-moving and construction equipment has been developed for virtually every type of condition encountered, with the result that the construction engineer today has at his command hydraulic buckets, clam-shovels and draglines, carryalls and loaders, rollers and conveyors, and other machines to meet the specifications of each particular assignment.[53] The increasing availability of more powerful and more versatile traction has also been important.

While these tools are common to many fields of construction, other types of equipment such as the canal stackers used in large construction projects, and plow ditchers employed in making smaller laterals and diversions, have been designed especially for canal construction. Notable progress was made in the early 1930's in the Boulder Canyon Project, when heavy mechanized equipment was devised to facilitate the construction of the All-American Canal to deliver water from the Colorado River to the cities of southern California. It was at this time that canal trimmers and canal framers were first employed.[54] Built around a large steel framework mounted on temporary rails on each bank of the canal, these machines provide the final shape or profile to the canal excavation, and line the bottom and sides with a layer of concrete in a continuous operation. As the work progresses, they propel themselves forward, being supplied with cement, aggregate, and water by auxiliary delivery equipment. In the construction of the 122-mile concrete-lined section of the Friant-Kern Canal in California's Central Valley, a huge canal trimmer was used to mix concrete, lay it over the surface of the canal excavation, and vibrate and smooth it to a finished surface.[55] Such trimmers have since become a standard part of large canal construction projects and have been equipped to handle a progressively greater volume of material.

Significant technical progress has been made in the last two decades in reducing the seepage losses involved in conveying water through canals. According to one recent estimate, nearly 35 million acre-feet, or about 40 per cent of the water diverted for irrigation in the seventeen western states, is lost before it reaches the farms where it is used.[56] In an intensive investigation conducted by the Bureau of Reclamation in the irrigated areas of Utah, Washington, and Wyoming, more than three-fourths of the conveyance losses, which constituted a slightly greater

---

[53] See pp. 221-27, "The Development of Earth-moving Equipment."

[54] Nadeau, *op. cit.*, p. 222.

[55] O. W. Israelson, *Irrigation Principles and Practices* (2nd ed.; New York: John Wiley, 1950), p. 73.

[56] A. R. Robinson and Carl Rohwer, "Measurement of Canal Seepage," *Proceedings, American Society of Civil Engineers,* 81, Separate No. 728 (New York, 1955), 1.

share of the total water entering canals, was attributed to seepage.[57]

By providing canals with relatively impermeable linings or otherwise treating the canal bottom and sides, these losses can be reduced greatly and the efficiency of the project accordingly increased. Additional benefits from the use of linings are found in the prevention of water-logging and other damage by seepage to low lands adjacent to the canals, reduction in maintenance costs (especially through elimination of weeds, retardation of moss growth, and curtailment of erosion), a greater safety margin against breaks due to rodent activity and other causes, and improvement in the general hydraulic properties of the structures.[58]

Canals can advantageously be lined or otherwise treated to make them less permeable to the point at which the value of water saved and other benefits equal or exceed the costs of treatment. However, the technical solution is complex, and at present there is no one best technique of reducing seepage. A number of practices exist, each of which is applicable under suitable conditions.

Of the many materials available for lining canals, concrete has been one of the most important. Where the danger of frost damage is great and high tensile strength is required, the concrete has usually been provided with steel mesh reinforcement. In the last two decades, however, a two- to four-inch lining of plain concrete, involving substantially less cost, has proved acceptable in many installations,[59] and precast concrete slabs with tongue and groove joints have also been effectively employed.[60] The installation of concrete linings in smaller diversion canals and laterals was facilitated prior to World War II by the development of the mechanical concrete laying machines described on page 252.[61]

Considerable seepage loss is experienced in concrete lining installations if the expansion joints which are left between the slabs are not

[57] Alfred R. Golzé, *Reclamation in the United States* (New York: McGraw-Hill, 1952), p. 408. For conveyance losses of specific canals in the West, consult Ivan E. Houk, *Irrigation Engineering* (New York: John Wiley, 1951), Vol. 1, pp. 388-91.

[58] U. S. Department of the Interior, Bureau of Reclamation, *Canal Linings and Methods of Reducing Costs,* Washington, 1952, pp. 3-5; Israelson, *op. cit.,* p. 72; Golzé, *op. cit.,* p. 406.

[59] U. S. Department of the Interior, Bureau of Reclamation, *ibid.,* p. 5.

[60] Israelson, *op. cit.*

[61] Allen S. Parks, "Lining Irrigation Canals," *Compressed Air Magazine,* 57 (1952), 279.

adequately waterproofed. Synthetic mastics and metal and rubber expansion strips have been developed which ensure watertight joints, and the adherence and flexibility of various asphaltic compounds used for this purpose have been greatly improved.

Linings of various earth materials involve far less construction expense than those consisting of concrete, and under proper conditions they are very effective. For many decades Bentonite, a talc-like clay mineral which expands from sixteen to thirty-five times its original volume when wet, has been used for lining canals. This material can be applied by mixing a small amount with the top layer of earth in the canal bottom or by spraying or injecting dilute Bentonite suspensions under pressure.[62] A considerable amount of experimental work with other comparable techniques has been carried on in recent years. For example, in central Nebraska it has been found that two-thirds of the seepage loss through porous soils can be eliminated by consolidating them with an injected silt slurry.[63]

One of the simplest ways of making canals less permeable, if the proper type of soil is present, is by heavy compaction after the excavation has been completed. Under some conditions, heavily weighted sheeps-foot rollers are able to establish a compacted layer of earth 6 to 8 feet thick in the canal bottom and sides. In some instances this has reduced seepage losses in amounts comparable to those obtained by the use of concrete, at one-half to one-sixth the cost.[64]

Plasticized asphaltic materials, in addition to being employed for sealing joints and in repairing leaks, are used in prefabricated membrane linings. The latter, averaging one-half inch in thickness and coming in sheets sometimes 100 feet in length, are laid over the canal bottom; a highly impervious lining is obtained after the joints are sealed. To provide protection against abrasion and other mechanical damage, the lining may be covered with a layer of sand or gravel. More recently, less expensive asphalt compositions have been developed which may be pneumatically sprayed onto the soil surface of the canal.

[62] "Control of Canal Seepage with Bentonite Improved," *Engineering News Record,* 152 (April 1, 1954), 54-55.

[63] G. E. Johnson, "Stabilization of Soil by Silt Injection Method," *Proceedings, American Society of Civil Engineers 1953,* 79, Separate No. 323 (New York, 1953), 18.

[64] T. V. D. Woodford, "Lower Cost Irrigation Canal Linings," *Engineering News Record,* 151 (November 5, 1953), 35-40; and J. J. Waddell, "Low-cost Compacted Earth Lining Shows Low Seepage Losses in Friant-Kern Canal," *Civil Engineering,* 23 (1953), 684-87.

Plate 15. Front view of a lining machine used to install concrete lining in a main canal; U. S. Bureau of Reclamation Columbia Basin project, near Stratford, Washington.

A typical composition contains 12 per cent emulsified asphalt and 3 to 5 per cent Portland cement mixed in sand and water. After it solidifies, it is usually covered with a layer of sand or gravel.[65] The cost is one-third to one-fifth less than that of the prefabricated asphalt membrane. Pneumatically applied mortar, sprayed asphalt, poured asphalt and tar, and other materials have also been used. Because little equipment is required, this technique is very mobile and can be economically employed where obstructions are present. It is also advantageous in that the amount of material applied can readily be altered. A more recent innovation involves use of polyethylene films. Such plastic linings, covered with a protective gravel fill, have completely eliminated seepage losses in a Utah test, and appear to be economically practical.[66] Various chemical earth stabilizers and chrome-lignin binders appear promising in the experimental stage.

For many of the older projects where seepage losses are high, these new canal lining techniques are among the most promising means of increasing the usable water supply. In 1940 there were 125,000 miles of irrigation canals and laterals in the western states, of which less than 5,000 miles were lined.[67] By 1957 the proportion of lined structures had probably grown several times, although precise figures are not known. All of the major canals delivering municipal water are lined, and more than 600 miles of irrigation canals were lined by the Bureau of Reclamation between 1947 and 1952.[68] In addition, the United States Department of Agriculture through 1954 had aided individual farmers in lining substantially more than 1,000 miles of privately owned laterals and diversion ditches.[69]

Today all new canals are lined or treated in some way, and an increasing number of those already in existence are being modified in order to capture a part of the water now lost in transit. Costs have varied from less than $0.50 per square yard of treated surface to more than $2.00,[70] depending on the materials used and conditions encoun-

[65] "Asphalt Gunite is Used to Line Canals," *Engineering News Record*, 157 (September 13, 1956), pp. 35-38.
[66] "Plastic Films in Farm Applications," *Agricultural Engineering*, 37 (1956), 743.
[67] Golzé, *op. cit.*, p. 406.
[68] U. S. Department of the Interior, Bureau of Reclamation, *op. cit.*, p. 2.
[69] U. S. Department of Agriculture, Agricultural Conservation Program, *Summary for 1954*, Washington, 1955.
[70] T. V. D. Woodford, *op. cit.*, p. 35; J. J. Waddell, *op. cit.*, p. 687.

tered. With the exception of canals lined with asphaltic sheets and plastic films, seepage cannot be entirely eliminated. Even a good concrete lining may show losses as high as 7 cubic feet of water per day for every 100 square feet of canal bottom. However, conveyance losses have been substantially reduced, and it appears that further savings of water can be effected as new materials and techniques become available.

TUNNEL CONSTRUCTION

The practice of tunneling has a number of important applications in water resource development. Several of the oldest tunnels constructed in America were driven for canal routes.[71] Today, however, tunneling has become especially useful in transporting municipal water supplies, in harnessing the hydroelectric potential of waterways, and in diverting water for irrigation.

Innovations in technology have greatly increased the feasibility of such projects over the last few decades. One of the earliest water tunnels in the United States was a 3,000-foot bore made during the 1880's to deliver water for irrigation to the San Bernardino Valley in southern California. The city of Los Angeles constructed a much larger tunnel in 1905, when it drilled five miles through the Tehachapi Mountains to carry water from the Owens River to the Los Angeles area.[72] Although pneumatically powered percussion drills had been introduced in the 1870's, the removal of muck, or the accumulated debris following blasting, was essentially unmechanized. Machinery for rapid underground loading and hauling of broken rock was developed and improved in the first few decades of this century. One of its first major uses was in the construction of the two six-mile tunnels which were built by an improvement district in central Colorado. After the "pioneer bore" had served as access to construction in the larger main bore which became the Moffat railroad tunnel, it was completed as a water tunnel to tap the headwaters of the Colorado River and divert its flow under the Continental Divide. The entire project was completed in 1928.

Important innovations in rock tunneling practice were made in the

[71] Archibald Black, *The Story of Tunnels* (New York: Whittlesey House, 1937), pp. 18-20. In addition to the canal tunnels referred to in this source, mention can be made of the tunnel constructed in the early part of the nineteenth century on the Chesapeake and Ohio Canal.

[72] Nadeau, *op. cit.*, pp. 8-9, 47.

Plate 16. Tunnel construction: a "jumbo" platform shown in position for tunneling operations, Trinity diversion project, California.

Plate 17. Tunnel construction: boring machine used in preparation for blasting operations; U. S. Army Corps of Engineers Fort Randall Dam, South Dakota.

construction of Hoover Dam. Started in 1931, two giant diversion tunnels were driven, each possessing a diameter of 50 feet after a concrete lining had been applied. To speed the drilling operation, use was made of a large framework of platforms mounted on a track. As many as thirty drillers could simultaneously attack the tunnel face from this frame. After the face was drilled, the platform was withdrawn to permit blasting and muck removal. These platforms, soon known as "jumbos," with mechanization and other improvements have been extensively employed in all large tunneling operations.

The work capacity of pneumatically driven drills, which have been used for several decades, has steadily increased. The tungsten carbide steel bits which are inserted in them have been greatly improved. For example, small openings now are included in the bit, through which compressed air, or air and water, may be blown. In this way the rock grindings are carried away from the cutting surface of the bits. The improvement of these tools also has increased control over blasting effects. Because of the accuracy the improved tools make possible, the firing pattern can be engineered with far greater certainty. Over-breakage, the material removed by blasting in excess of the "payline" (the intended limit to be cut), consequently has been minimized, and substantial cost reductions have been achieved.

The present capacity for tunnel boring rests in large part upon systems of muck removal which can expeditiously handle large volumes of material. The introduction of electric and air-driven "muckers" has improved working conditions because the air-pollution hazard is minimized in comparison with the employment of internal combustion engines.

A recurrent problem in tunnel construction projects has involved handling the large volumes of water that are often encountered. High-capacity pumping equipment to remove the water, and grouting material to seal off porous formations have reduced the severity of this problem.

The effect of these technical innovations can be seen in several different types of projects in which tunneling has been extensively used. When completed, the 44-mile West Delaware water tunnel, from the Delaware River to New York City, will move about one-fifth of the future supply of this large municipal water system. The cycle of drilling, blasting, and muck removal in boring this 11-foot tunnel was progressing in 1957 at an average rate of more than 50 feet a day.[73] A significant innovation in lining the earlier 25-mile East

[73] "44-mile Water Tunnel Hits Its First Snag," *Engineering News Record,* 157 (September 27, 1956), 30-32.

Delaware tunnel was the use of a 1,000-foot-long car or bridge, which was hoisted through the tunnel as work progressed, to make possible continuous roof lining with quick-setting concrete at a speed of almost one-fifth of a mile per day.[74]

In numerous hydroelectric installations, penstocks and tailraces have been built by tunneling. One of the most spectacular was the construction of twin tunnels 5.5 miles long and 51 feet in diameter for a new hydroelectric plant at Niagara.[75] In some instances, such as the nearly 10-acre excavation for the Kemano power plant inside Mount DuBose in British Columbia and the Chute-des-Passes plant in Quebec (in construction in 1958), entire power stations have been built underground.

A tunnel more than 13 miles long was constructed as a part of the Colorado-Big Thompson reclamation project in northern Colorado. It is the longest tunnel ever driven to carry water for irrigation, and makes possible the diversion of runoff collected on the Pacific slope of the Continental Divide through the mountains to the Atlantic side. The completed lined tunnel has a diameter of almost 10 feet, possesses a carrying capacity of 850 acre-feet a day (192,500 gallons per minute), and supplies irrigation water to an area of 610,000 acres.[76] More than 300,000 acre-feet were delivered from the western slope to the eastern slope through this tunnel during the 1954 crop year.[77]

The progress already made in transporting water through large and long tunnels is dwarfed by the plans for some projected developments. In the proposed Pennsylvania mine drainage project, a 300-mile tunnel is contemplated. Drainage from hundreds of mines would flow at 600,000 gallons per minute to the Susquehanna River.[78] Of equally

[74] "East Delaware Tunnel: Gallery of Up-to-date Tunneling Techniques," *Engineering News Record,* 150 (March 26, 1953), 32-36.

[75] For details of the tunneling operation involved in this project, see J. R. Glaeser, "Ontario Hydro Blasts Out Niagara Power Tunnels of 51-ft. Diameter," *Civil Engineering* 23 (1953), 676-80.

[76] For a short description of the tunneling operation involved in this project, see L. A. Luther, "Big Thompson Project Nearing Completion," *Compressed Air Magazine,* 57 (1952), 339-43.

[77] W. E. Blomgren, "Water Supply Procedures for Water-Resource Projects in the United States," *International Commission on Irrigation and Drainage, Annual Bulletin, 1954,* pp. 4-10.

[78] Willard W. Hodge, "Waste Disposal Problem in the Coal Mining Industry," *Industrial Wastes: Their Disposal and Treatment,* Willem Rudolfs, ed. (New York: Reinhold, 1953), pp. 312-418. Also, S. H. Ash, *et al., Flood Prevention in Anthracite Mines, Northern Field Anthracite Region of Pennsylvania Project*

Plate 18. "Mechanical mole" used for tunnel boring in soft rock formations; Oahe project, Missouri River, South Dakota.

large dimension are proposals to divert water from the Colorado River at Marble Canyon through a 143-mile tunnel to Granite Reef Dam in the Phoenix area. Another proposed project is the diversion of some 1,200,000 acre-feet of water annually from the Colorado River at Bridge Canyon into central Arizona through a 77-mile tunnel. The scope of these plans indicates the breadth of planning and interbasin diversion of water made possible by the tunneling techniques that now are standard practice.

5. Electricity Transmission[79]

The transportation of electricity over distance has been a problem particularly associated with hydroelectric development. Good hydro sites, even of modest natural dimension, are determined by topographic, geologic, and hydrologic conditions. More often than not they tend to be physically separated from the localities, or even the regions, where people are found in greatest numbers, where the optimum industrial sites are to be found, and where both the potential and effective demand for electricity is greatest. The problems of overcoming distance perhaps are most pronounced in connection with the best use of the sites of largest potential—those capable of development to a half-million kilowatts or more. For these sites it is obvious that the greater the distance that electricity can be transported (or transmitted), the more effective electrical service from the site can be. The increasing reliance of the United States on steam-electric power, although reducing requirements for very large flows of water, does not free power plants from riparian locations.

A second important reason for increasing effective transmission distance of electricity is the economic desirability of interconnecting load centers (foci of electricity demand) with each other, and with a group of generating plants on different sites. Different industries vary in their

No. 1 (Bulletin No. 545, 1954); *Flood Prevention in Anthracite Mines, Western Middle and Southern Fields, Anthracite Region of Pennsylvania Project No. 2* (Bulletin No. 546, 1955); *Flood Prevention in Anthracite Mines, Northern Field Anthracite Region of Pennsylvania* (Bulletin No. 547, 1955), U. S. Department of the Interior, Bureau of Mines, Washington.

[79] For a detailed map showing the principal electric utility service areas in the United States, see map in pocket at the end of this volume.

load characteristics,[80] industrial load may differ from municipal and rural load, municipality may differ from rural area, and municipalities from each other. The characteristics of generating plants also may differ from each other, although hydro plants within a given region tend to have less variance than do load centers. However, because of the seasonality of water supply in most parts of the country, considerable fluctuation in seasonal (or longer term) generating capability may occur. Therefore, the interconnection of a number of generating sites with each other, and with thermal power plants (which do not have seasonal limitations) will tend to produce a system with prime (continuously dependable) generating capacity of far higher proportion than if individual sites were separately exploited. Interconnection also can smooth the curve of incident demand where load centers of different types are brought into a single system and the sharing of capacity reserves among several systems is made possible.

But interconnection also requires the spanning of distance. Accordingly, the first problem for technology in electrical transmission was that of overcoming distance in matching supply with effective demand. Electricity is lost in transmission, especially at low voltages. Through technological advances the economic range of transmitting electricity was increased by raising transmission voltages and reducing normal line losses. Centers of demand thereby were brought within reach of the economical generating sites. As this was done, interconnection of a number of generating sites with each other and with a number of load centers also became possible.

A second problem was that of overcoming natural hazards to high-voltage transmission. Lightning and icing in particular presented problems for the maintenance of high-tension lines.

A third problem was that of keeping the costs of high-voltage transmission equipment, particularly line costs, within the limits of economic feasibility.

TECHNIQUES AND EQUIPMENT RELATED TO USE
OF INCREASED VOLTAGE

Much attention was given to the problem of transporting electricity for a number of years prior to 1890, but it was not until 1891 that "mobility" was given to electric power, when a 110-mile transmission

[80] Hour and season peak demand, total amount of electricity required, relation of prime to secondary power, etc.

line was installed in Germany between Lauffen and Frankfort.[81] This was made possible by the introduction of alternating current, which could be transformed to higher or lower voltages as required. By the turn of the century, the practice of electric power transmission was firmly established in the United States, especially in the Pacific Coast states. Among the lines built early were those from Willamette River Falls to Portland, Oregon; from Santa Ana Canyon to Los Angeles; and from Niagara Falls to Buffalo, New York. However, the voltage that could be carried on these facilities was limited to 40 kv.

When the pin type porcelain insulators, originally used on power lines and towers were replaced by suspension type insulators in the first decade of this century, it became possible to increase transmission voltages above the previous maximum of 60 kv. This marked the beginning of the development of the high-voltage, long-line transmission network of the present. In the following years, further improvements in insulators were made, high-voltage, high-capacity transformers were developed, and line corona losses were sufficiently minimized so that operation at 220 kv. was attained. By 1926, more than a thousand miles of 220-kv. lines had been installed (table 25). By the mid-thirties

TABLE 25. *Circuit Miles of Overhead Electric Line of 22,000 Volts and Above in Service, 1926–54*

Voltage category	Circuit miles in service			
	1926	1935	1948	1954
from 22 through 28 kv.	19,238	27,804	21,123	21,042
from 33 through 42 kv.	31,220	47,443	48,640	56,953
from 44 through 60 kv.	21,062	27,284	27,529	30,874
from 66 through 100 kv.	20,757	36,207	44,260	57,243
from 110 through 165 kv.	14,785	29,224	44,526	67,463
220 and 230 kv.	1,153	2,306	4,035	6,601
330 kv.	0	0	0	668
Total	108,215	170,268	190,113	240,844

Sources: Edison Electric Institute, *Electric Industry in the United States—Statistical Bulletin for the Year 1955* (Statistical Bulletin Number 23), New York, 1956; and *idem, The Electric Light and Power Industry in the United States* (Statistical Bulletin No. 4), New York, 1937.

---

[81] Hans Thirring, *Power Production: The Practical Application of World Energy* (London: Harrap, 1956), p. 147.

most of the high-voltage lines in the Atlantic Seaboard states were still equipped for maximum loadings of 110 kv., while in the Midwest 132 kv. was common and in the area of the Tennessee Valley 154 kv. was being employed.[82]

Studies which preceded installation, in the late 'twenties, of the transmission lines to deliver power from Hoover Dam to Los Angeles— a distance of some 240 miles—led to a number of important developments, such as the high speed, 3-cycle circuit breaker used for switching. Further improvements have made possible the very large prompt-acting 330-kv. circuit breakers, capable of carrying and switching currents of 2,000 amperes. Operating efficiency and service have been greatly improved as more reliable systems to control and regulate alternating frequency have been perfected[83] and transmission losses have been steadily reduced. For example, in the large system of the Tennessee Valley Authority transmission losses currently account for no more than 6.5 per cent of the total power generated.[84] This is a major reduction when compared to earlier transmission losses of 15 per cent or more.

AVOIDANCE OF NATURAL HAZARDS

Two of the most serious natural hazards to effective long-distance transmission of electricity in the United States have been lightning strokes and the icing of lines and insulators. The first of these to be successfully treated was the lightning problem. From 1921 to 1925, important investigations were made on the occurrence and characteristics of lightning and its effects upon the transmission of electricity. Equipment to reduce the hazard of its effects, such as ground wires, counterpoises, expulsion tubes and ultra-rapid reclosure systems, subsequently were developed and installed.

In the northern two-thirds of the country the collection of ice or

[82] Philip Sporn, "Recent and Past Progress in Power Transmission," *Electrical Engineering*, 74 (October 1955), 880.

[83] John Zaborszky and Joseph W. Rittenhouse, *Electric Power Transmission: The Power System in a Steady State* (New York: Ronald Press, 1954); chap. 10 describes load frequency regulation techniques. Additional information on significant electrical advances may be found in: Eugene Ayers and Charles A. Scarlott, *Energy Sources: Wealth of the World* (New York: McGraw-Hill, 1952), p. 251.

[84] TVA, *Annual Report, 1956*, p. 35.

glaze on the conductors is a hazard even more widespread than lightning. The weight of the ice on the conductor and the supporting towers, augmented by a natural vibration effect, may cause great structural damage. As large interconnected transmission networks have developed, it has become possible to inhibit ice formation and to melt it by remote control from a central station. This is accomplished by connecting numerous transmission lines in series so that there is a substantial total impedance in the circuit. Then when line voltage is connected to ground, the conductor is heated electrically and freed of its ice load. First practiced in 1938, this technique has been further refined, especially by the development of sensitive instruments for measuring the decrease in carrier signal strength, thus making possible the detection of ice formation.

LINE CONSTRUCTION COSTS

The advantage of higher operating voltages is based on the fact that the load-carrying ability of a transmission line is proportional to the square of the operating voltage. Although costs of transmission lines and terminal equipment rise rapidly with voltage increase, they are more than offset by the much greater load-carrying ability. Therefore the capital expenditure per kilowatt-hour of electricity transmitted a given distance has decreased as the voltage increased, so long as the full capacity of the high-voltage line was used.[85] In a survey conducted a few years ago, average construction costs per mile of a steel tower double circuit 220-kv. line was ascertained to be $33,000, and annual operating expenses amounted to $290 per mile. This compared to an initial cost of $6,500 per mile for a wood pole, single circuit 66-kv. line, for which operating expense was $95 per mile.[86] Although the high-voltage line required five times the investment and had three times the operating cost of the low-voltage line, its power transmission capacity would be about twenty times as large.

The use of aluminum conductor for high-tension lines has proved to be less costly than the previously preferred copper. It is almost as efficient a conductor, and its lighter weight permits greater economy in

[85] D. T. Braymer, "Extra High Voltage," *Electrical World*, 146 (July 9, 1956), 82-83.
[86] Federal Power Commission, *Electric Utility Cost Units, Transmission Plant*, Washington, 1951.

the construction of tower supports for the line. An illustrative installation is the 167-mile line from Grand Coulee Dam to Everett, Washington, where about a thousand miles of steel-reinforced aluminum conductor were used in place of copper, with major savings.

There is, however, ample opportunity for further application of technology to line construction and maintenance. Mounting construction costs have been a recent special problem. The cost of constructing transmission lines and substations doubled within the fifteen years 1940-55.[87] The over-all costs of delivering power from distant hydro installations is a factor in determining the relative advantage of hydro power development or thermal power installations near the load centers. Transmission costs therefore have direct bearing on water development for these purposes.

PRESENT TRANSMISSION DISTANCES AND COSTS

The technical events in all three of the areas illustrated above have greatly increased the economic range for transmitting power. The transmission line which was installed between Hoover Dam and the population centers in southern California in the early thirties was a pioneer among the longer distance facilities, but since then many lines between 200 and 300 miles in length have been completed. Today the lowest cost for energy transmitted a distance for fifty miles theoretically is attained with facilities equipped for 230 to 330 kv. Greater distances justify higher voltages; 330 to 440 kv. is optimum at a distance of a hundred miles and 400 to 500 kv. at twice this distance.[88] Voltages of 440 kv. are technically feasible. A 400-kv. transmission line is already operating in Sweden, and a proposed line in Utah and southern California may also be designed for 400-kv. It appears likely that it will soon be possible to install 500-kv. duplex conductor lines.[89] Transmission at such high operating voltages probably will not be economic unless the total system power exceeds 1 million kilowatts.

In 1955 more than 11,000 circuit miles of new transmission lines at

[87] H. D. Hunkins, "The Future of Power Transmission in the West," *Electrical Engineering*, 74 (1955), 1070.

[88] "Review of Early Performance of 330 kv," *Electrical World*, 145 (February 27, 1956), 25.

[89] Announcement by the U.S.S.R. indicated that a 500-kv. line would go into operation in 1958.

66 kv. and above were installed, the average cost exceeding $50,000 per mile,[90] and construction has since been maintained at about this rate.

The achievement of technology in this field may be partly measured in transmission costs. Although the distances spanned by transmission lines have greatly increased, costs of transmission have not changed significantly. The transmission expense incurred per kilowatt-hour of electricity marketed is about the same as it was twenty years ago, although materials, equipment, and construction costs have risen sharply (table 26). These rising costs began to show their effect after 1944, however, and it appears that transmission expense will probably rise along with other power development costs.

TABLE 26. *Transmission Expense per Kilowatt-hour of Electricity Sold in 1937–54*

1937	.30 mill
1940	.24 mill
1944	.22 mill
1947	.26 mill
1950	.26 mill
1954	.28 mill

Sources: Federal Power Commission, *Statistics of Electric Utilities in the United States: 1955,* Washington, 1956, p. xxiii; and *idem, Statistics of Electric Utilities in the United States: 1947,* Washington, 1948, p. xv. Figures include maintenance, depreciation and amortization, salaries and wages associated with transmission, insurance, and taxes.

DISTRIBUTION OF DISPERSED LOAD

The major developments in transmission technique have been the connection of large-scale generating units with distant markets for power. However, transmission techniques that have facilitated service to geographically dispersed load also have been important. Through these means the effective service territory of generating systems in the United States has been extended so that all but the most sparsely settled and developed areas are within economic reach of generating plants. These developments apply mainly to farm service, although they have bene-

[90] "Electrical Industry Statistics," *Electrical World,* 145 (January 23, 1956), 131.

fitted all rural settlements and their economic pursuits.

In 1935 less than 11 per cent of American farms were receiving central station electric service.[91] However, the construction program which was initiated and financed by the federal Rural Electrification Administration, created in that year, rapidly altered this situation. As of 1950, 999 new systems and 1,018,336 miles of line had been energized and 3,251,787 new consumers had been connected through sponsorship and aid given to local effort by the Rural Electrification Administration.[92] An important part of the REA success was achieved by reducing the cost of the installation of rural transmission facilities. Lighter construction, assembly line production methods, and standardization of equipment were a few of the measures which enabled new lines to be built under REA supervision before the war at an average cost of $825 per mile, including overhead.[93] In many cases, the costs of REA installations were only half those previously prevailing for comparable facilities. The activities of the REA have unquestionably spurred the privately owned utilities in this field, with the result that by 1954, 93 per cent of American farms had been electrified.[94]

6. Dispersal of Demand for Water as an Influence on Quantitative Inequalities of Demand and Supply

The problem of meeting large-scale water demand may in part be met by judicious geographical distribution of that demand as well as by developing transport and delivery facilities which alter its geographical availability. For example, planning practices which affect the location and density of urban settlement can be seen as influencing the geographical requirements for domestic water supply. Although dispersal of water demand is only one of many factors involved in urban and suburban development and growth, this effect is realized where nodes of settlement are scattered. With the rapid accelerating outward growth of American cities, some reduction in the areal intensity of

[91] U. S. Department of Agriculture, *Report of the Administrator of the Rural Electrification Administration: 1950,* Washington, 1950, p. 13.

[92] *Ibid.,* p. 10.

[93] U. S. Rural Electrification Administration, *1939 Report of the Rural Electrification Administration,* Washington, 1940, p. 100.

[94] U. S. Department of Commerce, Bureau of the Census, *Statistical Abstract, 1956,* p. 645.

demand for water appears to have occurred, especially in the last fifteen years. The consequences of such planning proposals as Ebenezer Howard's "Garden City," developed and made popular in the United States by Lewis Mumford; Clarence Stein's "Regional City"; or Frank Lloyd Wright's "Broadacre City" would be of this character, spreading the incidence of demand geographically and thus providing a larger area from which to obtain water supplies.[95] Indirectly, any technical change which makes the dispersal of industry possible, like the development of improved systems of communications and transportation, also fosters a spreading of the demand for water. Accordingly it can lessen the necessity of moving water from water-surplus areas to those of short supply.

Probably the most prominent and repeated examples of planned demand dispersal, however, have occurred within industry, and particularly the electro-metallurgical industry. The enormous amounts of power required by the aluminum industry in reducing alumina to aluminum (approximately 25 per cent of the cost of the aluminum ingot) have placed a premium on low-rate electric power for this purpose. Both through private industry and federal government planning, aluminum reduction plants until recently were dispersed in areas which had locational disadvantages both from the point of view of market for the metal, and the supply of alumina or bauxite. The location of several plants in the Pacific Northwest, and the large facilities of the Aluminium Company of Canada in Quebec and British Columbia all have illustrated conscious dispersal of this demand for electric power to sites near hydro generating facilities. Since 1955, however, policy for new plant location has changed somewhat in response to decreased thermal generating costs and the increasing remoteness of low-cost hydro generating sites. Some of the most recent reduction plant locations have been in the midst of the marketing area of eastern United States, where ample coal for thermal power generation is available at low cost.

[95] The "Garden City" idea was first advanced by Howard in 1898 in *Tomorrow: A Peaceful Path to Real Reform* (London: S. Sonnenschein & Co.). Its impact and subsequent development in England have recently been reviewed by Lloyd Rodwin (*The British New Towns Policy* [Cambridge, Mass.: Harvard University Press, 1956]). Mumford has discussed this development in terms of the American scene (*Green Memories* [New York: Harcourt Brace, 1947]). Stein has summarized his pioneering endeavors in this field in *Toward New Towns for America* (Liverpool, England: University Press of Liverpool, 1951). Frank Lloyd Wright first formulated his conception of the "Broadacre City" in *When Democracy Builds* (Chicago: University of Chicago, 1945).

SUMMARY COMMENT

It already is apparent that the techniques which permit the realization of scale economies can be extremely important in water development. The impact of some, like methods of handling materials and the manufacture or treatment of the materials themselves, already has been felt forcibly wherever water resource development has been undertaken within recent years. No project of any scope would be started without planning for their benefits. These are techniques whose application is simply related to their dependability and economy. The application of other equally well-proven techniques, however, is much more complex. Their application not only is dependent on perfection of the technique and its economic viability, but also on legal, political, and other features of the cultural environment within which development is undertaken. The application of integrated water control, large-scale integrated electricity distribution, and regional planning for demand-supply adjustments are of this character. No other group of techniques so well illustrates the principle that technological perfection does not necessarily imply ready or widespread adoption of a technique. The technique must be compatible with other features of the social environment in which it is applied before universal adoption is found. The extent of application of these techniques will be treated in Section IV of this study.

CHAPTER 10

Techniques Extending the Physical Range of Water Recovery —

FOUR CASES

Techniques Relating to Ground-water Exploitation: (1) Drilling Techniques—rotary drilling; the continuous flight auger. (2) Increasing Rates of Water Flow in Wells—acidizing and fracturing of wells, the horizontal collector. (3) Pumping Improvements. *Geographic Extension of Large-scale Surface-water Exploitation:* (4) The Case of Malaria Control. *Summary Comment.*

In general, techniques extending the physical range of water recovery appear to be more in the domain of the future than a part of the past and the present. They accordingly make up an important part of the discussion in Part III (chapters 13, 14, and 15). Nonetheless the past offers some illustrations of the meaning of extending the physical range of water recovery. Extension of man's capacity to exploit water may be made within the tropospheric part of the atmosphere, into geographical environments hitherto unavailable for exploitation (like the oceans), or into the rock formations underlying the surface of the land. The illustrations here selected are principally from the latter field, that of ground-water exploitation. One additional case is an extension into the new geographical environment of hitherto unusable surface waters.

The four cases are well-drilling techniques, increasing rates of water flow in wells, pump improvement, and malaria control techniques for reservoirs.

TECHNIQUES RELATING TO
GROUND-WATER EXPLOITATION

An overwhelming percentage of the total number of individual water facilities in the United States comprises ground-water installations. On the basis of 1949-50 data it has been estimated that 91 per cent of all installations for water withdrawal of any kind in the United States were for the withdrawal of ground water.[1] Sixty-four per cent of the farms having some irrigation used ground water, and 80 per cent of all rural domestic water use (except irrigation) in 1950 was from ground-water withdrawal.[2] In 1950, about 25 per cent of the irrigated acreage of the country was served by ground water, which supplied 23 per cent of the total water used for irrigation.[3] Of the total United States water withdrawals, ground water accounts for only about 18.5 per cent of total use.[4] The geographical pattern of ground-water use is one of wide dispersal, but most installations are small, shallow, and of relatively primitive character. Ground water therefore serves in an important way to meet dispersed demands on a small scale. Over the greater part of the country very little more than the surface of the enormous amounts of water stored within rock formations has been tapped.

Among the reasons for this limited exploitation is the somewhat erratic distribution of large-volume, freely flowing aquifers. Shallow-lying aquifers are widely found, particularly in unconsolidated sediments, but knowledge of the extent, internal characteristics, and manner of exploiting formations for a larger flow is imperfect. The situation has been summarized by Sayre:

> Ground-water investigations often are very complex, time-consuming, and, in some places, costly. They include not only geologic and geophysical mapping to determine the location, extent and thickness of water-bearing formations, but also hydrologic mapping and physical studies to determine the permeability and porosity of the rock, the amount of water that enters the aquifers, the direction and rate of movement of the water and the areas of recharge to and discharge from the aquifers. Inasmuch as water beneath the land surface is invisible, the relevant data are often determined by indirect methods

[1] Louis Koenig, "Ground Water—The Impact of Recent Technological Advances on its Utilization," Resources for the Future, Inc., unpublished manuscript, December 1956, p. 25.

[2] *Ibid.*, pp. 25, 24.

[3] *Ibid.*, p. 6.

[4] *Ibid.*

such as electric and gamma-ray logging and examination of the geologic strata in drilled wells, as well as by actual pumping tests.[5]

There have been other discouraging physical features about ground-water exploitation on large scale; the necessary expenditure of energy to lift water to the surface of the ground, for example; the small volumes of flow generally experienced in most wells; the tendency in some areas toward a higher mineralization than in surface flows; and the limitation of ground water to withdrawal uses only. The practicality of multiple use found in surface-water exploitation thus is usually absent.

It follows that any technical change facilitating the discovery of underground water, or increasing the physical depth of drilling, decreasing the cost of drilling or of lifting the water to the ground surface, or increasing the flow available for withdrawal, may offer some opportunity for extending the physical range of ground-water use. The first and newest of these techniques is described in chapter 15; the other three are discussed in this section as examples of well-developed technology in this field.

1. Drilling Techniques

The process of sinking a hole into the crust of the earth for the collection of fluid or gaseous materials has changed greatly since application of mechanical power toward this end within the last hundred years. Drilling methods are of interest because they have been associated with deep penetration into the rock materials of the earth's crust, and therefore into greater masses of water. Among the many drilling methods that have been developed over the years since the advent of mechanical power, rotary drilling and the continuous flight auger have been selected for brief description. The continuous flight auger is of particular interest because it enables the sinking of reasonably shallow wells very cheaply; its use therefore bears upon the problem of meeting dispersed demands for adequate water supply.

ROTARY DRILLING

The common method of drilling for water has been with a percussion bit. In percussion drilling a heavy bit is raised by a cable and

---

[5] A. N. Sayre, "Ground Water," *Scientific American,* November 1950, p. 18.

dropped in the drill hole, pulverizing any consolidated rock material. Water is kept in the hole and, when several feet of rock or sediment have been penetrated, the bit is withdrawn on the cable and a "bailer" is lowered to the bottom to remove the cuttings. This is repeated over and over again until the desired depth is reached.

In recent years, however, another method of drilling has been coming into use in many areas of the country. In rotary drilling, which long has been employed by the petroleum industry, a cutting bit is fastened to the end of drill pipe which is rotated in the well. To remove the cuttings a fluid (mud) is pumped down the drill pipe and is forced upward through the annular space between the drill pipe and the sides of the well; this not only removes the cuttings, but cleans the bit face and helps support the walls of the hole against caving.[6] While percussion drilling with cable tools still is more suitable for some rock formations, where it can be used rotary drilling tends to be faster and cheaper per foot of hole. It is particularly suited to drilling in "sticky" and unconsolidated formations, which comprise the most common material to be penetrated.

An innovation in rotary drilling recently introduced in the petroleum industry is air or gas drilling. In this procedure drilling mud is replaced with air or gas traveling at high velocities (3,000-6,000 feet per minute). The air or gas serves the same purpose as the drilling mud and allows notable increases in penetration rate, although it does not support the sides of the hole.[7] Air drilling is applicable to shallow wells and to moderately soft formations. For example, it has been used in drilling 400-600-foot deep, 6¼-inch diameter water flood wells in Oklahoma.[8] These wells were drilled, perforated, cleaned, cemented, and completed at the rate of one each 24-hour day with a crew of two per shift. The operation was cheaper than with cable tools and required only one-third of the time. It is probable that this case represents a penetration rate (during actual drilling operations) of at least 75 feet per hour. In another instance of water well air drilling in a harder formation the penetration rates in limestone are reported to have been as high

[6] In the so-called "reverse circulation" method, the flow of drilling fluid is in the opposite direction; i.e. down the annular space, and upward through the drill stem. Both this method and the original method of rotary drilling were used in water drilling before they were used in the petroleum industry.

[7] Phillip L. McLaughlin, "Many Drilling Advancements Likely," *World Oil,* April 1954, pp. 149-54.

[8] L. L. Brundred, "Air Drilling Shallow Water-Flood Wells," *Oil and Gas Journal,* June 11, 1956, pp. 144-47.

as 25 feet per hour (six times as fast as with cable tools).[9] It seems possible that air drilling may find an important place in sinking water wells, accentuating the trend to rotary drilling and lower well costs. It also is possible that further technical improvements in drilling developed by the petroleum industry, such as the magnetostrictive drill and the turbo drill,[10] may point the way to improved drilling techniques for water wells. Both operate on the principle of shortening the line of mechanical power transmission to the drill, thereby reducing power losses and accelerating tool removal for changing bits and other operations. The prime mover in each instance is near the drilling face. Application of the present magnetostrictive and turbo drill equipment to water well drilling is not foreseen, but the development of these techniques shows that further drilling improvements may be expected. Eighty per cent of the oil wells drilled in the Soviet Union in 1955 are reported to have been sunk with turbo drills, which were perfected in that country.[11] American application, in the petroleum industry, has commenced only recently.

THE CONTINUOUS FLIGHT AUGER

A much more modest piece of equipment promises wider application in water well drilling than the more spectacular depth drilling techniques now being experimented with in the petroleum industry. The continuous flight power auger has applications not only in drilling shallow water wells, but also in prospecting for and evaluating aquifers. Hand augers have been used for many years to obtain subsurface samples of soil for agricultural, highway, and foundation work. Power augers also have been used for test work and for sinking wells. These augers are rotated in a hole until the bit is full and then are pulled and unloaded. This process, repeated many times, sinks a hole and allows samples at any depth up to the limit of the equipment, in 1957 about 300 feet.

[9] Schramm, Inc. (advt.), *Water Well Journal*, July 1956, p. 27.

[10] These two machines are suspended deep in the well by cables, the magnetostrictive being powered by electricity transmitted to the motor at the bottom of the hole, and the turbo drill being driven by a small, powerful water turbine supplied with drilling mud through a flexible pipe from the drilling rig.

[11] J. H. Thacher and W. R. Postlewaite, "Turbo Drills, Past and Present," *Monitor*, I, No. 47 (November 1956) 5.

The continuous flight power auger has been increasingly used since 1951. Where earlier power augers had only a few flights of lifting spirals, this newer type has a continuous spiral extending from the bottom of the hole to the surface; cuttings are carried to the surface in the manner of a screw conveyor. Sections of the auger may be added to the length already in the hole as depth increases. This simple equipment is portable, usually being mounted on light trucks or jeeps. It can be operated by one man who can be trained in a few days. Drilling costs are 25-50 cents per foot. The continuous flight power auger is used regularly to 100 feet and functions satisfactorily as deep as 200-300 feet. Formations for which it is suitable include sand, gravel, clays, shales and slates, sandstones, limestones, and other sedimentary formations which are not massively cemented with silica and do not contain boulders approaching the diameter of the hole.

Under favorable conditions in unconsolidated formations, continuous flight auger units have bored six inch holes to 80 feet in less than one hour from the time of arriving on the property, with a single operator. The same operator, using the same rig, can set casing (of a special type) and complete the well ready for pumping in one additional hour.[12] In view of the traditional small scale of water well drilling, this equipment, with its cheapness, portability, low "set-up" time, and ability to drill without mud or water, may be an answer to more widely dispersed ground-water use.[13]

2. Increasing Rates of Water Flow in Wells

Water flow into most holes of water well size is usually not rapid because water underground normally must move through an intricate path of pore-sized openings in rock materials. While this rate may be sufficient to support the many small withdrawals for domestic purposes, flow adequate for larger demands, including those of irrigation and industry, is not commonly encountered. Only under unusual conditions of hydrostatic pressure (e.g. artesian flow) within an aquifer, or high

[12] Data on continuous flight auger drawn from personal communications to Louis Koenig from Mobile Drilling, Inc., Indianapolis, Ind., and from Leo Hough, Louisiana State Geologist, Baton Rouge, La.

[13] Ninety per cent of the wells drilled in the United States average less than 150 feet deep, and cost less than $800.

porosity (as in sands), are large-volume flows normally found. Consequently, techniques of increasing the rapidity of water flow into wells, and therefore increasing the possible rate of withdrawal, may make ground water physically and economically available for uses which otherwise could not be served. This is of particular importance to industrial use, where the low and constant temperature of ground water makes it the preferred source for industrial process cooling. Techniques of special interest in this connection are acidizing and fracturing of rock and use of the horizontal collector.

ACIDIZING AND FRACTURING OF WELLS

Acidizing and fracturing of oil wells to increase production capability has been practiced for some time in petroleum exploitation. Acidizing is the practice of pumping acid through a well bore into a formation. By a dissolving process the acid enlarges and interconnects the pores of the rock, thus increasing the permeability and yield of the producing formation. Conditioned and inhibited hydrochloric acid is used for limestone formations and hydrofluoric acid for rocks containing silicates. Fracturing consists of physically breaking the formation and making cracks or holes in it either by applying fluid under high pressure or by perforating with "bullets" or shaped charges. In acid-fracturing, the two techniques are combined by using acid as the fracturing liquid. This method supplements the dissolving action of the acid by a fracturing or cracking action due to the hydraulic pressure and the abrading action of sand used in the pressure fluid to prop open the fractures. The pressure is applied by pumping the acid into the well casing. It can be applied to various horizons by sealing off sections of the well with temporary plugs of various kinds.

Acidizing of water wells was first used about 1946, but the technique was not perfected until 1952. Acid-fracturing was introduced in 1954.[14] Acidizing can be used in unconsolidated as well as consolidated formations and in materials other than limestone, but fracturing may not be particularly useful in unconsolidated formations. By 1956 only a few hundred wells had been acidized, and only a dozen or so fractured.[15]

[14] Ivan R. Bielek (Dowell, Inc.), personal communications to Louis Koenig, 1956. Also, *idem.,* "Successful Acidizing," *Water Well Journal,* August 1956, pp. 9, 24, 26, 29.
[15] Bielek, "Successful Acidizing," *op. cit.*

Recent reports on well acidizing (and some fracturing) have shown very satisfactory results in application of these methods to forty wells in Missouri (table 27). Acidizing also has been applied to Illinois, Indiana, Kentucky, and Tennessee wells in sand, gravel, or limestone.

TABLE 27. *Results of Acidizing Forty Missouri Water Wells*

Per-cent apparent increase in production	Number of wells
0-50	7
50-100	7
100-200	9
200-500	10
500-1,000	3
Above 1,000	4

Source: John G. Grohskopf (Missouri Geological Survey), personal communication, 1958.

Acidizing techniques are not well standardized, largely because it is not yet clear how much acid is required. Practice varies usually between 35 and 100 gallons per foot of face in a 12-inch well (six to seventeen times the volume of the well bore opposite the face).[16] The full life of increased flow from well treatment by these methods is not yet known but it is at least three years, and the decline in yield of an acidized well is known to be less than for a nonacidized well pumped at the same rate.[17]

The cost of acidizing and acid fracturing varies according to quantity of acid employed, dependent in turn on type of formation and feet of aquifer face. In a few moderately deep wells in Missouri, the cost of acid fracturing was about one-fourth the cost of a new well. One new 850-foot well yielded 36 gpm (gallons per minute) before acid fracturing. After acid fracturing the capability was 160 gpm.[18] Thus the

[16] Brantley Myers (Dowell, Inc.), personal communication to Louis Koenig; John G. Grohskopf, "Acid Rescues a City Well," *Water Well Journal,* May-June 1955, 3 pp. reprint; Bielek, "Successful Acidizing," *op. cit.*

[17] Bielek, "Successful Acidizing," *op. cit.*

[18] Myers, *op. cit.;* Grohskopf, *op. cit.;* Finley B. Laverty and Herbert A. van der Goot, "Development of a Fresh Water Barrier in Southern California for the Prevention of Sea Water Intrusion," *Journal, American Water Works Association,* 47 (September 1955), 886-908.

owner had developed the temporary equivalent of three and one-half additional wells for about one-fourth the cost of a single well. In a rough appraisal of the comparative value of acidizing, it is assumed that there would be a linear decrease in productivity of the acidized well to the level of the nonacidized wells in three years, whereas the nonacidized wells are assumed to maintain constant production. The 350 per cent initial increase in productivity therefore might average 175 per cent increase over the long term, or the equivalent yield of nearly two additional wells; and the one well might be re-acidized six or seven times over a period of about twenty years for the investment in equivalent capacity in new wells.

However, an acid-fracturing cost as low as one-fourth new well cost is only reached in rather deep wells. For example, use of the method in wells less than 300 feet deep, with a 20-foot aquifer face does not appear economic (table 28). Since the average municipal and in-

TABLE 28. *Minimum Depths of Wells for "Economic" Acid-Fracturing*[1]

50-ft. face—limestone	540 ft.
50-ft. face—nonlimestone	890 ft.
20-ft. face—limestone	310 ft.
20-ft. face—nonlimestone	480 ft.

[1] Assumes an average municipal and industrial well cost of $16 per foot of bore, and use of 100 gallons of acid per foot of aquifer face.

dustrial well is only 240-265 feet deep, a majority of wells are of such depth that present cost of acidizing would be large compared with the rather low cost of a new well.

The method is probably not yet economic for rural domestic wells, but some domestic and irrigation operators effect a compromise by using acid in a crude manner to surge their wells.[19] Further developments applicable to smaller wells may be possible if oil well servicing companies consider the potential in water well servicing of sufficient value to develop cheaper techniques for this market. There are about

[19] Myers, *op. cit.*

14 million operating water wells in the country[20] compared to 600,000 oil wells.

It seems likely that acidizing and fracturing will have reasonably wide applicability in maintaining municipal and industrial wells, and larger irrigation wells. The impact on ground-water use should cause a decrease in the 400,000-500,000 replacement wells drilled each year.[21] Secondly, acidizing and fracturing should reduce the cost of water well investment. Finally, they should have the effect of increasing the over-all efficiency of pumping. Since the efficiency of a pump is at a maximum only when operating at the design conditions of head and flow rate, decreasing well yields decrease pumping efficiency. Use of these treatments at proper intervals would maintain well yield nearly constant, which in turn should increase the over-all efficiency of pump operation.

THE HORIZONTAL COLLECTOR

Fracturing increases the capability of a well by increasing the porosity of the water-bearing material. Other methods, applicable only to unconsolidated aquifers, have been used to increase capability by enlarging the well surface exposed to the water-bearing material.

In the gravel wall technique, for example, the well is drilled oversize, either throughout the length of the hole or by under-reaming of the aquifer face. The space unused for casing is filled with gravel. Another method of achieving the same result is to surge the well (i.e. apply liquid under pressure) during completion of construction. Surging washes the unconsolidated aquifer material surrounding the bore and sluices the fine particles into the well, from which they are removed by pumping. The coarser particles are thus left in the form of a gravel pack around the well screen.

These techniques result in an effective enlargement of well diameter, but well diameter in itself has surprisingly little effect on yield. Doubling the diameter will increase the yield by only 10 per cent. The main effect of these techniques is to reduce the velocity of approach into the well, which normally causes a pressure drop large enough to reduce

---

[20] Walter L. Picton, "The Water Picture Today: A National Summary of Ground Water Use and Projection to 1975," *Water Well Journal,* April 1956, pp. 10, 25, 26, 29.

[21] *Ibid.,* for replacement wells data.

head at the well face. Such drops in pressure not only result in increased drawdown (thus lowering well production), but also tend to cause chemical deposition from many well waters. Deposits may reduce the porosity of the surrounding formation, and even of the well screen itself, thus lowering water yield.

Another method of increasing well capability is the infiltration gallery. This system was employed in some of the ancient water works of the Near East and Mediterranean lands, but is not used in the United States. It consists of a pipe, tunnel, or a gravel-filled trench sunk into the aquifer. If constructed from the surface, infiltration galleries must be of limited depth; if tunneled from the bottom of a shaft, they are expensive. However, in modern times an ingenious method has been developed for constructing the equivalent of infiltration galleries, or horizontal wells, relatively inexpensively. This device is known as the horizontal collector.

In constructing a horizontal collector a shaft 13-20 feet in diameter is sunk into the ground as a caisson. Depths of 200 feet are possible, but the maximum practical depth under normal conditions is about 150 feet and the majority of present installations are less than 120 feet. From the bottom of the shaft large perforated pipes or screens are forced into the formation by hydraulic jacks. As they are installed they are surged (i.e., liquid is applied under pressure) to remove the fine material and form a gravel wall, typically 4 feet in diameter around each screen pipe. The screens may be 8-24 inches in diameter and can be projected as much as 300 feet horizontally. As much as 3,000 lineal feet of screen may be projected from a single caisson. Such a collector would correspond to a vertical well having 3,000 feet of face exposed to the aquifer. Moreover, in the horizontal well every foot of screen is subjected to the full hydrostatic pressure of the overlying water, whereas in a vertical well only the bottom of the screen is under the full head. The yield per foot of drawdown for horizontal wells is therefore much greater than for vertical wells. Since the velocity of water approach is exceedingly small, reduction of formation porosity is very slow and exceptionally long service life is obtainable.

The horizontal collector method was invented and patented in the early 1930's, and has been promoted by only one company since that time. The collector is an economic advantage usually only for water supplies larger than about 1 million gallons a day. In 1956, there were 138 units in operation, concentrated mainly in the Ohio and central Mississippi valleys. A study of 40 installations built since 1946 shows a construction cost range of 1.3 cents to 10 cents per daily gallon production rate, averaging less than 4 cents. This total includes shaft, horizontal laterals, slabs, superstructures, pumping and electrical equipment.[22] Hence, an installation having a 25-million gallon daily capacity might involve an investment of $1 million.

Where it is usable, the potential impact of the horizontal collector on ground-water utilization is great and comes about chiefly because its large capability makes it an ideal unit for induced infiltration. It is most useful in aquifers composed of unconsolidated rock materials. Horizontal collectors minimize the frequent disadvantage of ground-water supplies for municipal purposes—lack of flow in large volume. At the same time, they preserve the usual advantage of ground water in requiring no expensive filtration plants. The industrial preference for ground water and the increasing number of large water users suggest that horizontal collectors may have wider future application in ground-water exploitation. In 1957 there were estimated to be about 1,800 communities and 2,200 manufacturing establishments in the country having a daily water requirement of 1 million gallons or more.[23] While not all of these are accessible to adequate ground-water supplies, or to unconsolidated aquifers, a substantial number are so located. The collector is likely to have a place in serving these demands as they increase further under the impact of a larger, more urbanized, more industrial population.

3. Pumping Improvements

While improvements in the techniques of sinking wells and collecting ground water do affect water costs and physical availability, costs of lifting water from aquifer to ground surface probably are even more influential over the long run in determining the extent of ground-water use. Consequently, improvements in the technology of lifting water are of prime significance in the exploitation of ground water. Some of the more striking correlations in ground-water use increases are with technical changes in methods of supplying energy. As far as water use is concerned, these are good examples of "windfall" technical bene-

[22] Data on the horizontal collector from trade literature and personal communications to Louis Koenig, 1956, from Ranney Method Water Supplies, Inc.
[23] Koenig, *op. cit.*, p. 121.

fits. For instance, the practice of pumping has been continually improved by increase in the efficiency of pump motors, internal combustion engine improvement, and decreasing cost of electric power.[24] The possible areas for pumped wells have been enlarged through the extension of rural electrification.

Distinct from these gradual and windfall benefits, a clear example of the impact of a technological advance occurred with the introduction of the turbine pump. The effect was similar in many areas of the country where the depth to available water began to outdistance the maximum lift of suction pumps, about 25 feet. In some instances shafts were dug so that the pumps might be placed below the ground surface to reach water at greater depths. The practical limit of such shafts proved to be about 25 feet, making it possible to reach water at depths of 50 feet.

About 1937 the vertical turbine pump was finally perfected and commercialized. The turbine pump changed the pumping depth limit greatly. This pump is designed as a cylinder of small diameter containing a vertical turbine which may be sunk to the bottom of the well and which can operate below the water level. It is driven by a motor or engine at the surface by means of a long rotating shaft.

The effect of this development is well illustrated in the High Plains of Texas. There, the first irrigation was from shallow wells within the suction lift of centrifugal and piston pumps. As use of these pumps continued for irrigation from a water body having negligible replenishment, water levels over the years dropped beyond their reach. The introduction of high-capacity deep-well turbine pumps allowed a phenomenal expansion of irrigation in these areas. In the sixteen years following introduction of this pump the number of wells multiplied twenty-five times in the Texas High Plains[25] (figure 25).

While it greatly extends pumping depths, the vertical turbine pump does have a depth limitation. As the depth of the pump setting is increased the diameter of the drive shaft must be increased. This, together with bearing wear from misalignment in crooked and off-vertical holes, sets a practical depth limit on their use. A vertical pump shaft

[24] Carl Rowher, *Design and Operation of Small Irrigation Pumping Plants* (Circular 678), U. S. Department of Agriculture, Washington, 1943.

[25] Jack R. Barnes, personal communication to Louis Koenig, 1956; E. R. Leggat, *Summary of Ground Water Development in the Southern High Plains, Texas* (Bulletin 5402), Texas Board of Water Engineers, 1954; and *idem* (Bulletin 5410), 1954.

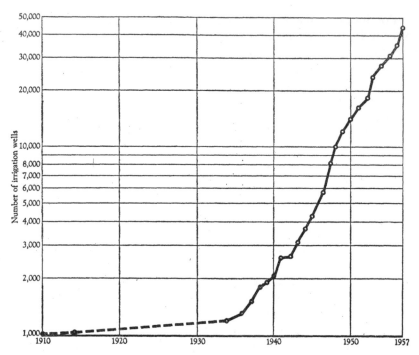

Figure 25. Increase in Irrigation Wells in the Texas High Plains.

for 500-foot depths is as much as 2¼ inches in diameter. Smaller shaft sizes are limited to 300- or 400-foot depths.[26] This will not be of great importance in the High Plains, where most of the water lies within 400 feet of the surface.[27]

In other sections of the country shallower aquifers are becoming exhausted and it is necessary to draw water from greater depths. At these depths the handicap of a long drive shaft becomes great. To circumvent the disadvantages of the long drive shaft, submergible pumps were introduced. But motor and pump are contained in a cylindrical structure which is installed inside the well bore and under the surface of the water. The only connections with the top of the well are the cables for supporting the unit and supplying electric power to it. In

[26] Catalog of the Aurora Pump Division, New York Airbrake Co., Aurora, Ill., December 1, 1954, Section 102.5.

[27] "Thickness of the Ogalala Formation" (map), *The Cross Section*, November 1956, pp. 2-3.

addition to great depth capacity, they may be used in crooked holes, are quiet in operation, and may be installed in locations subject to flooding.

Although submergible pumps were first developed about forty years ago, their underground use was limited to the oil industry for pumping from deep wells, and for pumping water for water flooding. It was not until 1948 that such pumps became available commercially for general water well service.[28] This recent application was facilitated by the development of a small-diameter pump, of about four-inch size.

Conditions other than depth led to their increasing use even at depths within the range of the vertical turbine pumps. This is indicated by the growth in submergible pump sales in the past few years:[29]

> 1953—34,785 units
> 1954—43,362 units
> 1955—51,583 units
> 1956—62,875 units
> 1957—65,196 units

The most popular size of these pumps is a small ½-horsepower unit. Seventy per cent of all units sold are 1 horsepower or less.[30] Water and brine pumping installations range in depth from 100 to 5,000 feet and in capacity from 5 to 3,500 gallons per minute.[31]

A slight disadvantage of the submergible pump is that gasoline or natural gas power cannot be used unless an electric generator also is purchased. In general, gas pumping costs about 80 per cent as much as electric pumping in areas where deep wells have been of particular significance.[32] Nevertheless submergible pumps now account for one-

[28] J. Carle and H. Worthington, "Development and Application of Submergible Pumps," *Water Well Journal,* March-April 1953 (reprint); "Necessity to Remove Great Volume Favors Use of Reda Pump," *Oil and Gas Journal,* February 24, 1938 (reprint); "Reda Submergible Pumps and Waterflooding," *Producers Monthly,* April 1954, p. 73.

[29] U. S. Department of Commerce, Bureau of the Census, *Facts for Industry—Farm Pumps* (Series M35G—07), Washington, April 16, 1958.

[30] Pump Committee, Manufacturers and Suppliers Division, National Water Well Association, "The Driller Should Know Best," *Water Well Journal,* October 1956, pp. 9, 18.

[31] Reda Pump Co., trade literature and catalogs.

[32] Rowher, *op. cit.;* Underground Water Commission (of Arizona), *The Underground Water Resources of Arizona,* Tucson, 1952; Texas Agricultural Experiment Station, *Cost of Water for Irrigation on the High Plains* (Bulletin 745), College Station, Tex., 1952.

fourth[33] of all deep well pumps sold and this ratio is expected to rise within the next few years.[34] The result will be cheaper water for large and small users, and extension of the available economic supply of ground water.

GEOGRAPHIC EXTENSION OF LARGE-SCALE
SURFACE-WATER EXPLOITATION

Past technology has had some bearing on extending the physical range of surface-water exploitation as well as on the exploitation of ground water. An interesting example has been the extension of large-scale, surface-water development into geographical areas where diseases borne by water-bred insects present serious hazards for public health. This problem first was met on a regional scale in the program of the Tennessee Valley Authority, which is described here.

4. The Case of Malaria Control

The malaria control activity of the Tennessee Valley Authority was initiated in an area where malaria was prevalent in endemic form. The hazard of this disease was recognized as one of the leading health problems of the region, where one-third of the population in some sections suffered from malaria. It was also realized that the creation of a system of artificial lakes or impoundments, as contemplated in the program of the TVA, would greatly aggravate the problem unless effective steps were taken to control the spread of malaria. Today the area of impounded water within the TVA exceeds 600,000 acres and a shoreline of more than 10,000 miles has been created.[35] As the dimensions of the undertaking contemplated for the Tennessee River Basin began to unfold, it became apparent that the cost of larvicidal

[33] U. S. Department of Commerce, *Facts for Industry—Farm Pumps, op. cit.*

[34] Pump Committee, *op. cit.*

[35] O. M. Derryberry and F. E. Gartrell, "Trends in Malaria Control Program of the Tennessee Valley Authority," *American Journal of Tropical Medicine and Hygiene,* 1 (1952), 500.

measures, until then the principal means of malaria control in such areas, would be prohibitive and that new types of measures were required.[36] As a result of the attention centered upon this problem, a considerable amount of basic information regarding malaria control has been obtained, and new techniques and measures have been devised.

Epidemiological evidence indicates that only one species of mosquito, *A. quadrimaculatus,* is of importance as a malarial vector in the Tennessee Valley area. There are four stages in the life cycle of this and other mosquitoes, the egg, the larva, the pupa and the adult winged insect or *imago.* The grown male feeds on plant nectar or other sweet or fermenting substances, but the females must obtain a blood meal before they can produce eggs. *A. quadrimaculatus* usually breeds in fresh, clear, quiet, slightly acid to alkaline bodies of warm water of a more or less permanent nature which contain an abundance of vegetation or floating material. The propagation of this mosquito is dependent upon the presence of living plants or plant products, such as seeds, twigs, leaves, or brush, which either float upon or protrude through the water surface. The larvae which hatch from the eggs are aquatic and free-swimming, feeding primarily upon green algae, diatoms, and flagellated protozoans. Water temperature is an important factor in their development, the optimum range being between 85° and 90°F. Two weeks is generally required for this species to develop from egg to adult. Although the effective flight range of *A. quadrimaculatus* is usually considered to be one mile, studies have shown that not more than one-quarter of the adults fly farther than half a mile from their breeding place.[37]

The successful propagation of the anopheles mosquito thus depends upon numerous environmental factors such as temperature, vegetation, and other conditions determining the suitability of breeding places, as well as the availability of nearby blood supplies. Conditions in the Tennessee Valley are such that the season of active propagation extends only from the middle of May to the middle of September.

However, the transmission of malaria among a human population does not depend entirely upon the presence of anopheline mosquitoes.

[36] E. L. Bishop, "Some Health Implications of Regional Water Control with Special Reference to Malaria," *Proceedings of the Conference of the Health Officers, Nurses and Sanitation Officers of Alabama, 1939,* Montgomery, Ala., 1939.

[37] U. S. Public Health Service and Tennessee Valley Authority Health and Safety Department, *Malaria Control on Impounded Waters* (Washington: U. S. Government Printing Office, 1947), especially pp. 260, 264.

Such mosquitoes must come into contact with persons whose blood contains viable malaria parasites, and climatic conditions must subsequently be favorable to the complete development of the malaria parasite within the mosquito. Only then can a person whose physiology permits the development of malaria parasites be infected and the disease spread.

There are three ways of attempting to control this health hazard: (1) measures directed toward the elimination of the mosquito; (2) measures aimed at the prevention of contact between man and the malaria mosquito; and (3) measures directed against the malaria parasite. From the beginning of the TVA malaria control program, primary emphasis has been given to measures of the first type, involving the control of mosquito propagation on impounded waters either by preventing the mosquito from breeding or by destroying it in the larval stage. To accomplish this, the principal measures practiced have been reservoir preparation, water level management, permanent shoreline improvement, shoreline and drainage maintenance, and larviciding.

Preparation of a reservoir basin prior to impoundage is recognized as fundamental to adequate mosquito control. By clearing the area, most of the water surface of a reservoir is rendered unsuitable for mosquito propagation. After the initial removal of perennial vegetation, a final conditioning operation is carried out in which all plant growth within the zone of normal fluctuation of water level during the mosquito breeding season is cut, piled, and burnt.[38] Great importance is attached to the impoundage of reservoirs as early as possible after cessation of the mosquito breeding season, in order to allow floating debris to become waterlogged and sink or to be stranded by wave action.[39] The advantage of low-stumping and arboricidal treatment in the elimination of woody growth has been clearly demonstrated, and power rakes and other heavy equipment have been extensively employed in the fall preparation of a reservoir before impoundage.

Unquestionably the most important techniques that have been incorporated into the TVA's malaria control program have centered upon the management of the level of impounded water. When the water is dropped below the limits of shore-line vegetation, the eggs and larvae are stranded, or drawn away from protective vegetation so that they are exposed to wave action and to predators like minnows, or other

[38] W. G. Stromquist, "Engineering Aspects of Mosquito Control," *Civil Engineering,* 14 (1944), 432.
[39] U. S. Public Health Service and TVA, *op. cit.,* pp. 10, 96.

natural enemies. Such management also facilitates the application of larvicides where their use proves necessary.

Wheeler Dam was the first installation in which provision was made for this activity, the gates having been built one foot higher than originally planned in order to make possible the seasonal fluctuation of the reservoir level for malaria control.[40] A "malaria surcharge" of one foot has been adopted as standard design for all subsequent main river projects. This flood surcharge is used in advance of the mosquito season to strand drift and flotage, which can be collected and destroyed after the reservoir is returned to normal high water level. The water level is next held relatively constant during the period of early spring growth to retard the development of marginal vegetation and thus provide a cleaner shoreline during the active mosquito breeding season. Current practice then calls for lowering the water level approximately one foot and returning it to the original level at weekly or ten-day intervals, after the commencement of the initial period of moderate mosquito production (see figure 26).

It has been discovered, however, that vegetation tends to encroach upon the area dewatered during the cyclical fluctuation. This condition can be countered by allowing the general water level to recede from 0.1 to 0.2 foot per week during the period of greatest mosquito production. By combining such a gradual seasonal recession of the water level with the cyclical fluctuation, the shoreline can be maintained in advance of encroaching vegetation during the mid-season period when marginal growth begins to invade the zone of fluctuation and maximum mosquito production is approaching. As the water level is allowed to recede during the later part of the mosquito season, the water at low elevation is thus removed from contact with the encroaching band of marginal vegetation.

Water level regulation has become the most effective single mosquito control measure in the TVA area. Of the twenty-four reservoirs which the TVA owns, such management alone provides satisfactory mosquito control on sixteen reservoirs, while on the others seasonal larvicidal operations and other measures must also be scheduled.[41]

The program of permanent shoreline improvement, aimed at eliminating the most serious mosquito breeding areas, consists mainly either of diking and dewatering, or of deepening and filling. With the first,

[40] W. G. Stromquist, "A Partnership in Malaria Control," *The Southern Medical Journal,* 34 (1941), 837.

[41] Derryberry, *op. cit.,* p. 501.

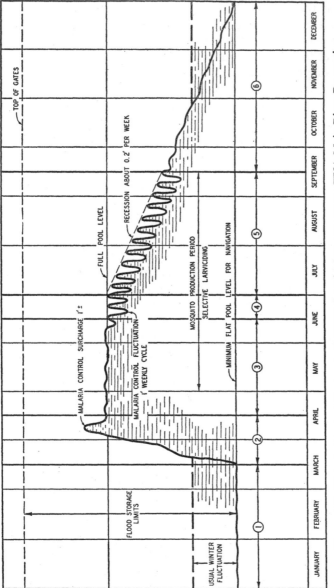

Figure 26. Malaria Control Features of Water Level Management on TVA Main River Reservoirs.

1. Low Winter Flood Control Levels: controls growth of submerged aquatics; permits marginal drainage and herbicidal operations.

2. Early Spring Filling: retards plant growth. Surcharge: strands drift above full pool level.

3. Constant Level Pool: provides long-range plant growth control.

4. Cyclical Fluctuation: destroys mosquito eggs and larvae.

5. Fluctuation and Recession: destroys eggs and larvae; reduces breeding area; provides clean shoreline.

6. Recession to Winter Levels: permits fall shoreline maintenance and improvement operations.

dikes are built to separate large shallow areas from the larger impounded bodies of water, and pumps are installed to remove the water which accumulates behind them. The second practice involves deepening the lower half of shallow mosquito breeding areas and placing the excavated earth on the upper portion of the area. Dewatering projects have generally proved unsatisfactory in terms of both mosquito control effectiveness and operating and maintenance costs, but excellent results have been achieved in the deepened and filled areas, in most cases without supplementary larviciding. More than 18,000 acres have been scheduled for the first type of treatment, while deepening and filling has been or will be practiced in about three-fourths of this area.[42] Shoreline and drainage maintenance constitutes another important aspect of the control program; measures vary greatly in accordance with the character of the growth to be removed and other factors, but a number of innovations, such as the use of 2,4-D as a herbicidal agent and dynamite in maintaining ditches, must be mentioned.

Although the main emphasis of the TVA's anti-malaria activity has been upon biological or naturalistic methods of countering mosquito production, various chemical control measures have been employed, and larviciding, the most important, continues to be required. Since DDT became available during World War II, like other agencies the TVA has experimented with the use of aircraft for applying insecticides and special spraying equipment has been developed. Study has also been directed to the apparent resistance which anopheline mosquitoes acquire to DDT.[43] As other aspects of the control program have been completed, the area being treated by larvicidal spraying has progressively been reduced. However, the extent to which this measure is practiced varies considerably from year to year, depending upon the prevailing conditions.

All of the measures mentioned are aimed at controlling mosquito production. Attention has also been given to preventing contact between man and the anopheline mosquito. Limited use has been made of the residual spraying of DDT, and special malaria control easements have been procured in several sections of the Kentucky Reservoir, prohibiting night-time occupancy of structures near the water during the anopheline breeding season.

It is clear that there is no single best method to be pursued in a

[42] *Ibid.*, p. 502.
[43] *Ibid.*, p. 505.

malaria control program within an extended river basin such as that of the Tennessee. The TVA program has included a variety of different measures, in combination or alone, within a given stretch of water. In the area impounded by the Kentucky Reservoir, for example, provision was made for the application of six different types of control measures, three permanent in character and three requiring annual repetition.[44]

The effectiveness of the program carried out by the TVA is suggested by the threefold reduction in the anopheline mosquito population achieved in the area of all reservoirs between 1945 and 1951, and the even more dramatic decrease in the incidence of malaria parasitemia among inhabitants of the region. In 1934, when the program was first initiated, about 28 per cent of the blood slides of human subjects examined for malaria parasites were found to be positive, while less than 1 per cent have been positive on all surveys since 1941.[45] By June 1956, the seventh consecutive calendar year was concluded in which not a single new case of malaria could be traced to mosquitoes breeding on the waters of the TVA system. This record is of further interest in the light of malaria control costs, which have declined greatly since 1947 when the TVA's expenditures for this purpose reached a peak of $1,220,040. For the fiscal year 1956 they were $414,291 for the entire system.[46]

At the same time, the control program is significant because it has been successfully formulated as part of a multiple-purpose undertaking. Management of the water level of TVA reservoirs is not determined solely in terms of a single purpose but is made to accord as fully as possible with the needs of all uses. Initially it was feared that malaria control activity within the TVA area would have an unfavorable effect upon desirable aquatic life and upon migrating waterfowl. Over the years, however, it was found that the interests of wildlife conservation need not be jeopardized by malaria control. Special measures have been devised to reduce the area of conflict. Illustrative is the refilling of areas dewatered for mosquito control during the summer, in order to provide autumn migrating waterfowl with resting and feeding grounds,

[44] E. L. Bishop and F. E. Gartrell, "Permanent Works for the Control of Anophelines on Impounded Waters," *Journal of the National Malaria Society*, 3 (1942), 211-19.

[45] Derryberry and Gartrell, *op. cit.*, p. 506.

[46] Tennessee Valley Authority, *Annual Report, 1947, Annual Report, 1956* (Washington: U. S. Government Printing Office).

and to improve wildlife food supplies.[47] This has demonstrated how the conflicts which arise between development purposes can be minimized, and acceptably reconciled.

As a means of extending the range of water development in the Tennessee Valley, malaria control has been of major significance. In spite of its manifold benefits, this river development program could not have been undertaken had it resulted in serious hazard to health. Successful control of malaria in this area and actual improvement in health conditions have therefore effectively permitted this major extension of the range of surface-water exploitation for multiple-purpose use.

SUMMARY COMMENT

Until a relatively few years ago the principal water resources available for use or development were those which man had used for thousands of years—soil moisture, surface streams, and shallow groundwater deposits. All are highly dependent on the regional incidence of natural precipitation. Impoundment and large-scale development of surface waters, furthermore, was sharply limited in parts of the world where insect vectors of tropical diseases were prevalent. Within the last thirty years notable progress has been made in the direction of opening tropical and subtropical areas to large-scale, surface-water development. The depth range and the geographical spread of groundwater exploitation also has been increased by proven techniques of recovery. On the whole, however, technical "breakthroughs" have not yet opened water resources in the directions where major extensions of the physical range of water recovery are to be found. These are in the waters of the oceans, the deep-lying aquifers, and the unprecipitated moisture of the atmosphere. Such techniques may be a part of the emerging technology, promising but not yet widely proven, which is to be treated in the succeeding section of this study.

---

[47] Stromquist, *op. cit.*, pp. 837-38; U. S. Public Health Service and TVA, *op. cit.*, pp. 341-48.

PART III

EMERGING TECHNOLOGY
OF POTENTIAL IMPACT ON WATER USE

CHAPTER 11

Technical Activities in Progress and
Related to Water Use —

INTRODUCTION

The United States still is faced with problems of adjustment to the technical progress of the past. This adjustment presumably will extend into the future. At the same time, water management and development also will be faced with the now evolving technology of the future. If efficient development alert to exploiting all the technical opportunities is sought, the problems brought by technology are compounded. An evaluation of the directions in which the newer technical activity is moving therefore is essential to the story of technology in water development.

Some of the recent technical events impinging upon water development and use are of revolutionary scope in their potential effects upon organized economic endeavor. A prominent example is the impact of industrial use of nuclear energy materials upon both the flow and withdrawal uses of water. The weather modification activity now being undertaken also is potentially revolutionary.

Methodical evaluation of the newer technology is particularly desirable because popular, and even political, evaluation of nascent techniques with revolutionary promise tends toward cycles of quick optimism and following disillusionment. Orderly progress toward flexible, responsive, administrative structure for planning, experiment, and development is difficult and sometimes impossible in such an atmosphere. Projections in the following descriptions of evolving techniques therefore are directed toward a balanced analysis on the basis of existing information.

Relatively little allowance is made for "mutational" or presently unfore-seeable technical events which, nonetheless, are likely to appear. The objective is description and evaluation of several prominent technical activities in the light of existing information, looking toward the problems they create for future water development. These problems, as they relate to administrative organization for water development, are considered in chapter 22.

The techniques described in this section are chosen from all of the principal categories used to classify the past technical events described in chapters 5-10. *Techniques increasing or decreasing water demand, and those promoting the scale economies* are here illustrated by analyses of: (1) new developments in the generation of electric power from conventional and nuclear fuels, and (2) the relation of waste disposal in the atomic energy industries to water requirements (chapter 12). Three descriptions are chosen from the group of *techniques extending the physical range of water recovery,* one each for the three large areas of unexploited water—sea, atmosphere, and underground stored waters. The techniques described and evaluated in this instance are the use of saline waters (chapter 13), weather modification (chapter 14), and those applying to the discovery and investigation of ground-water reserves and movement (chapter 15). *Techniques extending the services afforded a given unit of supply* are represented by three descriptions, one each from industry, agriculture, and water management: the recycling or reuse of water in industry, the budgeting of water use in agriculture, and the reduction of reservoir evaporation losses (chap-ter 16).

The subjects discussed in this section in places may not seem sharply defined from the case descriptions presented in Section II. While for some, like the impact of nuclear fuel use, and the disposal of nuclear waste products, the cleavage of past and future development is very definite, in other cases, like the industrial recycling of water, past employment of the technique has been carried on over a long period. Some techniques thus are evolutionary and do not represent a sharp break with the past. Nonetheless, their major impact is more a matter of the future than of the past or present. This undoubtedly is true for industrial water recycling. The examples have been chosen not only for their possible future significance in the United States economy, but also because their major growth is a matter of the future. Thus their sharpest impact upon the problems of administrative organization for water development still is in the future.

For each subject dealt with in the following five chapters a description of the technique or techniques is given, along with brief evaluation of the stage of development which each has reached. Also included are evaluations of the prospective effects of the developing technique upon water demand and supply, the likely effects upon cost and capital requirements for water supply, the likely impact upon multiple-purpose planning, the effect on water users generally, and the impact upon the demand for materials and services connected with water development.

CHAPTER 12

Developing Techniques Likely to Influence the Rate of Water Exploitation or the Nature of Water Demand

(1) Recent Developments in Thermal Methods of Generating Electricity —increase in electricity demand; relation between thermal and hydro-electric power capacity and potential; the place of atomic energy in electricity generation, and its relation to use of conventional fuels; new technology in steam electric generation based on conventional fuels; generation by internal combustion engine and gas turbine; water use in steam power generation; effects of new steam generation techniques on water withdrawal and disappearance; effects of new steam generation techniques on multiple-purpose planning, development, and use of water. (2) Peaceful Uses of Atomic Energy as They Affect Water Supply and Use—reactor operation for electric energy production; isotope production and use; uranium ore processing; water use in spent fuel processing; disposal of radioactive wastes in water. Summary Comment.

The use of nuclear reactions for energy and other purposes is probably the foremost technical event likely to affect the future use of resources. Because of its potentially revolutionary influence upon the economy and technology of the country, some review of its likely effects upon water requirements is essential. In this connection, the two principal aspects of peaceful uses of nuclear energy materials lie in generation of electric power and in waste disposal. However, the application of nuclear technology to electric power generation cannot be discussed intelligibly without concurrent discussion of other means of thermoelectric power production. Accordingly the two cases chosen to illustrate the effects of developing technology on water requirements are thermal methods of generating electricity, and the relation of nuclear energy to water demand, including waste-movement and waste-disposal requirements.

1. Recent Developments in Thermal Methods of Generating Electricity

The single most important flow use of water is in hydroelectric power generation. The single most important industrial withdrawal use is for thermal (or steam) electric power generation. Four events in the use and generation of electricity are of significance to the future industry and its relation to national water requirements: (1) the greatly accelerating demand for electricity throughout the country, (2) shift of the burden of supply heavily toward thermal power generation subsequent to the development of the low-cost prime hydroelectric sites, (3) recent improvements in thermal power generation from fossil fuels, and (4) the current experiments with nuclear power plants.

These events already have had some influence on water requirements, and undoubtedly will have still further effects as the nation's future generating system continues to change. Attention will be given here to the possible effects in water quantity requirements, water quality requirements, and in multiple-purpose development and management of streams.

INCREASE IN ELECTRICITY DEMAND

One of the most spectacular developments in the last two or three decades in the United States has been the rapid increase in the generation and use of electric power. With electric generation capacity and total power generated increasing at a rate of about 10 per cent per year, total United States generating capacity reached 115 million kilowatts in 1955, and generation totalled about 620 billion kilowatt-hours.[1] As of January 1, 1957, total installed capacity was 137 million kilowatts. Of the total, 82 per cent was generated in thermal power plants from

[1] Statistics on the electric power industry may be found in several publications, including: *Electrical World* (e.g. September 1956), Edison Electric Institute *Statistical Bulletin* (e.g. 1956), and Annual Reviews by the Federal Power Commission.

coal, oil, and gas.[2] Estimates in *Electrical World* show a 1975 generating capacity of 430 million kilowatts and a total generation of 1,970 billion kilowatt-hours.[3] The Federal Power Commission in 1955 estimated a 1975 generating capacity of 301 million kilowatts and a total generation of 1,354 billion kilowatt-hours.[4] It is anticipated that by 1975 1,200 to 1,600 billion kilowatt-hours should be generated in thermal power plants having a capacity of 250 to 350 million kilowatts.[5]

Although the exact size of electrical generation increases is not predictable, it is clear that the growth has been and will continue to be very substantial. Most of this increase will be in the form of power generated from the burning of fuels, and most of this by means of the steam power plant.

The reasons for this development relate to the increasing standard of living, population growth, and the development of new technology whereby electricity can be conveniently and advantageously employed in industry and the home. Growth of electrical appliances in the home (chapter 6, pages 137-46), increasing electrification in manufacturing plants through replacement of direct mechanical drives, and automation of industry are all responsible. Better home and street lighting also has contributed to this development. The steadily increasing efficiency with which electricity is generated has resulted in a gradual reduction in cost so that its use has gained at a greater rate than other sources of energy.

RELATION BETWEEN THERMAL AND HYDROELECTRIC
POWER CAPACITY AND POTENTIAL

Depending on year and season, between 10 and 20 per cent of the electric power generated in the United States is from hydroelectric stations. The remaining 80 to 90 per cent is generated from fuel-using plants of different types. In 1956, 19.2 per cent of the total installed

[2] *Ibid.* Also F. L. Weaver, "Prospects for Future Hydroelectric Development in the U. S. and What it is Likely to Cost," paper presented March 14, 1957, Fifth Atomic Energy in Industry Conference, Philadelphia, p. 1.

[3] *Electrical World,* September, 1956.

[4] Federal Power Commission, *Annual Review, 1955.*

[5] Joint Committee on Atomic Energy, *Peaceful Uses of Atomic Energy,* Report of the Panel on the Impact of the Peaceful Uses of Atomic Energy (*McKinney Report* [Washington: U. S. Government Printing Office, January 1956]), Vol. I, pp. 35-39, Vol. II, pp. 24-30. Predictions based on estimates in *Electrical World,* September, 1955, and report of Federal Power Commission, October 1955.

Plate 19. Cooling towers for water to be recycled at a steam electric generating plant. Zuni Electric Station (gross capability is 115,000 kw.), Denver, Colorado.

generating capacity of the country was hydroelectric and 18.3 per cent of the total generation from all types of plants was from this source.[6] The proportion of hydroelectric and thermal generation is expected to remain roughly the same during the next two or three decades, and new capacity is likely to be dominantly thermal. The heavy dependence of electricity consumers on power generated from fuel is clearly evident.

The position of hydroelectric generation in the total power supply picture can be further illustrated. The installed water power capacity in the United States was 26.3 million kilowatts in 1957; by 1980, a hydroelectric capacity of about 55.6 million kilowatts is expected to be in service.[7] If all the potential water power in the United States were developed regardless of cost, only about 113 million kilowatts of total capacity would be available, based on mean stream flow. Even with development of known storage sites, only about 64 million kilowatts could be available more than 50 per cent of the time.[8]

Although the total percentage contribution of hydroelectric power will remain small, the importance of this type of generation will be high in certain regions where good sites are available. For instance, 34 per cent of the total undeveloped capacity is at Pacific Northwest sites.[9]

A most useful future application of hydroelectric power is in the balancing of fluctuating loads of power distribution systems. Modern large-scale thermal power plants are most advantageously operated at fairly constant output. Increases and decreases in generation rates as system loads rise and fall are relatively costly where the standby capacity or spinning reserve must be steam units. However, if adequate water storage facilities and hydroelectric generating capacity are available, peak loads can be met by release of impounded water through these plants, which can respond rapidly and which have low operating costs when on a standby basis. A few situations may require the opposite type of co-ordination in management. Where reservoir control is impractical, the natural stream flow must be used, and "run-

---

[6] Weaver, *op. cit.*

[7] Weaver, *op. cit.*, p. 1; also other references previously cited.

[8] L. L. Young, *Developed and Potential Water Power of the U. S. and other Countries of the World, December 1954* (Circular No. 367), U. S. Geological Survey, Washington, 1955; Weaver (*op. cit.*) quotes the Federal Power Commission estimate of 113 million kilowatts total hydroelectric potential.

[9] Weaver, *op. cit.*, p. 3.

of-the-river" generation will be part of the base load of the system. Other water power-steam combinations include the carrying of base load by part of the hydroelectric plus all the steam capacity, the peak being provided by additional water power. Another schedule employs hydroelectric power for base load part of the year and steam for base load the rest of the year, peaks being carried alternately by the other source. Each hydroelectric power project and each generating system will have individual characteristics and a best method of operation, but general dependence on thermal power is expected gradually to increase as available low-cost hydroelectric potential becomes fully developed.

THE PLACE OF ATOMIC ENERGY IN ELECTRICITY GENERATION, AND ITS RELATION TO USE OF CONVENTIONAL FUELS

The place of atomic energy in electricity generation has already received extensive public attention. The promise held out for this revolutionary method of energy production has been so great that its position must be reviewed before use of conventional fuels may be considered or rejected as part of future technology in electricity generation.

A peaceful application of atomic energy which may have the greatest impact on the economy is the generation of electricity from nuclear fission reactions involving uranium and thorium. Although opinions vary widely as to the rate of development of atomic power in this country, there is almost universal agreement that a gradual increase in the percentage of the total electric generation capacity supplied by atomic energy will take place.[10] Considering the great expansion expected in total electric generation the conditions for assumption of some of this load by atomic power are ideal. The 1957 total United States electric generation capability of 137 million kilowatts is expected to rise by 1975 to a level between 300 and 450 million kilowatts.[11] This is approximately a threefold expansion. Estimates of total installed electrical capability in atomic power plants in 1975 range between 20 million and 45 million kilowatts.[12] Although these atomic power capabilities represent only 5 to 15 per cent of the expected total, annual additions of atomic electricity plants at the end of this twenty-year period possibly will be as much as 40 per cent of the total new capacity being built at that time.

[10] *McKinney Report*, Vol. I, pp. 35-39.
[11] *Ibid.*, Vol. II, pp. 24-30.
[12] *Ibid.*, Vol. I, p. 38; Vol. II, p. 27.

Most of the uncertainty in the rate of nuclear energy development is due to the difficulty in predicting costs of atomic power generation. In all atomic power plants now under construction or being planned, total costs of electricity are estimated to be considerably greater than costs of power from modern fuel-burning plants.[13] Not only are the fixed costs on the installation much greater, but the atomic fuel costs are not yet competitive. It is expected that both of these expenses will gradually decrease as a result of further development and practical operating experience. The rate at which these costs are reduced will in large measure control the proportion of new capacity that is based on atomic energy sources.

Numerous types of atomic energy reactors are in the research and development stages, and several other types are being planned or built for scheduled use in commercial atomic power plants. All are based on the principle of producing heat in a reactor through fission of uranium-235 (one of the natural uranium isotopes, present to the extent of 0.71 per cent in all uranium ores), the heat then being directly or indirectly transferred to steam for use in a conventional steam turbine— electric generator installation.[14] Thus, the nuclear reactor replaces the furnace section in a conventional fuel-operated electric generation plant. The factors which limit the efficiency of converting energy in steam at high pressure and temperature to electric power in conventional plants apply to atomic energy plants as well.

Because of radioactive elements produced in the nuclear reactor in the fission of uranium, this unit must be shielded from operating personnel and other equipment. However, the heat must be transferred from the reactor to steam, which has to leave the reactor zone and pass through a turbine. Since turbines require maintenance, radioactivity in the steam must be avoided. In most designs, therefore, heat is transferred first from the reactor core to a circulating fluid (liquid sodium, pressurized water, an organic liquid, an inert gas, or other

[13] L. H. Roddis, "Possibilities for Developing Economic Small Power Reactors," address to American Public Power Association Conference, Los Angeles, April 1956; also Joint Committee on Atomic Energy, *op. cit.*, Vol. 1, pp. 35, 38.

[14] The fundamentals of nuclear energy production and the important features of the several most important types of reactors are summarized by R. A. Charpie. "The Technology and Economics of Nuclear Power" (in the *McKinney Report*, Vol., II, pp. 31-42). Another review: A. M. Weinberg, "Fuel Cycles and Reactor Types," *Proceedings of the International Conference on Peaceful Uses of Atomic Energy, at Geneva, 1955,* Vol. III (New York: United Nations, 1956).

material) and then from the circulating fluid to steam in a separate heat exchanger. By this secondary heat exchange, steam free of radio-active impurities can be supplied to the turbine.

Atomic reactors operating at temperatures and pressures as high as those used in the most efficient and modern steam power plants are not yet feasible, because materials of construction combining properties essential in a reactor and properties needed for high pressure and temperature are not yet available. Difficult corrosion problems and strength factors dependent on atomic radiation have limited the maxi-mum working temperature and pressure of nuclear reactors.[15] As a result, over-all efficiencies of converting atomic heat to electric power do not approach those realized in best power plant practice. Typical values are about 25 per cent for atomic power plants in the design stage as compared with nearly 40 per cent in the best fuel-burning plants.[16] There is reason to expect improvement in atomic power plant efficiencies, as developments proceed, but they may never be fully as high as those in the best conventional fuel plants.[17]

A challenging potentiality in atomic power is the controlled nuclear fusion reaction of hydrogen to helium, with accompanying energy release. Although intensive research on this unsolved problem is being conducted by the Atomic Energy Commission (under the designation "Project Sherwood") and by scientists in other countries, no break-through has yet been reported which would assure the development of fusion atomic power for practical use.[18] An important step has recently been accomplished, however, in the momentary electro-magnetic pro-duction of the several million degrees temperature required for the con-trolled nuclear fusion reaction.

Another research effort with a highly desirable objective but without

[15] N. P. Jackson, "Atomic Energy and the American Economy," *Journal, American Water Works Association,* 47 (1955), 1139-47; also "Nucleonics," *General Electric Review,* 60 (January 1957), 36.

[16] J. A. Lane, "Economies of Nuclear Power," *Proceedings of the International Conference on Peaceful Uses of Atomic Energy, at Geneva, 1955,* Vol. I (New York: United Nations, 1956), pp. 309-21, 476.

[17] *McKinney Report,* Vol. I, p. 41.

[18] On May 9, 1957, a group of Texas electric utility companies contracted with the General Dynamics Corporation to support a major experimental program for investigating controlled nuclear fusion for peaceful use. (*New York Times,* May 10, 1957.) The announcement of this contract stressed a principal objec-tive of achieving an understanding of the reaction rather than the development of equipment for use of fusion reaction energy.

Plate 20. The Shippingport Atomic Power Station, Shippingport, Pa. The dark-colored building in the center houses the nuclear reactors. This station was the first atomic plant in the United States to be devoted exclusively to electricity generation. It is operated by the Duquesne Light Company. The nuclear section was built by the Westinghouse Electric Corporation under the direction of the Atomic Energy Commission.

very favorable prospects for early realization is the direct conversion of nuclear energy to electricity without intermediate heat production, engine operation, and generator use. Such a method for efficiently converting neutron flux to electron flow (electricity) would circumvent the inherent inefficiency of even the best heat engine system. It also might eliminate the need for water in the process.

In view of the problems of development still faced in the generation of electricity from atomic energy by uranium fission, and in view of the secondary position suggested for nuclear plants within the next two decades, even in optimistic forecasts, nuclear technology cannot now be expected to have a commanding influence on the pattern of electricity production for at least three decades.

NEW TECHNOLOGY IN STEAM ELECTRIC GENERATION
BASED ON CONVENTIONAL FUELS

The certain dominance of thermal means of producing electric power and the secondary position of nuclear power generating plants for some years in the future indicate that technical developments in the production of electricity from the fossil fuels are of vital interest in the generating techniques of at least the next twenty-five years.

The most significant technological development in steam power generation during recent years has been the steady increase in efficiency of converting the energy in fuels to electrical energy. In general, efficiencies have gradually increased through a large number of technical improvements, but a few developments of "breakthrough" nature have also appeared. Of these, the most important has been the introduction of extremely high-pressure, high-temperature boilers, culminating very recently in designs at pressures and temperatures above the critical point of water.[19] These developments have made operation at much higher efficiencies technically feasible. Thermodynamic limitations on the proportion of fuel energy which can be converted to electrical energy make it impossible to achieve exceptionally high efficiencies unless very high steam temperatures are employed. High pressure is also advantageous in increasing the conversion of fuel energy to electricity and in

[19] The critical point of water is the temperature above which no liquid water can exist, regardless of the pressure. This temperature is 705°F., and the critical pressure (pressure exerted by steam in presence of liquid at the critical temperature) is 3,206 pounds per square inch.

reducing equipment size per unit of work output. Recent developments in metallurgy have permitted the design and operation of steam boilers and turbines at temperatures well above 1,000 degrees Fahrenheit and at pressures exceeding 3,000 pounds per square inch.[20]

Of particular importance in the design and construction of boilers operating under such severe conditions has been the successful application of low-carbon, specially heat-treated austenitic steels in boiler tubing. Other improvements have been in the development of work-hardened superalloys, wrought superalloys strengthened by precipitation heat treatment, cast superalloys, titanium carbide ceremets, and molybdenum alloys.[21] Without these or other metallurgical developments, the improved power plant efficiencies would not have been possible. The best power plant designs as recently as ten or fifteen years ago permitted the generation of 1 kilowatt-hour of electricity with the combustion of about 12,000 Btu of fuel (corresponding to about 30 per cent energy conversion efficiency).[22] Now the most modern large plants are operating with heat rates of only 9,000 Btu per kilowatt-hour generated, and a new plant with a design heat rate of slightly over 8,000 Btu per kilowatt-hour will soon be in operation.[23] It is believed that by 1975 or 1980, heat rates as low as 7,500 will be obtained.[24]

Boiler operation at supercritical temperatures and pressures involves unique technological problems. Their satisfactory solution has been a major accomplishment. In addition to the metallurgical developments required, special techniques for boiler feed water treatment have had to be developed. The method of operation permits no separation of impurities in the production of the superheated and superpressure steam, hence the water must be of extremely high purity to prevent scale formation in the boiler, the turbine, and other facilities. In fact, complete demineralization of feed water is required.

Along with boiler developments, turbine improvements have also been realized. The use of extremely high pressure stages in the turbine

[20] R. G. Rincliffe, *op. cit.* Also J. H. Harlow, "Engineering the Eddystone Plant," paper presented at meeting of American Society of Mechanical Engineers, New York City, November 1956.

[21] "Turbines and Tomorrow's Metals," *Steel*, 136 (February 28, 1955), 88-90.

[22] R. G. Rincliffe, *Supercritical Pressures; Essential Forward Step in Power Generation* (Edison Electric Institute Bulletin 23), July 1955, pp. 208-12.

[23] I. Baumeister, "How Will We Lick Superpressure Problems?", *Power*, 98 (July 1954), 93-95; also Harlow, *op. cit.*

[24] *McKinney Report*, Vol. II, p. 15.

has required new designs involving different steam flow arrangements and turbine blade materials.

Developments leading to more efficient combustion of solid fuels also have contributed appreciably to the increase in power plant efficiency. Actual fuel losses have been reduced in modern power plants to negligible fractions of the energy supplied. Automatic stoker equipment, powdered coal firing, close control of air-fuel ratios, better combustion chamber design, and other improvements are responsible. Techniques for improving and using low-grade fuels also have been introduced. Beneficiation of low heating value sub-bituminous and lignite coals by a drying process followed by low temperature coking, has made steam power cheaper in sections of the country possessing these fuels.[25] Low-grade fuels also are used without beneficiation by mixing them with higher grade coal in the course of powdering boiler fuel.

Developments which have led to the practical design and construction of huge power generation units, now 250,000 kilowatts or more, have contributed indirectly to efficiency improvement. To some extent, the multiplying of these units into exceptionally large stations, some exceeding a million kilowatts capacity, can raise efficiency further. As plant sizes become progressively larger, the economics of recovering relatively small percentages of heat loss becomes more and more favorable. It therefore becomes practical to install equipment for obtaining even a few tenths of one per cent higher efficiency, whereas the costs of these improvements in smaller power plants could not be economically borne. The use of economizers, by means of which heat is recovered from flue gas by exchange with the incoming combustion air, and the installation of feed water heaters that employ flue gas for pre-heating feed water, are common practice in large new plants. The lower the final flue gas temperature, the higher is the over-all fuel conversion efficiency in the power plant.

Another advantage of the large-scale plant is increased economy through reduction in overhead, building cost, and other common costs per unit of capacity.

Two further developments of considerable importance in improving

---

[25] The process produces a char having a satisfactory heating value, together with a liquid tar by-product. A large installation of this type is in service at the plant of the Texas Power and Light Company in Rockdale, Texas. The process also may be readily adapted to the complete gasification of coal to produce a pipe-line fuel.

power plant efficiency are the successful application of steam reheat following partial expansion in the turbine, and the use of pressurized air in the boiler furnace.[26] By use of one or two stages of reheat, over-all thermodynamic efficiency in the heat engine cycle[27] can be increased, and at the same time mechanical difficulties due to moisture in the expanded steam can be greatly reduced. The latest practice involves two stages of reheat, that is, steam is withdrawn from a turbine after partial expansion, returned to a section of the boiler plant for reheating, and delivered again to the turbine for further expansion. Reheating a second time results in an additional increase in over-all efficiency.[28] The use of air at slight pressure in the fuel combustion process increases the efficiency of heat liberation and transfer to the boiler tubes.

GENERATION BY INTERNAL COMBUSTION ENGINE
AND GAS TURBINE

Approximately 82 per cent of the nation's electricity is generated in fuel-burning plants, and over 99 per cent of this is generated in steam power plants.[29] Nearly all of the small balance is generated by use of internal combustion engines, primarily diesel. A very small amount of electric power is generated in gas turbine plants, just recently introduced. However, practically no water is required in the operation of diesel engines and gas turbines.[30] The low water requirement of these power generation units is a distinct advantage in certain areas where water may be scarce. Processes depending on the combustion of liquid or gaseous fuel and the direct expansion of the resulting hot gases to drive recip-

[26] T. Kolflat, "Reheat Comes of Age," *Electric Light and Power,* 32 (September 1954), 88-91; also "Today's Station Designs: What's New?", *Electrical World,* 142 (October 18, 1954), 127-42.

[27] The heat engine cycle is the process by which heat transferred from fuel combustion products to a working fluid (steam) is partially converted to mechanical work by expansion of the steam in an engine (turbine). The useless heat energy remaining in the exhaust steam is discarded by condensing it with cooling water.

[28] "Efficiency Continues to Improve," *Electrical World,* 144 (October 17, 1955), 132-34.

[29] "53rd Annual Report on the Electrical Industry," *Electrical World,* 147 (January 28, 1957), 141.

[30] It is needed only for cooling certain engine components to prevent damage by overheating. Consequently there is only a small vaporization loss from a recirculating water system.

Plate 21. A 5,000-h.p. gas turbine, Cornudas Station, El Paso Natural Gas Company, Texas.

rocating or revolving machines coupled to electric generators appear to have a bright future in the production of power in small facilities (25,000 kilowatts or less). If present indications are borne out in practice, the gas turbine should largely supplant diesel engines in such applications.[31]

WATER USE IN STEAM POWER GENERATION

There are two needs for water in steam power plants. One requirement is for feed water which is supplied to the boiler and converted to steam for subsequent expansion in the turbines. The other and considerably larger water requirement is as a coolant by which exhaust steam from the turbines is condensed.

The purpose in condensing exhaust steam is to secure the lowest possible pressure (actually a partial vacuum) at the turbine exit. Because of this low exhaust pressure, the efficiency in converting heat to electrical energy is greater than if the exhaust steam were discharged directly to the atmosphere.

Since high-purity water is required in the boilers of modern steam power plants to prevent scaling and corrosion, treatment systems are practically always necessary for purifying the incoming boiler feed. The cost of this treatment and the availability of pure condensate from the turbine condensers dictate the recirculation of boiler feed water to the maximum practical extent. There is some unavoidable loss due to leakage and accumulation of impurities in this circulating stream, but the actual fresh-water requirements for boiler feed are very small. Although a modern power plant requires the conversion of about one gallon of water into steam in the boiler per kilowatt-hour generated, practically all of the steam is recovered as condensate in the condenser. Recirculation of all but 1 or 2 per cent lost and intentionally discarded permits power generation with a requirement for fresh or make-up feed water of .01 to .02 gallon per kilowatt-hour generated.[32] A modern 150,000-kilowatt power plant would therefore need 35,000 to 75,000 gallons of make-up feed water per day, with best recirculation practice. Although much more feed water is used in older plants, the trend is

[31] *General Electric Review,* 60 (1957) 20. Also "Turbines and Tomorrow's Metals," *Steel,* 136 (February 28, 1955), 88-90.

[32] J. G. Thon and G. L. Coltrin, "Morro Bay Steam-Electric Plant," *Proceedings, American Society of Civil Engineers,* 81 (Separate No. 737), July 1955.

toward low consumption of boiler feed water as these plants are replaced with more modern equipment.

By far the largest industrial water requirement is for cooling, and most of this is in steam power generation plants. Usage varies greatly from one plant to another, depending upon the extent of water recirculation. If there is no recirculation and the water temperature increases about 10 degrees in flowing through the condensers, approximately 100 gallons will be used per kilowatt-hour generated in the average power plant in use today. Small, older power plants require considerably more. With extensive recirculation of cooling water, however, only the water lost by evaporation in the cooling towers (plus a small make-up to compensate for other losses and purge) may be supplied. With efficient modern power plant practice involving about 9,000 Btu of fuel per kilowatt-hour generated, only 5 or 6 pounds of water need be evaporated per kilowatt hour generated.[33] Hence, the net consumption and withdrawal of approximately one gallon of water per kilowatt-hour generated can be considered a practical minimum. If the warmed condenser water is discharged into lakes without recirculation, the net loss or disappearance of water by evaporation from the warmed body of water is at least half that occurring in the average cooling tower system.[34] The evaporation rate in streams undoubtedly exceeds that in lakes. The range of daily water withdrawal requirements for a modern 150,000-kilowatt plant would be from a minimum of about 3 million gallons to as much as 300 million gallons.

Fortunately, quality requirements for cooling water are not severe, and many types of impure water, including brackish supplies and sea water, may be employed. A substantial part of electric power generated in coastal areas need not require fresh water for cooling purposes (chapter 13).

In contrast with the advantage of saline water use in coastal areas is the disadvantage encountered by power plants located on streams that are drawn upon heavily for industrial cooling purposes. Warm river water is a distinct handicap, because high power plant efficiency requires low cooling water temperatures. The concentration of industry in certain areas is leading to serious problems resulting from high river temperatures.

Waste disposal is not a problem related to water use in steam power

[33] R. G. Rincliffe, *op. cit.;* also "Kearney Plant Heat Rate Best Yet," *Electric Light and Power,* 32 (February 1954), 101-4.

[34] G. E. Harbeck, "The Use of Reservoirs and Lakes for the Dissipation of Heat" (Circular 282), U. S. Geological Survey, 1955.

plants, except as involved in waste heat removal. Gaseous and liquid fuels leave no residue whatever, and coal ashes are usually handled adequately by solids disposal techniques. Chemical wastes from water treatment operations are of minor significance. However, these statements do not apply to generation of energy with nuclear fuels. The possibility of radioactive contamination of streams presents such a serious hazard that an entirely different standard of purity is involved in such plant effluents.[35]

EFFECTS OF NEW STEAM GENERATION TECHNIQUES ON
WATER WITHDRAWAL AND DISAPPEARANCE

New techniques of steam power generation may affect water demand through a reduction in quantities evaporated, through change in requirements for water purity, and through reduction in withdrawals. The first change results directly from increase in thermodynamic efficiency; the second is caused indirectly by efficiency improvements; and the third results from recirculation as well as efficiency change.

Efficiency improvement reduces cooling water use, because less of the unused heat in the fuel must be discarded in cooling water when there is a lower heat input rate per kilowatt-hour generated. Most of the heat not converted to electricity is removed in cooling water, resulting in evaporation of that water. A lower heat rate will accordingly lower water requirements. If a heat rate of 12,000 Btu per kilowatt-hour prevails, approximately 7,500 Btu must be removed in cooling water, corresponding to the evaporation of about 0.9 gallon of cooling water. However, if the plant heat rate is as low as 8,000, only about 4,000 Btu would be discarded in the condenser and slightly less than one-half gallon of water will have to be evaporated to dissipate this heat.

Techniques of using cooling water can affect requirements both for quantity and quality of supply. Steam power plant location formerly was determined to a large extent by the availability of large quantities of water, or by access to high purity supplies for make-up water in a recirculation system. Technical improvements have practically eliminated these locational limitations. Large quantities of steam-electric power may now be generated with small withdrawals of even very low-quality water. The developments primarily responsible for these advantages are water treatment methods for removing minerals, and minimizing corrosion, scale deposits, and biological contamination. Better

[35] See section 2 of this chapter, pp. 324-32.

control of these difficulties has permitted substantial decrease in the proportion of water discard (technically known as "blow-down"), while maintaining satisfactory water quality in plant recycling equipment (see chapter 16).

Location restrictions and plant operational difficulties because of mineralization of boiler feed water also largely have been removed. The development of very high temperature and high pressure boilers, requiring nearly pure water, has been paralleled by better water treatment techniques. Close control of impurities in the circulating stream and the use of corrosion inhibitors in the feed water have yielded substantial operating benefits.

Contribution toward less severe locational requirements also has been made in the development of equipment which permits cooling with sea water or other water of high mineral content. Power plants located on the seacoasts may use sea water for condenser purposes, and even by distillation for boiler feed, if fresh water is unusually difficult to secure. Present techniques for handling sea water in power plant condensers are completely satisfactory in minimizing corrosion and biological fouling. Consequently, the growing scarcity and increasing temperature of fresh-water supplies in some areas are causing substantial increases in power generating capacity with salt-water cooling systems. Even larger proportions may be expected in the future. Increased preference for coastal locations and sea water use, even at the expense of longer power transmission lines, now seems likely. Former uncertainties in the use of sea water for condensing have been replaced by complete confidence in its suitability and continuous availability.

In spite of the remarkable increases in power plant efficiency and the resulting lower water requirements per kilowatt-hour generated, the demand for water in the generation of electric power will rise rapidly in the immediate future. The estimated 1975 national steam power generation, comprising 80 to 90 per cent of the total generation, is expected to be between 3 billion and 4.5 billion kilowatt-hours per day.[36] It is estimated that the present water use in steam electric power generation, about 60 billion gallons per day, will more than double by 1975 (131 billion gallons per day).[37] This rise will be accompanied by

[36] Computed from data in *McKinney Report,* Vol. ii, pp. 18-29.
[37] U. S. Department of Commerce, *Water Use in the United States, 1900-1975* (Business Service Bulletin 136), Washington, January 1956.

approximately a threefold increase in thermal power generation. The lack of proportionality is due to expected water savings resulting from increased power plant efficiency and more extensive recycle. This will be by far the largest industrial use of water in the United States.

These latter water quantities are total withdrawals, not disappearance. Extensive recycling could bring withdrawal down to levels approaching disappearance, now 1.5 to 2.0 billion gallons per day and expected to be 3.5 to 4.5 billion gallons per day in 1975. However, actual withdrawals are not likely to approach these low levels because of the lack of necessity in areas of plentiful water, including coastal regions where sea water is used for cooling. As nearly absolute minima, however, they represent useful figures for comparison. In regions of scarce water supplies withdrawal may be expected to approach disappearance in electric generating plants, roughly on a basis of a gallon of water per kilowatt-hour generated. Hence, a 100,000-kilowatt plant, operating at a 60 per cent load factor, could be expected to use about 1.5 million gallons of water per day. In some plants, particularly in the Rocky Mountain and Great Basin states, these low water withdrawal rates already prevail.

Power generation technology has developed to the point where practically no area need be without steam power facilities because of water scarcity. An example may illustrate this. A small community of 10,000 persons would use, on the basis of the national average (all uses), about 100,000 kilowatt-hours per day.[38] A minimum water requirement of about 100,000 gallons per day for a steam power plant would represent only about 7 per cent of the total water needs of the community, computed at 146 gallons per day per capita.[39] It can be seen, therefore, that if there is sufficient water for a modern community's existence there will be adequate supplies for power generation. Finally, for still smaller communities, the gas turbine power plant can make a further water saving in power generation. This unit requires practically no water for its operation, and where reasonably small quantities of power are required (about 25,000 kilowatts capacity or less), this type of plant may become dominant during the next two or three decades. The gas turbine will be a comparatively insignificant factor in the total

[38] Computed from 620 billion-kwh annual consumption and 175 million population.

[39] H. E. Jordan, "The Problems that Face our Cities," *Water, The Yearbook of Agriculture, 1955,* U. S. Department of Agriculture, Washington, 1955, p. 651.

power generation of the United States, but its freedom from dependence on water should make it an ideal source of power in water-scarce localities.

The high concentration of power generation plants in the Northeast, the Great Lakes region, and other areas where water supplies (including salt water) are generally plentiful indicates that there will be a considerable variation in water conservation practice between these areas and the regions where supplies are scarce. Hence, new technology in electric power generation has its main immediate influence on fuel use and economics rather than on water disappearance and conservation. As population and industrial activity increase in the more arid western states, and as competition for existing supplies in the eastern states becomes more severe, regional and local patterns of water use in electric power generation will differ more widely from the national average.

In summary, water demands will increase sharply because of power generation increase in the years ahead, but these water demand increases will be partially offset by savings resulting from higher power plant efficiencies and greater recirculation of water. Unlimited supplies of sea water for cooling purposes are available to a large part of the United States power generation capacity. Where water is scarce or will become scarce, extensive reuse can reduce requirements to a comparatively low level, and for small communities and small industries the gas turbine can free the power generation station from substantially all water requirements.

EFFECTS OF NEW STEAM GENERATION TECHNIQUES ON MULTIPLE-PURPOSE PLANNING, DEVELOPMENT, AND USE OF WATER

Beyond the direct effects of steam power generation upon withdrawal and disappearance of water, the effect of the new steam power technology upon multiple uses of water, both direct and indirect, must be considered. These conveniently may be reviewed by treating the effects upon hydroelectric development and operation, and upon other water uses. The new technology potentially can affect both the quantity and the type of water power developed, its methods of operation, and the investment costs which it can bear. The nature of hydroelectric development in turn may affect the position of other purposes in multiple-purpose development. Water withdrawal on large scale for steam power generation also may have direct effects on other water uses. All these may have bearing on the nature of multiple-purpose development.

Effects upon hydroelectric generation. A key question in all multiple-purpose use is the position of water power, about which there has been considerable speculation since the development of the nuclear reactor, and since the introduction of very large steam turbines. The question is of particular importance because of the dominating economic position of water power in all past multiple-purpose development. Reasons for a questioning attitude are suggested in these facts: (1) Hydroelectric power already is being generated at efficiencies above 90 per cent, with relatively little prospect for further improvement. (2) On the other hand, steam plant efficiency improvements actually are reducing thermal power costs. (3) Average capital costs of hydroelectric capacity are increasing because of the use of sites which are more expensive than formerly to develop.

Of the 86.9 million kilowatts of estimated undeveloped hydroelectric capacity in the United States in 1956, 29.3 million were authorized or licensed for construction, or in the state of application for development license. It is expected that all of this category will be developed in the reasonably near future, probably by 1980. Development of the remaining second category of 57.6 million kilowatts will "probably range from the feasible and competitive to the completely infeasible."[40] The entire 86.9 million kilowatts of undeveloped potential, however, is only about 31 per cent of the 280 million kilowatts additional generating capacity needed in the country by 1980.[41] A vast development of steam capacity therefore seems inevitable.

This suggests four questions: (1) Will the steam development at the high efficiencies and low costs now indicated eliminate the need for developing much of the second-category sites? (2) Will steam development change the manner of operation of all hydroelectric plants, including those existing, under construction, planned, and potentially developable? (3) Will priorities and values of other water uses in multiple-purpose developments increase enough to bear a larger share of project costs and thus make otherwise uneconomic water power competitive? (4) What effects are there likely to be in the use of water pumping-storage-hydroelectric generation techniques?

These questions are in large part answerable by considering the system relations of hydroelectric and the latest steam plants.

[40] Weaver, *op. cit.,* p. 8.
[41] *Idem.*

Power consumption in utility distribution systems generally is characterized by peak loads approaching double the average demand; generation systems must therefore be capable of producing nearly twice the power which they ordinarily supply. Modern steam power plants are not well suited to large variation in load and generation rate, whereas water turbines in hydroelectric stations are easily and efficiently controllable at various loads. Thus, if a generation system has both steam and water power facilities, and if the hydroelectric facilities have sufficient reservoir storage capacity or pondage, the steam plants may be operated to carry the base load, and the hydroelectric plants the peaking increment (see chapter 9, pages 248-50).

This new relation between steam and water power facilities brings an interesting cost factor into water development. Weaver has described it thus: ". . . . Energy costs per kilowatt-hour in fuel electric plants . . . vary inversely with capacity factor.* Thus, hydro power costing 10 to 15 mills per kilowatt-hour may be less costly than steam power used in the peak of the load, even though power from the steam plant might cost in the range of 6 to 8 mills per kilowatt-hour at system load factors." [42]

Where accessibility to fuel and the size of load call for the construction and operation of large steam power units, the development of relatively high-cost water power within reach of the system containing the large steam units thus may be feasible. At least some of the second-category undeveloped hydroelectric sites are therefore likely to be developed because of this relation. In addition, the changeover of existing hydroelectric capacity in similar situations to some peak load operation seems equally probable, as already has occurred on the TVA and some other systems.

Such changes in present water power facilities and in the design of new installations will generally require no increase in water storage volume, but additional penstock, turbine, and generator capacity will have to be provided if the available water is to be fully utilized in a peaking operation rather than in carrying base load. Although the higher capital cost and the lower load factor will result in higher kilo-

---

* This relation might be clarified thus: Energy costs per kilowatt-hour in fuel electric plants *increase* as the output percentage of capacity decreases. An exact inverse proportionality does not prevail because fuel cost is usually the major item in power generation cost, and it does not vary greatly per kilowatt-hour generated as capacity factor changes. (Authors' note.)

[42] *Ibid.,* pp. 7-8.

watt-hour costs, the cost increase may be less than for reserve steam capacity.

However, there are other situations where fuel costs are high and unusually favorable conditions exist for low-cost hydroelectric production at high load factor. This is mainly within the North Pacific drainage area, or the Pacific Northwest. In that area the past prevailing methods of hydroelectric operation are apt to continue longer than elsewhere. It is of interest that 38 per cent of the 86.9 million kilowatts undeveloped potential water power in the country, and 40.6 per cent of the expected total generation from this potential (358 billion kilowatt-hours per year) are within this Pacific drainage area.[43] However, at the rapid rate of development of favorable low-cost sites, it is not unlikely that even in this area, steam may become cheaper than the remaining undeveloped water power within twenty-five years.

The effect of operating hydroelectric facilities for carrying peak load rather than base load, is likely to be felt in other ways. A larger number of reservoirs are likely to be developed than otherwise might be the case, and pumped storage projects are likely to become more common. Conflicts between hydroelectric generation and other flow uses of water are likely to be intensified somewhat, since peaking operation causes less regular releases through turbines. Affected uses include direct recreational, fish and wildlife, waste-carrying, and downstream municipal and industrial supply. Where system operation is organized with both steam and water power, past troublesome conflicts between power and recreation, fish and wildlife, and withdrawal uses are not likely to disappear in the new pattern. Under circumstances of low priority for hydroelectric generation, devices of reregulation will often permit the use of conservation storage for best water power generation in an integrated system.

Operation of hydroelectric stations primarily for handling peak loads may also be in conflict with maximum flood control purposes. Periodic release of large volumes of water into a stream at flood stage could not usually be safely practiced. In small reservoirs, extreme daily changes in water level would compete with reserve capacity for flood control. Still another factor is the need, during flood seasons, of passing large volumes continuously. This requirement may dictate departure from hydroelectric operation only for peak loads, and a temporary conversion to base load operation.

[43] Federal Power Commission calculation, Weaver, *op. cit.*, p. 10.

Effects upon uses other than hydroelectric. Two indirect but important effects of the improved steam power technology upon water uses other than for water power merit consideration: (1) intensified competition for water by the expanded power generating capacity; and (2) the temperature increases of surface or ground waters resulting from greater waste heat disposal in power plant condensers. The latter may affect other industry, fish and wildlife, and waste carrying in streams. In the balance, however, it appears that the newer technology has the capacity not only for avoiding detrimental effects, but also for improving now unfavorable environmental conditions.

Past rapid growth in steam power generation has created the largest competitive industrial water demand. As technology has improved, power cost has decreased, use has increased, and water requirements have risen. Thus, in districts where water supply is limited, competition among water users for the existing supply has been accentuated by these developments. Something of a paradox appears here. Plentiful and cheap electric power is a strong aid to industrial activity in any area. In turn, additional water supplies are required for the industries themselves, and for the municipal systems associated with industrial concentrations. Thus the industrial growth stimulated by the availability of a good power supply might be expected to impair further development through severe competition for water supply.

Where local problems develop from competition for water withdrawal, existing or now developing techniques indicate satisfactory solutions. Recirculation methods (chapter 16) make possible very large reductions in water requirements. The effects of competition for water thus need not be heightened by increased power generation, where it serves as one base for expanded industry.

A detrimental effect of the rise in steam power generation is the temperature increase which may result in rivers used for condenser cooling. If cooling towers and reuse of condenser water are not employed, and if the normal river flow is small or if there are many users on the stream, highly undesirable temperature increases in the river may be encountered. The case of the Mahoning River in Ohio has been mentioned in chapter 2 (page 43).

Elevated water temperatures may be detrimental to other industry, to fish and wildlife, and to the capacity of a stream for handling wastes. Temperature rise is associated with a decrease in oxygen content and a resulting loss in capacity of moving water to oxidize waste materials which it otherwise could decompose. Oxygen content is also of direct importance to the survival of fish life in a stream, and the content of unoxidized waste in a body of water may affect both fish and wildlife unfavorably. In addition, high water temperatures impose cost handicaps on other industrial users of a stream for cooling purposes.

The return flow of cooling water into aquifers also may have undesirable effects, perhaps changing the permeability and hydraulics of the aquifer as well as its average temperature. However, comparatively little is known of the effects on ground water as yet.

As in the case of competition for withdrawal, undesirable temperature rises may be avoided or minimized by reuse of water within each generating plant, together with effective recooling in atmospheric cooling towers.

Summary comment on relation to multiple use of water. The newer steam power technology in almost every instance may be interpreted as having a favorable effect upon the development and management of multiple-purpose water facilities and multiple use of water in the United States. In spite of the vast expansion of steam power generating facilities now anticipated, steam plants are not expected to displace hydroelectric plants, because of the much enhanced value of water power for peaking purposes on systems within reach of both types of generation. These operating changes may intensify some conflicts between hydroelectric generation and other purposes served by storage facilities, as for example, fish and wildlife, waste carrying, and industrial supply. However, competition between withdrawals for steam power and other withdrawal uses may be expected to lessen or to disappear as reuse, recirculation, and low-quality water use increase in application.

2. Peaceful Uses of Atomic Energy as They Affect Water Supply and Use

While the use of atomic energy may grow more slowly than at first expected, the prospect for its widespread use in the economy of the United States at some time during this century appears very likely. Such a far-reaching development is certain to have significant effects on as basic a resource as water. Two general fields will be considered briefly here to illustrate the relation of atomic energy to water demand and supply: quantity and quality requirements of nuclear industry, and the relation of nuclear waste disposal to water use and supply.

In the succeeding discussion, two assumptions are made. One is that the principal process of producing energy from nuclear reactions for the next forty years will be by atomic fission and a steam engine cycle. The other is that the cost of electricity produced from atomic energy is likely to be about the same as the cost from modern steam plant generation using fossil fuels. Should a major scientific and technical breakthrough take place in the application of nuclear fusion processes to electric energy generation, both the outlook for costs and the problem of waste disposal might be radically changed. Other possibilities are the successful direct conversion of nuclear energy into electricity or the development of a gas turbine cycle powered by an atomic reactor. Either of these might materially reduce water requirements in nuclear-fueled power plants. However, planning of water resource use for the next twenty-five to forty years on the basis of the two stated assumptions appears reasonable at this time.

Water for processing, cooling, and waste disposal is and will be needed in the following activities concerned with applications of atomic energy: reactor operation for electric energy production, reactor operation for isotope production, the processing of uranium and thorium ores, and the processing of spent fuel from reactors. The principal problems and the most important water use are likely to be associated with reactor operation for electric energy production.

REACTOR OPERATION FOR ELECTRIC ENERGY PRODUCTION

Water quantity requirements for electric energy generated by fission processes are not likely to differ greatly from those of fossil-fueled plants. The similarity in design and operation of the steam cycle in conventional and atomic power plants is indicative of the comparatively small difference in water requirements of the two types. Water is used for the same purposes in each, namely as boiler feed and as a condenser coolant. The quantities of water required for these uses are discussed in the preceding section (pages 310-13). A practical minimum disappearance for conventional plants is about one gallon of water per kilowatt-hour generated, provided extensive recirculation of cooling water is employed. Average requirements may be about 50 gallons withdrawal per kilowatt-hour generated.[44] Slightly higher rates of withdrawal and dis-

[44] Based on 60 billion gallons daily withdrawal by utility plants in 1955 (U. S. Department of Commerce, *op. cit.*), and 430 billion kw-hours generated in 1955 in public utility steam plants (*Electrical World,* Annual Electrical Industry Forecast, September 1956).

Plate 22. Shippingport Atomic Power Station. Lowering a nuclear fuel charge into the reactor pressure vessel.

appearance may be required in atomic power plants because of the somewhat lower efficiency in converting heat to electrical power at the lower operating temperatures and pressures.

Although boiler feed water in atomic power plants does not involve any larger volumes than in conventional steam plants, purity requirements are exceptionally high. To maintain induced radioactivity in the circulating water and steam at an absolute minimum level, these fluids must be free of all dissolved and suspended impurities.[45] Complete demineralization of the original feed water and recirculation and continuous repurification for elimination of corrosion products are essential. Thus, fresh-water requirements in the steam and water circulation system are small simply because of the need for complete recirculation.

Presumably the same types of water suited to condenser cooling in conventionally fueled plants also will be usable in the condensers of atomic power plants. This means that in a "once-through" cooling system any sediment-free water may be used, but where recycling is employed additional treatment may be necessary. Water quality requirements therefore are not expected to be more rigorous for the atomic plants than they are in the fossil-fuel plants.

The possible indirect effect of atomic electricity generation on water use in conventional thermal and hydroelectric plants may be examined. For an integrated system of all three sources of energy, there is a reasonably clear indication that atomic plants, like the newest fossil-fuel plants, will be operated continuously at maximum capacity and load factor. Older thermal plants and hydroelectric plants having adequate water storage will be used for handling peak demands. Thus, each facility will be used in the manner best suited to its advantages and limitations. The combined effect on water demands should not differ appreciably from the pattern now developing with the fossil-fuel and hydroelectric systems.

Accordingly, the net effect on water requirements of supplementing conventional power plants by atomic power plants will be insignificant in the next few decades.

---

[45] A. E. Gorman and C. V. Theis, "The Treatment and Disposal of Wastes in the Atomic Energy Industry," *Water for Industry* (Publication No. 45 [Washington: American Association for the Advancement of Science, 1956]), p. 44. Also A. E. Gorman, "Disposal of Atomic Energy Industrial Wastes: Some Environmental Aspects," *Industrial and Chemical Engineering*, 45 (1953), 12.

ISOTOPE PRODUCTION AND USE

A second application or accompaniment of atomic energy development is the production of artificial radioactive elements, or isotopes, in nuclear reactors. Actually, this process takes place in all nuclear reactors, including those in which electric power is being generated. But in the production of plutonium from uranium for weapons the liberated heat is discarded in cooling water rather than utilized in power generation. Nuclear reactors may be used in similar manner for production of other radioactive isotopes for use in medicine, industrial tracers, food preservation, and many other applications.[46]

In the production of plutonium at Hanford, Washington, large quantities of Columbia River water are employed in cooling the reactors themselves.[47] A heavy local requirement for water is thus imposed. The location of the Hanford works was determined primarily by the availability of a very large and reasonably pure water supply. High purity is necessary to keep induced radioactivity in the effluent water at a safe level. This requires prior purification of the river water for removal of suspended matter.[48] Furthermore, effluent warm water from the reactors must be handled by special means so that the contamination can be rendered harmless before final discharge. This treatment involves retention for a period sufficient for short-lived radioactive isotopes to decay to safe levels, and for adequate dilution with additional river flow.

The future operation of nuclear reactors solely for isotope preparation will probably be of limited extent. With the exception of requirements for weapons and certain neutron-induced materials,[49] there should, in fact, be an adequate supply of by-product isotopes from power reactors. Therefore, the total demand for cooling water in the operation of special reactors apparently will not be severe, although the few installations required will have to be located on large water courses of acceptable purity.

Numerous uses of radioactive isotopes are already being made and

[46] *McKinney Report,* Vol. II, pp. 267-68.

[47] N. P. Jackson, "Atomic Energy and the American Economy," *Journal of the American Water Works Association,* 47 (1955) 1139-47.

[48] The dissolved solids in some streams might also have to be removed by chemical means if they otherwise were made dangerously radioactive by passage through the reactor.

[49] Elements made radioactive by exposure to neutrons in a nuclear reactor, rather than those resulting from fission.

still others can be foreseen. But any significant change in water demand and supply resulting either directly or indirectly from these applications does not appear probable, except in the disposal of radioactive wastes. This is a subject which merits separate discussion, on pages 324-32.

URANIUM ORE PROCESSING

The mining and milling of uranium ores are similar in many respects to operations in other metal-producing industries. Most of the important United States deposits of uranium are found in the Central Rocky Mountains, and adjacent Great Basin, where mining and milling facilities are now concentrated. The water requirements in these operations are not excessive, but in some regions where supplies are scarce judicious planning and conservation are required. The rapid expansion of small villages brought into existence by the discovery of uranium deposits and the setting up of milling facilities has occasionally caused water supply problems.

Uranium ores are generally reduced by one of several wet processes in which considerable quantities of water are used. In a typical mill about 3 tons of water are required per ton of ore treated; this corresponds roughly to 150 gallons of water per pound of uranium produced.[50] Although this is a large water requirement per unit of product, ranking about the same as aluminum, the tonnage production is comparatively small, and the water requirements of the entire industry are not significant in the total United States water demand. Even a large expansion in the mining and milling of uranium ores would not greatly affect the national demand. With total milling capacity equivalent to about 15,000 tons of ore per day[51] (in operation and under construction), water requirements would be only 10 million gallons. This may be compared with over 4.4 billion gallons per day in pulp and paper making (chapter 6, page 122).

Although the familiar stream pollution resulting from mill tailings can occur from uranium operations just as from other ore milling, the severity can be minimized by use of a tailings pond, retention basin, or other device. The wastes are not particularly dangerous, and the dis-

[50] H. L. Hazen, Inc., Denver, Colorado (uranium milling firm), personal communication, April 1957.

[51] From a quotation from Allan E. Jones, operations manager for the Atomic Energy Commission, Grand Junction, Colorado, in the *Denver Post* (Colorado), April 24, 1957.

posal problem in the industry is not great. Low normal stream flows in most of the region where mills are located necessitate practices which minimize pollution.

WATER USE IN SPENT FUEL PROCESSING

An important step in atomic power generation will be the recovery of unreacted uranium and the separation of by-products resulting from the atomic fission process taking place in the nuclear reactor. In contrast with the practically complete combustion of fossil fuels, only a small percentage of the uranium in a nuclear reactor can be utilized. This is because the products of the fission reaction build up to such a level that they diminish the net power output from the reactor, eventually causing cessation of operation if not removed. Therefore, maintenance of full power output requires periodic removal of the spent nuclear fuel elements and replacement with more uranium. Economy will dictate the processing of these spent fuel elements to separate and remove the fission products and recover the uranium for subsequent use.

Because of their extreme complexity, and because of the potential hazards in separating radioactive fission products from unreacted uranium, these processes will probably be carried on in separate facilities established expressly for this purpose. These chemical plants may be far from the power plants in which the nuclear fuels are being used, but even though highly radioactive spent fuels are transported to the reprocessing plants in shielded containers, shipping cost will be a very small fraction of total operating expense.

Water requirements in atomic fuel reprocessing plants are substantial but not excessive. In order to reduce the volume of radioactive sewage to a minimum, water intake is minimized. Even in the period several decades ahead, the reprocessing of spent nuclear fuels will therefore require relatively small quantities of water in comparison with paper making, steel manufacture, petroleum refining, and other heavy water consuming industries. The point of significance, however, is not the volume of intake water required, but the problems involved in handling the extremely radioactive sewage from these treatment plants.

DISPOSAL OF RADIOACTIVE WASTES IN WATER

In radioactive waste disposal and the potential hazards to water through radioactive contamination, the nation is faced with the most serious problem of waste disposal in its history.[52]

There are several major differences between radioactive wastes and other waste products such as industrial chemical wastes and municipal sewage. These differences appear in the degree of toxicity of the contaminants, the rate at which they can be naturally purified in streams, the methods for removal of wastes from plant effluents, and the final disposition of the removed waste materials. In each of these respects, radioactive pollutants are a much more serious and difficult problem than wastes previously or presently discharged into surface waters.

The principal potential source of radioactive pollution is the fission products of nuclear reactions. In the various types of nuclear reactors uranium spontaneously decomposes and liberates heat energy while doing so. The products of decomposition are other elements, many of which are radioactive. The type and degree of radioactivity differs a great deal among these materials, and the radioactive life is also greatly variable. Because the accumulation of these fission products beyond certain practical levels impairs both the performance and safety of the nuclear reactors, the nuclear fuel periodically must be removed and replaced by a fresh charge. In spite of high reprocessing costs, the recovery of unused uranium from these spent fuels appears an economic necessity, because of the relatively high percentage of unreacted uranium remaining in the "spent" fuel.

Because of highly specialized requirements in the chemical reprocessing of these spent fuels, present and foreseeable practice includes their transport to one of a few ideally located plants where spent fuels from numerous atomic power plants will be processed.[53] Reprocessing methods involve primarily chemical operations of dissolving the used reactor metallic mixtures and chemically separating the fission products from uranium. Uranium is then recovered, enriched with the fissionable uranium isotope U-235, and converted back to metallic fuel elements. Numerous reactions are involved in the separation of the various impurities. These fission products are the principal compounds which re-

[52] For general information on this topic, see: C. R. McCullough, *Safety Aspects of Nuclear Reactors* (New York: Van Nostrand, 1957); "Waste Disposal Curbs Nuclear Plants," *Electrical World,* 143 (April 25, 1955), 32; J. G. Terrill, D. W. Moeller, and S. C. Ingraham, "Fission Products from Nuclear Reactors," *Proceedings, American Society of Civil Engineers,* 81 (March 1955), Separate No. 643.

[53] B. V. Coplan and J. K. Davidson, "Reactor Fuel Processing—A New Chemical Business," *Chemical Engineering Progress,* 51 (November 1955), 493-96.

quire special disposal because of their high radioactivity. Although some have medical and industrial uses and are being further developed for application, most of this material must be disposed of as a worthless and troublesome by-product.

These process wastes may be of solid, liquid, or gaseous form. Their dangerous levels of radioactivity require very special handling. Some highly radioactive materials, like Strontium 90 and Cesium 141, could make a relatively large stream unfit for human use, even if discharged in minute quantities. For instance, less than 1/100 of a pound of Strontium 90 distributed in 75 million acre-feet of water would so affect the entire quantity of water.[54] In contrast with conventional industrial waste purification, where a 95 to 99 per cent removal of an impurity would be considered exceptionally satisfactory, a 99 per cent removal of a highly radioactive waste could be far from sufficient to permit discharge of the resulting effluent.

In handling the wastes from atomic fuel reprocessing plants, therefore, discharge of the long-lived, intensely radioactive fission products into water streams must be prevented. The quantities of water containing these materials are generally maintained at the lowest practical level in the plant, and finally must be handled in one of several ways. The method generally used is underground storage in large tanks, but this is not a permanent solution because some of the materials will require hundreds of years' storage before they can be released.

Underground tank storage of radioactive wastes is costly. In comparison with typical conventional sewage disposal costs of 2 cents per gallon, tank storage costs as much as $2.00 per gallon.[55] At an estimated 250,000-gallon daily waste production in 2000 A.D. (based

---

[54] Calculated from toxicity figure quoted by N. P. Jackson (*op. cit.*, p. 1145). The permissible increase in radioactivity in public water supplies, for unknown mixtures of materials is 1×10^{-7} microcuries per milliliter of water. (A microcurie is equivalent to the radioactivity from one-millionth of a gram of radium.) Thus radium itself would be present in maximum permissible quantity if about 4 ounces were distributed in 75 million acre-feet of water (1×10^{-13} gram/ milliliter.) Standards are reported in National Bureau of Standards *Handbook No. 52*, March 20, 1953. Seventy-five million acre-feet of water is equal to the total storage capacity of Lake Mead, Glen Canyon Reservoir (under construction, Colorado River), and Franklin D. Roosevelt Lake (behind Grand Coulee Dam).

[55] *McKinney Report*, Vol. II, p. 155. Also A. Wolman and A. E. Gorman, "The Management and Disposal of Radioactive Wastes," *Chemical Engineering Progress*, 51 (October 1955), 470-4.

on anticipated atomic energy generation and anticipated fission product concentration in waste water), an annual waste storage cost of nearly $200 million is indicated. To achieve the expected development rate for atomic energy, it is unlikely that use of this method can be continued indefinitely.[56]

A second technique for handling highly dangerous radioactive wastes is their deep-sea burial in containers, such as concrete tanks designed to contain the radioactivity for a great many years, or in less durable vessels expected to release the materials while some radioactivity still remains. In the latter case, the released impurities in theory would remain deep in the ocean and should not contaminate the waters and marine life near the surface. A supplementary method involves concentration of the radioactive sewage by evaporation or some other means, thereby reducing the total volume.[57] This process is expensive, however, even when the heat generated by the radioactive materials themselves is used in a self-evaporation system.

Another way to dispose of long-lived radioactive wastes is to place them in underground formations. Two variations are possible. In one, deep, isolated, and useless formations, such as those tapped by abandoned oil wells, abandoned mines, and salt-water aquifers of no conceivable value, can be used as receivers and permanent storage reservoirs for these wastes.[58] These disposal sites are of limited supply and questionable safety, however; seldom is there complete assurance against contamination of other formations by flow and percolation. Petroleum exploration and new production, for example, might be limited if wastes of this type existed in nearby formations. Nevertheless, where such formations can offer satisfactory isolated storage for several hundred years, they can be used.

A second type of underground disposal utilizes the absorption and exchange capacity in many types of soils, particularly clays and shales.[59] The contaminated waste is permitted to percolate through

[56] W. N. Mobley, "Fuel Reprocessing and Waste Disposal, Road-Blocks to Lower-Cost Atomic Power," *General Electric Review,* 58 (November 1955), 42-45; Terrill, Moeller, and Ingraham, *op. cit.*

[57] "Nucleonics," *General Electric Review,* 60 (January 1957), 36. Also C. E. Hirsch, "How Hanford Evaporates Fission Wastes," *Chemical Engineering,* 60 (November 1953), 184-85.

[58] Gorman and Theis, *op. cit.,* pp. 48-53.

[59] *Ibid.* Also Jackson, *op. cit.,* p. 1146; and W. J. Lacy and W. de Laguna, "Report to Division of Water, Sewage, and Sanitation Chemistry," American Chemical Society Convention, Miami, April 1957.

these formations, the radioactive impurities being absorbed and adsorbed in the soil and rock structures. When the water ultimately diffuses its way to formations which may then serve as ground-water supplies, the hazard can have been eliminated by a combination of waste removal and decomposition to less harmful elements.

Another proposed method of disposal is to transport the raw sewage or highly concentrated radioactive sewage to isolated desert or arctic burial grounds. But the costs of this procedure are obviously high, and the method cannot be considered particularly promising. Casting concrete blocks from Portland cement and radioactive waste water, for burial in land or sea, provides permanent, safe disposal, but also at considerable cost. Clays and other moisture-absorbing materials are being experimented with for use in a similar manner.

Conventional processes for treating sewage in water are not usually successful when applied to radioactive materials.[60] Biological and chemical poisons may be rendered relatively harmless by converting them to other forms, but a radioactive element has radiation hazard regardless of the chemical combination in which it appears. Only by its complete removal can the water be rendered safe, and even then the problem of its ultimate disposal still remains. A second problem in treating these radioactive impurities is the tendency of the added chemicals themselves to become radioactive by exposure to the primary impurities. Even though the primary radioactive element can be precipitated from water by chemical additives, by-products of this reaction may be too toxic to permit discharge.

Where the nature of the impurities permits, one successful method of removing radioactive contamination from water is by ion exchange.[61] Through the use of natural or artificial resins and chemical exchange compounds, such as those used in producing demineralized water for boiler feed, metallic ions are replaced with hydrogen, and nonmetallic ions with hydroxyl ions. These two substituted ions combine to form

[60] G. E. Eden, "Removal of Radioactive Substances from Water by Biological Treatment Processes," *Journal, American Water Works Association*, 46 (November 1954), Supplement 78. Also "Sanitary Engineering Aspects of Nuclear Energy," *Proceedings, American Society of Civil Engineers*, Vol. 81, No. 646 (March 1955), pp. 1-11.

[61] R. J. Morton and C. P. Straub, "Removal of Radionuclides from Water by Water Treatment Processes," *Journal, American Water Works Association*, 48 (May 1956), 545-58.

water, and radioactivity in the final effluent can be reduced to a very low level. The ion exchange materials then contain the radioactive elements and radicals, and instead of regenerating the exchanger by displacing the added ions with acids and alkalis, as is done in boiler feed treatment practice, the ion exchange chemicals are handled as solid radioactive wastes by one of the methods previously described. Some success has been obtained by using natural clays and shales in this same manner; mixing and agitating contaminated water with these finally divided materials has been found to remove as much as 95 per cent of many types of radioactive impurities.[62]

Coprecipitation is another method successfully applied in removing certain radioactive compounds from water. The formation of insoluble sediments in water by addition of suitable chemicals can trap or chemically combine with some of the radioactive compounds and permit their removal by settling and filtering. The lime-soda water-softening process has been experimentally used for this purpose.[63] Coagulation by conventional chemical floculation and by phosphate coagulation can also be effective for partial removal of some elements under carefully controlled conditions.[64]

In spite of comparatively high percentage removals of radioactive impurities attainable by these several methods, none begins to effect the purification required if the original concentrations are high, in the range of one microcurie per milliliter (actually only 3×10^{-5} ppm of Strontium 89, for example).

Certain types of radioactive impurities in process waters may have shorter life and lower toxicity than those previously discussed. In these cases retention ponds, constructed to prevent leakage, can render the wastes sufficiently harmless to permit their subsequent release. The length of time necessary for retention depends on the radioactive life of the impurities involved. This method is employed at the Hanford works for decreasing the radioactivity of cooling water after its passage through the plutonium production reactors at that installation.[65] Low-level radioactive wastes resulting from fuel reprocessing operations can be treated in a similar manner.

In comparison with the natural purification which occurs with conventional types of organic sewage and with many inorganic wastes also,

[62] Lacy and de Laguna, *op. cit.*
[63] Morton and Straub, *op. cit.*
[64] *Ibid.*
[65] Jackson, *op. cit.*, p. 1143.

radioactive wastes undergo very little decontamination in streams.[66] Since radioactivity is independent of chemical combination, the toxicity remains in spite of oxidation, neutralization, or other reactions. The only potential beneficial effects are the retention time provided in flow downstream and the questionable one of possible ion exchange with shales and other rocks. A retention time of a few days, however, is not highly significant when radioactivity may last months or years; neither is ion exchange with rocks a favorable prospect if at the same time it creates undesirable contamination of the stream bed itself.

It is clear that atomic fuel reprocessing wastes must be handled by entirely different techniques from those customarily employed for disposal of conventional process wastes in streams and lakes. Normal operations must be arranged so that there can be no conceivable contamination beyond safe limits. Location of fuel reprocessing plants must be considered from standpoints of satisfactory ultimate disposal of the liquid wastes. Moreover, since none of the measures so far developed or proposed is entirely satisfactory over a long period of time, growth of the atomic power industry requires development of a much better method for ultimate waste disposal.

Estimates of the future volume of high-level radioactive sewage in terms of final concentrated liquid wastes, in the United States—60 million gallons a year by 1964 and 250,000 gallons per day by the year 2000[67]—make it clear that the lowest practicable fresh-water use will be demanded in order to minimize waste disposal difficulties.

Contingent operational accidents or emergencies. Even when the problem of radioactive waste disposal is safely and adequately provided for, the possibility of operational accident must be considered for the spent fuel processing plants, the retention sites for decay of radioactive waste products, and the reactors producing power or commercial isotopes. If any one of the above three installations were located in a flood-susceptible area, for example, and a major flood were to cause the release of radioactive material, extensive contamination might be caused within the affected valley. An explosion or other serious accident which released dangerous fission products also might affect water supplies, with possible serious consequences among municipalities and other water users.

In the case of spent fuel reprocessing plants and the retention sites,

[66] Eden, *op. cit.*
[67] Mobley, *op. cit.*

risk of contingent accident could be minimized by locating the plants in areas where in the event of accident contamination would be sharply localized. By way of illustration, such sites are much more likely to be found within the arid parts of the West than they are within regions with permanent streams and interconnected drainage. Within humid regions. coastal locations would obviously be preferable to locations midstream or upstream. The opportunities for localizing damage, however, are greatest where either aridity or frozen ground conditions make water flow small or nonexistent.

On the other hand, it is improbable that power reactors can be limited to such locations. Their freedom from dependence on large tonnages of fuel is, in fact, a potential economic advantage. Furthermore, the need for cooling water requires location on rivers, lakes, or seacoast where supplies are adequate. Usually, therefore, power reactors will be installed near the market to be served—in or near thickly settled areas. Possible accident therefore has much more capacity for causing damage than in the case of installations which may be located farther from industrialized or thickly settled regions.

Hundreds of pounds of radioactive fission products may be present in a large atomic power plant reactor, so the possibility of a major accident must be reckoned with. This could occur if only a few ounces of fission products were released into a stream, thereby heavily poisoning it with radioactivity. Where atomic reactors or other facilities for handling radioactive materials are located within permanent surface drainage areas with substantial dependent downstream settlement, precautions against contingent accident will be imperative. Such measures might include alternative sources of water supply, emergency systems of water supply, establishment of treatment procedures to eliminate toxicity, development of methods of decontaminating equipment, monitoring of water content, and other new services. New concepts of a safe water supply system may be necessary in these localities. While the number of reactor sites will be limited, the long life of radioactive materials makes the zone of damage risk extensive where the reactor site is upstream and the drainage permanent.

If special measures are required to guard against accidental contamination, investment costs of water facilities may be increased. Continuous monitoring of radioactivity in water is relatively simple and inexpensive. However, the construction and maintenance of emergency water supply facilities, systems of diverting contaminated water, and reserve or standby water supplies suitable for use in the event of primary source contamination will involve substantial expenditure. A significant addition thereby would have to be made to the already mounting costs of municipal and industrial supplies in some parts of the country.

Summary Comment

As they concern future water demand and supply, the techniques treated in this chapter, for peaceful use of nuclear reactions and for the production of electric energy from conventional fuels, may be summarized thus:

1) No extraordinary national water withdrawal demands are likely from these aspects of the emerging technology. Such demands will grow, as population, industrial activity, and electric energy demands continue to grow, but at no greater rate than is indicated by the technology of the past. However, this growth means definite pressures on water supply in some regions. These pressures should be more readily met, however, as a result of technical improvements permitting maximum water conservation in power generation.

2) Problems of nuclear waste disposal and the contingent accidental appearance of radioactive products in streams and ground water, on the other hand, must be considered an extraordinary problem of the future. It is certain that the waste-carrying capacities of inland water are not adequate to meet the demands that will exist for the disposal of radioactive wastes in any form. It is probable that the radioactive waste disposal problem will have to be met by methods as yet undetermined, or at least unproven, but almost certainly not by way of surface-stream or ground-water carriage. A great deal of research and engineering is required for the solution of numerous and complex problems of radioactive waste disposal. The possibility of accidental discharge of such wastes creates new problems for water development and water supply which will require new methods of planning, new operational tasks, new facilities, and probably increased cost of supply.

3) Should a process be perfected for peaceful use of the nuclear fusion reaction, most of the problems of radioactive waste disposal should disappear. However, no plans presently can be based on the emergence of such a process.

4) Generation of electric power by steam, whether from conventional fuels or from nuclear reactions, will introduce new relations for the

several purposes in multiple-purpose water development. The economic position of hydroelectric power is likely to be enhanced, because of its value for peaking purposes in systems where the base load is supplied by steam generation. Possibilities of hydroelectric generation conflict with other flow purposes may be increased somewhat as compared to the past, if a pattern of peaking operation of water power facilities is followed. Problems of minimum flow maintenance may require special provision, as less regular releases of reservoir water are made for hydro-electric generation.

5) Some regional variation in the pattern of water power use may be expected, particularly in the Pacific Northwest, where the fossil fuels are lacking and where low-cost hydroelectric sites have a large potential. There, the impact of steam power use will be felt somewhat more slowly than in much of the remainder of the country.

6) Use of nuclear power plants may produce altered patterns of hydroelectric operation and altered relations among multiple water purposes. The extent of this alteration may depend upon the manipulation of nuclear materials prices and the degree of subsidy given to them. In general, the changes in water use should follow the pattern being established by the recent developments in steam power generation and the increased dependence on that source for electricity.

CHAPTER 13

Expanding the Physical Range of Recovery —

FUTURE USE OF SEA WATER AND SALINE INLAND WATERS

(1) Demineralization of Saline Water—compression distillation and multiple-effect evaporation; solar distillation; electrodialysis; possible use; summary. (2) Industrial Use of Untreated Saline Water—technical developments; potential effects of industrial sea-water use.

The vast water supplies of the oceans have been a challenge for man's beneficial use at least since the beginning of irrigation agriculture. The challenge has been especially pronounced where semi-arid or desertic lands border immediately on ocean or sea, as on the shores of the Mediterranean, the borders of the Red and Arabian seas, the western coast of South America, and the Pacific coast of Mexico and the United States. Within the United States, the opportunity for ultimate use of these supplies and inland saline waters has beckoned more and more, as exploitation of fresh-water supplies in the western half of the country has approached marginal costs, or has been more openly subsidized from the federal treasury. Yet the actual use of salt, saline, brackish, and alkaline waters remains minor as compared to their availability, both in the United States and abroad. Aside from on-site marine uses (navigation and exploitation of animal life) and as a source of minerals (solar salt, bromine, magnesium, etc.), the total amounts of highly mineralized or moderately highly mineralized waters in use are small. As water for withdrawal purposes, sea water, and saline, brackish, and alkaline inland waters are much more items of future interest than of past record.

There are two ways of using mineralized waters for withdrawal pur-

334

poses on which most current interest centers. One is through converting them to fresh or near-fresh water by processing. The other is through the adaptation of equipment, or other paraphernalia of economic production, to the direct use of the mineralized water as it exists. Some aspects of the latter approach, particularly for agriculture, were discussed in chapter 7 (pages 148-61). Attention here will be centered first on the demineralization of saline waters, and second on their direct use in industry.

1. Demineralization of Saline Water

Conversion of salt water to fresh water by distillation has been practiced on a small scale for centuries, particularly on shipboard. But only during the last decade have sustained efforts been made to develop practical processes for supplying large quantities of demineralized water for municipal, industrial, or agricultural supply.[1] These efforts are directed principally toward developing demineralization methods of reasonably broad and general economic applicability. Of more limited current utility is a technically successful process already in operation at several foreign coastal sites where fresh water is unavailable and fuel from nearby oil fields or refineries is exceptionally cheap. Outputs of several million gallons of high-cost distilled water per day are being produced under these unusual economic circumstances. Research and development programs in saline water demineralization are being carried forward in the United States, principally under the auspices of the Office of Saline Water, United States Department of the Interior. In Europe and North Africa programs are being undertaken by co-operative groups formed under the Organization for European Economic Co-operation.[2]

Sea water is of course the major potential saline water supply for demineralization. Highly mineralized underground waters and moderate

[1] For a historical account of sea-water distillation development see: C. A. Hampel, "Fresh Water from the Sea," *Chemical and Engineering News,* 26 (1948), 1982-85.

[2] Summaries of all development projects under the Office of Saline Water are contained in reports of the U. S. Department of the Interior: *Saline Water Conversion* (Annual Report of the Secretary of the Interior for 1955); also *Saline Water Conversion* (Report for 1956 and Report for 1957).

to highly saline surface streams and lakes represent sources of possible utility in noncoastal areas. Finally, there are the mixtures of sea water and fresh waters, often found in tidal basins and rivers near the coasts, which show a considerable range of salinity.[3] Since all processes for demineralization which depend on removing the water from the salts, such as distillation and freezing, are relatively unaffected by the type and concentration of salts present, all sources of mineralized water can be treated with nearly equal facility by these processes. In a few methods, such as electrodialysis, the salts are removed from the water, and their composition and quantity is an important factor. As indicated below, this method appears best suited to the demineralization of brackish or moderately saline water rather than sea water or more concentrated salt solutions.

Concerted efforts to develop economical demineralization processes suitable for wide use were commenced by the United States Government only as recently as 1954, and none of these new developments has yet reached the stage of full-scale use. However, the studies have proceeded far enough to show that several processes appear potentially useful and that others are at least worthy of further investigation.[4] It must be recognized, nevertheless, that even when viewed in the light of further technical advances and resulting economies, these processes will probably never produce *cheap* water, and it is likely that demineralized sea water will be more costly than our present municipal supplies. On the other hand, costs should become comparable with those involved in providing new water supplies from other marginal sources, such as by very deep pumping, large diversion projects, and other means.

Several of the demineralizing processes now appear to have some

---

[3] Although there are no exact definitions in terms of salt content, water may be considered fresh where it has less than 500 ppm of dissolved solids; brackish water is represented by a range of mineral contents from 2,000 to 5,000 ppm; whereas sea water has salinity of 25,000 to 60,000 ppm, averaging about 35,000 or 3.5 per cent. Some underground sources have dissolved salt contents of 20 per cent or more. In general, water containing more than about 1,000 ppm of mineral matter may be termed saline, regardless of source (so defined in R. A. Krieger, J. L. Hatchett, and J. L. Poole, *Preliminary Survey of the Saline Water Resources of the United States* (Water Supply Paper 1374) U. S. Geological Survey, Washington, 1957. According to the U. S. Public Health Service, potable water should contain not more than 500 ppm dissolved solids, and must not contain more than 1,000 ppm.

[4] Reports of the U. S. Department of the Interior, *Saline Water Conversion,* as cited above.

Plate 23. Shuweik triple-effect salt water distillation plant, near Kuwait, Arabia. Five of ten units are shown; each can produce 120,000 gallons of fresh water per day.

degree of technical and economic superiority over others. Three are processes in which pure water is first evaporated from the salt water and then condensed: compression distillation, multiple-effect evaporation, and solar distillation. A fourth is an electrolytic process wherein the mineral constituents are removed from the solution by electrically attracting them through synthetic membrane films. Other methods, such as freezing pure ice from saline water and separating salts by osmotic membranes, are being investigated, and one of these conceivably may emerge as a commercially successful process.

COMPRESSION DISTILLATION AND
MULTIPLE-EFFECT EVAPORATION

Compression distillation and multiple-effect evaporation are improvements on the simple distillation process used for hundreds of years. They are based on the principle that mineralized water can be evaporated to form pure water vapor, the salts remaining behind in the boiling solution. The water vapor is then conducted through tubes which are externally cooled with any source of cold water, whereupon the vapor condenses to pure liquid water.

If the vapor can be condensed in such a manner that it gives up its heat to more boiling water (boiling at lower pressure and temperature), additional distilled water can be obtained per unit of energy input. In compression distillation, the vapor produced by boiling salt water in the evaporator is first recompressed slightly by mechanical energy. This compressed vapor then flows into the heating coils of the evaporator, where it condenses and produces more vapor. Actually, very little heat is required, and most of the energy is supplied as work to the vapor compressor. In this way, much more water is produced per unit of energy supplied than in a conventional evaporator. A new design, known as the "Hickman still," was undergoing pilot plant testing during 1958. Evaluations by Gilliland[5] and by Hickman[6] place the probable future total cost of water produced by this process (including amortiza-

[5] E. R. Gilliland, "Fresh Water for the Future," *Industrial and Engineering Chemistry,* 47 (December 1955), 2410-22.

[6] K. C. D. Hickman, "The Water Conversion Problem," *Industrial and Engineering Chemistry,* 48 (April 1956) 7A-20A. Also Hickman, "Centrifugal Boiler Compression Still," *Industrial and Engineering Chemistry,* 49 (May 1957), 786-800.

tion, electricity at 5 mills per kilowatt-hour, and all other costs) between 80 cents (Gilliland) and 20 to 60 cents (Hickman) per thousand gallons, equivalent to a price range of $65 to $260 per acre-foot. These figures may be compared with typical industrial fresh-water costs ranging from about 5 cents to 25 cents per thousand gallons. Small demineralizing units in individual households can provide another approach to the supply of water for residential use in areas where the most convenient sources are highly saline. An automatic compression distiller of the Hickman type, with a daily capacity of 350 gallons, has recently been suggested for this purpose.[7] Power costs are claimed to be about $1.50 per thousand gallons, based on a price of 2 cents per kilowatt-hour.

Multiple-effect evaporation involves the use of vapor from one evaporator as the heat supply to a second evaporator operating at slightly lower temperature and pressure. The condensed vapor is then recovered as distilled water. This process can be repeated several times in multiple units, and one pound of original steam can produce several pounds of distilled water rather than only about one pound produced in a single evaporator. Some large installations are already providing distilled water to industrial plants and communities in the Caribbean and Red Sea areas, and two new multiple-effect evaporation systems in Kuwait (Persian Gulf) and Aruba (Netherlands West Indies) have daily capacities above 2.5 million gallons of fresh water. Costs are predicted to be in the range from $1.75 to $2.80 per thousand gallons.[8] A new design, involving more than a dozen evaporating stages, will soon receive evaluation; exceptional economies may be found possible.[9]

Since the principal operating cost is fuel, total water production cost is dependent on its availability and price. At a typical steam cost of 40 cents per million Btu (equivalent approximately to fuel at 25 cents per million Btu), it has been estimated that with reasonable engineering developments and improvements, the minimum total cost of water produced in plants of millions of gallons daily capacity can be reduced to 40 to 60 cents per thousand gallons, or $130 to $200 per acre-foot.[10]

[7] "For 'Poor Water' Areas—Private Water Supplies," *Industrial and Engineering Chemistry*, 50 (February 1958), 28A.

[8] "Evaporation Plants Solve Water Shortages," *Chemical Engineering*, 63 (October 1956), 126-34.

[9] U. S. Department of the Interior, *Saline Water Conversion* (1956), *op. cit.,* pp. 14-16.

[10] Gilliland, *op. cit.,* p. 2415.

SOLAR DISTILLATION

Solar distillation of salt water is potentially attractive because no fuel or electricity is required and because areas of severest aridity generally receive abundant sunshine. At present, however, it appears no cheaper than conventional evaporation processes because the large required investment offsets energy savings. Considerable research and development is therefore being carried on, primarily in the direction of reducing capital costs of construction.

The type of solar distilling unit generally considered most practical is a glass-enclosed shallow basin in which salt water is warmed by direct heat from solar radiation being absorbed on the black bottom of the basin.[11] The warm water slowly evaporates, condenses on the lower sides of the glass covers, and runs into separate troughs for collection and storage. Unevaporated brine runs to waste. Other models having thin plastic films in place of glass, and sloping black fabric wicks in place of a horizontal basin are being tested.[12] The development of entirely new and durable plastic films may have a far-reaching effect on the feasibility of solar distillation. Field trial of a large glass-covered basin and testing of several other solar still designs are being conducted by the Office of Saline Water in an integrated development program at a sea-coast site,[13] and experimental application of small, family-size solar distillers is being made in Algeria.[14]

Cost predictions for solar-distilled water vary widely because they depend largely on estimates of plant construction costs as yet unexperienced. With an investment of about $1.00 per square foot of basin, a water production cost of about $2.00 per thousand gallons has been estimated.[15] With construction and operation economies believed ob-

[11] The process has long been known, and has occasionally been used on a practical scale. The oldest and largest installation was made in 1883 in the Chilean Andes, where very salty well water was solar distilled in units totalling a little more than one acre in area. Maximum water yields were about 6,000 gallons per day. Further details can be found in: J. Harding, "Apparatus for Solar Distillation," *Proceedings, Institute of Civil Engineers,* 73 (1883), 284-88.

[12] U. S. Department of the Interior, *Saline Water Conversion* (1956), *op. cit.,* pp. 18-29.

[13] *Ibid.,* pp. 19-22. Also *Saline Water Conversion* (1957), *op. cit.,* p. 3.

[14] Societé d'Etudes pour le Traitement et L'Utilisation des Eaux (C. Gomella, director), "Demineralisation des eaux saumatres par distillation solaire," *Proceedings of Meeting of Working Party No. 8 at Algiers, May 1955.*

[15] Gilliland, *op. cit.,* p. 2416.

tainable by use of a shallow artificial lagoon or basin, the investment in a very large plant has been estimated at 50 cents per square foot of area, and large-scale water production total costs of about 75 cents per thousand gallons ($245 per acre-foot) might be realized.[16] The average capacity of such a distillation unit in a sunny climate is about 5,000 gallons per day per acre of basin area.

ELECTRODIALYSIS

The newest promising contender for a place in large-scale salt water demineralization is the electrodialysis process developed by Ionics, Inc. (Cambridge, Mass.), and others.[17] By means of an electric current, the ionized mineral constituents of saline water are transferred through thin plastic membranes into other streams of salt water. Thus, one portion of the saline feed water is demineralized, and the balance becomes more highly saline and is run to waste. The process and equipment have been field-tested on a moderate scale (30,000 gallons per day from underground brackish water ranging up to 3,500 ppm salinity), and several commercial installations of capacity up to 30,000 gallons per day are now in operation.[18] A unit designed to demineralize 86,400 gallons of saline water per day (3,140 ppm) entered service late in 1957.[19]

The largest application of this process was scheduled for service in the Orange Free State, Union of South Africa, during 1958.[20] This installation will utilize a new, cheap type of membrane, to be replaced periodically, in the conversion of 3.6 million gallons of saline mine

[16] G. O. G. Löf, unpublished progress report to Office of Saline Water, U. S. Department of the Interior, 1955. Also D. S. Jenkins and E. H. Sieveka, "Saline Water Conversion," *Proceedings of the Western Development Conference* (Phoenix, Arizona, November 1956 [Palo Alto, California: Stanford Research Institute, 1957]).

[17] W. Juda, *Results of Selected Laboratory Tests on an Ionics Demineralizer* (Research and Development Report No. 1), Office of Saline Water, U. S. Department of the Interior, Washington, April 1954.

[18] T. A. Kirkham, "More Fresh Water Via Membranes," *Chemical Engineering*, 63 (October 1956), 185-89.

[19] W. E. Katz, "The Present Status of Electric Membrane Demineralization," *Chemical Engineering Progress* (April 1957), 190-4.

[20] O. B. Volckman and W. H. Moyers, "Ion Selective Membrane Research and Electrodialysis Engineering Development in South Africa," *Proceedings of the Symposium on Saline Water Conversion*, November 4-6, 1957 (Washington: National Research Council, 1958).

water per day to a usable municipal supply. Total costs are reported to be about 30 cents per thousand gallons.[21]

Principal costs of this process are for electric power and the replacement of membranes. It appears unlikely at this time that these items can be low enough when sea water is being demineralized to permit competition with evaporation methods. With only moderately saline underground or surface waters, however, demineralization costs should be attractively low. Typical costs of demineralizing near-worthless brackish water containing 3,500 ppm dissolved solids (about one-tenth the salinity of sea water) may be as low as 25 cents per thousand gallons in plants having daily capacities of several million gallons.[22] Another estimate places minimum present costs at about 50 cents per thousand gallons ($160 per acre-foot).[23] Still more recent evaluations[24] place the cost of demineralizing 1.5 million gallons of brackish (4,000 ppm) water per day to an acceptable quality (500 ppm) at 80 cents per thousand gallons. At a production rate of 80 million gallons per day, the cost should decrease to about 42 cents.[25]

POSSIBLE USE

As is the case with most industrial processes, the larger the installation the cheaper will be the product. To operate at a demineralization cost below $1.00 per thousand gallons, a sea-water conversion plant would undoubtedly require a capacity of at least a million gallons per day. Under favorable circumstances, such a plant would require about 10,000 gallons of fuel oil per day (multiple-effect evaporation plant), or 50,000 kw-h of electric energy per day (vapor compression plant), or 200 acres of solar distillation basin. With potential technological improvements, the energy requirements may be reducible (but at higher capital expense) to 2,000 gallons of fuel oil and 15,000 kw-h of electricity, respectively.[26]

---

[21] "Needed: An Extra 250 Billion Gallons of Water a Day by 1975," *Chemical and Engineering News,* 36 (March 24, 1958), 50-53 ff.

[22] Gilliland, *op. cit.,* p. 2422.

[23] Kirkham, *op. cit.,* p. 187.

[24] Jenkins and Sieveka, *op. cit.*

[25] U. S. Department of the Interior, *Saline Water Conversion* (1957), *op. cit.*

[26] Hickman, *op. cit.,* pp. 10A, 12A, 14A.

If one or more of these experimental demineralization systems becomes a successful and economical large-scale process for converting saline water to potable fresh water, the effects could be far-reaching. In arid and semi-arid United States costal areas, such as southern California and southwestern Texas, there are large population centers and industrial plants, many of which are facing immediate water shortages. To alleviate this situation, new supplies from distant rivers are being considered, at costs much above those of present sources. For example, a proposed project for bringing additional water to the Nevada desert from the Columbia River would require pumping energy of 50 kw-h per thousand gallons,[27] and would involve total costs of about 40 cents per thousand gallons.

Several saline demineralization processes show promise for operation at costs descending toward 50 cents per thousand gallons of water produced. These are being developed and improved, and within a very few years firm cost figures should be available. If preliminary indications are borne out in full-size tests, the effects of these developments on local water supply in critical areas should be important. The addition of new, comparatively high-cost water to existing supplies may be reconsidered, and plans for extremely expensive diversion projects may be reappraised. Costs cannot be expected to be so low, however, that existing developed municipal supplies could be economically replaced.

Large developed supplies of irrigation water are available in the West at prices of 1 to 5 cents per thousand gallons (equivalent to $3 to $15 per acre-foot). It is clear that unless great increases in the cost of irrigation water could be tolerated, demineralized sea water could not be used for that purpose. Except perhaps in special situations or for very high-value produce, agricultural use of demineralized water therefore cannot be foreseen at this time.

But as the cost of obtaining further water supply for industrial and domestic uses rises, demineralized saline water will become more and more competitive for these applications. Even now, if a municipality or industrial plant is considering augmenting its present water supply by establishing a new source at a cost as high as 50 cents per thousand gallons, demineralized salt water should certainly be considered as a possible supply. A recent example in the United States is the town of Coalinga, California, where potable demineralized water will soon be distributed from an electrodialysis plant supplied with saline ground

[27] *Ibid.*, p. 18A. Calculation includes amortization of investment and electricity costs of 5 mills per kw-h.

water. A dual distribution system will furnish potable and saline supplies to each user. There are numerous technical and economic problems associated with most of these new processes, but they have all been tested on small scale, some have been used commercially on a sizable scale for many years, and no insurmountable difficulties need be expected.

Although only a few American coastal cities have immediate need for additional water supply, many water-scarce inland municipalities are within feasible pumping distance from sea water. With nominal power rates, large quantities of demineralized sea water could be pumped about a hundred miles inland for 5 to 10 cents per thousand gallons. Another 2 to 5 cents would raise it a thousand feet. This transportation expense, although substantial, should not preclude serving these contiguous areas, and wider use of demineralized sea water could thus be made. Coastal sections of California and Texas seem particularly interesting territory where this new source ultimately could be made available within the 1,000-foot elevation in the coastal zones (figure 27 and table 29).[28]

In addition to supplementing water supplies in coastal and near-coastal areas, demineralization systems should become useful in improving the quality and usability of inland underground and surface brackish water. Assuming that satisfactory brine disposal methods will be developed, this application will result in increased local water supply in some of the areas of greatest need. The occurrence of underground and surface brackish waters having mineral content above 1,000 ppm is remarkably widespread (figure 28).

Capital requirements for water demineralization plants will be considerably higher per unit of capacity than for existing water supply systems, but will probably compare favorably with the investment in some new proposed diversion schemes. There may be considerable difference, however, in the ultimate investment requirements of the several types of demineralization plants. The solar distillation plant, for example, requires a particularly high capital outlay, but it operates with no energy cost. In this respect it is analogous to a hydroelectric instal-

[28] U. S. Department of the Interior, Bureau of Reclamation, "Potential Use of Converted Sea Water for Irrigation in Parts of California and Texas," *Saline Water Conversion Program Research and Development Report No. 3,* Washington, April 1954. Although the report and maps are based on potential irrigation use of demineralized sea water in these areas, regardless of cost, the data are also useful in showing *areas* of increased water needs, which could conceivably be met by demineralized sea water if available at reasonable cost for other uses.

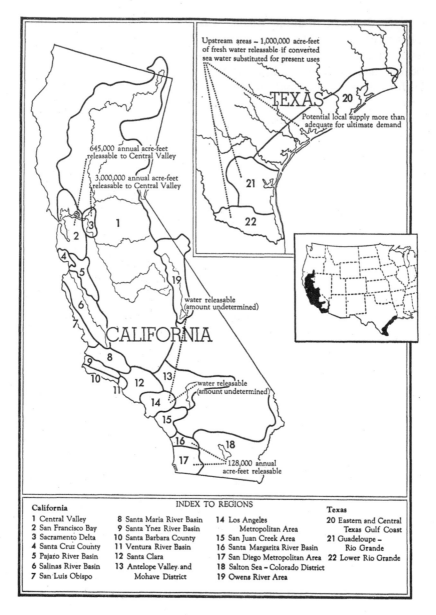

Figure 27. Regions of Potential Demand for Demineralized Sea Water for Irrigation. The figures in annual acre-feet indicate the approximate amount of fresh water potentially releasable to nearby areas if demineralized sea water were to be fully utilized in the coastal regions. (See also table 29.)

TABLE 29. *Regions of Potential Demand for Demineralized Sea Water for Irrigation*

Region	Feasible pumping elevation for demineralized sea water (feet)	Ultimate deficiency in local supply (annual acre-feet)	Estimated potential market for converted sea water (annual acre-feet)
California			
1. Central Valley	—	0	undetermined
2. San Francisco Bay	(¹)	1,330,000 to 2,400,000	1,000,000 to 2,000,000
3. Sacramento Delta	(¹)	0	3,000,000
4. Santa Cruz County	500	13,000	5,000
5. Pajaro River Basin	—	(²)	—
6. Salinas River Basin	500	(²)	—
7. San Luis Obispo	500	180,000	150,000
8. Santa Maria River Basin	500	80,000	80,000
9. Santa Ynez River Basin	500	80,000	60,000
10. Santa Barbara County	300	70,000	70,000
11. Ventura River Basin	200	5,000	5,000
12. Santa Clara River Basin	500	190,000	150,000
13. Antelope Valley and Mohave District	—	(³)	—
14. Los Angeles Metropolitan Area	2,000	2,100,000	2,100,000
15. San Juan Creek Area	1,000	140,000	135,000
16. Santa Margarita River Basin	1,000	1,116,000	160,000
17. San Diego Metropolitan Area	1,000		728,000
18. Salton Sea—Colorado District	—	(³)	3,500,000
19. Owens River Area	—	(³)	—
Texas			
20. Eastern and Central Texas Gulf Coast	—	0	—
21. Guadaloupe—Rio Grande	500	2,860,000	2,860,000
22. Lower Rio Grande	500	1,650,000	1,650,000

¹ Converted sea water would be usable at lower elevations. Maximum elevation not determined.
² Currently planned fresh-water developments adequate for anticipated future demand.
³ Ultimate deficiency undetermined.

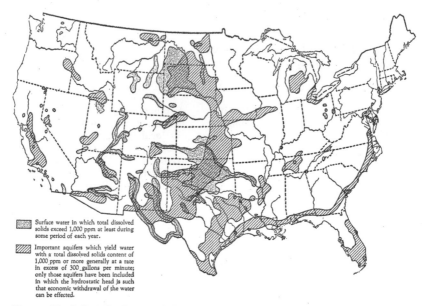

Surface water in which total dissolved
solids exceed 1,000 ppm at least during
some period of each year.

Important aquifers which yield water
with a total dissolved solids content of
1,000 ppm or more generally at a rate
in excess of 300 gallons per minute;
only those aquifers have been included
in which the hydrostatic head is such
that economic withdrawal of the water
can be effected.

Figure 28. Saline Surface and Ground Water in the United States.

lation in comparison with a fuel-operated steam power plant. It is believed that the investment in very large sea-water demineralization plants can ultimately be reduced to about $1.50 per gallon daily capacity in compression distillation, $1.00 in multiple-effect evaporation,[29] $4.00 in solar distillation,[30] and $1.00 in electrodialysis of moderately saline brackish water.[31] These figures may be compared with typical costs of 40 cents to $1.60 for some large-scale diversion projects, based on pipe-line, pumping equipment, and power plant investment requirements per unit of water transporting capacity.[32]

Total costs of demineralized water have already been indicated as potentially competitive with those of proposed diversion plans in the

[29] U. S. Department of the Interior, *Saline Water Conversion* (1957), *op. cit.,* p. 14. Also "OEEC Team Reports on Water Demineralization Methods being Investigated," *Chemical Engineering News,* 31 (September 28, 1953), 4024.

[30] Löf, *op. cit.*

[31] Kirkham, *op. cit.,* p. 188.

[32] A tabulation of investment in several projects may be found in E. D. Howe, "Sea Water as Source of Fresh Water." *Journal, American Water Works Association,* 44 (August 1952), 690. (These two costs are for the Philadelphia—Delaware River project and the San Francisco Hetch Hetchy project, respectively.)

Southwest. Undoubtedly municipal and industrial water prices will rise sharply in these areas as the more expensive sources are "averaged in" with existing supplies. As developments bring demineralized water costs down toward and possibly below 50 cents per thousand gallons, use of this source must be considered in water-scarce coastal areas and regions of limited fresh but ample brackish water supplies. Variation in fuel and electricity costs in different localities will markedly affect demineralization costs, and possibly will dictate the choice of the process to be used. Where sunshine is plentiful and other energy sources are expensive, solar distillation will very likely be used. Cheap electric power will favor the compression distillation and electrodialysis methods. Low fuel cost should lead to application of multiple-effect evaporation plants.

Since the high cost of demineralized saline water will limit significant use to municipal and industrial consumers, the planning of multipurpose river projects should seldom be appreciably affected. The irrigation aspect of any feasible fresh-water project could not be economically challenged by any demineralization process now known.

Much more development work is needed in the demineralization field before large-scale applications can be made confidently. Basic data and the results of considerable research are available, but small prototypes of economical operating plants need to be constructed, tested, and improved before reliable economic and design data can be obtained. Considerable progress along these lines is being made in the United States. By 1959 several representative pilot plants were expected to be in operation.

Even substantial use of demineralized water cannot be expected to have large effect on the development and use of other water sources. Only in a few unusual circumstances, such as in projected expensive water diversion from a distant river system, might plans be altered by this development. In such a case the stream could be used elsewhere, possibly to better long-term economic advantage. In general, however, it appears that *after* all other practical means have been developed by a community, including feasible aqueduct and diversion systems, wells, sewage reclamation, and so on, saline demineralization would enter use. This entry would probably be made at the point of cost equality between the saline source and the most expensive alternate source. There is no basis for expecting that the cost of demineralized saline water will be low enough to eliminate development of all concentrated surface fresh-water supplies. Where fresh ground water is lifted much more than 1,000 feet, however, demineralized brackish water may become competitive, depending on well capacity and developments in pumping economy.

The construction of sizable demineralization capacity will involve an increase in energy consumption wherever utilized (except for solar distillation plants in areas of high insolation). It has been estimated that for an exceptionally efficient compression distillation plant of 10 million gallons daily capacity (municipal supply for 50,000 to 100,000 people), the minimum potential power requirement could not be less than 5,000 to 6,000 kilowatts. For a fuel-operated plant, the oil requirement might ultimately be reduced to 500 barrels per day.[33] Although these energy needs are not difficult to meet in most well-populated areas, they represent demands sufficient to require planning. The suppliers of construction materials, particularly of corrosion-resistant alloys and plastics, fabricators of chemical plant equipment, and other technical and construction fields will be well represented in these developments.

SUMMARY

It is reasonably certain that one or more demineralization methods will soon permit commercial production of potable water from sea water at costs substantially below $1.00 and possibly as low as 50 cents per thousand gallons ($165 to $330 per acre-foot). Technical development and engineering now in progress should bring this about within the period 1960-62. Even when realized, these costs clearly will be too high to permit use of the processes to supply water for irrigating common crops. However, in water-scarce municipalities and industrial establishments new fresh-water sources are being considered which may approach these levels of cost in the foreseeable future,[34] and there is every reason to expect that still higher fresh-water costs will be involved in the years ahead. Thus, in these situations demineralized water will become increasingly competitive.

Desalinization of moderately brackish surface and underground water by electrodialysis appears feasible at ultimate costs of 25 to 50 cents per

[33] Hickman, *op. cit.,* pp. 12A, 14A.
[34] Gilliland, *op. cit.* (p. 2411) states that large new supplies of water are being considered in southern California which would result in delivered costs of 50 to 60 cents per thousand gallons. San Francisco's Hetch Hetchy project, even in 1925, required an investment which places water costs at about 30 cents per thousand gallons delivered.

thousand gallons.[35] Assuming successful full-scale development of this process from its present semi-commercial stage, and assuming that satisfactory methods will be devised for the troublesome problem of brine disposal, practical applications in inland arid areas should result. Because this particular process can be operated more cheaply than distillation if the salinity is low, it is expected that it will be the least expensive method in such circumstances.

The most significant limitation to all these processes is the cost of energy, since a large, if not major, fraction of the cost of demineralizing saline water is energy as electricity or fuel (except in solar distillation). Most of the costs estimated herein have been based on 5 mills per kw-h for electric power and $2.00 per barrel for fuel oil. These represent favorable rates, but in those United States locations where higher energy costs prevail, demineralized saline water would be even more expensive than indicated.

Although large-scale use of some processes is already possible, further development and improvement of all potentially useful methods is needed before substantial application can be expected. To the extent permitted by available funds, such development work is being undertaken.

The program of the United States Office of Saline Water is currently directed along two principal lines: (1) continued research and investigation of potentially useful "new" methods of saline water demineralization; and (2) development of the four to six most completely tested methods to a stage at which plants may be operated large enough to permit reliable evaluation of costs.[36] Other processes may become of practical interest as these studies proceed, and some now considered potentially feasible may become less attractive.

A particular challenge for the future lies in economical reduction of sodium content in western irrigation waters. Even though true fresh water is not obtained or sought, every degree of sodium ion reduction in the presently employed irrigation waters can reduce water demands for

[35] Kirkham, *op. cit.*, p. 187. Also Jenkins and Sieveka, *op. cit.*

[36] U. S. Department of the Interior, *Saline Water Conversion* (1957), *op. cit.* The construction of at least five demonstration saline water demineralization plants in the United States was authorized by Congress in the fall of 1958 (Public Law 883, 85th Congress, 2nd Session, 72 Stat. 1706). Two of three sea water plants will have daily capacities of at least 1 million gallons, and one of two brackish water plants will produce at least 250,000 gallons per day. Types of processes were to be determined by the Department of the Interior.

leaching, lessen drainage problems, decrease salt additions to soil from the water, increase crop productivity, and decrease the costs of such production.

2. *Industrial Use of Untreated Saline Water*

For many years sea water has been withdrawn for direct use on ships and in coastal industrial plants, in applications where the salt content does not constitute a serious problem. Its principal use has been as a coolant in power plant condensers. In general, however, where fresh water is available it is preferred and used because of several disadvantages associated with the high salinity of sea water. The most serious of these problems have been in the corrosive action of sea water on metal surfaces and the fouling of heat transfer equipment by marine organisms and salt deposits.

In recent years, improvements in corrosion-resistant alloys have led to major reduction in the corrosion problems of sea-water use. As this handicap to sea-water use in industrial plants is minimized, greater employment of this source of cooling water in coastal areas becomes practical. Developments in water treatment to reduce the growth of marine organisms in retaining ponds, pipe lines, and principal equipment, also have contributed to greater use of sea water for cooling purposes.

Some progress also has been made in the use of sea water for other industrial requirements, such as in flushing or washing operations, where salinity can be tolerated. Sluicing furnace ash, handling logs and wood chips in pulp mills, and even laundering are among these numerous uses.

TECHNICAL DEVELOPMENTS

The most severe corrosive action of sea water is due to the galvanic effects between dissimilar metals in the presence of the saline solution. Even when only one metal is used throughout the system, nonuniformity in its composition, even on a microscopic scale, results in pitting and dissolving. For many years, admiralty brass—an alloy of copper, zinc, and tin—has been used for condenser tubes cooled by sea water. This material stands up reasonably well, but if other parts of the exposed system contain iron or steel they may suffer particularly severe attack because of the electrolytic action.

The cost of employing sea water in coolers and condensers has usually been high because of the expense incurred in maintenance and replacement. In recent years, however, various alloys highly resistant to sea-water attack have been developed. Most effective are several types of stainless steel containing both chromium and nickel, cast iron containing nickel, Monel metal, other copper-nickel alloys, and an aluminum-brass alloy.[37] Service life of many years can be obtained with such materials in contact with sea water.

Plastic pipe is a second development in the field of construction materials which is becoming important in sea-water use. Strong and durable polyethylene and polyvinyl chloride pipe are replacing iron and copper in handling sea water and other corrosive liquids.[38] A closely related new development involves lining steel pipe with plastic and bituminous coatings for corrosion protection. This is used where pipe sizes are too large for all-plastic installations.[39]

Another important technical advance brings about cathodic protection of iron and steel from sea-water attack through the use of renewable magnesium metal anodes placed in the equipment at suitable points. In this method an intentional electrolytic action is set up. The magnesium is preferentially corroded and dissolved, and the iron and steel thus are electrolytically protected from attack. This technique is proving particularly useful in condensers, pipe fittings, and other areas where iron or steel is exposed to sea water. The magnesium anodes must be replaced occasionally, but other maintenance is minimized.[40]

With these several methods of controlling corrosion, salinity in cooling water is no longer considered a serious problem. As a result, clean sea water is now generally a cheaper and better coolant than contaminated fresh water which requires considerable pretreatment.[41]

[37] W. B. Brooks, "Sea Water as Industrial Coolant," *Petroleum Refiner.* 33 (October 1954), 127-30; 33 (November 1954), 179-82.

[38] "Plastic Pipe Over the Hump," *Chemical and Engineering News,* 33 (July 25, 1955), 3062-64.

[39] Brooks, *op. cit.* Also R. B. Seymour, "Piping, Valves, and Ducts," *Industrial and Engineering Chemistry,* 47 (July 1955), 1335-42.

[40] Brooks, *op. cit.* Also D. P. Thornton, Jr., "Dow Conquers Sea Water Corrosion," *Petroleum Processing,* Vol. 7 (November 1952), 1640-43.

[41] "Leaving the Salt Behind; Pacific Gas and Electric Steam-Electric Generating Plant will Operate on Water from the Sea," *Chemical Week,* 75 (August 28, 1954), 52. Also A. D. Rust, "Use of Sea Water in Industry," *Southern Power and Industry,* 73 (September 1955), 70-76; and 73 (December 1955), 50-52.

Recently, large applications of sea water have been made in electric power plant cooling. Petroleum refineries, chemical plants, steel mills, and other industrial concerns also are making extensive use of sea water, thereby replacing an equivalent fresh-water requirement. Of the 110 billion gallons per day (338,000 acre-feet) withdrawn by industrial water users in 1955, about 19 billion gallons (58,500 acre-feet) were saline water. This does not include withdrawals directly from the ocean.[42] Including sea-water use, probably 20 per cent of the total industrial withdrawal already is saline water.[43]

In using sea water from most areas, means must be employed to prevent slimes, mussels, barnacles, and certain plants from fouling equipment. Chlorination has been the traditional preventive, but other recently developed chemical treatments now are used very effectively. With small concentrations of certain organic compounds, such as sodium pentachlorphenate, and quaternary ammonium compounds, control of biological fouling can be maintained.[44]

Still another technical development is a potential water saver, although it has limited applicability. Several of the new synthetic detergents can be satisfactorily employed in highly saline waters for laundering. Unlike soap, which combines chemically with calcium and magnesium salts, sulfonate and phosphate detergents of some types can be used in sea water without particular loss in efficiency. Large commercial and industrial laundries could therefore employ sea water, and in many cases probably would do so if their supply systems did not generally come from municipal works.

---

[42] K. A. MacKichan, *Estimated Use of Water in the United States, 1955* (Washington: U. S. Department of the Interior, Geological Survey, Water Resources Division, December 6, 1956).

[43] In manufacturing alone, brackish water use in 1954 was about 14 per cent of the 32-billion-gallon daily withdrawal by manufacturers reporting intake greater than 55,000 g.p.d. These statistics represent more than 95 per cent of water used in manufacturing. Figures computed from data in U. S. Bureau of the Census, *U. S. Census of Manufactures: 1954* (Bulletin MC-209, Industrial Water Use), Washington, 1957.

[44] J. J. Maguire, "Biological Fouling in Recirculating Cooling Water Systems," *Industrial and Engineering Chemistry,* 48 (December 1956), 2162. Also H. J. Turner, *et al.,* "Chlorine and Sodium Pentachlorphenate as Fouling Preventives in Sea Water Conduits," *Industrial and Engineering Chemistry,* 40 (March 1948), 450-53.

POTENTIAL EFFECTS OF INDUSTRIAL SEA-WATER USE

Water withdrawn solely for cooling purposes in power plants and other industrial installations is consumed only to the extent that the extracted heat evaporates it. Nevertheless, substantial savings in fresh water can be made by sea-water substitution. The net evaporation of roughly 1 gallon of water per kilowatt-hour generated in a modern steam electric plant,[45] whether from a recirculating water system or from a river or lake in a once-through system, means an annual water disappearance of 3 to 4 billion gallons in a 500-megawatt power plant. If such a supply must come from sources having competitive uses, particularly in water-scarce regions, the advantages of sea water are clear. In some situations where once-through use of fresh cooling water is practiced or planned, at least 50 gallons of total fresh-water withdrawal per kilowatt-hour generated could be replaced by sea water or brackish water. Given a coastal location, therefore, even a comparatively small 100,000-kw plant could eliminate its annual fresh-water withdrawal of 30 to 50 billion gallons by substitution from a saline source.

An additional advantage in salt-water use lies in the conservation of pipe-line capacity and of ground-water supplies for demands that can be met only by fresh water. The location of manufacturing and power plants thus need not be tied to extreme dependence upon plentiful fresh water. The Morro Bay electric generating plant on the central California coast is an ideal example of this freedom. There, the sea provides not only the cooling water but also the source of boiler feed water in a demineralization system.[46]

In planning water supplies in coastal locations, historical averages on water consumption per capita, including all industrial and power uses, should be critically evaluated in the light of future substitution of sea water for fresh water in most large-scale cooling applications. Although industry's total water requirements are not significantly affected by substitution, its needs for fresh water could be materially reduced if sea water were employed wherever possible.

Effects on capital requirements and operating costs in industry. Sea

[45] At a modern, but not the very best, "heat rate" of 10,000 Btu of fuel value per kilowatt-hour generated, approximately 6,000 Btu must be discarded in the cooling water, resulting in the ultimate evaporation of about 1 gallon to maintain constant temperature.

[46] "Leaving the Salt Behind . . . .," *op. cit.*

water for cooling in manufacturing and power plants is generally more costly to use than an available, clear, uncontaminated fresh-water supply. More expensive construction materials are needed to withstand the corrosive action of sea water, and chemical disinfection of the water is required to prevent biological fouling. However, the over-all cost of sea-water use may be much lower than that of a contaminated fresh-water supply requiring treatment, or of one drawn from many deep-well sources, or of surface supplies from large stream diversion projects. The capital requirements for providing a saline supply, including the special equipment costs, would ordinarily be higher, but operating costs could well be below those of fresh-water systems. Each geographical site will require specific comparison among the several alternatives.

Effects on other water sources and uses. Water planning in coastal and near-coastal areas can be geared to extensive use of sea water for cooling and condensing purposes in large industrial plants if fresh water is in short supply or if its cost is considerably above the industrial average for the particular type of product. Since much of American industry is heavily concentrated close to the seacoasts, greater use of sea water can effect major fresh-water savings. As industry expands and demands grow, even those East Coast areas where fresh water is now plentiful are likely to find supplies insufficient to meet all competing needs. If, wherever practical, industry uses sea water, more fresh water will be released for those industrial and municipal purposes which demand a nonsaline supply. Where acute shortages are imminent, as they are in southern Texas and southern California, the release in this way of much-needed fresh water for other uses can have substantial general benefits.

Effect on distribution methods and designs in plant and outside plant. Usually several types of water are involved in supplying water to manufacturing and power plants. Potable supplies, process water, fresh cooling water, and sea water are commonly piped around an industrial coastal plant for meeting the various requirements. Where it is not already part of the supply, the use of sea water merely adds one type of service to an existing system. Increased sea-water utilization will necessitate larger piping and pumping facilities within plants, but will reduce the use of cooling towers and other fresh-water facilities. It is quite possible that plants located many miles from salt water will be served economically by private, public, or co-operative sea-water pipe lines, pumps, and canals. But this is in the future; in 1958 there were no examples of this practice.

Plate 24. Morro Bay Power Plant, California. Salt water is used for cooling in this plant.

Effects on demand for alloys, plastics, and other materials. As the use of sea water in industrial and power plant equipment increases, the demand for numerous alloys resistant to sea-water corrosion will unquestionably rise. There should also be some stimulus to the market for plastics and other synthetic materials capable of withstanding attack by sea water. These secondary effects may not be of major importance in view of the importance of these same materials for many other uses. But as the alloys, in particular, are improved, and especially as their cost is reduced, the further use of sea water in feasible applications may be expected.

Further possible developments and general evaluation. There are numerous other hypothetical uses for sea water as a replacement for fresh water in domestic and industrial applications. But their practicality as a means of substitution is remote. For example, a complete dual fresh-water–salt-water supply system in a large city would usually be too expensive under present conditions, even though reduction in fresh-water use of possibly 50 per cent might be achieved. Clean sea water could be used in home laundries, bathrooms, and possibly for some kitchen needs. In commercial and industrial establishments, flushing and washing operations of all kinds, cooling, and nearly all needs except those involving addition of water to a product could be met by sea water. It is not unlikely that these developments ultimately will take place, but only after all available fresh-water supplies have been fully developed, even at considerable expense. If, for example, the substitution of sea water for fresh water in such applications would permit the population of a coastal city to double where otherwise it would have to remain approximately stationary, these changes might take place. Their likelihood would depend on the relative costs of demineralizing sea water in large central coastal plants as compared with the expense and inconvenience of the dual water system. At present, demineralization costs are high and the dual system would usually be cheaper. Future developments may reverse this situation.

Even in the immediately foreseeable future, however, fresh-water savings are readily and reasonably obtainable. Many organizations are providing for sea-water use in new installations and are even substituting sea water for fresh water in water-scarce areas. Potential use of sea water by coastal industries to the maximum practical extent can result in significant and important fresh-water savings in the areas concerned.

CHAPTER 14

Expanding the Physical Range of Recovery —

INCREASING SOIL MOISTURE, RUNOFF, AND PERCOLATING WATERS THROUGH WEATHER MODIFICATION

Nature of Clouds and Conditions Necessary for Seeding. Methods of Seeding—dry ice seeding from airplanes; silver iodide smoke from ground generators; modification by water spraying. Commercial Weather Modification. The Advisory Committee on Weather Control. Results of Weather Modification Experiments and Commercial Cloud-seeding Programs. Control of Floods, Hail, Lightning, and Other Elements of Weather. Potential Effects of Weather Modification on Water Supply and Demand—on storage and distribution systems; on multiple-purpose planning; on the development of other water sources; on agricultural practices; on water system costs and capital requirements; on needs for basic data and further research and engineering; on legislation and administration. Summary and General Evaluation.

Artificial precipitation of moisture from the atmosphere has been a tempting subject for scientific attention because the atmosphere is naturally the most mobile part of all the earth's water supplies. It also has been attractive because recharge of water into the atmosphere is automatic, consequent upon the evaporating and vegetative transpiration effects of energy received from the sun. One of the most beneficial water projects conceivable would be a redistribution of precipitation from the atmosphere either geographically or in time, or both. Probably no basic change in the rate of total moisture moving into and out of the atmosphere in the hydrologic cycle would be possible without significant changes in the solar energy received. But temporal or geographical redistribution of rain, snow, and hail for some time has appeared to be a problem worthy of scientific attention.

The moist air masses moving from the Pacific Ocean eastward across the North American Continent go on from the East Coast with more than three-fourths of their original moisture. The potential for additional precipitation within United States territory is therefore immense.[1] In addition, extensive recirculation of water from the land to the atmosphere makes it possible to produce precipitation several times, even from the same masses of air and vapor.[2]

Even more significant than these estimates of total and re-evaporated supplies of atmospheric moisture is the low "efficiency" of natural precipitation. It has been estimated that even a heavy storm precipitates only 0.5 per cent of the overhead moisture in the storm area.[3] If relatively modest increases in this yield of water can be artificially induced, highly rewarding benefits will be obtained. For instance, a 15 per cent increase in total precipitation could mean an additional 100,000 acre-feet of water per year on a 1,000-square-mile watershed having an annual precipitation of only 15 inches. If this water were to fall in the areas where it is most needed in western and southwestern United States, its monetary value at $20 per acre-foot, would be $2,000 per square mile in agriculture, or, at 15 cents per thousand gallons, nearly $5,000 per square mile of municipal supply watershed.

[1] Studies by G. S. Benton, M. A. Estaque, and J. Dominitz, *Scientific Report No. 1, Contract AF-19 (122)-365* (Johns Hopkins University Department of Civil Engineering for Geophysics Research Division, Air Force Cambridge Research Center), show a 1949 water balance on the North American Continent of 91 cm. atmospheric moisture entering, 76 cm. leaving, and 15 cm. liquid runoff.

[2] Recent analysis of tritium contents of water in the Mississippi Valley, by F. Begemann, and W. F. Libby, *Continental Water Balance, Ground Water Inventory and Storage Times, Surface Ocean Mixing Rates and World-Wide Water Circulation Patterns from Cosmic Ray and Bomb Tritium* (Report No. AFOSR:TN-56-561, University of Chicago to Air Force Office of Scientific Research, November 15, 1956), shows that approximately two-thirds of the annual precipitation in the Upper Mississippi Valley is ocean water and one-third is re-evaporated land precipitation. Of approximately 100 cm. of ocean water vapor inward transport per year, 52 cm. precipitated in the valley, leaving 48 cm. of vapor to be joined by vapor from the land area, totalling 72 cm. outward transport. "Internal" recirculation involved 49 cm. annual evaporation, of which 25 cm. reprecipitated and the balance remained as vapor. The over-all difference of 28 cm. appeared as runoff.

[3] I. P. Krick, "Rain Increase Projects in Relation to Water Resources," *Water and Sewage Works,* 99 (July 1952), 260-1.

In addition to moisture redistribution, the prospect of weather modification has been attractive because of other possible benefits. These include the reduction of hailstorms and resulting damage, decrease in lightning storms and consequent reductions in multimillion-dollar forest fire losses, modification and diminution of hurricane intensity and damage, and reduction in frequency and severity of floods.

In spite of the size of potential benefits from weather modification, it is only within the last decade that some promise of success in this area has been reportable.

Nature of Clouds and Conditions Necessary for Seeding

The principle upon which induced precipitation depends is artificial nucleation. This is manipulation of the mechanism by which the tiny liquid droplets forming a cloud combine into particles of water or ice large enough to fall from the cloud as rain or snow.

A cloud is actually a mass of air in which a very large number of microscopic liquid water droplets are distributed, too small to fall through the surrounding air. Some high-altitude clouds are composed of tiny ice crystals. So-called warm clouds exist at temperatures above 32° F. whereas super-cooled clouds, the more common type in the United States, both in winter and summer are partially or entirely at temperatures below the freezing point of water. In tiny droplets, however, water will not start to freeze at temperatures as high as, or even considerably below, 32° F. In fact, liquid water droplets can exist at temperatures down to -39° F. without freezing. Below this temperature, however, the water droplets cannot exist as a liquid, and ice crystals spontaneously form.

In the process of precipitation from a super-cooled cloud, unless the temperature falls below -39° F., tiny foreign solid particles apparently must be present for initiating the coalescence and freezing of the liquid droplets. These naturally-occurring smoke and dust particles, containing mineral matter from the earth, meteorites, or the sea, serve as nuclei on which condensation or freezing is initiated. The effectiveness of these particles apparently depends upon their composition, size, and shape. Some are able to start ice crystal formation at comparatively high temperatures, close to 32° F., while others require much lower temperatures to be effective. After ice starts to form on these nuclei, the crystals grow by collection of liquid and vapor from the rest of the cloud, and when they are large enough to fall through the air mass they descend toward the earth. If the air temperatures are below freezing throughout their travel, they fall as snow. If warm air is encountered, they melt and fall as rain. Occasionally the lower air layers are so warm and dry that the drops revaporize and no precipitation actually reaches the ground.

A second rain-forming process is of greater importance in tropical and subtropical climates. Under proper conditions, the droplets in a warm cloud can start to coalesce or the vapor can start to condense, apparently on initiating nuclei of somewhat different character. By condensation of vapor and collision with other droplets, the liquid drop grows to a sufficient size for it to fall.

The atmosphere always contains many types of dust at various concentrations and altitudes; where the air has moved long distances over land areas these particles are usually present in sufficient quantity to initiate precipitation if other necessary conditions prevail. Thus, if a cloud containing an adequate supply of liquid droplets exists partly or wholly in a super-cooled condition, and if the nuclei are present in a form and concentration sufficient to cause ice crystal formation at the lowest prevailing temperature, precipitation usually will be naturally initiated. The process is apparently so sensitive to small changes in conditions, that a rather delicate balance of these factors must be present for precipitation to be initiated and completed. If the nuclei are not in sufficient abundance or activity, or if the cloud properties are not in the unstable range, precipitation will not be started and no moisture will be received at the ground. For instance, the Pacific air masses over western United States are thought to contain fewer effective precipitation nuclei than the air masses over the eastern part of the country, where they have crossed extended land area.[4]

One further important property of clouds which affects the natural and artificial nucleation process is the extent to which vertical air movement influences cloud formation. Vertical air movements are particullarly encountered over mountains, or where heating of the land surface causes local convection. In so-called orographic clouds, resulting from deflection of air upward by mountains, the cooling of warm moist air by adiabatic expansion as it rises results in condensation of moisture droplets (figure 29). An orographic cloud, having considerable vertical air

[4] V. J. Schaefer, *Ice Crystal Nucleation in the Free Atmosphere*, mimeographed summary of paper presented at Conference on Scientific Basis of Weather Modification Studies, Tucson, Arizona, April 1956.

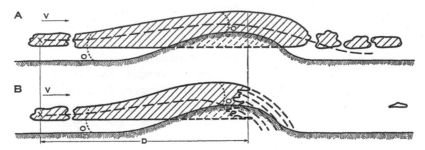

Figure 29. Formation of Snow in Orographic Clouds. The two usual results of introducing ice crystals or ice-forming nuclei at point X (several hundred meters above ground) are shown above. In A, orographic lifting carries the crystals upward, but if the distance, D, is too small or the mean wind speed, V, is too great, the crystals will not grow large enough to fall from the cloud; hence, they evaporate in the lee of the mountain. If D and V are such that there is an interval of 25 minutes or more for crystals to grow, snowfall will occur as shown in B.

movement, can be contrasted with the layer type or stratiform cloud formed at the junction of warm and cool air masses having little vertical movement. Orographic clouds are characterized by wide temperature variations throughout their height, usually being well above 32° F. at the base and below this temperature at the top. Thunder clouds are formed by small, rapidly rising air columns above heated land areas, and have somewhat similar temperature variations with elevation.

These natural conditions suggest the following possibility: if a cloud does not have sufficient quantities of ice-forming nuclei, but if other conditions such as temperature, air motion, water droplet size and distribution are satisfactory, the artificial addition of crystallization nuclei may initiate the precipitation process. The potentialities of increasing water supplies by introducing such particles into suitable clouds are therefore interesting.

Methods of Seeding

Late in 1946, Vincent Schaefer and his co-workers at the General Electric Company clearly demonstrated that super-cooled clouds in a laboratory cold box and in the atmosphere could be modified by artificial means so that precipitation would take place. This experiment and many others following it, conducted by General Electric research teams and others, have led water planners to hope that important increases in precipitation might be brought about in areas where additional water is needed.[5]

DRY ICE SEEDING FROM AIRPLANES

The first experiments were with dry ice (solidified carbon dioxide). Schaefer and Vonnegut in "Project Cirrus" (1947-48) demonstrated that certain types of clouds, particularly stratiform, could be induced to precipitate their moisture as snow by seeding them with particles of dry ice discharged from an airplane flying through or over the cloud.[6]

The success of dry ice seeding depends on the fact that a cloud of liquid droplets cannot exist at temperatures below -39° F. The unstable or super-cooled liquid condition therefore is upset if even a small part of the cloud is cooled below this temperature. Ice crystal formation is then initiated in the very cold part of the cloud, below -39°, and the crystals grow as previously described. Hence, if a very cold substance like dry ice, having a temperature of -80° F., is scattered in reasonably small particles through a super-cooled cloud, even at temperatures far above -39° F., small zones will be cooled below -39° F. where the dry ice comes in contact with air and moisture. Ice formation will be initiated in these zones, and the nucleation and crystal growth process will begin. Snow or rain may then occur if there is sufficient circulation to bring additional moisture-laden air into the seeded zone.

Methods of introducing dry ice into super-cooled clouds have generally involved distribution of small pellets from airplanes flying over or through suitable cloud formations. Particularly successful nucleation, accompanied, however, by very little precipitation, have been obtained

[5] Among several reviews and historical accounts of weather modification experiments, the following may be of special interest: V. J. Schaefer, "The Economic Aspects of Experimental Meteorology," *UNSCCUR Proceedings: Vol. IV, Water Resources* (New York: United Nations, Department of Economic Affairs, 1951, pp. 2-27); V. J. Schaefer, "Induced Precipitation and Experimental Meteorology," *Transactions, New York Academy of Science,* Ser. II., 12 (June 1950), pp. 260-64; B. Vonnegut, "Cloud Seeding," *Scientific American,* 186 (January 1952), 17-21; "Project Cirrus—The Story of Cloud Seeding," *General Electric Review,* 55 (November 1952), 8-26. "Tomorrow's Weather," *Fortune,* 47 (May 1953), 144-49; mimeographed papers of the Conference on Scientific Basis of Weather Modification Studies, Tucson, Arizona, April 1956.

[6] B. S. Havens, *History of Project Cirrus,* Report No. RL-756 (Schenectady, N. Y.: General Electric Co., July 1952).

with stratiform clouds, and some orographic clouds have similarly been seeded. Dry ice particles also may be scattered from unmanned balloons.

SILVER IODIDE SMOKE FROM GROUND GENERATORS

The other common method for artificially inducing precipitation is by introduction of solid nuclei for ice crystal formation in a cloud at temperatures above -39° and below +32°. The most effective nucleating agents appear to be those having a crystal structure similar to that of ice; and of these, silver iodide is exceptionally good. This crystalline solid can be converted to a very fine particulate smoke in a fuel-operated smoke generator on the ground. If there are rising air currents to carry the smoke upward, the silver iodide particles can be naturally transported into the cloud region. If temperatures and droplet concentrations are within suitable limits, and if an adequate supply of active silver iodide particles is present, ice crystal formation will be initiated. Other smokes have also been used with varying degrees of success. Copper sulfide has produced some favorable results.

The principal problem is one of introducing a sufficient number of active silver iodide particles into the portion of the cloud where they will be effective in causing ice crystal formation. The transport of the smoke by uncontrollable air currents is uncertain, and there is some decrease in effectiveness of silver iodide when exposed to sunlight, presumably because of partial photochemical action. The rather critical conditions required for start of precipitation, some of which are not clearly known or understood, make the success of a silver iodide cloud-seeding operation unpredictable.

"Project Cirrus," the federal government "Cloud Physics Project" (Weather Bureau, Air Force, and National Advisory Committee on Aeronautics), as well as workers in Australia, and several private weather modification contractors, began efforts to seed clouds with ground-based silver iodide smoke generators between 1948 and 1950.[7]

[7] In addition to references previously cited, the following apply more particularly to these later projects: T. J. Henderson, "Increased Interest in Weather Modification," *Edison Electric Institute Bulletin,* 21 (November 1953), 433 ff; R. K. Linsley, "Artificial Precipitation Control," *Proceedings, American Society of Civil Engineers,* 80 (November 1954), 1-19; E. G. Bowen, "Australian Experiments in Artificial Rainmaking," *Bulletin, American Meteorological Society,* 33 (June 1952), 244-46.

Although spectacular results often were achieved with the silver iodide experiments, there was considerable doubt that precipitation could be increased significantly and controllably by this process.[8] A common criticism of the claims for success was that the conditions chosen for individual experiments were those which usually should have resulted in natural precipitation without cloud seeding. Since the variability in cloud conditions is so great with respect to both time and place, the selection of normal or control conditions with which to compare experimental results was and still is difficult. Because the quantitative effects being produced by seeding are small in comparison with natural variations, statistical analysis of a large number of experiments was seen to be the best evaluation method. It was commenced during this period.

MODIFICATION BY WATER SPRAYING

Warm clouds are a comparatively minor source of precipitation in continental United States, particularly in the regions where supplementary moisture is most needed. This source of potential increased supply has nevertheless been investigated in a series of carefully controlled experiments in cloud modification by spraying water from an airplane into warm cumulus clouds. The results have indicated that precipitation can frequently be induced by artificial nucleation of this type.[9] Under the proper conditions, the spray droplets apparently collect more moisture from the much smaller cloud droplets by collision and by a vaporization-condensation process. The presence of droplets of various sizes leads to growth of some at the expense of others, resulting finally in rainfall.

Commercial Weather Modification

Despite the lack of agreement on the significance of the silver iodide experiments, commercial rainmaking started on a substantial scale about

[8] H. G. Houghton, "An Appraisal of Cloud Seeding as a Means of Increasing Precipitation," *Bulletin, American Meteorological Society*, 32 (February 1951), 39-46. Also, "Statement of the Council of the American Meteorological Society Regarding Artificial Precipitation," *Bulletin, American Meteorological Society* 34 (May 1953), 218-19.

[9] H. R. Byers, "Results of the University of Chicago Cloud Modification Experiments," Conference on Scientific Basis of Weather Modification Studies, Tucson, Arizona, April 1956 (mimeographed).

1950. Several commercial operators, working under contract with farm organizations, municipalities, and other groups, began extensive cloud-seeding programs. Most of them have operated in California, Oregon, New Mexico, Texas, Oklahoma, Colorado, and other western states where the water supply is most critical. During 1953 and 1954 commercial cloud-seeding operations continued and expanded. A particularly significant program of commercial cloud seeding was conducted in the Bishop Creek watershed, California, by the California Electric Power Co.[10] During the same period the federal government continued its investigations by establishing the Artificial Cloud Nucleation Project under Weather Bureau jurisdiction near the Washington coast, and the Bonneville Power Administration Project in eastern Washington and northern Idaho.

The introduction of commercial operation has intensified the controversial technical aspects of weather modification efforts. Even more, it has raised troublesome legal and administrative questions. For example, the possibility that cloud seeding and artificial precipitation in one state might reduce the amount of natural precipitation occurring in a down-wind state has raised the question of the need for regulating cloud-seeding programs. Possible liability of the various parties in case of artificially and accidentally induced floods and other disasters also is a consideration.

The Advisory Committee on Weather Control

In order that considered action on these questions might be taken, the President appointed a special Advisory Committee on Weather Control, and in July 1954 it commenced "a complete study and evaluation of public and private experiments in weather control for the purpose of determining the extent to which the United States should experiment with, engage in, or regulate activities designed to control weather conditions."[11]

The committee gathered data from all organizations engaged in cloud-seeding activities, made a statistical analysis of results, examined them in

---

[10] F. Hall, J. J. Henderson, and S. A. Cundiff, *Cloud Seeding Operations in the Bishop Creek, California Watershed,* U. S. Weather Bureau Research Paper No. 36, Washington, 1953.

[11] Public Law 256, 83rd Congress, First Session (67 Stat. 559).

the light of existing knowledge of atmospheric physics and meteorological phenomena, and investigated the legal aspects of the problems. On February 8, 1956, the committee's first interim report was submitted to the President with a recommendation that its activities be extended for two additional years.[12] This recommendation was approved and enacted, thereby permitting the committee to complete several studies before its disestablishment December 31, 1957. Conclusions and recommendations for subsequent governmental activities in this field were presented in the final report released in January 1958.[13]

Results of Weather Modification Experiments and Commercial Cloud-seeding Programs

The results of weather modification efforts have aroused one of the greatest scientific controversies of recent times. On the one extreme are claims of the weather modification contractors to the effect that their activities have directly caused heavy increases in precipitation.[14] At the other extreme are conservative meteorologists who feel that there are insufficient data for appraisal of the results, or that if there are any effects, they must be small.[15] The views of the groups employing the

[12] Advisory Committee on Weather Control, *First Interim Report, Advisory Committee on Weather Control* (Washington: U. S. Government Printing Office, February 1956), pp. 1-11.

[13] Advisory Committee on Weather Control, *Final Report of the Advisory Committee on Weather Control,* Vols. I and II (Washington: U. S. Government Printing Office, December 1957). Volume II of this publication is probably the most comprehensive and important reference on cloud-seeding technology and evaluation. It also contains an extensive bibliography.

[14] R. D. Hoak reports that 68 per cent increase in Denver water supply has been claimed: "Greater Reuse of Industrial Water Seen," *Chemical and Engineering News,* 33 (March 28, 1955), 1278-82. Increases in rainfall and snowpack of 20 to 40 per cent have frequently been claimed as resulting from commercial cloud seeding in various sections of western United States, as for example, by R. Bollay, "Clouds with Silver Lining," *Electrical West,* 110 (March 1953), 73-76; and in "Demands for Rainmaking Services Seen Increasing," *Engineering News,* 150 (April 9, 1953), 26.

[15] K. A. Brownlee, "Statistical Tests by the Method of a Control of the Rainmaking Hypothesis: Some Alternative Viewpoints," Conference on the Scientific Basis of Weather Modification Studies, Tucson, Arizona, April 1956 (mimeographed). Also, T. H. Evans, "Present Status of Rain Making," *Public Works,* Vol. 84, No. 53, pp. 53 f; and R. K. Linsley, "Artificial Precipitation Control," *Proceedings, American Society Civil Engineers,* 80 (November 1954), 1-19.

"rainmakers" range between these extremes, most of them apparently feeling that some beneficial effects have been secured.[16] Still another position is held by some of the pioneering physicists in this field, who claim that major cyclical changes in precipitation patterns over thousands of miles have been due to cloud-seeding experiments.[17]

Just as evaluation of the early experiments was hampered by the extreme natural variability in weather conditions, it has been difficult to distinguish effects which might be due to commercial cloud-seeding efforts. Precipitation apparently cannot be artificially induced unless all the atmospheric conditions required for natural rainfall or snowfall prevail, except sufficient active crystallization nuclei. Thus determination as to whether occurrence or absence of precipitation was due to, or in spite of, the cloud-seeding effort remains a key problem.

Although much cloud seeding has been performed during the past ten years, only a small fraction has been for the particular purpose of obtaining scientific data on which to base conclusions as to its effectiveness. Most of the operations have been by commercial groups whose primary purpose was to increase the available water supply, rather than to test the methods. By use of statistical techniques, however, analysis of a considerable number of commercial cloud-seeding operations has been possible. These analyses have generally involved a comparison of precipitation occurring in the artificially seeded target area with that presumed to occur naturally in an unaffected area normally having similar meteorological conditions. With a sufficient number of seeded and unseeded storms and measurements of resulting precipitation, the influence of artificial seeding may then be determined. For the results to be significant, however, the normal variation in precipitation between the stations compared must be smaller than the difference shown in the seeded and unseeded storms.

Several experimental projects devoted specifically to determining the effectiveness of cloud seeding have been undertaken by the United States

[16] A. Cage, "Artificial Rain Making Pays Off," *Electrical West,* 104 (May 1950), 60-62; "Denver Bank Pushes Weather Control," *Business Week,* June 17, 1950, p. 90; "Rainmakers Thrive While Scientists Decide Results," *Business Week,* March 15, 1952, pp. 108-10.

[17] I. Langmuir, "Widespread Modifications of Synoptic Weather Conditions Induced by Localized Silver Iodide Seeding," *Science,* 112 (October 20, 1950), 456. "Seeding and Peaking; Committee to Look Into Matter of Cloud Seeding," *Scientific American,* 190 (February 1954), 46-47.

Weather Bureau and others in this country and abroad.[18] From an analysis of the results of these research studies and many commercial cloud-seeding projects, the Advisory Committee on Weather Control made an investigation of the degree to which precipitation had been artificially affected. Progress reports by members of this committee and by others at the Conference on the Scientific Basis of Weather Modification, in April 1956, were among the first significant evaluations of the quantitative effects of cloud seeding, but there was a considerable area of disagreement among those intimately concerned with these studies.[19] Precipitation increases in southwestern United States, Australia, and other areas were postulated, but the statistical analyses were based on limited data.[20] The experiments in Australia, in which clouds were seeded with silver iodide from an airplane, showed the possibility of a 30 per cent increase in precipitation, but further results are needed for dependable conclusions.

The most comprehensive statistical analysis of cloud-seeding operations thus far conducted was sponsored by the Advisory Committee on Weather Control.[21] Eleven sustained commercial cloud-seeding projects in the United States and one in France were used in the evaluation. Regression lines were based on recorded precipitation in 5,096 storms in five orographic, three semiorographic and four nonorographic areas. Comparison of precipitation measurements in 386 seeded storms showed increases of 14 per cent in the orographic and semiorographic areas,

[18] R. J. Boucher, *Operation Overseed,* Advisory Committee on Weather Control, *Final Report,* Vol. II, pp. 126-36; S. Petterssen, "A Brief Survey of the Artificial Cloud Nucleation Program," summary of paper presented at Conference on the Scientific Basis of Weather Modification Studies, Tucson, Arizona, April 1956 (mimeograph); Byers, *op. cit.;* F. A. Berry, "Small-Scale Cloud Seeding Experiments," *Bulletin, American Meteorological Society,* Vol. 38, No. 2 (1957), p. 92.

[19] Summaries of papers presented at Tucson conference by H. C. S. Thom, K. A. Brownlee, H. R. Byers, G. W. Brier, and others (mimeograph).

[20] J. V. Hales, et al., "Cloud Seeding Evaluation in Utah," and E. G. Bowen, "An Aircraft Seeding Operation in the Snowy Mountains Area of Australia," Conference on the Scientific Basis of Weather Modification Studies, Tucson, Arizona, April 1956 (mimeographed).

[21] H. C. S. Thom, *An Evaluation of a Series of Orographic Cloud Seeding Operations,* Advisory Committee on Weather Control Technical Report No. 2 (Washington: U. S. Government Printing Office, 1957). A description of the method of evaluation is presented in Thom, *A Statistical Method of Evaluating Augmentation of Precipitation by Cloud Seeding,* Advisory Committee on Weather Control Technical Report No. 1, 1957. These two reports are also contained in Vol. II of the Advisory Committee's *Final Report,* pp. 25-49 and pp. 5-24.

and no significant increase in the nonorographic regions. The statistical analysis shows the 14 per cent figure to be significant at the 0.01 level of probability. The committee concluded that "[cloud seeding] in western United States produced an average increase in precipitation of 10 to 15 per cent from seeded storms with heavy odds that this increase was not the result of natural variations in the amount of rainfall."[22] The committee stated also, "In nonmountainous areas, the same statistical procedures did not detect any increase in precipitation that could be attributed to cloud seeding."[23] A second evaluation of a more limited area in South Dakota was also conducted by the committee.[24] Results of this analysis, however, were inconclusive except in the presentation of data on the relative risk of seeding during specified months of the growing season.

There is probably more disagreement on the controllability of precipitation than on the quantitative effects of cloud seeding. The extent to which silver iodide nuclei, in active form, reach the precipitation-forming zones in the clouds is not clearly known. The duration of a storm which might result and its lateral travel are also uncertain. Even with successful seeding the storm may yield little precipitation, particularly if circulation and inflow of moist air is inadequate. Questions as to the possibility of overseeding a cloud, actually interfering with or inhibiting precipitation, have not yet been completely answered. However, a recent comprehensive study of this factor indicates that overseeding is highly unlikely with conventional silver iodide smoke generators.[25] The exact conditions required for the most effective precipitation with a smoke-type nucleating agent or with dry ice are still to be determined.

It is generally agreed that if appreciable precipitation increases may be directly obtained by cloud seeding, the cost of a typical commercial program would be amply justified.[26] With a rainfall increase of, say, 10

[22] Advisory Committee on Weather Control, *Final Report,* Vol. I, p. vi.

[23] *Ibid.*

[24] Gerald D. Berndt, *An Evaluation of Commercial Cloud Seeding Operations Conducted During the Summer Months in South Dakota,* Advisory Committee on Weather Control Technical Report No. 5 (Washington: U. S. Government Printing Office, 1957). Also contained in Vol. II of the Advisory Committee's *Final Report,* pp. 69-85.

[25] Boucher, *op. cit.*

[26] No detailed analysis of the economics of weather modification had been made by 1958. However, the *Final Report of the Advisory Committee on Weather Control* treats the problems of making such an analysis and its content. See E. A. Ackerman, "Design Study for Economic Analysis of Weather Modification," Vol. II of *Final Report,* pp. 233-45.

Plate 25. A cloud-seeding experiment in progress, Manila, P. I. First signs of rain may be seen in the form of streaks below the base of the clouds.

Plate 26. Showers from a cumulus cloud after seeding, Manila, P.I.

per cent in an agricultural area where normal precipitation is inadequate, cloud-seeding expenditures of a few cents per acre could possibly yield crop value increases of several dollars per acre. More water for power generation, municipal supply, and industrial use could have even greater value, at small cost.[27] Cloud-seeding programs from airplanes rather than from ground generators, if found to be exceptionally effective, might justify the costs of this more expensive procedure.

One of the most conservative positions on the scientific aspects of weather modification and control has been taken by the Council of the American Meteorological Society.[28] In its statement of Policy on Weather Modification, April 30, 1957, there are nine points relating to (*a*) the qualitative effects and limitations of seeding various types of clouds witht dry ice, water drops, hygroscopic particles, and silver iodide crystals, and (*b*) the needs for further research in this field. Two particularly significant points are:

6. Evaluations performed by independent agencies have yielded reasonably convincing evidence of increases of precipitation due to the operation of ground-based silver iodide generators only for operations conducted in cold weather in regions where forced lifting of the air over a mountain range is an important factor. No convincing evidence has been presented which indicates that ground-based silver iodide seeding affects the amount or character of the precipitation over flat country. This does not prove that there are no such effects but suggests that if present they are too small to be detected by statistical analyses of data available to this date . . . .

8. Present knowledge of atmospheric processes offers no real basis for the belief that the weather or climate of a large portion of the country can be significantly modified by cloud seeding. It is not intended to rule out the possibility of large-scale modifications of the weather at some future time, but it is believed that, if possible at all, this will require methods that alter the large scale atmospheric circulations, possibly through changes in the radiation balance.

Although there now is general agreement that positive effects can be secured and that precipitation can be increased under favorable circumstances, the limitations of the methods, the quantitative results and their prediction and control are not yet known with any degree of certainty.

[27] For some cost and benefit information, see: C. Gardner, Jr., "Hauling Down More Water from the Sky," *Water, The Yearbook of Agriculture, 1955,* U. S. Department of Agriculture, Washington, pp. 91-95; Hall, Henderson, and Cundiff, *op. cit.;* and California State Water Resources Board, *Weather Modification Operations in California* (Bulletin No. 16), Sacramento, Cal., 1955.

[28] American Meteorological Society, *Statement on Weather Modification,* news release May 10, 1957. Also contained in *Final Report of the Advisory Committee on Weather Control,* Vol. II, p. 422.

Considerable analysis of most of the major cloud-seeding activities in
the United States shows that in orographic storms average precipitation
increases of about 14 per cent have been obtained, but effects over
regions distant from mountains are uncertain, and probably small.

The Advisory Committee on Weather Control stated in its final report that only by understanding the physics of cloud formation and precipitation can the potentialities of weather modification be fully developed. [29] This approach is generally agreed to be the most promising and
essential to successful application. Among its recommendations, the
Advisory Committee advocated government performance and support of
meteorological research, co-ordinated by the National Science
Foundation.[30]

Control of Floods, Hail, Lightning, and Other Elements of Weather

In addition to the possibilities of augmenting precipitation, there are
indications that other weather factors might be artificially controllable to
some extent. The control or moderation of floods by *reduction* in precipitation or by its redistribution may become possible, although there
is no clear technical evidence that these effects can be obtained. The
severity and frequency of hailstorms can possibly be minimized by
heavy seeding of thunderclouds to prevent their growth to large proportions and to reduce the degree of super-cooling. A statistical method
of evaluating results from hail suppression experiments has been suggested by Thom.[31] In the course of developing this method, Thom found
that the data employed (Magadino plain, Ticino, Switzerland) gave
some tentative indications of positive results from hail suppression
experiments. The possibility that hail damage might also be increased
on occasion rather than decreased also has been suggested.[32] In its final
report, the Advisory Committee on Weather Control stated that "hail
frequency data were completely inadequate for evaluation purposes and
no conclusions as to the effectiveness of hail suppression projects could

[29] Advisory Committee on Weather Control, *Final Report*, Vol. i, p. vii.

[30] *Ibid.*, Vol. i, p. viii.

[31] H. C. S. Thom, *A Method for the Evaluation of Hail Suppression*, Advisory
Committee on Weather Control Technical Report No. 4, 1957. Also found in
the Advisory Committee's *Final Report*, Vol. ii.

[32] H. Weickmann, "Physics of Cloud Modification," *Bulletin, American
Meteorological Society*, 38 (February 1957), 92-93.

be reached."[33] Reducing the severity of tornadoes and hurricanes through control of the conditions leading to their development or through modifying them after development, should be fruitful areas of future research. Knowledge in all these fields is very limited, and much research will have to be completed before the practical possibilities can be well evaluated.

With the exception of precipitation and hail control, lightning suppression has received more attention than other potential applications of cloud modification. Preliminary results are encouraging although not conclusive. [34] In most of the western forest lands, lightning is the greatest fire hazard, particularly because it often strikes important forests without accompanying rainfall. Investigations are in progress relating to the atmospheric conditions under which these storms develop and to the possibilities for their artificial control.[35]

On a much vaster scale are suggestions that major climate modification in large areas of the country might be effected by "steering" air masses through artificial means such as cloud seeding.[36] This very long-range prospect is not now supported by data permitting even preliminary appraisal. However, since there is little doubt that weather can be modified intermittently in a limited area, the possibility remains that large-scale changes might be brought about by extensive cloud-seeding programs or by other methods which might alter atmospheric circulation patterns.[37]

Potential Effects of Weather Modification on Water Supply and Demand

The foregoing discussion of the status of weather modification experiments and their evaluation shows that appraisal of the benefits first requires establishing some criterion of potential increase in precipitation.

[33] Advisory Committee on Weather Control, *Final Report*, Vol. i, p. vii.

[34] I. P. Krick, "Value of Weather Modification Operation to the Pulp and Paper Industries," *Pulp and Paper of Canada*, 56 (1955), 263 ff.

[35] V. J. Schaefer, "Jet Streams and Project Skyfire," *Transactions, New York Academy of Science*, 17 (1955), 470-75. Also studies in Advisory Committee on Weather Control, *Final Report*, Vol. ii, pp. 105-25, 273-82.

[36] G. H. T. Kimble, *Scientific American*, 182 (April 1950), 48-53. Also W. J. Pierson, "Large Scale Control of Weather by Introduction of Sublimation Nuclei into the Atmosphere," *Transactions, New York Academy of Science*, Ser. ii, 12 (June 1950), 268-70.

[37] American Meteorological Society, *Statement on Weather Modification, op. cit.*

Although there is no unanimity of agreement on the quantitative effects, the consensus is approaching a middle ground between the extreme views.

First, without claiming an exact quantitative influence, but stating instead an arbitrary figure based on latest analyses of cloud-seeding results, it will be assumed that annual precipitation can be increased 10 per cent in localities where orographic cloud formation commonly occurs. These are mainly over the Cascade Mountains, the Sierra Nevadas, the coastal ranges of the Pacific Coast, and the mountains of the Great Basin of western United States. The annual increase selected is effectively larger than indicated by the 10 to 15 per cent increase found in the statistical analysis of seeded storms, because it is based on precipitation over the entire year, part of which is not ideally suited to cloud-seeding programs. However, improvements in knowledge and techniques may bring the annual average increases up to the present level of effectiveness of seasonal and periodic seeding projects.

Second, in the absence of indications to the contrary it will be assumed that artificially induced precipitation in one region does not reduce precipitation down-wind from that region. There is a question as to whether such effects occur or not, but in view of the small fraction of atmospheric moisture actually precipitating in a storm, it seems reasonable to assume that this effect is limited.

Third, in view of somewhat more limited but reasonably consistent data on cloud-seeding attempts under nonorographic conditions in the absence of pronounced convection circulation, it will be assumed that no practical increases in precipitation by present seeding methods can be effected. This assumption may ultimately prove unduly pessimistic. If even a slight percentage increase in precipitation might be artificially obtainable, the actual quantities of water could be substantial, and their economic effects very beneficial.[38]

Fourth, it will be assumed that there is no economical way to cause artificial precipitation from warm clouds, such as commonly occur over Gulf coastal regions and in low-altitude summer clouds throughout the United States. Although they can be successfully seeded and precipitated by salt particles and water sprays distributed from airplanes, the cost as yet appears to be excessive in comparison with the benefits.

Finally, the precipitation occurring from convective storms caused by atmospheric thermal gradients, particularly over the Great Plains,

[38] See Berndt, *op. cit.,* pp. 41, 56.

will be assumed unalterable. The great variability in formation and location of these storms, the frequency with which natural precipitation occurs from them, and the improbability that artificial nucleation would significantly affect the total precipitation even when successful seeding is accomplished, leads to this assumption. Hailstorm and lightning reduction by suitable early seeding of storms of this type may become practical, but the influence on total rainfall in a sizable area is not expected to be large.

In light of these assumptions, and considering the topographic conditions of the country, it is evident that the major potential benefits from cloud seeding for precipitation increases would accrue in western United States (figure 30). Eleven western states contain perhaps 95 per cent of the area in the United States with atmospheric and geographic conditions under which the outlook for benefits from seeding might be considered somewhat optimistic. Assuming a 10 per cent increase in precipitation where orographic storms predominate, it is suggested that an increase in annual runoff of about 15 million acre-feet in these states might be possible (table 30). Some further benefits through soil moisture additions from increased precipitation also would be possible, but are not susceptible even to rough estimate at this time.

The eleven states include about a fifth of the total land area of the United States. However, nearly 90 per cent of the total estimated runoff increase under the assumptions made would occur within five of these eleven states: California, Oregon, Washington, Idaho, and Colorado. The major potential benefits from cloud seeding thus appear to be much more sharply regionalized than is popularly understood at the present time. At the same time, since much of this area is in need of additional water for irrigation, municipal supply, and industrial use, increased precipitation and runoff can be highly valuable.

Although there appears to be little promise of substantial effect on precipitation outside of the western region of the United States, there may be some indirect benefits in adjacent areas through augmentation of stream flow on the eastern slope of the Rockies. Increased withdrawal for irrigation, municipal supply, and industrial use in states downstream might thus be possible. Based on the best available preliminary information, full-scale weather modification efforts would have appreciable effect on total water available for use in the United States, but the increased supply would be most significant in one region.

Another regional benefit in the western states would be to agriculture not under irrigation. Additional precipitation, particularly in the winter

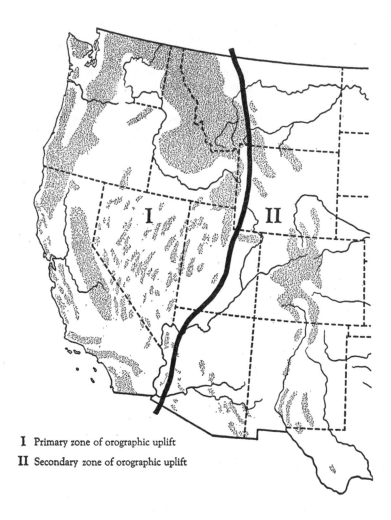

I Primary zone of orographic uplift

II Secondary zone of orographic uplift

Figure 30. Potentially Favorable Areas for Artificial Increase of Precipitation. Areas delineated are those in which the general topography appears to be favorable for AgI seeding. However, the effectiveness of seeding at any particular site depends upon the orientation of the target area to the prevailing moisture-bearing cloud masses, the frequencies of winter storms, special local physiographic features, and other factors. The general rugged characteristics of the terrain of the western states also may foster the formation of cloud systems favorable for seeding in the areas between those shown.

TABLE 30. *Estimate of Potential Precipitation Increases by Artificial Cloud Seeding in Western States*

State	Total area (acres)	Annual precipitation (inches)	Estimated part of annual precipitation from orographic clouds (per cent)	Estimated possible maximum increase in precipitation by artificial seeding		Estimated possible maximum increase in runoff by artificial seeding[1]	
				(inches)	(1,000 acre-feet)	(inches)	(1,000 acre-feet)
Arizona	72,901,760	12.5	20	0.3	2,300	0.02	100
California	101,563,520	20.9	50	1.0	8,500	0.4	3,400
Colorado	66,724,480	16.5	40	0.7	3,900	0.2	1,100
Idaho	53,476,480	17.2	60	1.0	4,500	0.5	2,200
Montana	94,168,320	14.8	30	0.4	3,100	0.1	800
Nevada	70,745,600	8.6	20	0.2	1,200	0.02	100
New Mexico	77,866,240	14.3	20	0.3	1,900	0.02	100
Oregon	62,067,840	27.8	40	1.1	5,700	0.6	3,100
Utah	54,346,624	13.2	30	0.4	1,800	0.05	200
Washington	43,642,880	32.0	50	1.6	5,700	1.0	3,600
Wyoming	62,664,960	14.6	30	0.4	2,100	0.1	500
Total	760,168,704				40,700		15,200
Total five Pacific Coast and Rocky Mountain states[2]							13,400
Percentage in five states							88

[1] Computed by use of natural ratio of runoff to precipitation.
[2] Washington, Oregon, California, Idaho, Colorado.

and spring, would yield substantial benefit in the value of increased agricultural yield. Here again, because of highly variable atmospheric conditions, this supply of water would not be dependable. But although no advantage may be gained in the driest years and little if any improvement realized in the wettest years, benefits should be obtainable under the more "normal" conditions. In the wheat-growing areas of the Northwest, for example, a one-inch increase in precipitation, obtained at times when most needed, might produce a yield improvement of 1 to 5 bushels per acre. Much western dry farming, however, takes place in regions relatively unfavorable for artificial cloud seeding, at a distance from the mountainous terrain necessary to provide orographic influence. Clearly, then, the prospects for precipitation increase in a locality must be appraised by examination of the conditions in the particular area.

At the local (individual watershed) level, the possible effects of artificial cloud seeding on the supply of water can range from zero to perhaps a 15 per cent increase. Such extremes could be found respectively from the arid or semidesert climates of southern Arizona and New Mexico to the ridges of the Sierra Nevada, the Cascade Mountains, and the Rocky Mountains, where ideal orographic conditions exist. In the planning of water development and use in the years ahead, consideration of the potentialities of artificial cloud seeding and weather modification in each particular area will have to be made. Such planning would involve not only organization of the cloud-seeding program itself, with respect to scheduling, locations, extent, etc., but also consideration of storage facilities, multiple-purpose use, priorities, and many other factors.

It cannot be assumed that *dependable* increases in water supply can be obtained by artificial means. Unless sufficient moisture-laden air under the right atmospheric conditions is available in the region, artificial seeding will not be effective. In periods and regions of drought, usually characterized by very little cloudiness, major precipitation increases would be unlikely. Unfortunately, the potentialities and limitations of these weather modification techniques are not yet sufficiently clear to permit long-range planning with a satisfactory degree of confidence. As the data become more available and reliable, particularly with respect to the locality for which plans are being made, steps for developing and using any potentially available additional water can be more definitely taken. The following topics appear appropriate for further consideration: potential effects of successful weather modification on available concentrated water supplies and on soil moisture; relation to the management of storage and distribution systems, relation to capital requirements for water development, relation to multiple-purpose planning, and relation to the development and management of other sources of water.

POTENTIAL EFFECTS ON STORAGE AND DISTRIBUTION SYSTEMS

Substantial increases in precipitation in many instances would require additional water storage facilities. In other situations, where reserve capacity is already available, a percentage increase of the expected magnitude might not require additional storage. In many of the areas where artificial seeding should be most successful, reservoirs rarely have been filled in recent years. This is due not only to low precipitation, but also to an increased rate of water withdrawal on a schedule not originally planned when the reservoirs were constructed. As population grows (rapidly in some of the eleven states) industrial needs for water increase; agricultural demands become greater than the supply, greater fractions of the average stream flow are withdrawn, and the amount of water available for storage is reduced below that originally planned. If, therefore, precipitation could be increased in a fairly regular seasonal pattern, new reservoir requirements would be minimal. Since there are indications, however, that artificial precipitation will be more successful in wet seasons and years than in dry periods, water storage requirements may generally become greater. But the relative success of augmenting winter, as contrasted with summer, precipitation has not yet been determined with sufficient accuracy to permit evaluation of this factor. Further data are required. Some additional reservoir capacity will probably be needed, but the extent of this additional requirement can be determined only after further data on precipitation patterns are obtained and analyzed, bearing in mind the existing storage and use factors in each area concerned.

POTENTIAL EFFECTS ON MULTIPLE-PURPOSE PLANNING

Since the probable limits of precipitation increase obtainable by weather modification efforts are considerably less than the seasonal and annual fluctuations in precipitation, it is unlikely that this factor would have any major effect on the planning of multiple-purpose projects.

In the planning of unbuilt reservoirs, power plants, and other multiple-purpose projects there will be a question as to whether the design data should include artificially induced precipitation and consequent runoff. Generally, it probably will be found that the additional water would not be a major factor in planning reservoir capacities and sizes. These maxima would be based on the expectable heavy natural precipitation in wet years, with the addition of relatively small secondary increases from artificial seeding.

In the consideration of minimum and total annual stream flows, where favorable atmospheric and geographic factors prevail, it also may be desirable to design for flows made higher by runoff from successful cloud-seeding programs. Distribution of water for power, industrial, municipal, and agricultural use would appear to be most simply based on consideration of total stream flow, whether caused by artificial or natural means. Evaluation of project costs and benefits during the planning stage would logically include the use of water expected to be produced by artificial cloud seeding, provided that past experiments in the area have shown reasonably conclusive and favorable results.

POTENTIAL EFFECTS ON THE DEVELOPMENT
OF OTHER WATER SOURCES

Moderate increases in precipitation in limited geographical and atmospheric situations cannot be expected to displace the water from other sources. In an expanding economy with steadily growing water requirements, the development of all economical sources will certainly proceed. The undependable nature of this additional source is perhaps the most potent reason for concluding that it will not displace another more reliable supply. In effect, it will be a bonus, sometimes obtained under certain favorable conditions. Since it may rarely be available in naturally dry areas, and rarely in more humid areas in dry years, other sources can be expected to be developed to their fullest practical extent regardless of weather modification potentialities.

POTENTIAL EFFECTS ON AGRICULTURAL PRACTICES

The modest size and relative undependability of results from cloud seeding make it improbable that there will be any substantial modification in agricultural practices, even in areas ideally suited to weather modification. Where dry farming is practised, crop growth depends on soil moisture from rain and snow; if precipitation is less than the minimum requirement of the particular crop, failure will result. The characteristic extreme fluctuations in precipitation will apparently not be reducible by artificial cloud-seeding activities. Even if a 15 per cent increase to a normal annual precipitation of 15 inches could be depended upon, the additional moisture would not permit any considerable modification in types of crops or agricultural practices.

Responses of irrigation farming to weather modifications may be greater than those of dry land farming. Although it also is unlikely that a cloud-seeding program would result in appreciable changes in this type of agriculture, precipitation increases in the watershed areas could directly benefit the agricultural users through improved crop yields where irrigation deficiencies frequently exist. It has been shown that during the growing season maintenance of optimum soil moisture for typical field crops is very important in achieving high yields.[39] For example, the yield of corn can be doubled by even a small increase in available moisture at the right time.[40] The possibilities for benefits of this type are greater than in nonirrigated farming because much western irrigation farming is located downstream from the better cloud-seeding areas and because reservoir storage makes possible optimum application of the additional water, regardless of the period of collection.

Each watershed and agricultural area must be individually examined for its adaptability to weather modification efforts. Experience is not yet sufficient to permit reliable evaluation of potentialities in a great many regions of the west, but as more data become available the possibilities can be assessed. Where the circumstances appear favorable, planning in irrigation districts and dry farm counties logically will include consideration of the probable costs and benefits of cloud-seeding programs and the desirable extent of such seeding.

Preliminary indications that the frequency and severity of hail might be controlled suggest the possibility of an ultimate reduction in high insurance premiums and high losses due to hail damage. But the data are too limited for appraisal of the potentialities at this time.

[39] Rhoades and Nelson, "Growing 100-Bushel Corn with Irrigation," *op. cit.*, pp. 394-400.
[40] C. W. Thornthwaite and J. R. Mather, "The Water Budget and its Use in Irrigation," *Water, The Yearbook of Agriculture, 1955*, U. S. Department of Agriculture, Washington, pp. 346-57. See also chapter 16, pp. 442-51.

POTENTIAL EFFECTS ON WATER SYSTEM COSTS
AND CAPITAL REQUIREMENTS

Experience with cloud-seeding programs has indicated that the capital requirements and operating costs for augmenting precipitation are very small in comparison with the value of the increased water supply. Commercial cloud-seeding firms have been undertaking projects at contract prices ranging from 1 cent to 5 cents per acre annually. These costs reflect total expenditures for materials, labor, and fixed charges on equipment, plus the profit of the weather modification contractor. It is clear that direct investment and operating costs in weather modification are of minor importance. Unless large increases in cloud-seeding intensity and frequency per unit area are found desirable, program costs may be expected to decrease even further.

The indirect costs of weather modification programs lie in any necessary increase in reservoir capacity and distribution facilities. In many situations capacities may already be ample. In others, enlarged facilities may have to be provided in order that full advantage of the additional precipitation can be obtained. Each situation will have to be examined, particular attention being given to expected seasonal precipitation patterns, annual fluctuations, available storage, and the need and cost of supplemental storage. If diversion works might be required to bring runoff from favorable cloud-seeding sites into another area where the water is needed, these investment and operating costs will have to be considered along with the costs of the weather modification program. If new storage and transmission facilities are required, their annual costs might be larger than the direct cloud-seeding program costs.

POTENTIAL EFFECTS ON NEEDS FOR BASIC DATA
AND FURTHER RESEARCH AND ENGINEERING

Disagreement on the quantitative effects of cloud seeding and doubts of dependable benefits show clearly the need for much more basic data and technical development. The Advisory Committee on Weather Control, the sponsors of cloud-seeding operations, and the weather modification firms themselves, are analyzing or have studied the results of these projects. The United States Weather Bureau and professional agencies in other countries also are carrying on work in this field. Basic investigations in cloud physics are limited, however. Additional research in the fundamentals of cloud formation processes and in the mechanisms of natural and artificial nucleation is a major requirement. There are possibilities for improvements in seeding agents and techniques which might materially increase the success of weather modification. Fundamental studies in cloud physics may open the way for systematic solution of many problems in the artificial production of precipitation.

Results of laboratory and field tests under the best control possible, and with randomized choice of operational procedure, are much needed. Information is necessary on types of atmospheric and geographic conditions suitable for cloud seeding, types of clouds amenable to modification, the best materials and methods for seeding the clouds, seasonal distribution of additional precipitation, hail and lightning suppression, increase or decrease of precipitation down-wind from the target area, and numerous other factors. Complete evaluation of field results in many areas will also require more extensive data on rainfall, snowfall, snowpack, snow moisture content, and runoff, and data on all other weather variables.

POTENTIAL EFFECTS ON LEGISLATION AND ADMINISTRATION

Some of the most complex problems raised by the possibilities of weather modification are legal and administrative. By 1957, fourteen states had passed laws providing for regulation of weather modification activities.[41] Five explicitly claim the ownership of water in the atmosphere above the state and possible additional precipitation on the land.[42] These laws may intensify problems of control and allocation where interstate water regulations exist, such as interstate water allocation, or river regulation compacts and power agreements. Many problems

[41] Arizona, California, Colorado, Florida, Louisiana, Massachusetts, Nebraska, Nevada, Oregon, South Dakota, Utah, Washington, Wisconsin, and Wyoming. New Hampshire, New Mexico and Oklahoma have given notice of state interest in weather modification by collateral legislative action. A full discussion of this question is presented by J. C. Oppenheimer in Advisory Committee on Weather Control, *Final Report*, Vol. II, pp. 211-32; and a summary appears in Vol. I, pp. 23-26.

[42] Colorado, Louisiana, Nebraska, South Dakota, and Wyoming. *Ibid.*, Vol. I, p. 24.

arising from conflicting interests have yet to be solved. Vacation and recreation establishments already have raised objections to additional man-made summer rain, and there have been protests against excessive winter snows attributed to artificial cloud seeding.[43] In the five court cases accepted for trial through late 1957, the trial courts found for the defendant weather modifiers in three cases; one case had not yet come to trial; and another had met with procedural delays.[44] Apparently, in 1957, legal restrictions had not inhibited weather modification operations to any notable degree. However, at that time no court case had reached an appellate stage.

Possibilities of conflicting interest undoubtedly exist although weather modification operations have not yet emphasized them. For instance, if river flow in a dry season is not adequate to satisfy the established rights of certain users, the additional flow caused by artificial cloud seeding operated by a low-priority user may have numerous claimants. Even though proof now appears difficult to establish, down-wind claimants of damages almost certainly may be expected to air their grievances in court. Interstate involvement and permanent federal government interest in the field are most likely. There appears to be ample Constitutional power (interstate commerce, and in the interest of national defense) for federal government participation in regulation, or in other weather modification activity, as needed.[45] In its final report, the Advisory Committee on Weather Control recommended enactment of legislation (S. 86, 85th Congress) authorizing the National Science Foundation to promote and co-ordinate technical studies in this field and recommended also that specific authority to obtain full information on weather modification activities be conferred on that organization. It was felt that procurement of information of this type is prerequisite to legislative actions of regulatory nature.[46]

[43] "Rainmaking Gets Go-Ahead in New York Court Case," *Engineering News Record*, 144 (May 18, 1950), 24. Also "Heavy Snows Spur Attack on Rain Makers," *Engineering News Record*, 147 (July 5, 1951), 26. At least one court case has been tried in which an injunction against cloud-seeding operations by a recreational interest was denied (Slutsky *vs.* City of New York, 97 N.Y.S. 2nd 238, 1950). See Oppenheimer, *op. cit.*, Vol. II, p. 214.

[44] Oppenheimer, *op. cit.*

[45] J. C. Oppenheimer, "Policy Considerations in Weather Modification," *State Government*, 30 (May 1957), 109-13.

[46] Advisory Committee on Weather Control, *Final Report*, Vol. II, pp. 217-18, and Vol. I, pp. 26-27.

Summary and General Evaluation

If the conclusions of the Advisory Committee on Weather Control are generally substantiated in practice, the potentialities of weather modification programs in suitable areas and under proper conditions are significant in the future water supply of the western states. Present indications are that substantial precipitation increases can be secured only in orographic situations where the cloud-forming processes lend themselves to artificial nucleation. These circumstances are encountered principally within eleven of the western states in a scattered and irregular pattern. They are of most importance to total runoff in the five states of California, Oregon, Washington, Idaho, and Colorado.

When more data are available on the conditions necessary for initiating and controlling precipitation, and on the quantitative effects of cloud seeding, it will be possible for planners in individual areas to evaluate cloud-seeding programs as a means for increasing the water supply. The potential local benefits in many instances appear substantial. If the indications are favorable, the potential value of the increased precipitation would far outweigh the cost of producing it. As yet, however, only fragmentary studies of the economic benefits from weather control have been made. The Advisory Committee on Weather Control was able to include only preliminary planning for such study in its 1953-57 program.[47]

The newness of this development indicates the probability of rapid technological change. Thus water resource planning should include, at all times, review of current work in this field.

[47] E. A. Ackerman, *op. cit.*

CHAPTER 15

Expanding the Physical Range of Recovery —

DEVELOPMENT OF TECHNIQUES FOR
THE DISCOVERY AND EVALUATION
OF GROUND-WATER DEPOSITS

Surface Geophysical Reconnaissance Methods—seismic refraction and
reflection; resistivity surveys; induced electrical polarization. Surface
Techniques Other than Geophysical—aerial photograph interpretation;
climatological interpretation. Subsurface Geophysical Methods—resistivity logging and spontaneous potential logging; gamma ray logging
and neutron radiation measurement; temperature logging; borehole
diameter logging; flow meter logging; fluid conductivity logging.
Prediction of Aquifer Productivity and Storage Capacity. Appraisal of
Natural Underground Movements of Water. Chapter Summary.

Technology can turn potential resources within the earth into employed
resources, but first the extent and quality of the potential resources must
be appraised. In the case of water, movement and change over time
also must be evaluated. For surface-water supplies these attributes are
fairly readily ascertained. The situation is not as favorable for water
existing under the land surface.

The technical problem of locating and appraising underground water
supplies is not only inherently more difficult than for surface water but
it also is much more costly. In spite of cost, a considerable amount of
effort has been directed to this task over the years. A number of
recently developed techniques and analytical tools are facilitating the
discovery of ground-water supplies. They are also being employed to
make a far more accurate description and analysis of ground-water

characteristics and conditions of occurrence than has been available in the past. The nature of these techniques and their value in expanding the physical range of ground-water recovery are the subjects of this chapter. They are considered to be significant because (1) our knowledge of ground water is very incomplete, and (2) the magnitude of ground-water resources is vast (see chapter 2, pages 24-28).

While considerable progress has been achieved, all of the existing methods of exploring for ground water and investigating the character of its occurrence have definite limitations. Short of actually drilling into the earth, there still is no entirely reliable way to locate ground water, to ascertain its quality, or to determine the producing capacity of the deposits that are found. The geophysical techniques employed for this purpose provide only indirect measures of the relevant factors. Each of the techniques makes available data from which geological conditions can be inferred. These conditions include the texture, stratigraphy, and structure of the geologic formations within which ground water occurs. The techniques also provide indications of pertinent aspects of ground-water occurrence, movement, and quality. When several of these techniques are employed in combination, however, the comprehensiveness and accuracy of such analysis are substantially enhanced. It has become possible greatly to increase the probability of locating ground water through such reconnaissance activity. Perhaps even more important, however, is the use of these techniques in evaluating ground-water deposits once they have been located.

The development of geophysical methods of mineral prospecting and exploration is of comparatively recent date. It was only thirty years ago that oil was first discovered through the use of these new tools of investigation. Since then, primary emphasis has been given to improving the means of detecting petroleum deposits and many of the more valuable minerals. The potentialities for ground-water investigation of some of these new methods have been realized only recently. In the future they can be expected to assume much greater importance.

The measuring techniques used by the geophysicist in studying geological structure in the field and in exploring for subsurface natural wealth are based upon detecting differences or anomalies of such physical properties as density, magnetism, elasticity, electrical conductivity, radioactivity, and temperature within the earth's crust. When used for analyzing ground-water resources, these measurements can be interpreted to provide indications of the character of the prevailing geological structure, the various types of underlying rock, and the existence of water. Occasionally it also is possible to infer the porosity of a formation, its water content, and the chemical quality of its water. However, additional information regarding the hydrology of the area under analysis is necessary in order to interpret the data correctly and to establish the limits of usefulness of the deposit. Therefore, applied geophysics, properly used by the ground-water hydrologist, is a powerful tool, but only one of several upon which the hydrologist relies.[1]

There are two groups of applicable geophysical techniques: the surface methods, and the subsurface techniques. Surface methods other than geophysical, like aerial photograph interpretation, also may be useful in ground-water investigations.

Surface Geophysical Reconnaissance Methods

Several geophysical methods have proved useful in ground-water exploration and study. Seismic methods, electrical resistivity measurement, and induced polarization are good examples. They are termed surface methods because the measuring instruments may be above ground or in comparatively shallow holes.

SEISMIC REFRACTION AND REFLECTION

In the seismic refraction and reflection techniques, a small charge of dynamite is exploded on the earth's surface, and the travel time is recorded for the resulting direct and reflected sound or shock waves to reach instruments located on the ground at known distances. Since the speed of a wave depends upon the nature of materials through which the wave passes, it is possible to obtain relevant geological information from these measurements. The waves travel comparatively rapidly through solid igneous rock, but move much more slowly through unconsolidated materials. Of special interest for water exploration is the fact that wave velocity is higher in moist than in dry formations. The greater the contrast of the elastic properties which govern seismic wave velocities within the sub-surface formation, the more clearly can the different formations and their boundaries be identified.

[1] David K. Todd, "Investigating Ground Water by Applied Geophysics," *Proceedings, American Society of Civil Engineers,* 81 (Separate No. 625, 1955), 2.

When shock waves reach an interface between two materials having no appreciable difference in elastic properties, such as that marked by a water table, the wave which approaches at the critical angle of incidence will move along the interface. As the wave travels along this boundary, a series of waves is transmitted back into the unsaturated layers above. At any point on the surface the first wave will arrive either directly from the shot point or from a refracted path. By measuring the time of the first arrival at varying distances from the shot point, information can be obtained regarding the distance of these interfaces below the surface.[2]

Because seismic refraction data are most reliably interpreted when homogeneous subsurface strata are encountered, this method is of assistance in ascertaining water table levels which constitute planes of regular configuration. However, the actual presence of ground water is often difficult to determine by this technique since the velocities from saturated and unsaturated zones frequently overlap. Nonetheless, it has been used extensively in New England for locating ground-water supplies.[3] Elsewhere, this method has been of most use in conducting rapid and economical reconnaissance surveys to eliminate areas unfavorable for test drilling. It also is an aid in mapping water table levels where the forthcoming data can be supplemented with additional information. Seismic refraction techniques have found some application as an engineering aid as well, especially in evaluating the adequacy of proposed dam foundations.

RESISTIVITY SURVEYS

A surface reconnaissance technique which has proved even more important in ground-water exploration involves measuring the electrical resistivity of the subsurface strata. This method is the one most frequently used in water prospecting. Measurements are taken of the amount of resistance a unit volume of a rock formation or other underground material presents to the passage of an electrical current. Four surface electrodes may be used, two for measuring and two "charging" electrodes. In one arrangement they are placed on the ground in a straight line with the charging electrodes located at the two extreme ends

[2] Todd, *op. cit.,* p. 5.
[3] D. Linehan and S. Keith, "Seismic Reconnaissance for Ground Water Development," *Journal of the New England Water Works Association,* 63 (1949), 76-95.

of the line. Current is applied to the charging electrodes and the resistance values of the intervening earth materials determined through the measuring electrodes. As the lateral spacing between the electrodes is increased, the resistance is affected to some extent by conditions at greater depths. Vertical exploration thus can be undertaken by horizontal movement of surface equipment. In relatively porous earth materials, resistivity is controlled more by the amount and the chemical character of the water held in the formation than by the properties of the rock itself. Especially in the case of aquifers consisting of unconsolidated material, it is the resistivity of the ground water which is dominant. From the resulting records of the subsurface resistivity conditions at selected points, it is possible where conditions are favorable to prepare iso-resistivity maps, and from them to outline localities or areas underlain by permeable water-bearing material.

Although the interpretation of subsurface conditions from resistivity-depth curves is a complex and frequently difficult problem, several methods of making such inferences have been developed. Some prior information on the rock formations is required for the most reliable data interpretations. Numerous resistivity investigations have been carried out in many different regions of the United States. They have been used in locating perched water tables, alluvium filled valleys, and water "pockets" in weathered crystalline rocks. They also have located various geologic features that affect the occurrence of ground water, such as faults, fracture zones, and dikes. Such surveys have been used in delimiting underground bodies of salt water and in determining both the vertical and the horizontal extent of sea water intrusions.[4] This is possible because resistivity measurements are affected by the degree of salinity of the ground water. This method is also being used in locating areas of extensive seepage along canals[5] and in determining the leakage potential of dam sites and reservoir locations. Well drilling on recommended sites generally has confirmed the usefulness of the technique. In one case, successful wells were developed in three out of

---

[4] See, for example J. H. Swartz, "Resistivity Survey of Scofield Plateau," Territory of Hawaii, Division of Hydrography, Bulletin No. 5, pp. 56-59, in H. T. Stearns, *Supplement to the Geology and Ground Water Resources of the Island of Oahu, Hawaii,* Honolulu, 1940.

[5] C. N. Conwell, *Application of the Electrical Resistivity Method to Delineation of Areas of Seepage* (Geological Report No. G-114, U. S. Department of the Interior Bureau of Reclamation), Denver, Colo., 1956.

four trials, while in another area the accuracy of the resistivity survey finding, as verified by test drilling, appeared to be higher than 90 per cent.

INDUCED ELECTRICAL POLARIZATION

It appears probable that the technique known as induced electrical polarization, introduced and first developed at the New Mexico Institute of Mining and Technology, will become a useful geophysical aid in prospecting for ground water from the surface.[6] As in the case of the resistivity technique, an electrical current is passed through the earth between a system of electrodes spaced on the surface at known intervals. In this instance only a straight line configuration of the electrodes is used. For induced polarization, measurement is made of the decay rate of the reverse polarization current which occurs between the electrodes after the charging current is cut off. By increasing the lateral spacing between the electrodes, deeper penetration of the subsurface is obtained, making it possible to accumulate data for a profile of any depth to the working limit of the method. The maximum polarization voltage is low, however, and it decays with great rapidity. As a result, high-speed detecting equipment is required. While no entirely satisfactory explanation of this polarization effect has been made, its empirical utility in providing indications of the water content of various formations has been experimentally demonstrated. Water has been detected by this method as far as 200 feet below the surface.[7] If measurement of the induced polarization effect can be developed as a suitable field technique, it should prove especially advantageous because the induced voltage varies directly with the permeability of the material tested, its water content, and its quality. This fortunate relationship is in contrast to the resistivity method in which water quantity and quality increases have opposing effects on the resistivity measurement.

[6] Victor Vacquier, *Prospecting for Ground Water by Induced Electrical Polarization* (Socorro, N. M.: New Mexico Institute of Mining and Technology, Research and Development Division, 1956). Also Victor Vacquier *et al.*, "Prospecting for Ground Water by Induced Electrical Polarization," *Geophysics*, 22 (July 1957), 660-87.

[7] P. R. Kintzinger, New Mexico Institute of Mining and Technology, personal communication, December 1957.

Surface Techniques Other Than Geophysical

AERIAL PHOTOGRAPH INTERPRETATION

Recent reports indicate that another new technique of ground-water reconnaissance appears to have promise when used under the proper conditions.[8] Aerial photographs, when interpreted by skilled personnel, have been effectively employed in evaluating the probability of locating ground water in certain areas. The information regarding vegetation, land use, topography, and drainage features which can be obtained from aerial photographs is useful in this respect. Aerial photographs can also be employed to identify the textures of surface materials, which have considerable interpretative significance. Furthermore, water-deposited material such as alluvial plains, terraces, outwash deposits, granular morainal areas, and other potential sources of ground water such as sandy lake beds, all possess specific patterns recognizable from aerial photographs. This technique probably is applicable primarily to the more humid areas where fairly shallow aquifers exist, but it affords an inexpensive means of ascertaining the general ground-water conditions of a broad area. Test well sites can accordingly be located in what appears to be a favorable locality for water prospecting, and costs may thereby be reduced.

CLIMATOLOGICAL INTERPRETATION

The information stemming from another line of endeavor—the application of climatological analysis to hydrologic problems—is also proving to have increasing significance in evaluating the ground-water potential of different localities. Data derived from this source are especially useful in ascertaining the capacity of an area to yield ground water over an extended period of time. When any region is viewed as a self-contained unit, it is clear that withdrawals on a sustained yield basis cannot exceed the amount of water made available through infiltration into the ground. As knowledge regarding the disposition of the moisture provided through precipitation becomes available, determination of the maximum

[8] R. H. L. Howe, H. L. Wilke, and D. E. Bloodgood, "Application of Air Photo Interpretation in the Location of Ground Water," *Journal, American Water Works Association,* 48 (1956), 1380-90.

values for ground-water recharge is possible, regardless of geologic condition. The water budget described in chapter 16 (pages 442-51), may be applicable in ground-water evaluation also.

Subsurface Geophysical Methods

As has been indicated, several surface reconnaissance techniques can be of assistance in locating areas where ground water is probable. Far more definitive and reliable information, however, is obtainable through various subsurface geophysical methods generally characterized as "logging" techniques. Among them are electrical resistivity logging, spontaneous potential logging, gamma ray and neutron radiation measurement, temperature logging, borehole diameter logging, fluid conductivity logging, and flow meter logging. Like the surface methods, the several logging techniques require supplementary data before reliable interpretations can be made. These data usually are in the form of physical samples of the underground formations.

Since the data utilized in this type of analysis are obtained at various depths below the surface, it is necessary that boreholes or some other type of access into the ground should be available. Where wells do not already exist, test wells are often drilled so that logs or records of changes in subsurface material at various depths may be taken. Exploratory drilling has been facilitated in recent years by the development of a continuous flight power auger which can be used in sedimentary formations to sink holes up to 300 feet in depth. This machine is described more fully in chapter 10 (pages 273-74). Because drilling costs are low, the continuous flight auger appears to possess considerable value both as an exploratory tool in itself and in facilitating the collection of further information from logs of various types.

RESISTIVITY AND SPONTANEOUS POTENTIAL LOGGING

Of the various subsurface geophysical techniques, electrical logging is at present the most important. Where this method is employed, a log is recorded of the apparent resistance and spontaneous potential of the formations penetrated by a drill hole. Resistance and resistivity are measured by sending an electrical current into the wall of the hole and measuring the potential drop between the point of current application and various systems of electrodes, some or all of which may be in the borehole, at a fixed distance from the point of current application.[9] The drop in electrical potential, or voltage difference varies directly with the resistivity of the material through which the current passes. For a water-saturated formation, voltage drop is a function of the volume, salinity, and distribution of the water it contains. Because different sediments can be characterized by the type and the abundance of openings containing water, the resistivity data obtained from an electric log can be used to make inferences regarding the subsurface lithology and to delineate the formation boundaries.

These logs, however, have greater utility for evaluating the chemical quality of water in granular aquifers and for estimating the porosity of the water-bearing material. Although their use in this respect is limited to beds of sandstone, sand, gravel, and other unconsolidated materials, this does not appear a serious limitation. Probably more than three-fourths of the ground water annually withdrawn in the United States is obtained from such sources.[10]

Micrologging is the name given to a recent refinement of this technique in which the electrodes are more closely spaced, the separation not exceeding an inch. Due to the resulting shallow penetration, determination of the properties of a formation filled with drilling mud becomes possible. These measurements may then be used in interpreting the data obtained from the normal long electric log to estimate the porosity of the formation.

Spontaneous potential logging involves measurement of the electric currents flowing at various elevations in a well. These currents are generated primarily by two effects. One, called electrofiltration or streaming potential, is a function of the differential hydraulic head that causes fluid to flow between a borehole and an adjacent permeable formation, while the other, identified as the electrochemical potential, is a function of the difference in the concentration of ions in the mud or water in a borehole and in the water in an adjacent permeable formation.[11] Passage of water from the drilling mud or slurry through the mud cake takes place only when the formation being logged is permeable. If some of the mud filtrate is to pass through the mud cake and invade the formation, its fluid pressure must also be higher than that of the

[9] P. H. Jones and H. E. Skibitzke, "Subsurface Geophysical Methods in Ground-Water Hydrology," *Advances in Geophysics,* Vol. III (New York: Academic Press, 1957), pp. 243 ff.
[10] *Ibid.,* p. 260.   [11] *Ibid.,* p. 248.

formation. Although the electrokinetic effect of such filtration generates a potential, the electrochemical potential created at the boundary between the water in the formation and the drilling fluid, as a result of their ionic concentration differences, is much greater. These potentials, which are additive, generate currents which increase when the electrochemical differences of the boundary become more marked, as when a saline water interface is encountered. The difference in electric potential between a point on the surface and the formations traversed by the borehole is then used in appraising the chemical quality of the water.

Appraisals of quality are possible by the resistivity method also, because the resistivity of water varies with the concentration and mobility of ions it contains. Mobility of ions in turn depends upon molecular weight and electrical charge. For example, the ion mobility of chloride is several times that of carbonate. Information regarding the chemical composition of ground water therefore can be inferred from the differences in resistivity measurements as well as obtained from the spontaneous potential logs.[12] Typical graphs empirically relating ion concentrations to total dissolved solids have been prepared for some major aquifers to serve as aids in making such estimates.[13]

Generally, resistivity logging and spontaneous potential logging are accomplished simultaneously. The same electrodes are used for both and the logs are often plotted on a single base line. The combined use of spontaneous and resistivity data facilitates identification of the lithology of the strata penetrated by the borehole, and it provides a fairly accurate and detailed record of the boundaries between the different formations. These electric logs are being used to define the depth and thickness of aquifers, to appraise the chemical characteristics of known ground-water supplies, and to observe important relative changes which occur within them over time.

Electric well-logging studies have been undertaken in most states where serious ground-water supply problems have been encountered. In the San Joaquin Valley of California probably half of the rotary drilled wells are now electrically logged at the time of drilling.[14] The information obtained is used to determine the placement of perforations in well casings, to ascertain formation types, to correlate formations from well

[12] Todd, *op. cit.*, p. 7.

[13] P. H. Jones and T. B. Buford, "Electric Logging Methods Applied to Ground Water Exploration," *Geophysics*, 16 (1951), 115-39.

[14] F. L. Bryan, "Application of Electric Logging to Water Well Problems," *Water Well Journal*, 4 (1950), 3-7.

to well, and to indicate the probable location of additional aquifers.

Although considerable skill and experience are required to interpret the resistivity curves, they are proving to be useful in locating the boundaries between fresh and saline water. Extensive surveys for this purpose have been conducted in Louisiana. Where fresh ground water is protected by conservation laws, logging is employed to determine where casings are required to prevent saline intrusions.

In addition to its role in evaluating ground-water supplies, electric logging is also widely employed in well-drilling operations to help determine when drill stem tests to sample the water should be conducted. If well drilling is interrupted for testing purposes only when the logs indicate favorable conditions, drilling costs can be minimized.

GAMMA RAY LOGGING AND NEUTRON RADIATION MEASUREMENT

Next to electric logging, the most important subsurface geophysical methods in ground-water investigations depend on measurement of the intensity of gamma ray and neutron radiations. Although almost all of these techniques were originally developed for the purpose of petroleum prospecting, several radioactivity well-logging methods have already proved of considerable value in the water field. It seems quite clear that their utility will increase in the future. One of their most valuable features is the fact that measurements can be obtained through cased wells. This is an important consideration, since electric logging is applicable only in uncased holes and therefore cannot be employed in areas underlain with unconsolidated materials where only cased wells are available for exploration.

Prior to World War II, geophysical measurement of radioactivity centered upon logging or recording the intensity of natural radiation emitted by the rocks of the earth's crust. The objective in natural radioactivity logging is to measure the emissions of radioactive substances in the rock formations adjacent to the walls of the drill hole, and to plot the intensity of the radiations at successive depths.[15] Generally, naturally emitted gamma rays are measured, the intensity of which may be computed by use of an ionization chamber, a Geiger-Mueller counter, or the more sensitive scintillation counter. Minute quantities of radioactive material are found in all types of igneous, metamorphic, and sedimentary rocks. Different rocks show characteris-

[15] V. J. Mercier, "Radioactivity Well Logging," *Quarterly of the Colorado School of Mines: Subsurface Geologic Methods*, 44 (July 1949), 345-59.

tic radioactive values which, when accurately measured, may identify the formations encountered and their stratigraphic characteristics. However, because the ranges in radioactivity of several types of rock overlap, the correlation of lithology with radiation intensity is far from perfect. Radioactivity logging in which natural gamma ray intensity is measured also provides no indication on the fluid content of rocks. It is useful mainly in identifying formations and in determining their boundaries, as well as for stratigraphic studies based upon analysis of several bore holes in a given locality.[16]

Within recent years natural gamma ray logging has been supplemented by induced radiation logging. The techniques, which induce and measure the passage and capture of fast neutrons within rock formations, may have particular application to aquifer evaluation. Information missing from the results of natural gamma ray measurement may be obtained in this way, including data on the fluid content of rocks.

Two radiation logging techniques employed in ground-water studies make use of a source of fast neutrons, which is lowered into a bore hole along with specially constructed detection equipment. The neutrons emitted from this source penetrate the casing, if the well or bore hole is cased, and radiate outward into the formation under investigation.

One of these techniques is known as a gamma-neutron log, the other as a neutron-neutron log. Both depend upon the consequent action of induced fast neutrons (figure 31). In the first of these methods, use is made of the intensity of the secondary gamma radiation emitted when neutrons are captured by atoms within the formation. In the second, the number of slow neutrons which reach the radioactive detection element in a specified time is recorded. The amount of gamma radiation from neutron capture is in large part a function of the number of hydrogen atoms present in the formation. So is the extent to which neutrons lose their energy and are slowed as they collide with the nuclei of the material through which they travel. Since water is one of the principal substances containing hydrogen within the earth's crust, inferences can be made from the measurements obtained by these logging techniques regarding the water content of the subsurface formations.[17] It has been reported that when the results of these two logs are com-

---

[16] W. E. Jackson and J. L. P. Campbell, "Some Practical Aspects of Radioactivity Well Logging," *Petroleum Technology* (Technical Publ. 1923), American Institute of Mineral and Metallurgical Engineering, 1945, pp. 241-67.

[17] They have been more widely used in petroleum exploration.

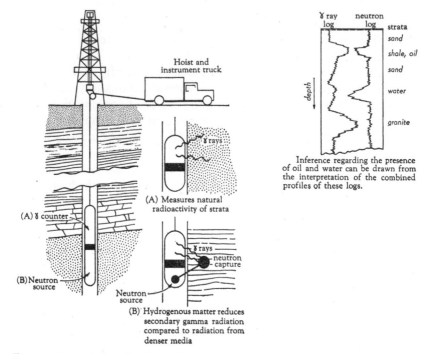

Figure 31. Radiation Well Logging.

bined they possess interpretative significance comparable to that afforded by the electrical log.[18]

While neutron logging can indicate the probable water content of a
formation, it cannot distinguish between permeable and impermeable
formations. For instance, both a saturated shale and a saturated sandstone might give the same measurable secondary gamma radiation.[19]
However, the former is of no value as a source of withdrawable water,
but the latter is often an excellent aquifer. Neutron logging therefore
is best used in conjunction with other forms of logging which can
distinguish permeability, like the spontaneous potential log, and measurement of natural gamma radiation.[20]

[18] Jones and Skibitzke, *op. cit.*, p. 274.

[19] Robert V. West, Jr., "How to Interpret Logs Quantitatively and How Radioactivity and Induction Logging Work," *Oil and Gas Journal*, April 2, 1956,
pp. 115, 116, 119, 123.

[20] Louis Koenig, "Ground Water—The Impact of Recent Technological Advances on its Utilization," Resources for the Future, Inc., unpublished manuscript, December 1956, pp. 59d-59e.

TEMPERATURE LOGGING

The significance of temperature in ground-water hydrology has long been recognized, but measures to make possible the quantitative evaluation of temperature effects are a comparatively recent development. As the practical depth of ground-water withdrawal continues to increase, however, various methods of temperature logging are likely to gain greater application.

Temperature is in itself an important factor in appraising ground-water resources, due to its influence upon the viscosity of water and upon the effective permeability of an aquifer—the latter of significance in the ultimate specific capacities of developed wells. Since electrical resistance of solutions is a function of temperature, electric logs of wells containing highly mineralized water may yield misleading information if the temperature is not known. Where this problem is serious, it has been found advisable to correct the resistivity values that are recorded in electric well logging by computing them at a standard temperature. Because accurate data on borehole temperatures are necessary for such computation, the logging of temperature has become comparatively common practice.

Below the zone of seasonal fluctuation the thermal gradient, or increase in temperature with depth beneath the surface, also serves as a valuable indicator of geophysical conditions. This is the case because the rate of radial heat flow from the high temperature core of the earth is essentially constant. Differences in the temperature gradients through the rocks of the earth's crust thus reflect the properties of thermal conductivity of successive formations, their textural characteristics, and their porosity and moisture content. Both the absolute value of thermal conductivity and the discontinuities in depth-temperature gradients, indicated by these measurements, are of use in interpreting the stratigraphic and structural features of the subsurface formations. The possibility of obtaining information from the extremely small differentials in temperature existing between short vertical distances in the subsoil structures appears especially promising in the case of the recently developed surveying technique known as the "delta-log." This method requires the use of very sensitive equipment, able to measure changes as small as 0.001°F. Particularly advantageous is the fact that the values obtained reflect a passive response to normal thermal gradients and an active response to abnormal ones. Differential temperature logging is also useful in that, like radioactivity logging, it can be practiced in cased holes; moreover, it is far less costly.

The differential temperature log can be employed most effectively together with the conventional depth-temperature gradient log. These techniques have proved useful in identifying water from several aquifers in a single borehole, and in relating the occurrence of water in one hole with that of another. In the future they should be useful in areas underlain by consolidated rock aquifers, where no other logging device is found effective, and where knowledge is sought of fissure location or solution opening from which a well draws water.[21]

BOREHOLE DIAMETER LOGGING

Borehole diameter logging has been utilized for several decades. It has proved useful not only in helping to resolve various problems of well construction and drill stem testing, but also as an aid in determining lithologic sequence and in delineating the areal extent of various formations. The logs depend upon measurement of the variations in the diameter of a rotary drilled bore, obtained today by electrical components which are adapted for use as an adjunct of the conventional electric logging equipment. A borehole varies in width because the intersected strata are not worn away evenly by the abrasive action of the drilling tools or by subsequent erosion and hydration caused by the drilling mud. As a consequence, a well profile reflects, in addition to the action of the drilling tools, the relative hardness, cementation, fluid content, and permeability of the rock formations that are penetrated; and it reflects as well the extent to which the rocks dissolve in, or become hydrated by, the drilling mud. The information thus obtained regarding the characteristics of the formations, and regarding the resistance they afford to erosion and solution, serves as a useful supplement to other logging data.

FLOW METER LOGGING

Flow meter logging is a technique which is used to record the velocity and the direction of flow of water within a well. With current meters similar to those used in surface stream gauging, aquifers tapped by cased wells having multiple screens can be evaluated by this method. Leaks in cased wells and permeable zones penetrated by

[21] Jones and Skibitzke, *op. cit.,* p. 284.

uncased wells also can be identified.[22] The method is employed to measure either the natural or the induced vertical rate of flow of water in the borehole, which varies with the rate of inflow or loss of water through the borehole wall. Such logging aids in identifying each of the aquifers tapped by a well and in determining their static hydraulic head and other characteristics. Relevant information can be obtained from these logs not only on the thickness of aquifers but also on their specific capacities.

FLUID CONDUCTIVITY LOGGING

In fluid conductivity logging, the electrical conductivity or conductance of the borehole fluid is recorded. These properties are the reciprocals of resistivity or resistance, and are obtained through the use of equipment of similar design. They assist in determining the depth of salt-water leaks in cased artesian wells, and the depth and relative artesian head of salt-water aquifers penetrated by uncased wells. Because of its comparative simplicity, low cost, and reliability in identifying intrusions of saline water, it has become a widely applied technique. The logs, when correlated with the spontaneous potential curves of electric logs, are valuable in determining the salinity of the water in the penetrated aquifers.

Prediction of Aquifer Productivity and Storage Capacity

One of the crucial scientific tasks in the field of ground-water hydrology involves the prediction of attainable withdrawal rates. Significant advances have been made upon this problem, particularly since the development, more than two decades ago, of the Theis nonequilibrium formula for determining the transmissibility and storage capacity of an aquifer.[23] Prior to this, the rate of recharge of an aquifer was determined by empirically ascertaining the point of equilibrium at which the rate of withdrawal as a result of pumping became equal to the rate of recharge as indicated by constant water level in the producing well. Where adjacent observation wells are available, the effect which this equilibrium water level has through its cone of

[22] *Ibid.,* p. 289.
[23] Koenig, *op. cit.,* pp. 64-65.

influence upon the level of the other wells, yields information regarding the permeability or transmissibility of the formation. Another old method determines average withdrawal capacity of an aquifer by calculating the hydrologic balance each season for several years. This involves knowing the total actual withdrawals, the recharge from precipitation, the uses of the water (needed in establishing return flow), and the ground water level.[24] This method obviously requires long-term measurements. It is well illustrated in the program of ground-water study which has been in progress in Nebraska since 1930. Even there, however, precise balances over extended areas are still to be determined. In 1957 it was thought that such balances might be possible in "the not-too-distant future."[25]

While application of the Theis formula[26] involves a procedure similar

[24] Blair Bower, personal communication, January 1958.

[25] E. C. Reed, "Ground Water," paper presented at *Nebraska Water Conference,* February 28-March 1, 1957.

[26] The Theis nonequilibrium formula is written:

$$(h_e - h) = \frac{114.60}{T} \int_{\frac{r^2}{t}}^{\infty} \frac{e^{-1.87r^2S/Tt}}{1.87r^2S/Tt} \, d\left(\frac{r^2}{t}\right)$$

Where:

h_e = elevation in feet of original water surface above bottom of aquifer

h = elevation of water surface at any point when pumping

T = coefficient of transmissibility in gallons per day per foot under gradient of one foot per foot

r = distance from center of pumped well, corresponding to h

t = thickness of aquifer

S = coefficient of storage, volume of water yielded per volume of formation unwatered by a one-foot lowering in water table or head

The formula is applied by pumping one well of a group at a constant rate Q and measuring the water level (h) in it and in a number of observation wells at distances (r) at various times (t). Evaluating the integral makes it possible to determine, prior to the state of equilibrium between withdrawal and recharge, the value of S and T, which are characteristics of the formation. When S and T are known, the drawdown in the pumping and observation wells, the contour of the water surface, and the radius of influence may be predicted. The contour of the water table at any time under any schedule of pumping thus may be determined, as can the total available water in the aquifer and the capacity of the aquifer for recharge.

Sources: Koenig, *op. cit.,* pp. 65a-65c; H. E. Babbitt and J. J. Doland, *Water Supply Engineering* (New York: McGraw-Hill, 1955); C. E. Jacob, "Flow of Ground Water," *Engineering Hydraulics* (Hunter Rouse, ed. [New York: Wiley, 1950]).

to that employed in the equilibrium method, it differs in that the data regarding changes in well levels may be utilized as soon as pumping is commenced, before the equilibrium condition has been attained. This is of considerable importance, since extended periods of pumping are often required to reach equilibrium, and if no recharge occurs this condition is never attained. This method also supplies much more reliable information regarding the permeability of a water-bearing formation, as well as making possible the determination of its storage capacity. As a result, the amount of water that will be yielded for each foot of drop in the water table or artesian head can be predicted.[27]

Development of the Theis nonequilibrium formula constituted a major engineering advance. It not only provided a much more reliable technique for determining the capacities of ground-water occurrences, but also opened up an entirely new understanding of underground water movement.[28] It was decisive in arriving at the present capacity to treat ground-water flow as an engineering problem. With this important tool, the technical possibility of developing ground-water supplies to design criteria, so as to insure their maximum usefulness, has been greatly enhanced.

Appraisal of Natural Underground Movements of Water

Even with the useful Theis formula, there remain important gaps in present knowledge of the movement of water underground. Although many deposits of water below the vadose layer may be essentially static, lateral and vertical movement of water occurs in most aquifers of likely economic significance. The characteristics of this motion, accounting for the recharge and release of the water supplies stored underground, is understood in a general way. However, more exact information regarding the volume, rate, and direction of this flow would greatly help to ensure the optimum use of ground-water resources.

The suggestion that the movement of ground water can be traced by adding foreign substances, which can then be followed, was first made many decades ago. Since then, numerous studies have been conducted

[27] D. F. Petersen, "Hydraulics of Wells," *Proceedings of the American Society of Civil Engineers,* Vol. 81, No. 708 (1955).

[28] H. E. Babbitt and J. J. Doland, *Water Supply Engineering* (New York: McGraw-Hill, 1955).

in which various chemical dyes are used as tracers.[29] However, such compounds as fluorescein, one of the most extensively used materials, can only be effectively employed in highly permeable granular strata or in fissured or cavernized formations. They are also subject to other limitations.

Since World War II, attention has turned to the use of radioactive isotopes for this purpose. Such tracers can be detected in very low concentrations and provide the basis for extremely sensitive measurements. Several radioactive tracers have already been utilized and the results indicate some promise, especially for Iodine-131 and for tritium, the radioactive isotope of hydrogen.

Due to the special characteristics of tritium,[30] it appears probable that analysis of ground-water movement may be facilitated by its use. The potentialities of tritium in this field are being actively investigated. Because of the close identity of tritium with hydrogen, it is capable of combining with oxygen to form water molecules which in most respects are indistinguishable from the hydrogen-water molecule. Minute amounts of tritium are constantly being formed in the upper atmosphere, where it reacts with oxygen, the resulting water molecules descending to the earth in the course of precipitation. Tritium has a radioactive half-life of 12.5 years, indicating that half of the atoms will decay in that time. Before the first thermonuclear bomb test (November 1952), the concentration of tritium present in the surface waters of the earth was thought to remain constant, because the rate of decay or disintegration is continually balanced by the rate of replenishment. The tritium concentration in underground water may be substantially less than that in surface waters, if there has been a delay of months or years in the infiltration of precipitation or surface waters into the aquifer. It therefore was thought that the age of ground water, or the length of time it has been underground, might be determined on the basis of the intensity of tritium radioactivity. Such age measurements undoubtedly will have some applicability in the future, although the manner of determination now appears more complicated than originally envisioned.[31]

[29] W. J. Kaufman and G. T. Orlob, "Measuring Ground Water Movement with Radioactive and Chemical Tracers," *Journal, American Water Works Association,* 48 (1956), 559-61.

[30] W. F. Libby, "The Potential Usefulness of Natural Tritium," *Proceedings, National Academy of Sciences,* 39 (1953), 245-47.

[31] Wallace de Laguna, "The Use of Tritium for Determining the Age of Ground Water," *U. S. Geological Survey Water Resources Bulletin,* August 10, 1957, pp. 29-31.

Tritium perhaps holds most promise as an artificially introduced tracer element which could be employed to determine the source, direction of flow, and rate of natural recharge of aquifers; to evaluate the movement of water through the vadose layer; to study the relation between water age and depth; and to investigate more definitely the potentialities of artificial recharge. If some technical obstacles in the utilization of this technique can be surmounted, it seems certain that an additional analytical tool will be made available.

One step toward further evaluation of tritium in such applications was taken in a study of the capacity of soil minerals to adsorb tritium water.[32] Selective adsorption of tritium water in comparison with ordinary water would greatly reduce the value of tritium as a quantitatively usable tracer in ground water. However, such selective adsorption is thought to be of negligible occurrence. When conclusive information of the behavior of tritium in the presence of soil materials is available, determination of its value as a tracer of water movement should be possible.

Chapter Summary

Many geophysical methods developed in the last few decades are useful in ground-water exploration and appraisal. There is no single technique which is universally applicable or which provides the basis for obtaining all of the information pertinent in such analysis. Rather, each is limited both in terms of the interpretative results which can be derived from it and with regard to the geologic conditions and other circumstances under which it can be employed. In most instances, the reliability of the information obtained is additive, and it is consequently desirable to employ several techniques simultaneously. Not only is confidence in the results thereby increased, but the range of usable information is augmented. While it can be concluded that there are no serious limitations upon current technical capacity to locate and analyze ground-water occurrences, questions must be raised with regard to the practical availability of these techniques. The prevailing cost of adequate electric logging, for example, may run to about half the average cost of domestic wells.[33] Application of the other techniques also is costly. In almost

---

[32] Resources for the Future project, completed, 1958.
[33] Koenig, *op. cit.*, p. 60.

every instance the equipment must be manned by skilled and experienced personnel; careful comparative studies are then required for practical and reliable results. These costs are amply demonstrated in the experience of propecting for petroleum, a much more valuable commodity than water in our present economy, per unit of weight or volume.

The costs of prospecting and evaluation are not serious where deposits of large volume, like the deposits of the High Plains, occur at relatively shallow depths (i.e. within 500 feet of the surface of the land). Costs also are not a serious deterrent where there is need for accurate information on the capacity, behavior, and probable life of a large aquifer serving a concentrated demand, as in the vicinity of a large metropolitan area. There is thus little doubt that these techniques of exploration and evaluation will be applied successfully and profitably to the known, but incompletely evaluated aquifers. Included among them will be: (1) all from which significant amounts of water are withdrawn to meet metropolitan and industrial water needs; (2) the bolson, alluvial fan, and lava bed aquifers of the West; (3) the Atlantic and Gulf Coastal Plain sediments; (4) the known major aquifers of the Great plains; and (5) the Mississippi and other alluvial valley sediments of the eastern states. Planning for the use of these aquifers on a more permanent basis should be possible, almost certainly with the addition of substantial water supplies not now exploited.

Beyond the benefits from more precise definition of the known aquifers of some capacity, the potential additions to our water supply from application of these techniques of exploration and evaluation are doubtful. Their cost makes them of limited value in locating and evaluating low-capacity aquifers which can meet dispersed local demands. At the present time farm and nonfarm rural wells comprise 96 per cent of all ground-water installations. A great majority of these, and of the annual new construction, is made up of wells having a daily yield of about 400 gallons. Most of them also are owned by families, and pumped for domestic use.[34] In candor, it must be admitted that dowsing (or "water witching") is still holding its own in the popular mind as a technique of exploration for ground water on this scale. In places, however, the new techniques have been applied successfully in locating even small domestic wells. This has been the case with respect to seismic prospecting in New England. The surface techniques may be of most assist-

[34] Koenig, *op. cit.*, pp. 25, 28.

ance in this respect because they are the lowest in cost. However, their applicability is almost certain to be geographically limited.

Rural ground-water use cannot be dismissed as unimportant on the grounds of small installations. About 50 million American people are served by rural and suburban wells.[35] This is 30 per cent of the total population of the country.

Perhaps of even greater interest is the capacity of the new techniques to explore for and evaluate water deposits lying deeper within the earth than most of the important known aquifers. Experiences with exploration for and discovery of petroleum illustrate that the technical capacity for such discovery and evaluation certainly can be developed, even if it does not exist at this time. There is little doubt that the physical range of our knowledge and definition of ground-water deposits can be expanded greatly. However, practicality will place some limits on the depth of such exploration and evaluation. The practical limit appears to be the maximum depth from which water has been pumped for a large-scale use and as a marginal unit of supply. This is about 1,000 feet, as practiced in recent years in the Salt River Valley of Arizona.

Between 500 to 600 feet and 1,000 feet there is ample room for exploration and discovery. Beyond the 1,000-foot depth, however, energy and capital costs of lifting the water to the surface appear to be high enough to make present exploration of limited interest and limited economic importance for agricultural supplies and small-scale wells (e.g. for domestic supply) unless there are further favorable changes in energy costs or the efficiency of pumping equipment. A different situation may be foreseen for productive deep-lying aquifers which could be exploited for municipal and industrial supply. In these latter instances, high exploration and drilling costs can become minor factors if water production is high. Total costs of lifting water even as much as 2,000 feet would not exceed present total water costs in many localities, provided the aquifer can support a well with a million gallon per day capacity.

The future value of the new techniques may be summarized thus:

1) They will be of considerable value in delimiting, understanding the behavior of, and defining the characteristics of known aquifers.

2) The techniques will be of limited value in appraising the small-scale supplies needed to meet additions to dispersed rural demands. Surface techniques may be of some local assistance in such appraisal.

[35] *Ibid.,* p. 20

3) They will be valuable in further exploration for ground water within the economic range of pumping (about 1,000-2,000 feet, depending on productiveness of the aquifer).

4) The desirability of their use in exploration for deep-lying deposits will vary considerably by region. Further exploitation of ground-water deposits must be weighed against other potential sources of new supply. Throughout the East deep-lying deposits may be judged to have little economic significance indefinitely in the future, except in a few localities of extremely intensive demand. Within the West the prospects of desalinization, weather modification, and long-distance transportation of water must be considered as alternatives to use of deep-lying ground water. Competition between desalted sea water, augmented precipitation, and deep ground water may be particularly evident in all the Pacific Coast states. Within the Rocky Mountain states (Idaho, Montana, Colorado, in particular) weather modification promises some cheaper supplemental water. Within the Great Basin and the Great Plains, however, there may be reason for careful assessment of the underlying ground waters, to considerable depth. It is within these regions that the new techniques may have their most important exploratory application in the near future.

CHAPTER 16

Extending the Use of a Given Supply —

THREE CASES FROM INDUSTRY AND AGRICULTURE

(1) Techniques of Industrial Recycling and Reuse of Water—industrial water demand; extent of water reuse in industry; types of industrial water reuse; specific examples of water reuse in manufacturing industry; effects of industrial water reuse; summary statement of future prospects. (2) Water Budgeting in Irrigation. (3) Reduction of Reservoir Evaporation Losses. Chapter Summary.

While some of the more spectacular potential achievements of technology relate to expanding the available supply of water, less conspicuous measures directed toward extending the service afforded by a given unit of supply may have even greater significance. In view of the general agreement on the prospect for much increased water use and withdrawal by manufacturing industry, possible measures for the reuse of water within industrial plants seem especially appropriate for examination. However, savings of water within agriculture may be aided by new techniques also. These are illustrated by a scheduling procedure which has been proven in some irrigation enterprises: water budgeting. Finally, some new techniques of conservation may be applicable wherever water is stored for use. These are illustrated by evaporation control methods now under experiment. The industrial case will be discussed in greatest detail because the potential water savings through reuse now appear to be both clear and of large size. Eventual results in the other two cases are somewhat more speculative at this time.

1. Techniques of Industrial Recycling and Reuse of Water

Of the estimated 738,000 acre-feet of water withdrawn per day in the United States in 1955 (excluding hydroelectric use), nearly one-half the total (about 358,000 acre-feet, or roughly 115 billion gallons per day) were used by industry.[1] Most industrial use is for cooling purposes and waste disposal, neither of which causes sizable disappearance of water. Water in these uses is a carrier for discarded heat and dissolved or suspended matter from industrial operations. Primarily because of the great quantities of water involved, the prospects of extensive industrial water recirculation and reuse deserve exceptional attention. Possibilities also exist for some reuse of municipal water, but municipal use has a comparatively small volume.

There are two ways in which industry uses water several times: on-site reuse, occurring within a single plant or works; and return flow use, which occurs commonly along every stream from which heavy withdrawals are made.

On-site water reuse may be expressed quantitatively as the total rate of water supply to all parts of a plant, over and above the intake rate, as a fraction or percentage of that intake rate.[2] Thus, 50 per cent reuse means that 50 per cent more water is being supplied to the assembly of units in a plant than is being taken into the plant water system. For

[1] K. A. MacKichan, *Estimated Use of Water in the United States—1955* (Circular No. 398), U. S. Department of the Interior, Geological Survey, Washington, 1957, pp. 1-18. Also in *Journal, American Water Works Association*, 49 (1957), 369-91.

[2] The relationship between intake or withdrawal (W), reuse (R), total water requirements (T), outflow or discharge (O), and disappearance (D), may be illustrated thus:

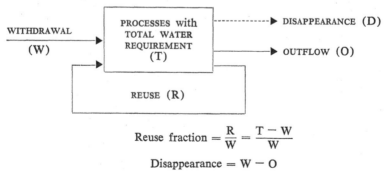

$$\text{Reuse fraction} = \frac{R}{W} = \frac{T-W}{W}$$

$$\text{Disappearance} = W - O$$

every 2 gallons of water entering the plant from the water source, 1 gallon is being recirculated, thus providing 3 gallons for use in plant equipment. Two gallons would be simultaneously discharged in waste streams, vapor, or in the plant product.

Where stream flow is adequate and contamination from upstream industries is not unduly severe, many plants reuse water which has had prior use by other factories upstream. The effluent from one plant thus serves as a partial water supply to another. In regions where stream flows are small, as in most of the western half of the country, there are fewer examples, first because of much lower industrial concentration, and second because flows are often too low for adequate dilution of the waste from one plant. This water therefore cannot serve as a satisfactory supply for other plants downstream.

The techniques hereinafter discussed are exclusively those of on-site reuse. Return flow use requires no special water recycling facilities and does not depend on any substantial new technology. As explained subsequently, its continuance in the next several decades largely depends on widespread application of on-site reuse or other treatment.

INDUSTRIAL WATER DEMAND

A comparatively small number of industries account for a large proportion of industrial water use.[3] Not only do these relatively few industries use large amounts per unit weight of product, but large plants and production involve heavy total water demands on any given site. Some of the largest unit water requirements are found in pulp and paper mills (76 gallons per pound of product),[4] petroleum refineries (38 gallons per gallon of crude oil processed), acetate rayon plants (100 to 200 gallons per pound of product), textile mills (70 gallons per pound of woolen cloth), synthetic rubber industries (90 gallons per pound of product), and steel mills (30 gallons per pound of finished steel).[5] In

---

[3] A survey of industrial water use by The Conservation Foundation showed that only 6 per cent of the plants consumed 80 per cent of the total industrial water used. National Association of Manufacturers and The Conservation Foundation, *Water in Industry* (New York, 1950), p. 10.

[4] The data in this paragraph are on a "once-through" basis, with no recycling, except where total daily use figures are presented.

[5] Calculated from U. S. Bureau of the Census, *U. S. Census of Manufactures; 1954* (Bulletin MC-209, Industrial Water Use), Washington, 1957. Other data, illustrating the range of consumption within individual industries are given in the

the case of steel, for example, total water withdrawal in the United States was about 7.7 billion gallons (23,600 acre-feet) per day in 1954 (table 31.) Withdrawal for the production of chemicals and related products

TABLE 31. *Withdrawal of Water and Recirculation in Some Manufacturing Industries—United States, 1954*[1]

(Total per year in billions of gallons)

Industry	(1) Total water intake	(2) Brackish or salt water intake	(3) Water required if no water was recirculated or reused
United States Total	11,757	1,622	21,906
Pulp, paper, paperboard and products	1,607	135	4,129
Pulp mills	755	85	1,957
Paper and paperboard mills	748	47	1,968
Chemicals and allied products .	2,810	513	5,004
Petroleum refining	1,220	565	4,083
Rubber products	113	5	170
Blast furnaces	836	66	1,104
Steel works and rolling mills ...	1,976	55	2,496
Primary aluminum	263	2	341
Food and kindred products	590	60	1,248
Textile mill products	273	3	307
Lumber and products, except furniture	133	26	155

[1] Source: U. S. Bureau of the Census, *U. S. Census of Manufacturers: 1954* (Bulletin MC-209, "Industrial Water Use"), Washington, 1957. The census of industrial water use was made only among those establishments which indicated that gross intake was 20 million gallons or more in 1954. Electric utility plants (steam or hydro) are not included.

following references: H. E. Hudson, Jr., and J. Abu-Lughod, "Water Requirements," *Water for Industry* (Washington: American Association for the Advancement of Science, 1956), pp. 12-22; Ohio Water Resources Board; H. E. Jordan, "Industrial Requirements for Water," *Journal, American Water Works Association,* January 1946; Iron and Steel Institute; K. S. Watson, "Need for Water Management Program in Industry," *Journal, American Water Works Association,* 47 (1955), 973-81; National Association of Manufacturers and Conservation Foundation, *op. cit.*

TABLE 31 *(continued)*

By Regions and Industry Groups[2]

(Total per year in billions of gallons)

Industry	(1) Total water intake	(2) Brackish or salt water intake	(3) Water required if no water was recirculated or reused
New England Region:			
Total	578	74	877
Food and kindred products	21	11	26
Pulp and paper	249	2	487
Chemicals	41	21	50
Petroleum and coal	22	20	24
Petroleum refineries	7	—	8
Primary metal industries	47	4	51
Delaware and Hudson Region:			
Total	1,092	303	1,969
Food and kindred products	67	22	368
Pulp and paper	93	4	255
Chemicals	237	93	333
Petroleum and coal	261	133	483
Petroleum refineries	248	132	465
Primary metal industries	252	28	285
Blast furnaces	43	—	47
Steel mills and rolling mills ..	153	(*)	172
Chesapeake Bay Region:			
Total	551	180	873
Food and kindred products	33	5	50
Pulp and paper	109	5	225
Chemicals	144	15	264
Petroleum and coal	60	55	61
Petroleum refineries	36	35	37
Primary metal industries	157	100	184
Blast furnaces	66	59	70
Steel mills and rolling mills ..	87	41	107
Eastern Great Lakes–St. Lawrence Region:			
Total	1,435	11	2,189
Food and kindred products	23	2	32
Pulp and paper	110	1	223

[2] Regional boundaries are shown on figure 32, "Census Bureau Drainage Basin Regions," p. 415. Where significant data was readily available information for major industries within industry groups is shown. The group designated as "Primary metal industries" includes the processing of metallurgical coke.

(*) indicates a brackish or salt-water intake of less than one-half billion gallons.

TABLE 31 *(continued)*
By Regions and Industry Groups[2]

(Total per year in billions of gallons)

Industry	(1) Total water intake	(2) Brackish or salt water intake	(3) Water required if no water was recirculated or reused
Eastern Great Lakes—St. Lawrence Region (cont.)			
Chemicals	352	3	580
Petroleum and coal	107	2	212
Petroleum refineries	49	2	140
Primary metal industries	590	1	656
Blast furnaces	218	2	231
Steel mills and rolling mills ..	268	1	288
Ohio Region:			
Total	2,058	14	2,962
Food and kindred products	46	1	68
Pulp and paper	54	2	148
Chemicals	495	1	733
Petroleum and coal	153	7	291
Petroleum refineries	51	3	141
Primary metal industries	1,185	2	1,458
Blast furnaces	349	2	457
Steel mills and rolling mills ..	685	—	834
Cumberland Region:			
Total	16	1	38
Food and kindred products	1	(*)	3
Pulp and paper	1	—	1
Chemicals	14	—	32
Petroleum and coal	—	—	—
Primary metal industries	—	—	—
Tennessee Region:			
Total	228	3	338
Food and kindred products	2	(*)	5
Pulp and paper	37	2	82
Chemicals	152	2	193
Petroleum and coal	1	(*)	3
Primary metal industries	16	1	28
Southeast Region:			
Total	746	116	1,624
Food and kindred products	40	2	65
Pulp and paper	329	94	899
Chemicals	102	9	147
Petroleum and coal	19	1	31
Petroleum refineries	5	1	6

TABLE 31 *(continued)*
By Regions and Industry Groups[2]

(Total per year in billions of gallons)

Industry	(1) Total water intake	(2) Brackish or salt water intake	(3) Water required if no water was recirculated or reused
Southeast Region (cont.)			
Primary metal industries	68	4	245
Blast furnaces	25	4	89
Steel mills and rolling mills ..	37	—	130
Western Great Lakes Region:			
Total	1,539	18	2,063
Food and kindred products	69	1	90
Pulp and paper	142	7	270
Chemicals	167	4	257
Petroleum and coal	185	2	309
Petroleum refineries	136	1	220
Primary metal industries	780	2	898
Blast furnaces	121	—	123
Steel mills and rolling mills ..	617	2	689
Upper Mississippi Region:			
Total	540	7	1,021
Food and kindred products	86	1	224
Pulp and paper	114	1	228
Chemicals	38	1	80
Petroleum and coal	24	2	139
Petroleum refineries	19	1	126
Primary metal industries	100	2	134
Blast furnaces	6	—	11
Steel mills and rolling mills ..	64	—	116
Lower Mississippi Region:			
Total	324	11	685
Food and kindred products	41	6	49
Pulp and paper	31	2	75
Chemicals	41	2	169
Petroleum and coal	67	2	238
Petroleum refineries	66	2	235
Primary metal industries	136	—	136
Missouri Region:			
Total	571	4	1,210
Food and kindred products	60	1	92
Pulp and paper	12	1	26
Chemicals	443	2	944
Petroleum and coal	28	3	108
Petroleum refineries	28	3	102

TABLE 31 *(continued)*

By Regions and Industry Groups[2]

(Total per year in billions of gallons)

Industry	(1) Total water intake	(2) Brackish or salt water intake	(3) Water required if no water was recirculated or reused
Missouri Region (cont.)			
Primary metal industries	18	—	19
Steel mills and rolling mills ..	12	—	12
Arkansas, White, Red Region:			
Total	177	16	1,077
Food and kindred products	12	—	19
Pulp and paper	56	11	210
Chemicals	37	1	187
Petroleum and coal	34	2	593
Petroleum refineries	28	1	585
Primary metal industries	26	1	48
Blast furnaces	3	—	3
Steel mills and rolling mills ..	19	—	31
Western Gulf Region:			
Total	1,112	636	2,989
Food and kindred products	13	(*)	36
Pulp and paper	29	1	213
Chemicals	471	324	907
Petroleum and coal	383	241	1,494
Petroleum refineries	385	241	1,480
Primary metal industries	190	69	271
Steel mills and rolling mills ..	16	11	37
Colorado Region:			
Total	16	—	53
Food and kindred products	2	—	4
Pulp and paper	—	—	—
Chemicals	3	—	11
Petroleum and coal	—	—	—
Primary metal industries	10	—	35
Great Basin Region:			
Total	34	6	144
Food and kindred products	4	—	5
Pulp and paper	—	—	—
Chemicals	4	4	—
Petroleum and coal	9	2	51
Petroleum refineries	2	—	35
Primary metal industries	—	—	—
Blast furnaces	3	—	12
Steel mills and rolling mills ..	8	—	69

TABLE 31 *(continued)*

By Regions and Industry Groups[2]

(Total per year in billions of gallons)

Industry	(1) Total water intake	(2) Brackish or salt water intake	(3) Water required if no water was recirculated or reused
Pacific Northwest Region:			
Total	411	44	966
Food and kindred products	23	1	30
Pulp and paper	224	9	740
Chemicals	27	7	31
Petroleum and coal	1	(*)	5
Primary metal industries	38	5	58
California Region:			
Total	329	179	828
Food and kindred products	47	7	82
Pulp and paper	17	—	47
Chemicals	42	26	86
Petroleum and coal	162	136	473
Petroleum refineries	158	135	437
Primary metal industries	11	1	61
Steel mills and rolling mills ..	8	—	34

Figure 32. Census Bureau Drainage Basin Regions.

was another 7.7 billion gallons per day; for pulp and paper, about 4.4 billion gallons (13,500 acre-feet) per day; and petroleum products, 3.4 billion gallons (10,800 acre-feet).[6] Most of the water used by industry is not actually consumed but is returned to streams and lakes, and occasionally to underground sources.

By far the largest user of water is the electric power industry, where withdrawal averages about 65 gallons per kilowatt-hour generated in fuel-burning power plants. This water use is almost entirely for cooling purposes, and is consumed only to slight extent, the balance generally being returned to streams and lakes. Total withdrawal for steam-electric power production in 1955 was about 72.2 billion gallons (222,000 acre-feet) per day, constituting more than three-fifths of all industrial uses.[7] According to a survey by the Federal Power Commission, 65.5 billion gallons per day were used for this purpose in 1954.[8] (Table 32.)

TABLE 32. *Withdrawal of Water and Recirculation in Thermal-Electric Power Plants of 25,000-kw Capacity or More, United States, 1954*

(Total for year in billions of gallons)

Region	Total water intake	Brackish or salt water intake	Total water required if no water was re-circulated or reused	Water recircu-lated as a per-centage of total water intake (per cent)
United States Total	23,145	5,355	26,196	13.2
New England	1,119	700	1,135	1.43
Delaware and Hudson ..	3,182	1,886	3,182	0
Chesapeake Bay	1,469	366	1,470	0.07
Eastern Great Lakes— St. Lawrence	2,052		2,064	0.58
Ohio	4,128		4,487	8.7
Cumberland	12		28	133.3
Tennessee	917		917	0
Southeast	2,417	733	2,487	2.9
Western Great Lakes ...	2,780		2,780	0

[6] U. S. Bureau of the Census, *U. S. Census of Manufactures: 1954* (Bulletin MC-209, 1957).

[7] MacKichan, *op. cit.,* p. 11.

[8] Federal Power Commission, *Water Requirements of Utility Steam-Electric Power Plants, 1954,* Washington, 1957.

TABLE 32 *(continued)*

(Total for year in billions of gallons)

Region	Total water intake	Brackish or salt water intake	Total water required if no water was recirculated or reused	Water recirculated as a percentage of total water intake (per cent)
Upper Mississippi	1,816		1,884	3.7
Lower Mississippi	312	157	537	72.1
Missouri	437		603	38.0
Arkansas, White, Red Rivers	457	1	1,194	163.5
Western Gulf	577	276	1,508	161.3
Colorado	25		151	504.0
Great Basin	25		75	200.0
Pacific Northwest	5		5	0
California	1,415	1,236	1,688	19.3

Source: Unpublished summaries and files of the United States Federal Power Commission for the year 1954.

Although the principal industrial uses for water are for cooling and processing, distribution of these and other uses is considerably different in large and small plants (table 33.) Cooling water includes that used in equipment such as furnaces, reactors, refrigeration plants, and so on, as well as in the condensers of public utility and industrial power plants.

TABLE 33. *Estimated Percentage of Industrial Water Intake Used for Various Purposes*

Purpose	Plants withdrawing less than 10 million gallons per day (per cent)	Plants withdrawing more than 10 million gallons per day[1] (per cent)
Cooling	23	54
Process	36	32
Boiler feed	12	9
Sanitary and service	26	6
Other	3	4

[1] Total exceeds 100 per cent because of inclusion of reused water by some reporting plants.

Source: National Association of Manufacturers and The Conservation Foundation, *Water in Industry* (New York, 1950), p. 15.

Process water includes supplies for the washing of raw materials and products to remove impurities, for conveying materials either in suspension or in solution, for actual physical and chemical addition to products, and for many other uses.

EXTENT OF WATER REUSE IN INDUSTRY

Among the plants included in the 1954 Census of Manufactures, average water recirculation was about 86 per cent of intake.[9] Considerable variation among industries is characteristic: water recirculation by petroleum refiners averaged 235 per cent in 1954; pulp and paper, 157 per cent; chemicals, 78 per cent; blast furnaces and steel mills, 60 per cent; rubber products, 50 per cent; and textile mills, 12 per cent.[10] It is clear that water recirculation is being practiced by many users. Much of the water reused is for cooling, where the usual practice is a recirculation rate several hundred per cent of intake.

In the petroleum, chemical, and wood pulp industries, water recirculation has been used more extensively than by most industrial manufacturers. Among producers of chemicals and allied products using more than 20 million gallons of water per year, fresh-water withdrawals were about 46 per cent of the total water used in all plant processes in these industries. Brackish water constituted 18 per cent of the total intake and 10 per cent of the total water used, while recirculated water accounted for the balance of 44 per cent. Since there would be very little reuse of brackish water, the amount of recirculated water is about equal to the amount of incoming fresh water. In effect, therefore, each gallon of incoming fresh water is used twice before it disappears, runs to waste, or enters a chemical product. Reuse is even more intensive in petroleum refining. Among petroleum refineries with more than 20 million gallons of gross water intake annually, fresh water requirements were about 16 per cent of the total water used in all plant processes in these industries.[11] Brackish water intake constituted 46 per cent of the total intake and 14 per cent of the total water used, while recirculated water accounted for the balance of 70 per cent.

Unless there are some exceptionally difficult purification problems,

[9] U. S. Bureau of the Census, *U. S. Census of Manufactures: 1954* (Bulletin MC-209, 1957).
[10] Calculated from data in table 31.
[11] *Ibid.*

recovery and recirculation of large quantities of waste water and reduction of intake requirements is technically feasible in all industry. Whether an industrial establishment actually practices recirculation depends, however, on several factors, principally of economic nature. Because manufacturing costs are the prime consideration of industrial managements in their selection of materials and processes, the extent and type of water reuse in a particular situation depends almost entirely on the costs of recovering and recirculating waste waters in comparison with the cost of supplying the same quantity of fresh water. Because of the usual compact design of industrial plants, recirculation systems seldom require highly extensive and costly distribution facilities, but the cost of purification works, cooling equipment, and storage and pumping installations may be large or controlling factors in the choice of a water supply system. Thus, in locations where fresh water is scarce and expensive, or where its low quality requires costly purification, considerable reuse is usually practiced. Conversely, ample supplies of good-quality water available at low cost generally make reuse uneconomical.

Occasionally, recirculation practice has been dictated by corollary considerations, such as the necessity for complying with anti-pollution laws or the recovery of valuable materials from waste streams. When product recovery has led to water recycling, economics has again been the determining factor.[12] Legislation requiring a plant to cease disposal of its nuisance wastes in a stream or lake has led to higher expenses in some recirculation operations, but even these cases may become examples of maximum economy as fresh water costs rise.

Management's cognizance of costs and pollution regulations has been accompanied by the development of equipment to make the repurification of water reasonably economical. In the pulp and paper industry, for example, the development of economical continuous filters for the recovery of small amounts of waste fiber from large volumes of water has been an important factor in achieving the large recirculation rates now common in many plants.[13] There has been substantial progress in developing equipment that is unique for the industry and operation concerned. Frequently this involves special chemical processes. In the

[12] Of plants reporting waste water treatment in 1949, nearly one-third recovered useful by-products. National Association of Manufacturers and The Conservation Foundation, *op. cit.,* p. 41.

[13] H. B. Brown, *op. cit.,* p. 2154.

atomic energy industry, the development of highly effective ion exchange resins has made possible the complete demineralization of water before its exposure to radioactivity. Recirculation of this stream can then be practiced with minimum danger to personnel and facilities. (Chapter 12, pages 324-32, contains a more complete discussion of this process.)[14]

TYPES OF INDUSTRIAL WATER REUSE

Cooling water. The estimated 95 billion gallons (291,000 acre-feet) of water used per day for cooling in power plants and manufacturing establishments in the United States in 1955 represents withdrawal from streams, lakes, brackish supplies, and wells.[15] Most of this is used once for cooling and condensing purposes and is then discharged into surface streams and lakes. The balance is used as make-up water to recirculating cooling systems, part of which ultimately is discharged as liquid, the remainder being evaporated.

The objective of any industrial cooling operation is the transfer of heat from the material being processed and the processing machinery to the atmosphere or the earth. This can be accomplished by the direct use of air as a coolant, by "once-through" use of ground water returned to the aquifer, by the once-through use of surface water returned to a stream, or by recirculated surface or ground water. From the point of view of water use, the most conservational means are air cooling and once-through use of ground water returned to the aquifer. However, a ground-to-ground cooling process may involve sizeable pumping costs, and problems may arise because of the aquifer's response to the return of warm or hot water. Air cooling, particularly of high-temperature processes, until recently has been confronted with lack of suitable metals to facilitate heat transfer. Another handicap is the lower heat transfer rate to air and the greater surface required. Water therefore has been

[14] See also A. E. Gorman and C. V. Theis, *op. cit.,* p. 47. Also A. L. Biladeau, "Reuse of Cooling Water in an Atomic Energy Commission Installation," *Industrial and Engineering Chemistry,* 48 (1956), 2159-61.

[15] Estimated thus: 68.3 billion gallons for power plant cooling; 26.6 billion gallons for cooling purposes in other industrial establishments including air conditioning. The latter figure is 60 per cent of the total industrial use of water estimated by MacKichan, *op. cit.* p. 11. The first figure is from MacKichan, *op. cit.,* p. 11.

a preferred cooling medium because its use has permitted more economically constructed equipment. Moreover, its temperature is lower and less variable than that of the atmosphere. More effective cooling and fewer problems of operational adjustment are therefore encountered with water-cooled equipment. These are particularly important factors in the efficiency (heat rate) of a thermal electric power plant. Consequently, most industrial cooling has been by means of water, either in a once-through process or, more recently, in a recirculation system.

Where once-through surface-water systems are employed, or recirculated ground or surface water, or direct air cooling, the ultimate receiver of discarded heat is the atmosphere. If plentiful and cheaply obtained supplies of relatively cool water are available, there is little question that this source will be chosen for use in an industrial plant on a once-through basis. However, if reduction of withdrawal or lowering of initial water temperatures is sought, water recirculation or other plant cooling measures may be employed.

Cooling water recirculation is a relatively simple process. The water withdrawn from a stream or the ground first receives the discarded process heat. It is then pumped to cooling towers, where it is brought into contact with air. Part of the water thereupon is evaporated, transferring the process heat to the atmosphere. The minimum temperature to which the water can be cooled is known as the wet-bulb temperature, which depends on the temperature and humidity of the atmosphere. The wet-bulb temperature may be only a few degrees below atmospheric temperature when the humidity is high, or 25 to 30 degrees below in dry atmospheres. If a once-through surface-water cooling system is used, the warmed water entering lakes and streams evaporates more slowly, but ultimately transfers its heat into the same atmospheric receiver.

Cooling water recirculation does not reduce the over-all evaporation or disappearance of water. (See also discussion of water recirculation in thermal power plants, chapter 12, pages 309-19.) If the once-through cooling process is used, no water is actually "lost" at the plant, because it is simply warmed a few degrees (5 to 15 degrees increase in common practice) in passing through the plant. But this temperature increase results in slow evaporation of the effluent from a river or lake, the net result being vaporization into the atmosphere to an extent depending on several natural factors. At least half as much water vaporizes from the stream or lake as from a recirculation system removing an equivalent amount of heat.[16] In one case, evaporation takes place at the plant where cooling is performed, and in the other case it occurs in the effluent from the plant. It has been shown in chapter 12 that repeated heavy use of a stream for cooling purposes in once-through systems results also in excessive river temperature rise. This is particularly aggravated by quiet, deep, low-falling flow, where aeration is limited. Thermal "pollution" of streams may thus be a major problem, solvable at least in part by recirculation.[17]

Problems associated with the older natural-draft cooling tower recirculation systems have been the slow rates of evaporation in the towers, the energy required for continuous circulation of the cooling water, and corrosion and fouling from biological and inorganic contaminants. Speeding the rate of evaporation and reducing the over-all size and cost of the recirculation system have been accomplished by the introduction of forced-draft cooling towers, evaporative condensers, and sequential cooling systems.

Probably the most significant factor in the growth of recirculation cooling systems has been the simple and economical design of forced-draft cooling towers. These units are now so widely used that they are standard equipment in all large modern recirculation cooling systems. The additional power cost in blowing air upward through the tower counter-current to the descending drops of warm water is more than justified by the savings in construction costs.

Another new development in cooling equipment is the evaporative condenser. This unit combines the functions of the process cooler or condenser with the cooling tower for water recooling and recirculation. The heat from a large refrigeration plant, for example, may be removed by spraying water on the condenser tubes, exposed also to the air. Heat is dissipated largely by evaporation of a portion of the water into a natural or forced air draft, the unevaporated water being recirculated without storage. The consumption of water is no less than in a standard condenser and cooling tower system, but equipment economies may be substantial. They may permit recirculation where separate cooling tower costs would be so high that a once-through system would otherwise have to be chosen. Evaporative condensers are most advantageously

[16] G. E. Harbeck, "The Use of Reservoirs and Lakes for the Dissipation of Heat," U. S. Geological Survey Circular 282, 1955.

[17] E. W. Moore, "Thermal 'Pollution' of Streams," *Industrial and Engineering Chemistry,* 50 (1958), 87A.

used in small to moderate sized installations, whereas large cooling loads, as from a power plant, are more economically handled in separate condensers and conventional cooling towers.

Another type of recirculation system for process cooling involves the sequential use of water for cooling at several increasing temperature levels. In many industrial plants there are some cooling requirements which may be satisfied by water considerably above the lowest available temperature. In these situations, cold water can first be used in those operations where the lowest temperatures are demanded. The resulting warmed water is then used as a coolant for the higher temperature operations. In steel mills, for example, water may first be used in the power plant condensers where the lowest obtainable temperatures are desirable. The water may next be used for cooling process equipment operating at temperatures from 100 to 300 degrees F. Warm water from this operation may be used even further for cooling furnace walls and burner ports before its final discharge to cooling tower or surface disposal streams.

In this way, the same water supply can be used two or more times, its temperature progressively increasing without evaporation or appreciable loss. Ultimately, however, the total evaporation in a cooling tower or from surface disposal streams will be the same as though separate cooling supplies were used in each operation. Except for facilitating nonevaporative cooling, as explained below, the advantage is limited to permitting lower cooling tower capacity requirements and other water handling economies.

Improved methods for maintaining purity in recirculated water have led to significant reductions in withdrawal rates by increasing the extent of recirculation possible in cooling systems. Contact with air, particularly when warm, promotes the growth of biological impurities and causes corrosion of equipment by metallic ions, oxygen, and other impurities accumulating in the water. Better materials and methods for purifying the recirculated streams have resulted in cost reductions and greater water reuse. An additional result is the decrease in the necessary rate of purge or discard of water from the system to prevent excessive accumulation of impurities. Improved chemical germicides have been particularly effective in this respect. (See also chapter 12, pages 311-12.) The use of corrosion-resistant alloys in condensers and other heat transfer equipment has also permitted higher recirculation ratios. Instead of a once-through cooling water requirement of 50 to 150 gallons of water per kilowatt-hour generated, the intake fresh water can be reduced to as low as one gallon per kilowatt-hour generated.[18,19] This quantity is actually the water evaporated in the cooling tower plus the small discharge required to prevent accumulation of impurities.

Although none of the foregoing improvements in cooling water recirculation results in an over-all decrease in water evaporation, there are two other types of cooling which do reduce water disappearance. If the temperature of the used cooling water is substantially above atmospheric temperature, its heat can be dissipated without any evaporation simply by passing the water through an air-cooled heat exchanger without actual contact with the air stream. The recirculated water can thus be cooled to a few degrees above atmospheric temperature without any water loss. In many cooling operations in the metallurgical, petroleum, and chemical industries, temperatures of effluent cooling water are well above atmospheric, and considerable reduction in fresh-water requirements can be achieved by use of this indirect heat discard method. Even if the recirculated water is required at a temperature below atmospheric, much of the heat can be removed from the used coolant by nonevaporative air cooling. Only the residual heat between the atmospheric and wet-bulb temperatures need be removed by evaporation in cooling towers.

Unfortunately, cooling water for condensers in efficient electric power plants must be the coldest available, and always below summer atmospheric temperatures. Furthermore, its temperature increase in use should be relatively small, generally not more than 10 degrees. Provision is made for this by affording ample flow rates through the condensers. Steam power plants, although the largest water users, must therefore use evaporative cooling in any recirculation system. Another limitation to the use of nonevaporative water coolers is their cost, which is considerably higher than that of the evaporative, cooling tower type. Possible water savings must therefore be balanced against additional equipment costs.

The other technique for saving cooling water is the use of direct air cooling whenever possible and practical. Instead of using water cooling in operations at temperatures far above atmospheric, water often can be

---

[18] Calculated from best modern heat rates of about 9,000 Btu per kilowatt-hour. Data also presented by Hudson and Abu-Lughod, *op. cit.*, p. 19.

[19] The Federal Power Commission indicates, for example, that the average water intake per kilowatt-hour generated in Wyoming steam-electric power plants (1954) was only 0.8 gallon (*Water Requirements of Utility Steam-Electric Power Plants, 1954*).

replaced by air with substantial benefit. Especially where water is expensive, this type of cooling is receiving consideration and use. For example, in the study of equipment and process designs for shale oil production from Colorado oil shale, much attention is given to water requirements. The very limited water supply in the regions of the shale deposits places processes which employ air cooling at a considerable advantage.[20] Air-cooled condensers have been installed recently in oil refineries even where water was plentiful, because total costs were found to be lower than water-cooled condensers and additional water supply facilities.[21] However, general applicability of air cooling is restricted by equipment design requirements, usually more easily and economically met by water jackets, coils, and tubes.

The development of new alloys and ceramic materials resistant to oxidation and loss of strength at high temperatures, has made possible the use of air cooling in hot reactors and furnaces, where water cooling previously had to be used in order to minimize equipment damage. New cooling designs and techniques have led to use of dual cooling systems. Air may be used for the high-temperature part of the cooling requirement. Water is employed for cooling below atmospheric temperatures, or where a liquid coolant is indispensable. The petroleum refining industry has been a leader in applying these techniques.[21]

The relative ease with which cooling water can be recooled and recirculated to plant processes has resulted in widespread year-round use of recirculation systems except where fresh cold water is plentiful and cheap. Typical costs of operating a recirculating cooling water system range from 1 cent to 5 cents per thousand gallons.[22] Since the cost of pumping and purifying industrial fresh water varies from 2 or 3 cents per thousand gallons to as much as 25 cents, the economic advantages of recirculation can be substantial in suitable circumstances.[23]

[20] F. L. Hartley and C. S. Brinegar, "Oil Shale and Bituminous Sand," paper presented at Energy Resources Conference, Denver, Colo., October 1956.

[21] J. W. Thomas, "Economics of Air Cooling in Refinery Installations Where Water is Plentiful," paper presented at meeting of the American Institute of Chemical Engineers, Cincinnati, Ohio, December, 1958; D. H. Stormont, "Water Conservation," *Oil and Gas Journal,* October 27, 1949, p. 58; "How to Save Cooling Water: 2 Case Histories, *"Petroleum Processing,* 10 (December 1955), 1901-7.

[22] R. S. Aries and R. D. Newton, *Chemical Engineering Cost Estimation* (New York: McGraw-Hill, 1955), p. 172.

[23] E. R. Gilliland, "Fresh Water for the Future," *Industrial and Engineering Chemistry,* 47 (December 1955), 2410-22. K. S. Watson reports extensive cost

If fresh-water supplies are limited, recirculation may be mandatory because of insufficient volume even if total stream flow were utilized in a once-through system. Another situation occasionally dictating the use of recirculation is excessively warm fresh water. Since power plant efficiency is highest when water is the coldest, the use of cooling towers to reduce water temperature below that of the fresh supply may be advantageous. Recirculation is then employed.

Particularly in view of the new methods, reuse of cooling water and the practice of air cooling can be expected to increase steadily, both in number of installations and in extent where used. Rising costs and competing demands for fresh water will spur the growth of these conservation practices.

Recirculation of industrial process water. The term "process water," used in industrial operations, applies to all water which, for any purpose other than cooling, is brought in contact with the materials being manufactured or processed. Except for small amounts of water which may be vaporized or may enter a product either as part of a mixture or in chemical combination, process water is eventually discharged, usually containing waste materials from the manufacturing operations. This supply therefore has an important waste-disposal function, used process water actually being sewage. Just as municipal sewage can be treated and the water made reusable and even potable, industrial effluents can be purified for reuse. The problem basically is not different from municipal sewage disposal, since only the treatment processes vary as required by particular impurities.

Waste from industrial operations may be suspended matter, dissolved organic compounds, dissolved inorganic impurities, and bacteriological contamination. Most of the process water effluent comes from washing operations in which products, intermediates, and raw materials in gaseous, liquid, or solid form are freed of impurities by washing them. The process water supply must be of a purity sufficient to yield a product of the required composition. Reuse of the process effluent simply requires removal of accumulated impurities to a degree permitting use of the

data from the General Electric Company showing costs of industrial water from municipal supplies ranging from 10.5 cents to 25.6 cents per thousand gallons. Water from company wells averaged 3.7 cents, and from streams, after chlorination, 3.9 cents per thousand gallons. In the company's one treatment plant 1,000 gallons of process water cost 8.6 cents. Total costs of recirculated water (type not specified) are reported to be 3.6 cents per thousand gallons. Watson, *op. cit.,* pp. 973-81.

water again in the same operation or, if this is impractical, for reuse in operations having less severe purity requirements.

Until recent times, when process water became contaminated with plant wastes it was run to the nearest convenient stream for disposal without treatment of any kind. Three factors other than the water-saving incentive have led toward recycle of these waste streams. Probably the most important of these factors is legal restriction against excessive contamination of streams, measured both in quantity of waste and in concentration of objectionable or dangerous contaminants. Legislation enacted in many states has necessitated industrial sewage treatment, which in practice has required recirculation of waste streams in order to decrease their quantity and to reduce or eliminate some of the contaminants. If the effluent receives treatment such that it approaches or exceeds the purity of the plant water supply, recycling of this stream becomes practical and economical.

Occasionally impurities in waste streams have become recoverable at a profit. Loss of substantial quantities of plant products in waste water has frequently been reduced, thereby reducing the net cost of waste treatment. In other instances salable by-products have been developed from recovered waste material. Still another advantage in waste recovery is in the reduction of raw-material requirements, where the loss of valuable reagents in effluent streams may have been excessive.

The third indirect factor stimulating the recovery of process water is the fact that many industries require process water of higher purity than that of the available water supplies. Complete and frequently expensive water purification systems therefore are mandatory for the process water supply at these plants. In some cases it is only slightly more costly, and sometimes it is cheaper, to remove process impurities from the waste and recycle it to the operation than to purify the incoming raw water.[24]

Numerous examples of extensive water reuse originally adopted for one of the three purposes outlined above might be mentioned.[25] In the

[24] D. E. Noll and H. M. Rivers, "Reuse of Steam Condensate as Boiler Feed Water," *Industrial and Engineering Chemistry,* 48 (December 1956), 2146-50. Also M. Olchoff, "Conservation Program Halves Water Cost," *Factory Management and Maintenance,* September 1950, p. 120; J. N. Welsh, "How to Get the Most From Your Plant Water," *Mill and Factory,* 50 (February 1952), 102-5, and (March 1952), 113-15.

[25] Of 521 plants reporting waste-water treatment in the 1949 sampling survey, 30 per cent recovered useful by-products from the waste. National Association of Manufacturers and The Conservation Foundation, *op. cit.,* p. 41.

pulp and paper industry, regulations against disposal of "white water"[26] in water courses has led to increased yields of pulp products; and the recovery of waste pulping liquors, also legally imposed, has developed the chemical by-products based on lignins and other wood chemicals. (See also chapter 7, p. 180.) In the iron and steel industry, waste waters from coke plants have polluted streams and lakes with phenol, an odorous and toxic chemical. Pollution of municipal supply finally necessitated removal of phenol, the value of which has frequently paid for operating the treatment unit.

Because of the necessity of complying with pollution regulations, the feasibility of recovering by-products, and the stimulus to minimizing raw water purification costs, recycling of process-water streams has become relatively common practice, especially among large water users. But only in the last ten years or so has there been the additional incentive to reduce fresh-water demand in areas where supplies are limited. Wherever the quality and quantity of available water is becoming a significant factor in the location of a process industry, plant reuse is now an important consideration from the *water* standpoint also.

The technological changes that have contributed most to greater process-water recycling have been those developments which have so greatly increased industrial productivity. In other words, the demand for much more water has itself forced conservation and pollution abatement. Improvements in waste-water treatment methods and equipment have been directed toward reducing operating costs and in developing new methods for treating waste waters where standard techniques have not been successful.

Among the principal systems for waste treatment and water recovery, the simplest is the settling basin for removing suspended matter. A familiar example is the tailings pond for mine and mill wastes. Solids which will not readily settle may be removed by mechanical filtering, or in extreme cases by centrifuging. Sewage which contains decomposable organic materials can be satisfactorily treated in conventional equipment such as activated sludge tanks used in municipal disposal plants. Wastes from food processing industries are well handled in this type of system. Where highly acid or alkaline wastes are involved, neutralization, by addition of lime in the first case and sulfuric acid in the other, can be effected. Contamination from bacteria and other small organisms

---

[26] White water is the large volume of miscellaneous pulp and paper mill waste containing very fine wood pulp fibres, which are suspended in the water.

can be removed by chlorination or treatment with ammonium salts or phenolic compounds. In certain instances, water containing dissolved or suspended matter may be recycled without removal of the contaminants, if their presence is not detrimental to the product or process. The impurities therefore simply recirculate in a stream of relatively constant composition. Occasionally their presence is an advantage in that they may reduce the further dissolving and loss of process materials.

Some industrial water effluents must be treated by special processes for removal of chemical contaminants. Organic chemical wastes of certain types have presented exceptionally difficult problems. Individual plants and certain industries have unique purification requirements which must be met if recycling is to be successfully employed. Other plants may have unusually high purity requirements for fresh water. In the manufacture of nitrocellulose, for example, no dissolved iron compounds can be tolerated, and special water purification measures are used.

Some of the most important technological changes which are leading to water conservation in industry are those which promote frugal use of water in the processes requiring it. The first of several logical steps is discontinuing the use of pure water when a somewhat impure water can be substituted with equal effectiveness. Recirculated wash water having considerable concentrations of a process impurity can frequently be used without difficulty.

A second important conservation measure is the reuse of process water containing impurities followed by final processing with pure water in much smaller quantity. The major water requirement can thus be satisfied with a recirculated stream maintained at a uniform level of impurities by small continual fresh-water make-up and discharge. Further savings can be realized by using the effluent water from the operations requiring the pure supply, as the make-up stream to the recirculating water of lower purity. Thus, where 10 gallons of fresh water may formerly have been used to process 1 pound of product, perhaps only 1 gallon of fresh water and 9 gallons of recirculated water need be employed.

Repetition of washing or processing operations with several water streams of progressively increasing purity is frequently accomplished in what is known as counter-current operation. Fresh water is first used in treating the final products. Then, as its purity decreases it is employed in immediately preceding processes. As the water accumulates impurities, it is used successively in still earlier operations until it is finally brought into contact with the crude materials being processed. Water from this operation may then be purified and recycled.

Reclamation of sanitary sewage as industrial water supply. The use of a municipal sewage plant effluent by one or more industries to meet process and cooling requirements also can be an important conservation measure. This practice may relieve a municipal system of supplying fresh water to industrial plants, or it may reduce the competition for fresh water from surface or underground sources which plants and municipalities in the region are exploiting.

A well-designed municipal sewage disposal plant can be operated so that the effluent is as pure as the municipal supply. Suspended matter is completely removed, dissolved organic compounds are largely eliminated by fermentation, and bacteriological contamination is eliminated usually by chlorination. The possibilities for using this effluent as direct industrial water supply are sound and attractive.

Examples of direct supply of sewage effluents to industrial plants are still somewhat limited. The largest single industrial use of municipal sewage is in the Bethlehem Steel Corporation's installation at Sparrows Point, Maryland. The treated effluent from the city of Baltimore, a few miles away, supplies about 100 million gallons of cooling and process water per day to this iron and steel mill. Several smaller systems also have been put into operation in recent years.

As industries expand and as new firms move into areas of limited water supply and distribution facilities, industrial use of treated municipal sewage will certainly increase. The cost of this type of water supply may be substantially lower than the estimated cost of alternative supplies at the same point (table 34). The ease with which municipal sewage can be purified is a large factor in the potentialities of its use by industry. Although not yet universally practiced, municipal sewage treatment is becoming nearly a requirement, and with the added incentive of industrial needs for the effluent, these co-operative arrangements should become relatively common.

The feasibility of a plan for sewage reuse may often depend on the location of a potential industrial user in relation to the sewage treatment plant. The industry may not be able to make economical use of the sewage effluent if the plant is located at a much higher elevation or at a considerable distance from the sewage disposal plant. Although moderate pumping costs readily can be met, an industrial user located downstream from the city sewage disposal system obviously is in a more favorable position.

TABLE 34. *Use and Cost of Reclaimed Municipal Sewage at Selected Plants, 1955*

Location of user	Type of use	Approximate quantity used (acre-ft. per year)	Approximate cost (dollars per acre-ft.)	Approximate cost of alternative water (dollars per acre-ft.)
Grand Canyon, Arizona	Power plant[1]	200	$120	$650
Los Angeles, California	Sewage plant Hyperion	12,000	[2]	40
Baltimore, Maryland	Bethlehem Steel plant	[3] 60,000	[4] 4	33
Amarillo, Texas	Refinery, Texas Company	1,700	[5] 14	45
Big Springs, Texas	Refinery, Cosden	2,200	16	57

[1] Effluent also used for irrigation.
[2] Pumping and chlorination are only costs.
[3] Subsequently increased to approximately 100,000 acre-feet per year.
[4] Approximate cost does not include additional treatment and pumping, and amortization of $2 million investment.
[5] Does not include amortization costs paid by refinery which would raise cost to approximately that of city water at minimum use.

Source: Reproduced by permission of State Water Pollution Control Board, Sacramento, Calif., Publication No. 12, *A Survey of Direct Utilization of Waste Waters, 1955.*

The large magnitude of industrial water requirements relative to municipal use (the latter being nearly equal to the total volume of municipal sewage) must be recognized as a limitation to this application. Industrial water requirements are over 100 billion gallons per day, whereas public water systems supply less than 20 billion gallons. Daily industrial water demand may reach 245 billion gallons by 1975 in comparison with municipal water demands of 30 billion gallons.[27] Even if practically all municipal sewage were applied to industrial use, now or

[27] Walter L. Picton, *Summary of Information on Water Use in the United States,* Business Service Bulletin No. BSB-136, U. S. Department of Commerce, January 1956.

in the next twenty-five years, less than 20 per cent of industrial requirements could be met in that manner. But even a 10 to 15 per cent addition to scarce industrial water supplies in some western states can have a substantial effect on manufacturing in those areas.

Multiple recycle systems. In plants where maximum water conservation is practiced, not only one, but several complete recirculation systems may be in simultaneous use. Water of different type and purity moves in each set of reprocessing facilities. The supply of three or more types of water for the many uses in large plants is comparatively common. Thus, potable water, process waters of lower purity, and fresh and saline cooling water are often provided. Except for potable water and saline water, these supplies may be largely the treated and recirculated effluents from the processes where used. In other arrangements the effluent from one operation, with or without some purification, can be used as the supply water for another operation in which the contaminants are not objectionable. The pulp and paper industry has been particularly successful in using effluent water streams in this manner.[28]

In the planning of multiple water systems in an industrial plant, several negative factors must be balanced against the potential advantages. These are:

1) Separate piping, pumping, and storage must be provided for each type of water.

2) Separate purification and cooling equipment generally must be used for each system.

3) Intermittent and small-scale operations usually require disproportionately large water storage facilities.

Nearly all plants in which any recirculation is practiced maintain a separate cooling water recirculation system so that its treatment (other than recooling) can be minimized. Make-up is often fresh, clear water, not necessarily potable. If process waters are required, one or more systems may be employed containing water with various impurities and recirculating to various degrees. Make-up may occasionally be through addition of one stream to another, but usually by fresh supply. Potable water is usually from a municipal system, and sanitary sewage may often be handled by municipal facilities also. If sanitary sewage flow is large, industrial treatment works are often provided, the treated effluent then being suitable for recycle as process water or cooling make-up.

[28] H. B. Brown, "Conservation of Water in the Pulp and Paper Industry through Recycle, Re-use, and Reclamation," *Industrial and Engineering Chemistry,* 48 (1956), 2151-55.

SPECIFIC EXAMPLES OF WATER REUSE
IN MANUFACTURING INDUSTRY

The most significant applications of water reuse are being made in the industries requiring large quantities of water for processing purposes. The pulp and paper industry, for example, which ranks third in the demand for water by manufacturers, withdrew about 4.4 billion gallons per day (1954), or 13 per cent of all industrial water use in the nation, excepting that used in power plants. In areas where water supply and waste disposal are presenting difficulties, producers are reducing intake water to as little as one-fifth of total water use in all operations. These improvements have been due not only to enlightened policy, but also to development of waste recovery equipment and better co-ordination and integration of the many uses for process water in the manufacturing operations.[29] To produce 1 ton of bleached kraft paper from wood, at least 180,000 gallons of water are required by all the steps in the process.[30] The average withdrawal use of fresh water, however, is about 75,000 gallons, and in some mills less than 50,000 gallons of fresh water are demanded.[31] Corresponding figures for unbleached kraft paper are 140,000 gallons minimum total requirement as compared with 25,000 gallons average fresh-water demand; for newsprint, at least 110,000 total gallons are used, compared with an average supply of 25,000 gallons of fresh water.

In petroleum refining, the principal use for water is as a coolant, where its contamination is minimal. Since its recirculation can therefore be accomplished without undue difficulty, the petroleum refining industry has been notably successful in reducing its over-all demands for fresh water. An estimate of average fresh-water intake of about 770 gallons per barrel of crude was made ten years ago.[32] According to statistics for 1954, intake (fresh and brackish) was about 460 gallons per barrel of crude compared with a total requirement of 1,540 gallons.[33]

[29] H. W. Gehm and W. A. Moggio, "Conservation of Water in the Kraft Pulp and Paper Industry," *TAPPI*, 37 (March 1954), sup. 158A-61A.

[30] Brown, *op. cit.*, p. 2151.

[31] *Ibid.*

[32] Computed from figures in National Association of Manufacturers and The Conservation Foundation, *op. cit.*, p. 16.

[33] Based on 2.66 billion barrels of crude oil refined, 1,220 billion gallons water withdrawal, and 4,515 billion gallons total water used in 1954. Figures on petroleum from *Petroleum Facts and Figures* (New York: American Petroleum Institute, 1956), p. 199. Figures on water from U. S. Bureau of the Census, *U. S. Census of Manufactures: 1954* (Bulletin MC-209, 1957).

Even though these figures indicate substantial increases in industrial water recirculation, they do not show the potential extent of these practices. In the first place, the averages include consumption in plants not employing any recirculation as well as in those practicing extensive reuse. Furthermore, many of the largest plants are located where water is available in practically unlimited quantities, and recirculation has not become necessary. Examination of current water reuse practice in industrial plants now employing maximum recirculation is, therefore, a better guide to future industrial demands for fresh and reused water. In the petroleum industry, for example, actual fresh-water withdrawal in some refineries where a high degree of recirculation is practiced is as low as 35 to 70 gallons per barrel of petroleum, compared with the estimated once-through requirement of 1,540 gallons. Water is therefore used twenty to forty times in these plants.[34] Even though average water withdrawal for pulp and paper manufacture has been substantially reduced, considerably greater reductions have been made by individual plants. A progressive mill producing unbleached kraft paper was withdrawing only 21,000 gallons per ton of product, an intake requirement about 15 per cent of total normal demand.[35] A groundwood pulp mill located at Flagstaff, Arizona, withdraws only 2,000 gallons per ton, or about 5 to 10 per cent of typical water use for groundwood plants.[36] Future water demands in this industry could be met by an average supply of 20,000 gallons per ton and maximum reuse practice.[37]

In an unusual situation where water conservation has been an absolute necessity, the possibility of achieving major water savings in iron and steel manufacture has been demonstrated. Instead of a typical 40,000 to 65,000 gallons per ton, the Fontana Division of Kaiser Steel Company uses only 1,400 gallons, equivalent to a recirculation of 2,800 to 4,600 per cent, or reuse of water about thirty-five times.[38]

[34] J. L. Partin, "Water Conservation, A By-product of Industrial Waste Control," *Sewage and Industrial Wastes,* 25 (September 1953), 1050-59. Also R. N. Simonsen, "How Four Oil Refineries Use Water," *Sewage and Industrial Wastes,* 24 (November 1952), 1372-77.

[35] Brown, *op. cit.*

[36] J. M. Potter, Arizona Pulp and Paper Co., Flagstaff, Arizona, personal communication, June 1958.

[37] H. R. Amberg, "Reuse of Water in Pulp and Paper Mills," *TAPPI,* 38 (November 1955), sup. 154A-55A.

[38] Reported in numerous references, including Hudson and Abu-Lughod, *op. cit.,* pp. 12-22; and H. I. Riegel, "Waste Disposal at the Fontana Steel Plant," *Sewage and Industrial Wastes,* 24 (September 1952), 1121-29.

The objectionable and sometimes dangerous character of certain chemical effluents, and the occasional by-product value, have been important factors in the chemical industry's progress in water reuse. In water-scarce regions, however, the need for conserving the water itself has recently been felt by chemical producers. With high requirements for water purity in many chemical processes, it has sometimes been cheaper to purify and reuse effluents than to treat equally large quantities of impure raw water. The diversity of the chemical industry and the great number of products and processes involved make generalizations on potential recirculation extent of little value. Certain impurities, especially organic chemicals, are extremely difficult to remove from waste waters. Special treatment methods frequently can be developed for handling these troublesome materials. Some chemical manufacturing steps may be carried out with complete water recirculation, whereas others may not tolerate any recycling within economic limits.

EFFECTS OF INDUSTRIAL WATER REUSE

Regional and total water demands. The increasing application and extent of industrial reuse will have substantial effects on future industrial water demands. This may not superficially be evident because of industrial growth and the heavy water demand by these new and enlarged production facilities. But with daily industrial water requirement increasing from the present 116.7 billion gallons to a predicted 250 billion gallons[39] over the next twenty years, reuse will necessarily be practiced to the maximum practical extent.

Measured in total gallons released for other uses or for increased industrial application, reuse of water by industry should be one of the most important technological developments in the water resource field during the next ten to twenty-five years. The present average water reuse rate in manufacturing, nearly 100 per cent, (2 gallons used in processes per gallon of intake fresh water) may be expected to increase to at least 300 per cent, which will be only moderately higher than the present average practice in petroleum refineries. In many plants and processes, these reuse ratios have already been far exceeded. With increase in water recirculation of this size, twice the present output of

[39] Figure rounded because of the large number of assumptions underlying such an estimate.

goods would be possible without significant increase in industrial fresh-water demand. At the present rate of total water use in all manufacturing processes, this recirculation increase would release 15 to 20 billion gallons of water (46,000 to 61,000 acre-feet) per day for other needs while maintaining the same industrial output. If recirculation in utility steam electric power plants (13.7 per cent in 1954) increases even to 100 per cent (1 gallon recirculated for each gallon intake), another 30 billion gallons of water per day will become available.[40]

Increased water reuse also will make possible the establishment of more industry in water-scarce areas. Extreme water conservation measures can free even such large water users as steel and paper mills from requirements for location on sea coasts and large rivers.

It is evident from the general progress of industry, and from the dramatic success of certain plants and industries in conserving water by reuse, that future industrial requirements for fresh water will show a marked decrease per unit of product. Predictions of total industrial water requirements therefore cannot be based on past average use applied to expected manufacturing growth. Because of regional and local variation in water availability, and because the location of industries and the availability of water are not perfectly correlated, the extent of reuse will be highly variable, geographically, even among plants producing the same products (figure 33). Reuse in the southwestern states in relation to fresh-water use is already far more important than in eastern United States.

Most of the industries using large quantities of water have been located where supplies are ample, but as industrial concentration grows in these areas, water shortages are likely to develop. Some evidence already is at hand to suggest that reuse is increasing within the eastern and midwestern states. For instance, it has been reported that a selected canvass of industries in the state of Illinois showed no appreciable increase in total water use in that state between 1953 and 1956, although industrial production had grown considerably.[41]

Estimation of industrial water supply potential and individual in-

[40] Computed from data in Federal Power Commission, *Water Requirements of Utility Steam-Electric Power Plants in 1954.*

[41] Gilbert F. White, *United States Water Resources for the Future,* paper presented at Association of American Geographers Symposium on Resources for the Future, Cincinnati, Ohio, April 3, 1957. The survey was conducted by William C. Ackermann of the Illinois Water Survey Division.

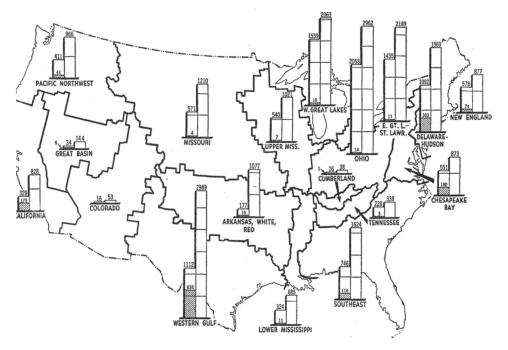

Figure 33. Industrial Water Withdrawal and Recirculation by Census Regions 1954. (Data from table 31. Shaded bar is salt water intake, entire left hand bar is total water intake, and right hand bar is water requirement if no recirculation had been practiced.)

dustry requirements will have to be undertaken in the light of reuse ratios of the particular industry, total plant capacity, present water costs and availability, and expected future water limitations. Planning will require close knowledge of the latest developments in water reuse in each industry, as well as reliable supply and cost information and growth potentials.

Costs and capital requirements. As recycling facilities are added to industrial water supply systems, capital requirements and operating expense must increase. The size of this increase depends upon the type of water being used, as well as upon the extent and type of reuse.

Within industrial establishments utilizing recirculated water of only one type, with fresh water being used as make-up to the system, storage and distribution facilities need be no more complex than in once-through systems. Where two or more types of process and cooling water are used in various plant areas, multiple systems for water distribution and sewage collection are required. Although purification and cooling of collected waste waters are usually performed in integrated facilities, water may also be recycled about a single operation, treatment being an auxiliary special step. Centralized treating, storage, and pumping facilities for several water types require substantial increases in piping and a heavier capital outlay and operating cost than do single water supply systems.

In spite of these necessary measures, capital requirements for recirculated water systems in industrial plants are not usually excessive in comparison with other service expenses and in relation to over-all investment. These systems, moreover, usually have long service life, low maintenance cost, and reasonable freedom from operational difficulties. Annual operating costs vary over a wide range, depending on the extent of recycling and the degree of repurification and effluent water treatment required. The great majority of cases should have costs within a range of 1 cent to 10 cents per thousand gallons of water recirculated. Simple cooling-tower recirculation systems with chemical control of corrosion and biological contamination usually involve total costs of less than 2 cents per thousand gallons.[42] Except where there is practically unlimited fresh-water supply needing no purification and only moderate pumping, this type of recirculation is usually cheaper than the fresh supply.

If chemical processes and physical separations are required in the purification of recirculated process waters, costs are considerably higher and dependent on the particular purification methods required.

Complete sanitary sewage treatment may cost from 2 cents to 6 cents per thousand gallons.[43] The costs of treated municipal sewage effluents usable by industry for process and cooling water are in the same range (table 34). Highly specialized sewage treatment usually is not required. Costs of alternate conventional supplies in each instance among the reported cases are considerably higher than the treated municipal waste water.

By comparison, investments in canals and long pipe lines are in the ranges of $1.00 to $3.00 and $10.00 to $15.00, respectively, per thousand gallons per day per mile, for flow rates on the order of 10 million gallons per day.[44] For a 10-mile pipe line and a 10 per cent total annual fixed cost, about 3 cents per thousand gallons would have to be charged against investment. When the costs of pumping, mainte-

[42] Aries and Newton, *op. cit.*

[43] G. M. Fair and J. C. Geyer, *Water Supply and Waste Water Disposal* (New York: John Wiley, 1954), p. 83.

[44] Calculated from latest available cost data.

nance, water rights, pretreatment, etc., are added, recirculation not infrequently will be found cheaper.

The exceptional cost-consciousness of industrial management assures the cheapest water source being utilized in plant operation. During the next few decades, wherever water is nearby, plentiful, and of good quality, once-through use will generally continue. The greater the expense of providing fresh water, including costs of pumping and purification and fixed charges on wells, reservoirs, canals, pipe lines, and other supply works, the greater will be the incentive to reuse part of the plant waste waters.

The notable exception to the cost-controlled decisions of management occurs when stream pollution laws may require expensive waste-treatment practices. Public relations policies may also dictate waste treatment, but in the long view even these decisions are based at least partly on the objective of maximum financial return to the manufacturer.

Downstream water users. The principal effects of industrial water recirculation on downstream users pertain to quality rather than quantity of water. Water disappearance by combination with plant products, by evaporation, and by seepage, if any, represents the difference between inflow and outflow. This *net* consumption is substantially the same whether reuse or once-through use is practiced. The additional evaporation in the cooling tower of a recirculation system is approached in the slower evaporation from the warmed effluent from a once-through cooling system. Hence, there is little difference in the volume of downstream flow a few miles below a plant, provided that intake and discharge are in the same stream.

Downstream water quality can and will gradually improve as water reuse replaces once-through practice. Pollution abatement is an accompaniment of water recirculation because the process for purifying water for reuse can be conveniently and economically used on the final effluent also. Since water purification and reuse involve *removal* of wastes otherwise discharged into streams, the growth of this practice will necessarily reduce contamination. Treatment of paper-mill "white waters" and sulfate pulping liquors illustrates the reduction of severe nuisance by extensive recirculation and effective removal of solids. Removal of chemical wastes or their neutralization by additives is rehabilitating deteriorated surface streams. Sanitary sewage purification is making some effluents substantially of the same quality as municipal supplies. All these steps will unquestionably benefit downstream industrial, municipal, and agricultural users.

Although stream quality should improve as water reuse rises, some of the final effluents from recirculation systems may contain dissolved solids, finely suspended matter, and other impurities too difficult and expensive to remove.[45] These materials might be undesirable downstream unless highly diluted by stream flow or unless subsequent feedwater purification can eliminate the objectionable contaminants. If there are many industries along a stream and each contributes small amounts of objectionable wastes, serious difficulties might develop. Where water supplies are limited, industrial effluents to streams will ultimately have to be substantially free of contaminants which might constitute a major problem to other water users. This quality requirement may occasionally be difficult and expensive to meet; nevertheless, maximum future water usefulness will depend upon it.

A major contribution of industrial reuse of cooling water will be in minimizing undesirable rises in river temperatures. If there are numerous once-through cooling water users concentrated in a relatively few miles of river valley, the 10 to 15 degree temperature rise in the flow through each plant handicaps downstream users and upsets the biological balance in the stream. (See also discussion and example in chapter 12, pages 318-19.) Reuse with recirculation of water through cooling towers immediately dissipates this heat into the atmosphere, thereby minimizing downstream temperature rise. It is not unlikely that legal restrictions on "thermal pollution" of streams may become as stringent as those concerning liquid and solid pollution.[46]

SUMMARY STATEMENT OF FUTURE PROSPECTS

The very heavy demands for water by industry and the rate at which industry is expected to expand will impose severe requirements on future water supplies. Many plants and industries have made substantial progress in the reduction of water demand per unit of product, through treatment and purification of water effluents to permit their reuse in

---

[45] A discussion of this factor may be found in "Water Conservation in Industry," Task Group Report, *Journal, American Water Works Association,* 45 (December 1953). A pertinent quotation is: "It must be anticipated that recycling techniques, which tend to concentrate the mineral constituents present in the intake water into a volume of water only 2-5% that of the original, may result in 'concentrative pollution' of the water source."

[46] Moore, *op. cit.*

plant processes. Where water has been scarce and expensive, remarkable savings in fresh-water requirements have been effected. In areas where there soon will be severe restrictions on new fresh-water supplies for expanded industrial requirements, these practices can be expected to increase substantially. In view of the large water requirements for many new products, the estimated industrial demand of 250 billion gallons per day by 1975 may be exceeded unless water reuse expands substantially.[47]

Limitations on the application of these methods vary greatly between industries, and the needs and incentives for water conservation differ considerably throughout the country. Wide differences in type and degree of reuse can therefore be expected. In general, economics will dictate the practice. Industry will use the cheapest source of acceptable water, whether fresh, partially reprocessed and recirculated, treated municipal sewage, sea water (where available and acceptable), demineralized sea water, or some other supply.

Considerable technical improvement can be expected in the treatment and purification of waste waters to make them more completely reusable. Each industry group and manufacturing plant will have certain unique water treatment problems as well as some in common. Recirculation of cooling water and control of its quality have been well developed and are widely practiced. Nonevaporative cooling is becoming significant and developments may be expected in this field. Removal of chemical and biological wastes from industrial process waters has advanced with the development of processes and equipment, and there should be further progress along these lines. It is likely that most water-using industries, in areas where new supplies are scarce and expensive, will find it economical to purify and recirculate nearly all their process and cooling water streams, thereby reducing fresh-water requirements to a small fraction of normal once-through use. As water scarcity moves into areas now well supplied, the importance of this practice in the over-all water economy of the nation will be great. A several-fold increase in the nation's industrial output is possible, at little or no increase in its total fresh water demand. Actually both will rise, but it is reasonable to expect that industrial production will rise at a considerably higher rate than fresh-water use.

---

[47] Assumes an increase of about 160 per cent in industrial production. This estimate reflects indications given in estimates of future productivity and gross national product. It is not to be taken as a projection, but rather as a probability.

2. Water Budgeting in Irrigation

Reuse of water already is common in western United States irrigation. This type of reuse is not the on-site type being developed within industry, but is the return flow reuse within a watershed. There also may be some recycling of ground water used for irrigation because of deep percolation back to the aquifer from which the supply is being withdrawn. But there appears to be little opportunity for reuse completely within the producing agricultural unit, or farm, in the manner of on-site industrial recirculation. However, the possibility of more efficient techniques of applying water to the production of crops shows promise of assisting water conservation within agriculture. A scheduling technique, water budgeting in irrigation, will be discussed as an example.

The same principle applied in multiple-purpose reservoir operation (chapter 8, pages 185-94) may be extended to other phases of water use. The economic advantage of accurate quantitative appraisal of water needs, equally accurate appraisal of the vagaries of supply, and subsequent synchronization of the two, appears in small-scale use as well as in major water regulation systems. The water budgeting technique is, in a sense, a "miniature" scheduling technique, but one with possibilities of wide application to irrigation.

Irrigation today is of greatest importance in the West. The provision of water to supplement that supplied by precipitation thus far has had only limited significance in the eastern half of the country, although it has long been employed in many localities there. But during the last decade information has increased as to the benefits that can be derived from irrigation in the East. (See also chapter 6, pp. 133-35.) Not only can losses be averted when conditions of severe drought are experienced during the growing season, but production in normal years can be substantially increased; the market quality, nutritional character, and dollar value of a crop can be materially improved.[48]

Portable sprinkler equipment has greatly facilitated the development of irrigation, particularly in the East.[49] This equipment also has been

[48] C. W. Thornthwaite, *The Place of Supplemental Irrigation in Postwar Planning,* Publications in Climatology, Vol. vi, No. 2 (Seabrook, N. J.: Johns Hopkins University Laboratory of Climatology, 1953), pp. 11-16, 27-28.

[49] Harry Rubey, *Supplemental Irrigation for the Eastern United States* (Danville, Ill.: Interstate, 1954). A more complete discussion of the development of sprinkler irrigation appears in chapter 6.

used to supply plant nutrients to crops through the addition of fertilizer to the irrigation water. The combined application of irrigation water and fertilizer, according to a recent review of the subject, opens up a new frontier for American agriculture.[50]

Deficiencies in soil moisture are experienced regularly at some time in virtually all of the agricultural regions of the United States. Although there are substantial returns to be derived from eliminating or minimizing them, the profitability of using portable irrigation equipment varies greatly with the extent and frequency of these deficiencies. Other factors must be considered in evaluating the practice, such as the availability of suitable supplies of water, pumping costs, and equipment costs. However, the value of irrigation is rapidly being demonstrated for field crops as well as for row crops, and its accelerated use in the humid areas of the country is one of the significant developments in present-day agriculture.

There remains, however, the crucial problem of determining how to make the optimum use of irrigation facilities. To employ them effectively, a farmer needs to know when his crops require additional water, and he has to have some basis for ascertaining how much to apply. While deficiencies of soil moisture seriously limit plant growth, an excess of water also is undesirable. Over-irrigation wastes otherwise usable water and causes unnecessary expense; it leaches plant nutrients from the soil, prevents proper soil aeration, and may contribute to erosion.

Agronomists generally agree as to the conditions of available soil moisture which are most conducive to plant growth. There are two "bench-marks" or soil moisture constants which have long been recognized: "field capacity," the maximum amount of water that can be retained by gravity and thus held in storage within the soil profile; and the "wilting co-efficient," or permanent wilting percentage. The latter is the soil moisture content at which the supply in the root zone is so low that plants cannot withdraw water from the soil. Wilting results, and the plants are unable to recover turgidity. Highest crop yields are reported to be obtained when the available soil moisture does not drop below 25 per cent nor rise above 90 per cent of field capacity.[51].

Although the optimum range of soil moisture conditions can be

---

[50] C. H. M. van Bavel and T. V. Wilson, "Evapotranspiration Estimates as Criteria for Determining Time of Irrigation," *Agricultural Engineering*, 33 (1952), 417-20.

[51] Thornthwaite, *op. cit.*, p. 22.

identified, the problem of determining the actual moisture conditions of the soil has persisted. Because this task is so difficult in practice, irrigation water is frequently applied in accordance with various rule-of-thumb guides. In many western irrigation districts, for example, irrigation may be scheduled so as to ensure the delivery of a specified amount of water to the root zone of the crop each week, irrigation water being used to compensate for the deficiencies of rainfall. However, such a procedure makes no allowance for the changes occurring in water need as the growing season progresses. More often, water is applied when signs of moisture deficiency are recognized in the growing plants. This procedure is also inadequate, for by the time plants begin to show moisture deficiency, they are already suffering and yields have accordingly been depressed.

Irrigation water is used most efficiently when its application is timed so accurately that imminent deficiency of moisture within the soil can be avoided. The time of water need, it follows, must be determined from the actual soil moisture conditions existing within a particular locality at a given time. Over the years, many techniques of appraising and measuring the moisture content of the soil have been developed. They range from examination of soil samples removed by augers or similar tools, to use of tensiometers of several types,[52] and of instruments employed to detect the speed with which fast neutrons emitted from a radioactive source pass through soil.[53] While several of these instruments have considerable significance as research tools, none appears to be both sufficiently reliable and simple to be of practical assistance to the irrigator.

In recent years an entirely different approach to this problem has been developed, by which commonly available climatological data are employed to determine the occurrence and the extent of deficiencies in the supply of soil moisture available for plant growth. During the last twenty years several variants on this approach have appeared, yielding similar results in practice. The principal differences among the techniques are those of converting the climatic data into estimates of soil moisture loss. The better known among the techniques are the Thornthwaite method, the Blaney and Criddle method, and the van Bavel-

[52] Tensiometers measure the electrical resistance of soils, which varies with their moisture content.

[53] H. R. Haise, "How to Measure the Moisture in the Soil," *Water, The Yearbook of Agriculture, 1955,* U. S. Department of Agriculture, Washington, 1955, pp. 362-71.

Penman method.[54] The succeeding description is based on the Thornthwaite method. Dr. Thornthwaite's study of the problem of such conversion is among the longest and most consistently applied in the United States.

Unless there is a large amount of deep percolation, the moisture in the soil can be regarded as a balance between precipitation input, and the outgo in evaporation from the soil surface, transpiration by plants, and runoff. The quantities of water added to the soil can be determined readily through rain gauge measurement, as can runoff through stream gauges. If the losses due to the combined effect of evaporation and transpiration can be ascertained with comparable ease, the available supply of moisture in the soil at any time could be determined by means of a simple accounting procedure. The resulting information regarding soil moisture conditions can then be used to schedule irrigation, water being applied in an amount sufficient to bring soil moisture back to the optimum level whenever it becomes significantly depleted.

The first step in the water budget technique is to calculate the amount of water that would be transferred from the soil to the atmosphere by evaporation and transpiration if abundant moisture were constantly available in the soil for the use of vegetation. This is *potential* evapotranspiration, in contrast to the moisture actually transferred by evaporation and transpiration. The rate of *actual* evapotranspiration depends upon five factors: weather, supply of soil moisture, plant cover, soil type and structure, and land management. However, the amount of *potential* evapotranspiration is estimated by analysis of only two climatic factors, the intake of solar radiation by plants and the soil surface, and to a lesser extent the temperature of the layer of air adjacent to the soil

[54] C. W. Thornthwaite and J. R. Mather, "The Water Budget and its Use in Irrigation," *Water, The Yearbook of Agriculture, 1955,* pp. 346-58; H. F. Blaney and W. D. Criddle, *Determining Water Requirements in Irrigated Areas from Climatological and Irrigation Data* (Technical Publication No. 96), U. S. Department of Agriculture, Soil Conservation Service, Washington, 1952; C. H. M. van Bavel, *Estimating Soil Moisture Conditions and Time for Irrigation with the Evapotranspiration Method,* U. S. Department of Agriculture, Agricultural Research Service and North Carolina Agricultural Experiment Station, Soils Department, Raleigh, N. C., 1957 (revised). Van Bavel has recently published several descriptions of the regional application of his method; e.g. C. H. M. van Bavel and J. J. Lillard, *Agricultural Drought in Virginia* (Technical Bulletin No. 128), Virginia Agricultural Experiment Station, Department of Agricultural Engineering and U. S. Department of Agriculture, Agricultural Research Service, Blacksburg, Va., 1957.

surface.[55] By use of an empirical formula, the changing rate of potential evapotranspiration throughout the year may be computed for any locality for which temperature records are available. Comparison of the amount of actual precipitation with the changing rate of potential evapotranspiration through the seasons, will then indicate the amount and the time of soil moisture surplus and soil moisture deficiency.

Although its application is subject to some uncertainties, the concept of potential evapotranspiration is proving to be a useful analytical tool. Because the rate of actual evaporation and transpiration is limited by the supply of moisture available in the soil, at some time during the year it is almost certainly less than the rate of potential evapotranspiration, the amount of the difference indicating the extent of the soil moisture deficiency. The situation in various parts of the country has been represented in synoptic form by Thornthwaite in what he calls the annual march of precipitation and potential evaporation (figure 34).

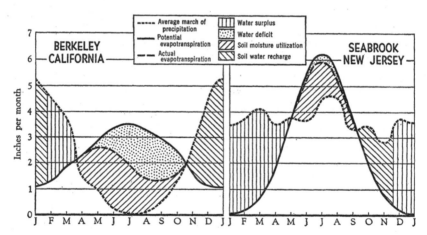

Figure 34. Annual March of Precipitation and Potential Evapotranspiration at Berkeley, California, and Seabrook, New Jersey.

At Seabrook, New Jersey, the potential evapotranspiration is negligible during the winter months, but it rises rapidly in the spring, reach-

[55] See C. B. Tanner, "Factors Affecting Evaporation from Plants and Soils," *Journal of Soil and Water Conservation,* 12 (1957), 221-27: ". . . . We should look for those factors which affect the latent heat exchange at the surface. This is evapotranspiration."

ing a peak of more than 6 inches in July, and then declining sharply in the autumn. The occurrence of precipitation is far more evenly distributed throughout the year. It can thus be seen that a water deficit occurs during the summer months in this area, while in the autumn water need falls below precipitation. After the excess rainfall has replaced the soil moisture previously used, a water surplus is experienced.

When the soil moisture is at field capacity, actual and potential evapotranspiration are equal, and the amount of precipitation in excess of the potential evapotranspiration rate then serves either to raise groundwater levels or is released as surface and subsurface runoff. When the potential evapotranspiration rate becomes greater than precipitation the following spring, soil moisture storage is drawn upon for actual evapotranspiration. However, the soil becomes progressively drier during the summer months. When the moisture content reaches a point where water is no longer available to plants in an amount necessary for optimum growth, actual evapotranspiration becomes less than potential evapotranspiration. This is the water deficit.

The climatic situation at Berkeley, California (figure 34) is quite different. There, little precipitation occurs during the summer, nearly all of the rainfall being experienced during the winter months. The annual water deficit is seven times as large as at Seabrook, New Jersey, totaling some 7 inches. The water surplus is only one-fourth as great, as seen from a comparison of the march of precipitation and potential evapotranspiration through the course of a year.

Encouraging results have been obtained in comparing the adequacy of the information made available by the water budget procedure with that from other sources. Thus the calculated values for potential evapotranspiration have been reported to agree closely with the amount of water disappearing ("consumptively used") in some irrigated areas. They have been tested against measurements derived from a number of specially designed transpirometers installed in various parts of the world, and from lysimeters operated by the Soil Conservation Service. When used to compute the theoretical runoff within different watersheds, the calculated values agree approximately with the results obtained by field measurements.[56]

This general procedure can be used as a practical aid in scheduling irrigation. Actual daily climatological values are employed to evaluate

[56] C. W. Thornthwaite and J. R. Mather, *The Water Balance,* Publications in Climatology, Vol. viii, No. 1 (Centerton, N. J.: Laboratory of Climatology, 1955), p. 51.

soil moisture conditions instead of monthly means of temperature and precipitation. A day-by-day inventory of the moisture in the soil can then be maintained by a simple accounting procedure in which the losses through evapotranspiration are debited and the additions through precipitation during the growing season are credited. The starting point for the "moisture account" may be taken as the soil moisture present at the beginning of the growing season for the particular crop. This may be at "field capacity," a condition when the soil contains no surplus of gravitational water and no deficit of capillary water. If precipitation occurs when the soil is at field capacity, the additional supply of moisture is rapidly lost by downward percolation, or lateral movement. The length of time surplus water is retained depends upon the amount of water and permeability of the soil. In the absence of precipitation, withdrawals from the balance begin to occur as the store of soil moisture is depleted by evaporation and transpiration. At this time, the amount of actual and potential evapotranspiration is the same. The balance is decreased by the amount of moisture loss, computed by a formula for potential evapotranspiration. When the soil begins to dry out, it becomes progressively more difficult for additional water to be expended by evaporation and transpiration.

Some studies have indicated that the rate of evapotranspiration for a particular crop and season is proportional to the amount of water remaining in the soil, but there is apparently considerable departure from this generalization. As withdrawals occur in the store of soil moisture, the actual rate of evapotranspiration generally declines, and always drops if the wilting condition is approached. In the event that the balance is fully restored by additions to the soil moisture supply, evapotranspiration again goes on at its maximum rate, while lesser additions from precipitation or irrigation may result in more moderate evapotranspiration rates.

The methodological basis of this technique for maintaining a daily water budget has here been outlined, but in practice the procedure has been simplified by the preparation of irrigation guides or tables for different localities which facilitate determination of the withdrawals from the soil moisture balance under varying moisture conditions. The resulting information is then employed directly in fulfilling an irrigation schedule (figure 35). This schedule indicates the amount of moisture to be applied to the soil and the timing of the application so that soil moisture deficiency does not drop below undesirable levels. As practiced, this technique of irrigation depends not only on water balance calcula-

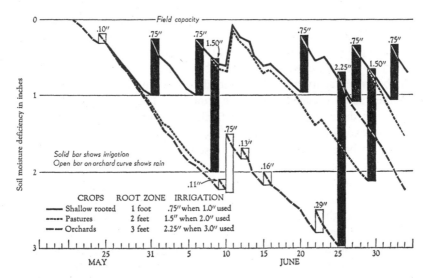

Figure 35. Typical Irrigation Schedule in the Humid Eastern States. For deep-rooted orchard crops, a predetermined maximum soil moisture deficiency of about 3 inches is seen to have been established. The dashed line, starting May 21 at field capacity (zero soil moisture deficiency) drops fairly steadily to June 10 when 0.75-inch rainfall causes an increase. Subsequent evapotranspiration and small precipitation additions occur until June 25, when the deficiency reaches 3 inches. Application of 3 inches of irrigation then raises soil moisture by 2.25 inches (balance is lost by runoff and surface evaporation), so that the deficiency is only 0.75 inch. The deficiency then starts to increase again. Shallower crops (solid line) are not permitted to sustain deficiencies greater than 1 inch, so more frequent irrigation in lesser quantities is scheduled as shown.

tions, but also on consideration of root depth of the crop being grown, soil type, and the economic and technical feasibility of irrigation practice. Irrigation water is applied whenever the soil moisture balance becomes depleted to the predetermined significant level, the water "inventory" thereby being increased by the appropriate amount. Shallow-rooted crops must be irrigated more frequently, but require smaller amounts of water than deeper-rooted pastures or orchards. The growth cycle of the crop is also a factor. Even under the same solar and temperature conditions, actual evapotranspiration varies with the changing moisture needs of a plant as it grows. An irrigation schedule should therefore be adjusted or compensated for these changes in water requirements.

Several additional environmental elements must be considered in employing this technique. The drainage characteristics of the farm land must be well understood, otherwise undetected additions to the soil moisture supply from adjacent localities might take place. It is also necessary to ascertain whether all of the rainfall which occurs when a field is below capacity actually is added to the supply of soil moisture, or whether some escapes as surface runoff.[57] The daily water budget procedure is deductive and hypothetical, and therefore judgment must initially be exercised in determining the appropriate parameters within which it can be employed. Properly used, it has already been demonstrated as a very useful tool. Where irrigation is scheduled to keep continuous account of the soil moisture, no serious moisture deficiency can develop in the soil to limit plant growth. Over-irrigation, damaging both crops and the soil and resulting in a wasteful misuse of water, also can be avoided.

The use of an irrigation scheduling technique can affect the productive employment of water wherever it is applied to crops. It is clear that relatively accurate estimate of the course of evapotranspiration, and its relation to the scheduling of water application, will be of interest in both eastern and western United States. However, the technique is of special interest in the East. The reason lies in the nature of water deficiencies in the two major geographical divisions of the country. The irrigated West is characterized by recurring seasonal deficiencies, with very little and uncertain supplies of moisture from the atmosphere during the growing season. In this situation irrigation may be planned on a seasonal basis, once the water requirements of the crop are known. It is likely that a fair adjustment already has been made in western irrigation areas between the water applied to crops and the potential evapotranspiration. Thornthwaite reported in 1953 a very close correspondence between the observed consumptive use in twelve western irrigation areas and the computed potential evapotranspiration.[58]

This does not mean an absence of overwatering in the West, for there are instances of applying larger amounts of water than needed by crops. However, this is known to be related in many cases to the need for preventing an accumulation of salts in the soil. The water budget or similar techniques, then, according to the available information, appear unlikely to have a large impact in the West.

[57] This occurs on occasion during intensive thunderstorms, when the top soil is dry and badly crusted.

[58] Thornthwaite, "The Place of Supplemental Irrigation in Postwar Planning," *op. cit.*, pp. 18-19.

Water deficiencies in the East, on the other hand, are of an entirely different character. They are of relatively short duration and occur within the normal growing season. They are not periodic in the sense of recurrence at specific dates, nor are they yet forecastable by the techniques of synoptic meteorology. It is within this situation that the water budget method is best adapted to profitable use.

The probability of overwatering is relatively great where irrigation is undertaken in the East without technical guidance. Expansion of irrigation in this region therefore can be undertaken in the future at the expenditure of less water, and presumably less cost, than otherwise might have been the case. The meaning of the technique in terms of water quantities obviously will depend upon the amount of irrigation which eventually is undertaken in the East. Studies of irrigation in the Mississippi Valley section of Arkansas have shown a decrease in irrigation water requirements where the water budgeting technique is used, as compared to previous practice without water budget scheduling.[59] Even though at present it is not possible to estimate the future quantities of water involved, the importance of the technique, or of improvements upon it, may be substantial.

3. Reduction of Reservoir Evaporation Losses

The steadily expanding reservoir capacity of the United States is not all gain for water supply. The extended water surfaces that have been created have an undesirable accompaniment in loss of impounded water by evaporation. Part of the vaporized water may return as rain (see chapter 14, page 356), but this will seldom occur in the region from which the water came. The permanent supply of the region thus falls short of its potential. Average gross annual losses through evaporation may be as high as 7 feet of water depth (figure 36). Under severe drought conditions, evaporation of 10 feet of water in a year may occur. In terms of total volume lost from artificial reservoirs in the United States, over 20 million acre-feet have been estimated to evaporate per year.[60] In the eleven western states, annual evaporation losses

[59] *Ibid.*
[60] H. R. Drew, "Evaporation Control Research in Texas," paper delivered at 26th Annual Convention, National Reclamation Association, Phoenix, Arizona, November 7, 1957.

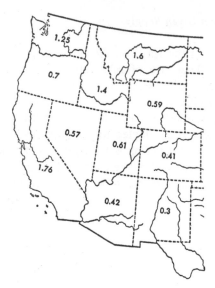

Figure 36. Evaporation Losses from Water Surfaces, Eleven Western States. (Figures are in millions of acre-feet per year.) In addition to quantities shown, evaporation from reservoirs and portions of streams coinciding with state boundaries are 0.5 from the Columbia River and 1.4 from the Colorado River. Evaporation of 0.1 million acre-feet from the Snake River on the Idaho-Oregon boundary is included in the Idaho total.

exceed 11 million acre-feet.[61] These water loss rates represent major volumes in comparison with many uses. The national evaporation loss exceeds all municipal public supplies, for example. In the eleven western states, use of water in manufacture could expand four times without exceeding reservoir evaporation in that area.

The source of energy for evaporation of water from reservoirs is the sun. Absorption of radiation in the water and on the reservoir bottom supplies heat which, on a sunny day, can vaporize 1 to 2 gallons of water per square yard of reservoir surface. The resulting vapor is then carried away by air movement. Although increases in water temperature, air temperature and wind velocity, and decrease in air relative humidity will temporarily increase evaporation rate, average water loss over a period of weeks or months is primarily limited by the solar radiation received and absorbed. However, indirect effects of temperature increase may be important in causing greater growth of algae and

[61] E. D. Eaton, *Control of Evaporation Losses,* Memorandum of the Chairman to Members of the Senate Committee on Interior and Insular Affairs, April 14, 1958 (Washington: U. S. Government Printing Office).

other materials in the water which can increase the fraction of the solar radiation absorbed.

In recent years, control of evaporation from these open-water surfaces has been seen as a tempting source of water not otherwise available. In achieving such control it has been clear that if the reservoir surface can be protected from the sun, the large evaporation losses which occur in regions of high insolation can be eliminated completely. Hence, in municipal supply systems, where the value of the water has been sufficiently increased by purification treatment, pumping, transmission, and other operations, expensive concrete or metal reservoir covers often can be economically justified. But the raw water evaporated from artificial lakes, like those formed by dams across canyons and ravines, does not have sufficient value to bear the very large cost of a structural cover. Where possible, of course, minimizing surface area by employing maximum practical water depths is advantageous. Exceptionally favorable conditions may eliminate all evaporation loss, by use of underground storage in well-sealed formations. Another possibility is the use of a nonvolatile immiscible liquid on the reservoir surface, creating another type of physical barrier between the water and the overlying air.

Although the basic principles of vaporization reduction, through the use of insoluble liquid films on water surfaces, have been known for at least thirty years,[62] it was not until about 1952 that substantial progress was made toward applying them.[63] Certain types of organic chemicals having molecules containing a long hydrocarbon chain terminating in an alcohol group, acid group, or some other radical compatible with water (hydrophilic group), will form an invisible, insoluble film only one molecule thick on a water surface. The rod-like molecules orient themselves vertically side by side, with the hydrophilic group downward toward the water and the hydrocarbon chain forming the new surface. Because of the extreme thinness of the film, a minute quantity of the added compound will completely cover a large water area. Moreover, these compounds have such low volatility and solubility in water that the film may be remarkably permanent.

Some aspects of this evaporation reduction process are not com-

[62] I. Langmuir and D. B. Langmuir, "The Effect of Monomolecular Films on the Evaporation of Ether Solutions," *Journal of Physical Chemistry*, 31 (1927), 1719.

[63] W. W. Mansfield, "Influence of Monolayers on the Natural Rate of Evaporation of Water," *Nature*, 175 (1955), 247.

pletely understood, but the essential effect of the film is to prevent the normal escape of water vapor molecules from the liquid surface by a "blanket" of foreign molecules. If the vaporization of water can thus be prevented, the additional requirements are: (1) the establishment and maintenance of a continuous film on the surface of the tank, pond, or large reservoir; and (2) the absence of undesirable side-effects, like toxicity to plants or animals later using the water, unpalatability, etc.

A heavy layer of oil also reduces evaporation loss, but it has several undesirable aspects. Passage of oxygen into the water and escape of carbon dioxide are so decreased that fish can be destroyed, organic wastes accumulated, and water quality seriously impaired. Tastes, odors, and even toxicity may result from dissolved fractions of the oil layer. Recreational use of the reservoir is largely destroyed. In contrast, small quantities of several relatively cheap chemical compounds will apparently produce protective monomolecular films without undesirable side effects.

The present technical status of evaporation control by monomolecular surface films is substantially as follows: (*a*) Evaporation can be reduced 50 per cent or more if a continuous film is present. Experimental field tests have generally shown reductions of 20 to 30 per cent.[64] (*b*) Side effects such as excessive water temperature changes, biological modifications, decrease in oxygen and carbon dioxide transfer rates, and detectable tastes or odors, are apparently absent. However, further information is needed to establish some of these factors with certainty. (*c*) The formation and maintenance of a continuous film on large reservoir surfaces are the principal problems in the practical application of the method, and are the subjects of current development studies.[65]

Hexadecanol (also known as cetyl alcohol), a 16-carbon, solid, fatty alcohol derived from animal fats or from petroleum by synthetic processes, has been found the most effective evaporation control agent. Several other compounds of similar structure also show satisfactory performance in laboratory and small tank tests. The 18-carbon

[64] Eaton, *op. cit.* Also W. W. Mansfield, *Summary of Field Trials on the Use of Cetyl Alcohol to Restrict Evaporation from Open Storages During the Season 1954-55*, CSIRO Division of Industrial Chemistry Service Nos. 74, 75, November 1955, Melbourne, Australia; W. T. Moran and W. V. Garstka, "Reduction of Evaporation through Use of Monomolecular Films," paper presented at Third Congress of International Commission on Irrigation and Drainage, San Francisco, Calif., May 4, 1957 (Washington: U. S. Bureau of Reclamation).

[65] Eaton, *op. cit.*

octadecanol and the corresponding fatty acids have been effective under carefully controlled laboratory conditions. Hexadecanol has repeatedly reduced evaporation from 10-foot outdoor tanks by approximately 50 per cent for about a week, and by more than 90 per cent in the first part of the tests.[66] The relatively short period of effectiveness on such small surfaces indicates the need for improvement in the film chemical itself or in the method of its application and maintenance. Efforts in these two directions are being made in the United States by several organizations, including the Bureau of Reclamation, the Geological Survey, and the Southwest Research Institute. They also are being made in Australia and Africa.

Preliminary tests in 1956 by the Bureau of Reclamation on a small lake in Oklahoma (Kids Lake near Oklahoma City) were conducted primarily to determine any effects on aquatic life, toxicity to higher animals, and taste or odor properties. In the summer of 1957, further studies on a 97-acre Colorado reservoir were directed primarily toward determining whether a film can be maintained on a sizable scale where wave action, biological effects, floating objects, and other natural phenomena are present.[67] Techniques are being investigated for establishing the film and replenishing the supply of the coating chemical to replace unavoidable losses. The design and location of suitable feeders for adequate but not excessive addition of the film-forming chemical are also being studied. Preliminary tests show favorable results by replenishment of the film with small beads of hexadecanol (less than ⅛-inch diameter) and by broadcasting quite small particles.[68] There are indications that the following problems may be encountered in application of the films: (1) sliding of the film from the water surface by formation of a coating on floating solid objects and upon the shore; (2) bacterial attack on hexadecanol; (3) slight evaporation and dissolving of the chemical used in the film. In addition, the effect of wind and waves is not fully known.

The 1958 program scheduled by the Bureau of Reclamation included

[66] B. W. Beadle and R. R. Cruse, "Water Conservation through Control of Evaporation," *Journal, American Water Works Association,* 49 (1957), 397-404.

[67] W. T. Moran, Bureau of Reclamation, Denver, Colo., personal communication, April 1957. Also Bureau of Reclamation, Division of Engineering Laboratories, "Special Investigations Memoranda Nos. 55-57" (Denver, Colo., 1955, 1956). Also U. S. Department of the Interior, Information Service, "Reclamation Tests Way to Prevent Evaporation Losses from Vital Reservoirs," news release for June 6, 1957.

[68] Beadle and Cruse, *op. cit.;* Eaton, *op. cit.*

studies of film application and maintenance techniques for the 1,000-acre Carter Lake in Colorado, and a full-scale evaporation control investigation on 2,500-acre Lake Hefner near Oklahoma City.[69] Preliminary results of the full summer project at Lake Hefner indicate a 9 per cent reduction in evaporation loss.

Biological effects of the surface films on small test lakes have been investigated by the United States Public Health Service. Although freedom from toxicity to animals and human beings has been established, and no deleterious effects on fish and plankton have been observed,[70] possible influence of the film and the chemical compound on natural bacteria and other microscopic life has not fully been determined. Since a slight unbalance in even a minor species might have a chain of effects with an ultimate impairment of water quality, large-scale use of the technique must await further test results. Even the mechanism of film action in preventing passage of water molecules while permitting passage of oxygen and carbon dioxide molecules is not clear. Also, there is the question of the effects of slight increases in water temperature due to solar heat absorption without normal evaporation. These fundamental problems are being investigated because the ultimate benefits of a successful evaporation control process can be of major significance. It is possible that in a few years several million acre-feet of water can be conserved annually by full application of a well-developed process.

It has been estimated that ponds not larger than 2 acres will require the addition of one-half to one pound of cetyl alcohol per month.[71] Tests by the Bureau of Reclamation on a reservoir of 150 acres involved the daily use of 0.2 to 0.4 pound per acre.[72] Actual requirements have not yet been satisfactorily determined, however. Based on data obtained in Australia, the cost of reducing evaporation from small ponds by 25 per cent may be in the range of $12 to $24 per acre-foot saved, where corresponding normal evaporation losses are from 8 feet to 4 feet per

[69] Eaton, *op. cit.*

[70] W. T. Moran, personal communication, April 1957.

[71] E. D. Eaton, "Control of Evaporation Losses," *op. cit.*, pp. 22-23. These data quoted by Eaton are from *Saving Water in Dams,* Leaflet No. 15 (Melbourne, Australia: Commonwealth Scientific and Industrial Research Organization, 1956).

[72] L. O. Timblin and Q. L. Florey, "Reservoir Evaporation Reduction Through Use of Monomolecular Layers," paper presented at 31st Annual Meeting, Rocky Mountain Section, American Waterworks Association, Santa Fe, New Mexico, September 25, 1957 (Denver, Colo.: Bureau of Reclamation).

year.[73] A more optimistic estimate is based on small-scale tests in Southwestern United States with cetyl alcohol, extrapolated to a monthly application rate of 2.2 lb. per acre. If this quantity proves adequate and effective, it is estimated that the cost of saving an acre-foot of water would be about $1.60, or half a cent per 1,000 gallons.[74] If processes are perfected and these preliminary estimates are realized in practice, this large addition to usable water supply in a wide section of the nation, actually where most needed, will undoubtedly be an economical "new" source.

In multiple-purpose reservoir developments, recreational, fish and wildlife, navigation, and flood control functions will be unaffected by application of evaporation control. All flow and consumptive uses will benefit, however. Hydroelectric generation will be increased in proportion to the added volume of water available; irrigation, municipal, and industrial users will benefit in similar degree. Still another potential benefit of this process may be in the reclamation of moderately saline lakes by reducing the evaporation-inflow ratio. If salinity in marginal or submarginal supplies can thus be decreased, not only will there be savings through evaporation decrease, but the entire reservoir capacity might, in several seasons, be brought into use.

The need for much more basic data in this field is clear. When the method is ready for general use, there will be a substantial demand for information on its practical application. Personnel in water management and operation will undoubtedly require special training in the necessary techniques, and it is not unlikely that inspection and maintenance of small reservoirs will have to be increased somewhat. An interesting effect, outside the water resources field, may be a remarkably sudden demand for a chemical in quantities of millions of pounds per year where past annual consumption, in cosmetics, has probably not exceeded a few thousand pounds.

Chapter Summary

The three illustrations here given—reuse, water budgeting, and evaporation control—suggest that comprehensive water development

---

[73] Eaton, *op. cit.*

[74] Beadle and Cruse, *op. cit.,* p. 399. This cost estimate is much below the reported Australian costs, and is undoubtedly based on expectations of increasing the effectiveness of the film and reducing the application rate.

planning and efficient management of water resources may well include the newer techniques that improve the using and storing of water. Substantial contributions to available water supply may be made no less through adding to the uses of a given supply, or restricting its disappearance, than through development of new supplies. Some of the cheapest water obtainable is promised by techniques like those here described.

The significance of these three illustrations does not rest in their individual promise alone; other techniques to extend the use of a given supply eventually may be developed. Some anticipation of these would seem warranted as part of efficient, comprehensive water development planning.

PART IV

ORGANIZATIONAL RESPONSES TO PROBLEMS AND OPPORTUNITIES FOR WATER DEVELOPMENT INTRODUCED BY TECHNOLOGIC CHANGE

CHAPTER 17

Administrative Organization and Past
Technologic Advance —

INTRODUCTION

The foregoing sections describing the nature of technical events influencing water development have provided a simplified framework for the process of matching water demand and supply. Technical change has been presented in some detail as a force influencing both water supply and water demand. Other forces influencing the nature of water demand also have been illustrated briefly (chapter 3), in particular the dynamic aspects of the country's population, and political factors. The institutional framework within which supply is matched with demand, however, is far from being simple, and has been referred to only incidentally.

The Institutional Framework

Products and services from a water resource, like those from any other instrument of production in a complex modern society, must be channeled through an institutional framework. Efforts toward developing further production from a resource to meet existing demand likewise must proceed within the same framework. The manner and effectiveness of matching demand with supply is much influenced by legal, political, and economic institutions and by administrative organiza-

461

tion.[1] The modified market economy prevailing in the United States, major private ownership of the means of production, constitutional law, division of United States territory into forty-nine state jurisdictions, prior appropriation and riparian water law, public utility regulation, and other important institutions all share in determining the manner in which demand and supply are equated. One part of this institutional structure of special importance to technical change is administrative organization for development and management. The value of technical effort toward solution of the problems of discontinuity of timing in supply and demand, geographical deficiencies of supply, and quality divergence between demand and supply, depends in part upon administrative organization to apply that effort. If technology is directed especially toward mastery of the physical environment, administrative organization may be seen as a means of facilitating or depressing the flow of services from the resource, which is part of the physical environment. Administrative organization also may be the means of directing the application of technology itself.

The place of administrative organization in matching supply and demand assumes an importance and complexity in water provision which does not prevail for most other products and services entering trade. The free market economy operates very imperfectly in the case of water because of the large size and the indivisibility of many units of supply.[2] Consequently, the political process appears to be needed in achieving timely development of supply for some water-derived services, and public administrative organization bears a responsibility which it otherwise would not have.

Introduction of political forces alongside the normal force of demand which might operate in the free market brings one face to face with the complexity of relating technology to administrative organization. Discretely considered, the problem of matching water supply and demand is one of adjusting volume, flow, quality, and unit costs of production and distribution to the volume, nature, intensity, and location of demand. Technology and administrative organization go hand in hand toward meeting this problem within a given cultural and physical environment. However, the political forces the problem brings into

[1] Local and regional cultural characteristics (other than political and legal) also may be strong influences on the manner of water use.

[2] John V. Krutilla and Otto Eckstein, *Multiple Purpose River Basin Development: Studies in Applied Economic Analysis* (Baltimore: The Johns Hopkins Press, for Resources for the Future, 1958), chap. III.

play may act also in the direction of *other* objectives through the *same* administrative organization. Other objectives may be actions in the interest of national security (either military defense, or safe-guarding the means of future resource production) and equity or preferential treatment in the distribution of benefits from production. Equity considerations in benefit distribution arise out of differences in demand: by geographical location, by purpose or function (i.e. industrial, domestic, transportation, agricultural, etc.), and by income and ownership class.

The multiplicity of objectives served by administrative organization for water development in the past has tended to obscure the relation which obviously exists between technology and administrative structure and technique. It is entirely possible that there may have been a "cultural lag" in the adaptation of American administrative organization for water development to the problems and opportunities brought by past technical events. It is even more possible that the emerging technology is mirrored by still less adaptation in our administrative structure. It is the purpose of the remainder of this study to examine these relations, insofar as they exist. In this analysis it is recognized that:

1) Other institutional factors than administrative organization also are important in channeling water supply and demand within the United States.

2) Administrative organization for water development serves not only the ends of adjusting water supply to the nature of water demand, but also objectives of national security, economic equity or preference, and other ends which may find convenient political expression through it. Technology therefore is only one of several factors influencing the pattern of administrative organization.

Recognizing these facts, *the analysis to follow must not be considered a complete answer to appropriate water development organization.* Neither must it be considered a complete analysis of the bearing of technical events upon the institutions which channel water demand and supply. Rather it will treat the organizational responses to the opportunities and problems presented by technology in the past, and the problems brought by the emerging technology, together with indicated responses thereto. The analysis is concerned primarily with the adjustment of volume, flow, quality, and unit costs of production and distribution to the volume, nature, intensity, and location of demand, and the unit benefits therefor. Insofar as they have validity, the facts brought forth can serve as data, along with data or other ends served by the same organization, for examination of administrative effectiveness.

The relation of technology and water development organization will be discussed against the background of the following features of the physical and institutional environment within which development has taken place:

1) The United States is a land of large river basins. Nearly three-fourths of continental United States lies within five major river basins and the Great Basin of interior drainage.

2) Generally speaking, the United States is divided into two great regions, one eastern and the other western, each having distinctive characteristics of water supply.

3) United States territory is similarly divided into eastern and western regions with distinctive characteristics of water demand, although the regions in this sense are less sharply contrasted than in the case of water supply.

4) The economy of the United States is dominated by judgments of the market place in the satisfaction of consumer demands. However, this process appears to be substantially modified in the provision of water and water-derived services, which depends in part upon socially based political or governmental decisions.

5) Private ownership of the means of economic production and distribution is dominant in the United States.

6) Constitutional law supplies a basic legal framework within which the organization of development efforts must proceed. Existence of the interstate commerce clause and other constitutional provisions have given a distinctive legal structure encouraging federal participation in some important phases of water development.

7) The territory of the United States is divided into forty-nine state jurisdictions which have little correspondence with the limits of the major drainage basins, or even with the limits of the important tributary basins. Furthermore, hundreds of county and special district jurisdictions over water exist within the states.

8) A further important institutional difference exists in that some states subscribe to the prior appropriation doctrine of allocating water among users, and others follow the riparian doctrine of reasonable use and continued natural flow. This division in legal point of view is a regional one, with West again distinguished from East.

The relation between technology and administrative organization will be considered within the limitations of these eight characteristics of the United States of the present day. Furthermore, it is fully recognized that technology can be only one of several final determinants of the administrative pattern.

What Administrative Organization Does

A second step in describing the relation of technology and administrative organization is to set forth briefly the functions of administrative organization in the type of resource development here being considered. Such organization mobilizes and directs group effort to the end of planning for development, realizing development, and operational management,[3] as follows:

1) *Planning*
   a) Supplying initiative toward the recognition of development problems.
   b) Policy formulation. Provision of specific interpretation of general statutes, or necessary legislative modification thereof.
   c) Applying all types of research techniques to the clarification and solution of development problems, i.e. planning analysis, programming, etc.
   d) Physical planning of development; co-ordination of planning for several purposes of development.
   e) Evaluating engineering and economic feasibility of development.
   f) Project scheduling.
   g) Review of experience, and evaluation of progress. Drafting of measures for necessary administrative or procedural reorganization or correction.

2) *Realizing Development*
   a) Securing or supplying entrepreneurship for development and later operational management.
   b) Securing necessary legal authorizations for development and subsequent operation of facilities.

---

[3] "Development" and "management" here and hereafter in this study are intended to mean efficient development and efficient management, that is, maximizing the net value of the products or services which can be made available from the resource under known techniques.

  c) Securing or supplying financing; administering disbursement of
  investment funds.
  d) Management and supervision of construction or other developmental activities.

3) *Operational Management*
  a) Necessary physical and mechanical operations.
  b) Sale of vendible products or services.
  c) Provision of debt repayment.
  d) Co-ordination of all operations with other public or private
  operations having impinging areas of interest.

These functions are applicable whatever the scale of the organization,
whether it represents a few families, a community, a district, a state, a
region, or the nation.

Application of Technical Phenomena to Administrative
Functions in Development

Five major types of technical changes have been discussed previously
in this study: techniques accelerating, and those decreasing water demand; techniques extending services afforded by a given unit of supply;
technical improvements promoting scale economies; and techniques
extending the physical range of water recovery. Reference to these five
classes of the manifold technical events of the past may assist in answering the questions which are at the heart of any analysis of the relation of
technology and administration: (1) Has the administrative organization prevailing in United States water development facilitated technical
advance? (2) In the course of past development, has our administrative organization given full opportunity for the application of past
technical advances? (3) What functional and geographical scope is
characteristic of the administrative problems created by technology, and
what scope do the administrative responses to these problems have?

As Charles McKinley has noted,[4] the first and second of these questions raise issues which are not peculiar to water resource functions nor
exclusively related to technical changes. They are part of administrative problems of larger scope and application: administrative leadership

---

[4] Charles McKinley, personal communication to the authors, 1957.

and administrative planning which facilitate adaptation and promote institutional efficiency in the face of a changing environment.[5] It is recognized that the responsiveness or rigidity of United States administrative organization in the presence of the particular problems of water resource development may have roots much deeper than the specialized water functions within our economy. Nevertheless they are specific questions which must be kept in mind in exploring the relation between technology and administrative organization for water development.

The second, third, and fourth groups of techniques, those decreasing demand, those extending water services, and those promoting scale economies, bear upon all three of the principal functions of administrative organization. The process of planning, the realization of development, and operational management all are effected by the use of devices like the scheduling techniques and integrated basin water control. Techniques extending the physical range of recovery have been of particular interest in past planning. On the other hand, the techniques accelerating the rate of exploitation have been of concern in development and operational management as well as in planning.

The five groups also have somewhat differing stimuli in the geographical environment of the United States. The application of the scale economies is especially interesting. The existence of large river basins predisposes the country to the use of scale economies when they are available for water development. On one hand, the arrangement of state boundaries, and the existence of two great regions which differ from each other in their patterns of water demand, water supply, and institutional characteristics, raise some formidable barriers to the use of these techniques. On the other hand, the existence of a scarce supply region with montane water sources (the West) has stimulated and favored the reception of techniques for extending water services, for decreasing water demand, and for extending the physical range of recovery. While techniques accelerating demand have affected the nation as a whole, they have fallen especially heavily upon the western region of scarce supply. Thus it would seem that the presence of a large region of scarce supply has given a peculiar cast to the development of water-related technology, and to the problems which technology has brought to administrative organization. They also have been affected by the flood disasters which large basins like the Mississippi are predisposed to

---

[5] See also P. H. Appleby, *Big Democracy* (New York: Alfred A. Knopf, 1945), pp. 105-11.

have. These will be described as they have related individually to planning, to the realization of development, and to operational management.

Finally, the impact of the five groups of techniques differs as far as the level of administrative organization is concerned. Most technical advance has presented increasing opportunities for administrative organization on the regional or national levels. The scale economies, extending the physical range of recovery, and the changes accelerating the rate of exploitation, all have affected community, locality, district, and state, but their important challenges have been to administration on the national and regional levels. Discussion in the succeeding three chapters accordingly will center upon the problems of the more comprehensive administrative levels.

CHAPTER 18

The Relation of Technologic Advance to Administrative Organization for Water Development Planning

The Administration of Planning and the Scale Economy Techniques—site identification and preservation; research and data collection for planning analysis; the organization of basin-wide or region-wide planning; planning for interbasin use of water and water-derived services. Planning Use Extensions of a Given Water Supply—(1) multipurpose planning; (2) other techniques extending the services of a given supply, or increasing availability. Extending the Physical Range of Water Recovery. Appraisal of Influences Accelerating the Rate of Exploitation. Promoting an Environment Favoring Application of Further Technical Improvements—the importance of qualified personnel; review of past experience.

This and the two succeeding chapters discuss separately the problems and opportunities which technologic advance has brought to the administration of planning, development, and operation of facilities for water use. The purpose is to analyze the relation which has prevailed between technology and administration, rather than to comment upon the conduct or policy of public and private agencies which have held important responsibilities in administering past water development and use in the United States. An insight into the process at work is sought, not a criti-

469

cal review of professional and political activities relating to past water development.[1]

In practice, the same administrative agency may have planning, development, and operational responsibilities, with one set of problems merging almost imperceptibly into the other. Planning cannot be divorced wholly from development and operation. Important operational assumptions always must be made in planning. Nonetheless, where analytical clarity is sought, as in this discussion, separate treatment ot these three functions of administrative organization seems desirable. That is undertaken in this chapter, chapter 19, and chapter 20. At the same time, a division of treatment in no way prejudges the suitability of integrating planning, development, and operation under the same professional administration.

The difficulty of separating planning and development is first encountered in examining the relation of technology and administration as they concern planning. The type of planning to be undertaken depends in part on the type of development that is feasible. Planning questions closely tied to the feasibility of development are considered by implication in the analysis of technology and development. Nonetheless there are certain planning actions which not only precede development but to which development has little "feed-back" relation. This discussion is addressed to these more independent aspects of planning for development and use.

The techniques promoting the scale economies and those extending the services of water supply in particular have presented an interesting set of problems for the administration of planning in this field.

The Administration of Planning and the Scale Economy Techniques

Four different types of problems were introduced by the scale economy techniques for administrative organization of planning: (1) identification and preservation of large-scale dam and reservoir sites; (2) data collection for planning analysis on a scientific basis; (3) the organization of planning on a basin-wide pattern; and (4) the organization of interbasin planning for development.

---

[1] The method here used is one of presenting illustrations, rather than an exhaustive treatment of all facets of existing water resource administration in the United States.

SITE IDENTIFICATION AND PRESERVATION

One of the first problems posed by technology for administrative organization of water development planning was the conservation of large-scale storage sites. As soon as the feasibility of large detention structures became evident, the question of site preservation arose. Even a rough reconnaissance of topographical and geological features of United States stream courses reveals definite limitations on the number of sites which permit the storage of water on a large scale. While the development of materials and engineering skills has permitted some adjustment to site conditions (as in the condition of foundation rock), not even the nuclear age has brought the prospect of creating a good storage site in the absence of specified natural surface conditions.

Three questions for administrative organization arose as the engineering feasibility of constructing large-scale reservoirs became apparent. The first was means of early identification of the sites; the second, regulation of construction so as to guarantee efficient use of the site's potentialities over the long run; the third entailed forestalling the economic encumbrance of sites which might make their later use unnecessarily or impossibly expensive.

Identifying the sites. The first of these questions was met in several ways. The possibility of using some of the sites for structures which might produce vendible services like electricity led to the identification and exploitation[2] of a number of sites by private enterprise. There can be no doubt that the organization provided in the electric utilities, and in some of the early large power consumers like the Aluminum Company of America, materially aided in the identification of the sites. Thus the Tennessee Valley Authority, when it commenced operation in 1933, found that private operations already had identified a number of sites on the river and its tributaries, including the important Fontana site. Private enterprise also had placed six major structures within the basin.[3] From 1900 onward considerable activity of this kind over both West and East resulted in many additions to knowledge of the sites.

[2] "Exploitation" here, as elsewhere in this study, has its original meaning of "use." No value judgment is attached.

[3] Hales Bar, Ocoee No. 1, Blue Ridge, Cheoah, Calderwood, and Santeetlah. The first three were sold to TVA by the Tennessee Electric Power Company; and the latter three are still owned by the Aluminum Company of America, but are integrated operationally in the TVA water control system.

Contributions toward this end also were made by the municipal water companies and irrigation districts. Thus part of the administrative answer to the problem of identification of sites came through private enterprise.

Private activity has been complemented, and in recent years overshadowed, by the surveys of public agencies. Agency surveys have been of special significance in identifying major storage sites. The earliest extensive work of this kind was undertaken within the Department of the Interior as it administered the Reclamation Fund from 1902 onward. Limited surveys of western streams for storage sites were undertaken by the Geological Survey in 1903 and subsequently,[4] and by the Reclamation Service in 1907 and subsequently.[5] The work of the Geological Survey has continued to the present day, particularly on the public lands. The Forest Service actively co-operated with the Geological Survey in early surveys for this purpose. Although commenced later, some of the most extensive surveys undertaken were made by the Army Corps of Engineers. The basins considered suitable for such surveys were first listed in the now well-known "308" report,[6] from which the later surveys take their name ("308 reports").[7] Since the original authorization was

---

[4] Even earlier appraisals of water development possibilities had been made in the Powell Irrigation Survey (1888-91), but apparently this Survey was not concerned with major storage sites so much as with identification of irrigable lands. The Act of March 3, 1879 (20 Stat. 377, 43 U.S.C. 31) establishing the Survey gave it responsibility for classifying the public domain lands as sites valuable for power purposes (Sec. 1), among other duties. If so classified by the Survey, lands of the public domain have been, and continue to be, withdrawn from all forms of entry under the public land laws.

[5] U. S. Department of the Interior, *Bureau of Reclamation Appropriation Acts and Allotments,* Washington, 1948, pp. 3-509.

[6] This report was prepared by the Corps of Engineers and the Federal Power Commission in response to a directive, given by the Congress to the Secretary of War and the Commission in 1925, jointly to estimate the cost of making surveys of those navigable streams and their tributaries "whereon power development appears feasible and practicable." (Act of March 3, 1925, Sec. 3, 43 Stat. 1186, 1190.) The Colorado River was excluded from this directive. (President's Water Resources Policy Commission, *Water Resources Law,* Vol. 3 of Report [Washington: U. S. Government Printing Office, 1951], p. 408.)

[7] House Document No. 308, 69th Congress, 1st session. Section 1 of the River and Harbor Act of January 21, 1927, directed that surveys be made as indicated in that document. The following streams were included in the list given in House Document No. 308 (pp. 3-4):

Streams draining to Atlantic Ocean north of Cape Cod as follows: St. Croix, Machias, Union, Penobscot, Kennebec, Androscoggin, Presumpscot, Saco,

Kennebunk, Salmon Falls, and Merrimack.

Streams draining to Atlantic Ocean between Cape Cod and New York Harbor as follows: Taunton, Pawtucket, Pawcatuck, Thames, Connecticut, Housatonic.

Hudson River and tributaries as follows: Mohawk, Hoosic, Batten Kill, Wappinger Creek, Wallkill, Kinderhook Creek.

Streams draining to Lake Champlain and Richelieu Rivers as follows: Poultney, Otter Creek, Boquet, Ausable, Saranac, Big Chazy, Winooski, Lamoille, and Missisquoi.

Raritan River.

Delaware River and tributaries as follows: Shohola Creek, Mongaup River, Neversink, Lehigh, Tohickon Creek, Neshaminy Creek, Perkiomen Creek.

Rivers draining into Chesapeake Bay as follows: Susquehanna, Pamunkey, Rappahannock, Occoquan Creek, Patuxent, Potomac, and James.

Streams draining to Atlantic Ocean south of Chesapeake as follows: Roanoke, Meherrin, Neuse, Tar, Cape Fear, Yadkin, Pee Dee, Santee, Savannah, Altamaha, Satilla, and St. Marys.

Streams except the Mississippi River draining to Gulf of Mexico, as follows: Suwannee, Withlacoochee, Apalachicola and tributaries, Mobile River System including the Coosa, Warrior and Tombigbee Rivers; Guadalupe, Calcasieu, Amite, Tickfaw, Tangipahoa, Chefuncte, Bayou Nezpique, Bayou Teche.

Mississippi River and minor tributaries as follows: Ouachita, St. Francis, Meramec, Illinois, Des Moines, Iowa, Wisconsin, Chippewa, and St. Croix.

Arkansas River and tributaries: White, Grand, Illinois, Petit Jean, Fourchee La Favre, and Poteau.

Ohio River and minor tributaries as follows: Tradewater, Wabash, Green and Barren, Salt, Kentucky, Miami, Licking, Guyandot, Big Sandy, Muskingum, Little Kanawha, Beaver, Monongahela, Allegheny.

Tennessee River.

Cumberland River.

Kanawha River.

Missouri River and tributaries as follows: Madison, Jefferson, Gallatin, Marias, Musselshell, Milk, Yellowstone, Little Missouri, Cannon Ball, Grand, Moreau, Cheyenne, Bad, White, Niobrara, James, Big Sioux, Little Sioux, Platte and Kansas, Osage and Gasconade.

Streams draining into Lake of the Woods and Hudson Bay drainage basin, as follows: Rainy, Big Fork, Little Fork, Vermillion, Kawishiwi.

Streams draining into Lake Superior as follows: Pigeon, Brule, Devil Track, Cascade, Poplar, Temperance, Manitou, Baptism, Beaver Bay, Gooseberry, St. Louis, Amnicon, Bad, Montreal, Sturgeon, and Carp.

Streams draining into Lake Michigan as follows: Wolf, Oconto, Peshtigo, Menominee, Manistique, Manistee, Muskegon, Grand, Kalamazoo, and St. Joseph.

Streams emptying into Pacific Ocean south of Columbia River as follows: Eel, Mad, Klamath, Sacramento, San Joaquin, Kern.

Columbia River and minor tributaries as follows: Cowlitz, Lewis, Willamette, John Day.

Snake River and tributaries.

made, the Corps has completed 217 surveys of major basins,[8] each being a "comprehensive investigation with a view to the formulation of general plans for their most effective improvement for the purposes of navigation, in combination with the most efficient development of the potential water power, the control of floods and needs of irrigation."[9] This has been a most important contribution to knowledge of the storage sites and other characteristics of American streams. Within its limited area of jurisdiction a notable survey also was made by the Tennessee Valley Authority, which later converted the main stem of the Tennessee River into a 627-mile unbroken series of reservoirs from Knoxville to the Ohio River. During their existence the National Resources Committee and the National Resources Planning Board (1933-43) also encouraged the detailed investigation of sites in the drainage basin surveys organized under their auspices.

Opinions have differed as to the engineering conclusions which should be drawn regarding some parts of the surveys of storage sites, and also on the timeliness of the surveys.[10] However, the question of identification of major storage sites appears to have been met with some thoroughness in the United States through a combination of private and public action. The prevailing administrative organization of water resource development thus appears to have been able to meet one challenge which technical advance gave to these agencies.

Streams draining into Pacific Ocean north of Columbia River as follows: Skagit, Snohomish, Stilaguamish, Puyallup, Chehalis.
Rivers in Hawaiian Islands.

In the "308" surveys the Chief of Engineers recommended, among other things, studies sufficient to determine: "The locations and capacities of reservoir sites . . . the location and practicability of dam sites . . . the capacities of power sites . . . the best plan of improvement for all purposes . . . the feasibility of the best plan of improvement." (p. 2.)

The authorization was broadened to include important economic data in the River and Harbor Act of August 30, 1935 (Section 6). It was also extended in the River and Harbor Act of May 15, 1928, which extended the work on the Mississippi River system.

[8] Communication from the Office of the Chief of Engineers, November 5, 1956.

[9] U. S. Army Corps of Engineers, North Pacific Division, *Review Report on Columbia River and Tributaries,* Portland, Ore., 1948, p. i-3.

[10] A Congressional bill providing for the survey of sites had been introduced by Senator Newlands of Nevada almost fifteen years earlier than the "308" authorization in 1927, but failed to pass, in part because of Corps of Engineers opposition.

Site identification for secondary structures, however, has not progressed as far. The lack of adequate coverage in both topographic and geological maps in the United States has made site reconnaissance an expensive procedure, dependent on field survey in many areas. Consequently, knowledge of the lesser sites is very fragmentary. This recently has been illustrated in the University of Kansas study of the Kansas River Basin.[11]

Regulation of construction to guarantee efficient use of dam sites. Identification of sites is only part of a planning process leading to development of a stream. Particularly in a country with an economic system like that of the United States, conflicts in objective may arise as to the way in which a site is to be used. The manner of development best suited to the needs of a localized corporation or a section of a state may not be the same as that yielding the greatest benefits to the nation, or to the region of the entire river basin. Within the last five years this has been dramatically illustrated in the Hells Canyon case on the Snake River (between Idaho and Oregon). Use of this key site on the Columbia River system was sought and obtained by the Idaho Power Company, whose best immediate interest was served by the construction of three power dams at periods corresponding to expected or actual load growth on their electricity distribution system. The best use of the site from the point of view of the northwestern region as a whole, and the highest long-range economic return to the nation, on the other hand, appears to have been an entirely different plan comprising one dam, or at most two dams.[12]

The controversy which developed in 1952 and later on the Hells Canyon case did not present an extreme contrast in benefits under the two different objectives, private and public, but it was illustrative of a situation within which extreme contrasts are possible. Many major multiple-purpose storage sites in the country could be or could have been used profitably for the single purpose of hydroelectric development. Such use, however, usually would eliminate the possibility of constructing economical multiple-purpose facilities on the same site or on a nearby site. Even small stream control works require substantial capital investment, usually amortizable in terms of several decades.

[11] University of Kansas, *The Kansas Basin: Pilot Study of a Watershed,* Lawrence, Kan., 1956.

[12] Krutilla and Eckstein, *op. cit.* See also Roy F. Bessey, "The Political Issues of the Hells Canyon Controversy," *The Western Political Quarterly,* 9 (1956), 676-90.

Construction of immediately profitable secondary works on or near a major storage site therefore may preclude later development of an important regulation facility as needs are foreseen. From the viewpoint of long-range planning there can be public interests which are not necessarily served by development with private or localized objectives. Even a major facility may be an underdevelopment of the full potentialities of a site, as is the case in Hells Canyon, and as appears possible for other sites on the Columbia River and its tributaries, like Nez Perce (Idaho-Oregon) and Paradise (Montana).

Regulation of site use accordingly requires administrative provision in proper water development planning.

A degree of site protection has been achieved through the protection of navigable waters from obstruction. The River and Harbor Act of 1899[13] and subsequent related legislation require approval of the Chief of Engineers, United States Army, for any construction of bridges, dams, dikes, causeways, or water diversions likely to affect the navigation capacity of a navigable stream of the United States. Since storage reservoirs have had a close relation to navigation capacity on a number of streams, some further protection against obstruction to good site use has been obtained. Such regulation continues to be administered by the Chief of Engineers, except in the case of the Tennessee River and its tributaries, which generally are under the jurisdiction of the Tennessee Valley Authority Board of Directors.[14]

Although all of the major storage sites in the Tennessee Valley have now been put to use, the regulatory provision of the TVA Act continues to be useful for secondary sites, channel protection, and protecting the integrity of the unified water control system. The powers delegated to the Chief of Engineers in the River and Harbor Act of 1899 and subsequently, as administered, continue to be useful adjuncts to the governmental machinery for site protection in the public interest.

Other early legislation came in the 1906 General Dam Act[15] and in the 1910 General Dam Act.[16] The first required the installation of

[13] Act of March 3, 1899, Secs. 9-20, 30 Stat. 1121, 1151.

[14] Act of May 18, 1933, 48 Stat. 58, as amended, 16 U. S. C. 831-831dd. Section 26a refers to the regulation of nonfederal construction on the river. Provision is made for appeal to the Secretary of War (Army) in the case of structures proposed for the Little Tennessee, a tributary. The foregoing reference to the U. S. code refers to the 1952 edition. All succeeding references to the Code are from the same edition.

[15] Act of June 21, 1906, 34 Stat. 386.

[16] Act of June 23, 1910, 36 Stat. 593.

fish passage and navigation facilities in dams placed on navigable streams; and the second required consideration of the relation of any such structure to "a comprehensive plan for the improvement of the waterway . . . with a view to the promotion of its navigable quality and for the full development of water power."[17]

Legislation of potential major importance in this function was included in the Federal Water Power Act of 1920, which called for federal licensing of nonfederal construction on navigable streams.[18] This provision has been administered by the Federal Power Commission since its inception. The Commission was given the responsibility for seeing that good use was made of sites on navigable streams proposed for private power development. The Federal Power Act of 1935, which superseded and included amended provisions of the 1920 Act, stipulates that an adopted project site must be one which in the judgment of the Commission will be:

> best adapted to a comprehensive plan for improving or developing a waterway or waterways for the use or benefits of interstate or foreign commerce, for the improvement and utilization of water-power development, and for other beneficial public uses, including recreational purposes.[19]

Further legislation toward these ends was included in the Flood Control Act of 1938[20] and in all subsequent flood-control legislation. This is the provision requiring installation of penstocks and similar facilities for later possible electric power production at all storage projects for flood prevention.

In a sense, site protection also has been achieved by federal construction. As the important water development programs of the federal agencies have proceeded, each has pre-empted sites for the construction of facilities fitting its program. Although there were some questions

[17] Sec. 1, 36 Stat. 593.

[18] Act of June 10, 1920, 41 Stat. 1063, Act of August 26, 1935, 49 Stat. 838, as amended, 16 U. S. C. 791a-825r.

[19] Sec. 10 (a), 41 Stat. 1063, 1068, as amended, 16 U. S. C. 803(a). The Commission's discharge of responsibilities delegated under these provisions in recent years has been criticized as inadequate, tending on occasion to be single purpose, as in the Hells Canyon case. Public power groups, in particular have disagreed with the Commission's interpretations.

[20] Act of June 28, 1938, Sec. 4, 52 Stat. 1215, 1216, 33 U. S. C. 701j. The penstocks provision in part may be traced to the activities of the Federal Power Commission, meeting responsibilities given it under the Federal Power Act.

as to results from the Corps of Engineers' program in the 1930's,[21] and regional differences in Corps' policy,[22] the large majority of federal facilities constructed in the last twenty years, including those of the Corps, has represented a real effort to find the most efficient use of sites in the interest of the general public. Particularly after the stimulus of the National Resources Planning Board efforts during the 1930's, and the examples of Tennessee Valley construction, the major construction agencies have had elements of comprehensive development in their planning. The Corps' program on the Columbia, that of the Tennessee Valley Authority on the Tennessee, and that of the Bureau of Reclamation in the Central Valley of California are all examples of conscientious planning and construction to get the most from the available natural sites in terms of multiple-purpose benefits.[23]

Forestalling the economic encumbrance of reservoir sites. Provisions for guaranteeing efficient physical use of dam sites has not been matched in the United States with preventive measures forestalling the economic encumbrance of good reservoir sites. A reservoir site may be made unusable, from an economic point of view, by intensive agricultural occupancy or industrial and commercial development. Once established, such activities may become impossibly expensive to relocate when reservoir use of the site is contemplated. Failure to provide for this aspect of planning may have been influenced by four things: (1) the

[21] There was a definite possibility, for instance, that a series of thirty-two low dams would have been constructed on the Tennessee, mainly for navigation, had the administration of the Tennessee River program remained under the Corps of Engineers. This was recommended (with provisions for substitution of high dams by private construction where desired by private interests) by both the District and Division Engineers in the "308" Report on the Tennessee (71st Congress, 2nd Session, House Document No. 328, p. 5). The recommendation was supported implicitly by the Chief of Engineers in the same document (p. 7). Under such a plan, sites since proven of great multiple purpose value would have been pre-empted by single-purpose construction. Plans for development of the lower Snake River had a somewhat similar history. Between 1933 and 1941 plans for ten low dams were seriously considered alternatives to the four-dam plan finally authorized. (See 73rd Congress, 2nd Session, House Document No. 190; and 75th Congress, 3rd Session, House Document No. 704.)

[22] In this connection question has arisen as to the plans recommended for New England streams by the Corps. See William E. Leuchtenberg, *Flood Control Politics: The Connecticut River Valley Problem, 1927-1950* (Cambridge, Mass.: Harvard University Press, 1953).

[23] Between 1953 and 1958 federal agencies have not always worked toward the same objectives in site protection. The Federal Power Commission's interpretation of the best use of the Hells Canyon site (three dams) differed from that of both the Corps and Bureau of Reclamation (one dam).

sparsity of settlement and economic use within a great many basins of the United States; (2) the fact that administration of real property regulation was in the hands of state or local authority, while large-scale river development is principally a matter of interstate or federal concern; (3) delays in federal or other legislative authorization and appropriations after initial planning; and (4) the difficulty of evaluating the prospective benefits of prospective water developments as against the immediate value of using potential reservoir sites as transportation and communication corridors; use for business, factory, and residential sites, and for still other purposes. At the time that land transportation facilities were located in major valleys, for instance, little was known of the possible water development of these valleys.

Fortunately, only a few major streams in the United States have been seriously encumbered by valley occupancy. The main stem Ohio River, the upper tributaries of the Ohio, and the middle reach of the Mississippi River are those most affected. It is likely that some potentially very valuable sites have become economically unusable because of valley occupancy along the Ohio by farms, factories, railroads, bridges, highways, residences, warehouses, and other commercial buildings. The million-kilowatt Dog Island site between Kentucky and Illinois may have been eliminated by such considerations, for example.[24] It is significant that the Corps of Engineers' comprehensive program for the Ohio River Basin, as that program was described in 1951, did not include a single major reservoir along the entire 981 miles of the main stem of the Ohio.[25] All major control works for the river were located on tributaries. Yet there is no lack of natural reservoir sites along the main stem of the Ohio.

For most of the remainder of the country possessing valuable reservoir sites, valley occupancy presents lesser but still significant economic handicaps to reservoir development. This is true for parts of the West, New England, the Lakes states, the Great Plains area, and the Southeast. In some of the sparsely settled sections of the West such encumbrance must be considered minor. In other areas the seriousness of the problem depends in part on the availability of alternative sites or alternative means of meeting the demand served by storage facilities.

As late as 1956 very little machinery existed within the United States

[24] Tennessee Valley Authority, Board of Directors, *Report to the Congress on the Unified Development of the Tennessee River System,* Knoxville, Tenn., 1936, pp. 68-71.

[25] President's Water Resources Policy Commission, *Ten Rivers in America's Future* (Vol. 2 of Report), 1951, pp. 638-40, and fig. 5.

to guide the interim use of potentially valuable reservoir sites beyond the actual physical development of the site itself. In some instances local zoning regulations may have provided regulation of flood-plain occupancy, and thereby incidental regulation of building on or other use of potential reservoir sites. However, these have been located mainly in small urban communities, and the net effect of such regulation has been small. Where specific important reservoirs have reached a stage of detailed planning in the Corps of Engineers' program for a river basin, the Corps occasionally has intervened to divert any major construction within the potential reservoir area. Such appeared to be the case in the Corps' opposition during 1956 to the construction of a Potomac Electric Power Company thermal electric generating plant along the Potomac River in Loudon County, Virginia. The proposed plant would have made infeasible the major regulation work planned for the Potomac, the Riverbend dam and reservoir. This effort on the part of the Corps at least temporarily forestalled construction within the Riverbend reservoir site. Other attempts by the Corps to forestall encumbering construction on reservoir sites have not always met with success. An example was the highway bridge across the Columbia River between Wasco County, Oregon, and Klickitat County, Washington, construction of which was commenced over The Dalles dam site in spite of the Corps' plans for use of this important site on the Columbia.[26]

In addition to the construction of physical facilities, another type of encumbrance frequently has forestalled construction of the best-suited works, and sometimes has paved the way for underdevelopment or outright encumbrance later. This is special-interest opposition to public action where adverse economic results are foreseen for the special interest. The Corps of Engineers often has been very sensitive to this type of opposition in the past, as in New England and in the case of the now famous Grand Coulee Dam. The Board of Engineers for Rivers and Harbors, mindful of a rival plan being favored by private utilities in eastern Washington State, once found the then proposed Grand Coulee Dam an infeasible project.[27]

In fairness it could hardly be maintained that the encumbrance of

---

[26] Relocation of the bridge was accomplished after Congress authorized the construction of The Dalles dam in 1950, but at an additional expense to the federal government of $1,994,000 for this specific purpose. (Communication from the Office of the Chief of Engineers, 1957.)

[27] See George Sundborg, *Hail Columbia* (New York: Macmillan, 1954), pp. 28-29, 121.

reservoir sites by other economic use was a pressing national problem in 1957. The major construction program of national scope likely to cause reservoir encumbrance was that for the National System of Interstate Highways authorized in 1956. It could be expected to raise this problem in at least some localities before its completion in 1975 or later. This in itself should be considered worthy of careful attention at the federal level. Nevertheless, little administrative organization had been developed for coping with the site problem as a whole, or even for meeting it with consistency. Aside from reservation of sites on public lands, the best that had been done was effective spot action by an agency with a project at stake which also had been planned in some detail. *Ad hoc* defenses of sites which had entered project plans offered the main procedure for site preservation. The problem is one of bringing zoning or other techniques of controlling land use to bear on the national interest in preserving good or unique sites for future valuable use. The need of regional determination of national interest is apparent.

RESEARCH AND DATA COLLECTION FOR PLANNING ANALYSIS

The scale economy techniques greatly increased the need for the collection and interpretation of accurate basic data on precipitation, hydrology, demographic movements, economic growth, and other items essential to effective development planning. A distinction here will be made between those aspects of data collection that are part of the process of drafting detailed plans for individual projects, and those essential to planning for the most effective large-scale development including basin-wide and interbasin potentialities. Large dams and large reservoirs of course brought with them the need for improved mapping of wider coverage and larger scale, more detailed structural geology data, data on sedimentation trends, and observation of the hydrologic and biological effects of large-scale impoundments. As might be expected, these new demands were met as part of the normal order of business by the organizations or agencies responsible for planning and constructing individual projects, like the Corps of Engineers, the Bureau of Reclamation, the Tennessee Valley Authority, or private corporations.

The possibilities of basin-wide integrated water management and interbasin diversions, however, also brought with them much more complex demands for basic data collection, some of which have not yet been fully satisfied. The succeeding discussion will treat these broader aspects of data collection, since they are of the most concern to administrative organization.

The scale economy techniques brought the following needs for data assembly and research: (1) precise and adequate geographical coverage for water yield data, and an understanding of the fluctuations thereof; (2) an understanding of the dynamic aspects of the economic and demographic environment, both national and regional, within which planning and development would proceed; (3) a precise observation of the effects of comprehensive development upon the quality and quantity of water available for each specific purpose.

The scale economy techniques, if employed in their fullest and most effective manner, treat entire regions. Accordingly the type of data that is entirely adequate for planning a single small or moderate sized project is little more than a beginning to the research and data collection necessary for profitable employment of the scale economy techniques. The chance, or automatic, adjustments which a regional economy will make to a single project are out of the question when basin-wide or interbasin development takes place. For many years this important point was ignored or obscured within the United States by those who favored "construction now" whatever inadequacies of basic data existed in planning for construction in a given basin.[28] Basic data collection and research for basin-wide development of water resources therefore must be judged by the clarity of its revelation of regional environment, trends, and potentialities.

Precision and coverage for water yield data. Relatively precise data on water availability (including quality and flow characteristics) is essential to well-planned regional or interregional water development. The more that is known of the hydrology of a basin in detail the more precise engineering may be in its project planning. To capture and make use of the abnormally large flow, to connect water surplus and water deficit areas, to equalize the availability over time, and to minimize the effects of temporarily intense natural shortages, call for accurate knowledge of hydrology. Quantitative knowledge of the hydrologic cycle in operation over an entire basin is necessary. This means knowledge of evapotranspiration and the water budget, the relation of ground-water storage and ground-water movement to runoff, and the hydrologic effects of land-use changes as well as good coverage of pre-

---

[28] See in this connection Senate Document 191, 78th Congress, 2nd Session, 1944, for an illustration of the attitudes and practices prevailing in major basin planning as late as the early 1940's.

cipitation receipts and stream flow. Furthermore, accurate estimates of the range of fluctuation to be expected for runoff within the physical or economic life of the facilities also are essential to informed planning of the fullest use.

The method of administering research and collection of data in this field in the United States has been a fragmented one. The principal agencies administering programs of data collection and research on a national basis are the Weather Bureau of the United States Department of Commerce and the Geological Survey of the United States Department of the Interior. Supplementing their efforts have been either localized or specialized observations by the Soil Conservation Service, the Forest Service, and other agencies of the Department of Agriculture; the Bureau of Reclamation; the Surgeon General; and the United States Army Corps of Engineers. Within the Tennessee Valley, the TVA may undertake any data collection necessary to the pursuit of its operational program.[29] Of these various programs, those of the Weather Bureau, the Geological Survey, the Department of Agriculture, and the TVA will be mentioned to illustrate the division of responsibility, and the prevailing organization of activity in this area.

Since the date of its establishment in 1890, the Weather Bureau has been charged with ". . . taking of such meteorological observations as may be necessary to establish and record the climatic conditions of the United States . . ." in addition to its duties of forecasting the course of weather.[30] Since 1938 provision also has been made for the operation of a current information service on precipitation and flood forecasts, according to the discretion of the United States Army Chief of Engineers and the Chief of the Weather Bureau.[31] These authorizations place the principal responsibility for the collection of data on precipitation, and the principal research on fluctuations in precipitation—long range and short range—in the hands of this agency. Its statutory responsibilities also lead to the conduct of observation and research on evaporation, transpiration, and snow melt. It thus is clear that the important basic data on probability and frequency relating to water supply and flow are to be collected and interpreted by the Weather Bureau. However, the immediate needs of forecasting appear to have dominated Weather Bureau activities from the beginning, and the important tasks of analyzing its enormous collection of precipitation

[29] Act of May 18, 1933, Sec. 22, 48 Stat. 69.
[30] Act of October 1, 1890, 26 Stat. 653, as amended, 15 U. S. C. 313.
[31] Act of June 28, 1938, Sec. 8, 52 Stat. 1215, 1226, 33 U. S. C. 706.

observations for probability and frequency have lagged behind the
techniques and machinery available for such analysis. The needs of the
Weather Bureau for serving other purposes than water development
(viz., aviation and many other pursuits) undoubtedly have conditioned its responsiveness to hydrologic data requirements. Appropriations toward more complete probability and frequency analysis have
been sought,[32] but they have not yet been made. The Weather Bureau
also has been unable to establish the complete network of precipitation
stations needed for good coverage in this country. In 1950 it was
estimated that an additional 4,456 precipitation stations, 242 evaporation stations, and 68 complete meteorological stations were required for
good coverage in the United States.[33] In 1955 the Presidential Advisory
Committee on Water Resources Policy again recommended the establishment of "about 5,000 additional stations, mostly precipitation . . ."[34]
A net addition to the network of 1,200 precipitation stations and 100
evaporation stations was made between 1950 and 1958. Weather
Bureau studies of the problem of measuring evapotranspiration also
must be regarded as inconclusive, from its own testimony.[35]

The Weather Bureau also was given responsibility in its 1890 Act
for "the gauging and reporting of rivers."[36] In practice, however, most
stream gauging in the United States has been undertaken by the
Geological Survey of the Department of the Interior, which never has
received a general authorization from Congress for the conduct of
this work. Commencing with the Sundry Civil Appropriation Act of
1894, Congress consistently has made funds available by yearly appropriations to the Survey for "gauging the streams and determining
the water supply of the United States, including the investigation of
underground currents and artesian wells in arid and semiarid sections."[37]

---

[32] See F. W. Reichelderfer, "The Appraisal of Water Resources in the United
States," in House Document No. 706, 81st Congress, 2nd Session, 1950, pp.
72-74.

[33] *Ibid.*, p. 72. These estimates were originally prepared under the auspices of
the Federal Interagency River Basin Committee. The meaning of the term "complete meteorological station" is subject to several interpretations, and was not
defined in the above reference. The Weather Bureau in 1958 had fewer "first
order" stations than in 1950, but it had a more efficient network than in 1950.

[34] The Presidential Advisory Committee on Water Resources Policy, *Water
Resources Policy* (Washington: U. S. Government Printing Office, 1955), p. 10.

[35] Reichelderfer, *op. cit.*, p. 73.

[36] Act of October 1, 1890, Sec. 3.

[37] Act of August 18, 1894, 28 Stat. 372, 398.

Aside from appropriation acts, the Geological Survey received authority for the acquisition of basic stream flow records from the Reclamation Act of 1902, by delegation. A 1942 statute also authorizes the Secretary of the Interior to acquire lands or the use thereof for the Survey's stream gauging activities.[38] The Survey now maintains in co-operation with the governments of forty-eight states, about 7,000 gauging stations, and conducts research on the occurrence, quantity, and quality of water throughout the nation. According to the testimony of a recent director of the Survey, this work "essentially . . . has been geared to immediate local problems and needs . . . comparatively little effort authorized directly to the Geological Survey could be turned to the broad interpretive evaluations of the water supplies available to meet shifting trends in agricultural and industrial development over the nation . . . or to needed research in the principles of water occurrence and movement . . . ."[39] At the same time the Survey's program has included investigations of the hydrologic cycle, and has made some important contributions to its description.

In 1950 the Director of the Survey called for a water inventory program which would include "Nation-wide coverage, and the maintenance thereafter of the minimum networks that will accurately record the changes with time of all elements in the hydrologic cycle." He added that "Existing networks of precipitation stations, stream gauging stations, observation wells, evaporation stations, water-quality stations, and stream sediment stations need considerable expansion and adaptation to serve this purpose." Also needed, in his view, was a "program of research on the effects of geologic and other factors that influence the origin, movements, occurrence, accumulation, availability, and other characteristics of water . . . ."[40] While improvements in the Survey's program have come since 1950, the situation described at that time for this key operation in basic data collection remains substantially the same. The Presidential Advisory Committee in 1955, for example, recommended an increase in numbers of stream flow measuring stations by 50 per cent, tripling the number of water quality stations, and quadrupling the number of stations for measuring sedimentation.[41]

[38] Act of December 24, 1942, 56 Stat. 1086, 43 U. S. C. 36b.

[39] W. E. Wrather, "An Accelerated Program for Obtaining Basic Data and Promoting Research on the Nation's Water Resources," in House Document No. 706, 81st Congress, 2nd Session, 1950, p. 81.

[40] *Ibid.*, p. 92.

[41] Presidential Advisory Committee, *op. cit.*, pp. 10-11.

The responsibilities assigned to the Department of Agriculture are important to precise knowledge concerning the hydrologic cycle in the United States. It is within this agency, particularly, that competence should be found to treat the hydrologic effects of land use and the mechanism and course of evapotranspiration. Basic research and data collection within these fields and others are undertaken in several of the bureaus of the Department, including those in the Soil Conservation Service, the Agricultural Research Service, and the Forest Service. Such activities are carried out under authorization of the Soil Conservation and Domestic Allotment Act,[42] the Bankhead-Jones Farm Tenant Act,[43] the Research and Marketing Act of 1946,[44] the Watershed Protection and Flood Prevention Act,[45] and other legislation.[46] The Department also has authority for the conduct of research on long-range weather forecasting, should it care to do so.[47]

The range of interest and activity within the Department of Agriculture is illustrated in the 1955 Yearbook which reports upon the water-related research and operational interests of the Department.[48] While a considerable amount of fundamental research on problems related to the hydrologic cycle is and has been conducted within the Department, the results and administrative orientation of the research tend to be in the direction of the Department's operational responsibilities. Contributions toward the development of scale economies in water development have been incidental, rather than a principal objective of the water-related research activities. The efforts of the Department appear to have been strongest in the fields of land-use effects upon hydrologic events, through the work of the Soil Conservation

[42] Act of April 27, 1935, Sec. 1, 49 Stat. 163, 16 U. S. C. 590a(1).

[43] Act of July 22, 1937, Secs. 31, 47, 50 Stat. 525, 531, as amended, 7 U. S. C. 1010, 1021.

[44] Act of August 14, 1946, 60 Stat. 1082, as amended, 7 U. S. C. 427 *et seq.*

[45] Act of August 4, 1954, Sec. 3, 68 Stat. 566 (Watershed Protection and Flood Prevention Act).

[46] The 1938 and 1944 Flood Control Acts (Act of June 28, 1938, 52 Stat. 1215, 33 U. S. C. 701b; Act of December 22, 1944, 58 Stat. 887, 889, 33 U. S. C. 701a-1) assign basic responsibilities for "investigations of watersheds and measures for run-off and water-flow retardation . . ." to the Department. See also Act of March 2, 1901, 31 Stat. 922, 926; and Act of February 29, 1936, 49 Stat. 1148, as amended, U. S. C. 590g(a), 590 i.

[47] See Reorganization Plan No. IV, Sec. 8, effective June 30, 1940, 54 Stat. 1234, 1236, 5 U. S. C. 133t note following.

Service, the Forest Service,[49] the Bureau of Plant Industry, Soils and Agricultural Engineering, and, indirectly, the agricultural experiment stations scattered through forty-eight states.

The Department of Agriculture appears to have given lesser attention to the problems of evapotranspiration in the hydrologic cycle. The needs for basic evapotranspiration data have had only minimal coverage. House Document No. 706 (81st Congress, 2nd Session) contained the following comments on United States knowledge of evapotranspiration: "The area of greatest deficiency in accumulated data and research in the hydrologic cycle is evapotranspiration . . . . The fragmentary character or unavailability of data on evapotranspiration leaves many areas without the basic data most needed in effective water resources management. . . . Very little has been done to relate these processes to projects of water resources management. . . . The determination of rates of evapotranspiration and of the amounts of water involved in the various processes of returning water to the atmosphere has not been made for any drainage basin. . . . No basis has been established for a comparison of the rates and amounts involved in evaporation and transpiration with the amounts involved in other phases of the hydrologic cycle."[50] These statements should not be interpreted as indicating complete inattention to the subject, for a number of investigators have been actively studying evaporation phenomena for twenty years or more. They have worked within the Soil Conservation Service and other units of the Department of Agriculture, the Bureau of Reclamation, the Geological Survey, and the state agricultural experiment stations.[51] However, most of this research has been highly localized, both in terms of area and in terms of the type of vegetative covering treated.[52]

It is of some interest that the data gap on evapotranspiration has received effective attention by private investigators. Noteworthy among

[48] U. S. Department of Agriculture, *Water, The Yearbook of Agriculture, 1955,* Washington, pp. 95-102, 144-51, 151-59, and 228-34.

[49] The conduct of the forest land-use experiments by the Forest Service at Coweeta, N. C., are notable examples of this type of work.

[50] *A Program to Strengthen the Scientific Foundation in Natural Resources,* House Document No. 706, pp. 62, 64.

[51] See U. S. Geological Survey, *Principal Federal Sources of Hydrologic Data* (Notes on Hydrologic Activities, Bulletin 8, Vol. 1, Inter-agency Committee on Water Resources, Subcommittee on Hydrology, Washington, 1956), pp. 2-3 for an account of research interests of the several federal agencies in this field, and of research in progress or completed by 1956.

[52] See chap. 16, *n* [54] for further references on this subject.

these has been the research undertaken by Thornthwaite Associates, both as to the processes of evapotranspiration, and the accumulation of data on local soil moisture balances.[53]

One important aspect of basic data coverage and interpretation is the regionalization of the data, in view of the definite regional (or river-basin) characteristics of water differences. It is obvious that the foregoing authorizations and activities have been organized on a national or interregional basis. Integration of all hydrologic data on a regional basis is done by the Soil Conservation Service for the small watersheds, or parts thereof, as they are investigated or treated in its program. It also has been undertaken by the Corps of Engineers for the Mississippi River and its tributaries on a continuing basis,[54] and in the "308" reports, which required the interpretation of all the available stream flow data. Regional integrations of data for its specific operational objectives have been made by the Bureau of Reclamation. A relatively broad authorization for this purpose is included within the 1939 Reclamation Project Act.[55] However, the broadest authorization for regional data collection, collation, and interpretation is contained in the TVA Act.[56] Under it, TVA has conducted both research and data collection which have produced one of the best-rounded physical research programs for any individual river basin in the country. The carefully planned hydrologic observation coverage and the field studies of the hydrologic effects of different types of land use have been noteworthy.

In addition to data interpretations on a larger regional basis, both collection and interpretation are needed on tributary basins. Some work of this type has been done in the tributary watersheds program of TVA, like that on the Parker Branch, North Carolina, and the Chestuee and Beech River watersheds, Tennessee. Of wider geographical but more narrow functional significance have been the research and data interpre-

---

[53] Some of Dr. Thornthwaite's work (see chap. 16, pp. 445-51) was commenced while he was an employee of the Soil Conservation Service, and he since has acted as consultant to the Department of Agriculture on these matters.

[54] R. S. Sec. 5252, from Res. of February 21, 1871, No. 40, 16 Stat. 598, as amended, 33 U. S. C. 4.

[55] Act of August 4, 1939, Secs. 9(a), 9(b), 53 Stat. 1187, 1193, 1194, 43 U. S. C. 485h(a), 485(b).

[56] Act of May 18, 1933, Sec. 22. "The President is hereby authorized, . . . to make such surveys of . . . said Tennessee Basin and adjoining territory as may be useful . . . in guiding . . . development . . . for the general purpose of . . . physical, economic and social development of said areas." (In practice delegated to TVA.)

tations undertaken under the Soil Conservation Service small watershed program, and other independent programs, like the work on the Umpqua River undertaken by Douglas County (Oregon) in co-operation with the United States Weather Bureau and the Geological Survey.[57] Such coverage, however, as yet applies only to a small percentage of the land surface of the nation.

Data on the dynamic aspects of the economic and demographic environment of water development. An understanding of the economic and social environment within which development is to take place is as essential to effective use of the scale economies as good physical data. Data on the trends of migration and population growth, the nature of industrial growth and migration, pertinent social trends (like shorter work hours), agricultural production and consumption trends, trends in energy consumption, the agricultural-services-industry balance in employment, the structure of political jurisdiction trends in international trade, and other matters are essential information for a well-planned development of large scale.

There are, of course, a great number of agencies within the United States collecting information of this kind. Such data have many other uses than water development planning. Each of the federal departments treating civil affairs within the United States collects such information, especially the Department of Commerce (which includes the Bureau of the Census), the Department of Agriculture, the Department of the Interior, the Department of Labor, the Department of Health, Education and Welfare, and a number of the independent offices and establishments (like the Federal Power Commission, the Federal Trade Commission, the Federal Reserve Board, the Council of Economic Advisers, and others). In addition, large amounts of data are collected and published through the efforts of public agencies in the forty-nine states, whose responsibilities parallel those of federal agencies within their territory of jurisdiction. There is added the significant and voluminous research and publication undertaken by university faculties, professional associations, research institutes with public objectives, and the work of privately owned utilities, other private industry, and trade associations. The volume of raw data on the economic and social environment within the United States is tremendous.

The question arises as to whether or not this great mass of economic, social, politico-geographic, and administrative data is adequate for the

---

[57] Charles McKinley, personal communication to the authors, 1957.

needs of water development planning. In sheer volume it may be more than enough, even though mostly collected with other purposes in view. However, an important difficulty arises in the use of this data, precisely because it is collected for other purposes.

To be of most value in water development planning, economic and social data must be organized and interpreted on a regional basis. As noted previously, water phenomena exhibit a high degree of regional differentiation. An understanding of the regional complexion in terms of its demography, industrial dynamics, agricultural trends, social welfare trends, employment characteristics, and the structure of political jurisdiction is essential to the projections on which sound planning must be based. Often data on a subregional scale also are needed.

Within recent years almost every federal water project or program statement has included some analysis of the economic, social, and administrative environment within the project or program proposed for operation. Thus the Department of Interior's Bureau of Reclamation included fairly elaborate background analyses in its project investigations for the Central Valley Project and the Columbia Basin Project.[58] Such analysis also was part of the recently authorized Upper Colorado River Basin Project.[59] The Corps of Engineers has undertaken the compilation of specific economic data on a river-basin pattern in connection with its "308" surveys [60] and, in a more limited manner, for its navigation projects [61] and in other connections. The Federal Power Commission has had authority since 1920 to make continuing studies of the relation of water resource use to industry, commerce, and electric power markets.[62] It also has broad investigative power to collect information necessary to its regulatory responsibilities.[63] The Tennessee Valley Authority Act contains probably the broadest authorizations for the compilation and interpretation of economic and social data on a regional basis. This Act empowers the President "to make such surveys of and general plans for [the] Tennessee basin and adjoining territory as may be useful . . .

[58] U. S. Department of Interior, Bureau of Reclamation, *Central Valley Basin,* Washington, 1949; *idem, The Columbia River,* 1947, p. 398.

[59] *Colorado River Storage Project,* House Document No. 364, 83rd Congress, 2nd Session (1954).

[60] House Document 308, 69th Congress, 1st Session (1926), p. 2.

[61] For instance, see Act of June 13, 1902, Sec. 3, 32 Stat. 331, 372, as amended, 33 U. S. C. 541.

[62] Act of June 10, 1920, Sec. 4 (a), 41 Stat. 1063, 1065, as amended, 16 U. S. C. 797(a).

[63] Act of August 26, 1935, Sec. 311, 49 Stat. 838, 859, 16 U. S. C. 825j.

in guiding and controlling the extent, sequency, and nature of development . . . for the general purpose of fostering an orderly and proper physical, economic, and social development of said areas."[64] This would seem to give ample latitude for the collection of basic data necessary in the effective development planning for this particular basin and the surrounding region. Such authorization is not paralleled on a continuing basis for economic and social data on any other region of the country. However, during the period 1934-43 the surveys and staff work of the National Resources Board, the National Resources Committee, and the National Resources Planning Board did provide a great deal of basic information of the sort here described. During this period an effort was made to collect adequate information on the regional, social, and economic environment within which water development was to be undertaken. The Board was discontinued in 1943.

Besides the authorized information gathering and interpretation of federal government agencies, mention should be made of the voluntary organization which has operated through the Interagency Committee on Water Resources in Washington, D. C., and its field committees like those on the Columbia and on the Missouri basins.[65] For their respective areas there is no doubt that the collection and interpretation of economic, social, and administrative data has been facilitated by the activity of these committees, even though it has been on a voluntary basis. Results from these committees, however, have depended upon the degree of enthusiasm and the extent of staff provided by the component federal agencies. During the period following the Second World War, for instance, the vigor of the Columbia Basin Interagency Committee owed much to enthusiastic support by the Department of the Interior for the field committee method, and to the staff and interests of the Bonneville Power Administration. The support of the divisional organization of the Corps and of the Department of Agriculture also were needed to achieve results from these committees, which were without staffs of their own. The annual regional program reports of the Department of the Interior, which relate departmental plans for resource development to regional economic trends, record the results obtainable from this type of organization. Those issued prior to 1953 are of special interest in this connection.

[64] Act of May 18, 1933, Sec. 22, 48 Stat. 69.
[65] The work of the Interagency Committees in part succeeded that of the regional committees of the National Resources Planning Board.

The activities of the private utilities within their respective service areas also deserve notation. Most utilities have effective means of collecting and interpreting data of use in promoting the economic development of their service areas. Since most of the utility service areas geographically are smaller than the areas covered by the scale economies in water development, only a few of these efforts are effective support for the scale economies in water development.

It must be admitted that, with the sole exception of the Tennessee Valley, the organization of research and the interpretation of data needed in planning the scale economies is deficient in the present United States approach to water development. One of the principal shortcomings is a lack of any provision for systematic, professional analysis of the regions of the United States. This is important because water, as previously observed, shows regional patterns of occurrence, and its full development must be organized on a regional basis.

Development consistently must contend with different types of regions which rarely have limits that coincide. A river basin is essentially a surface-water or run-off region. Other important physical aspects of water development have differing regions, like ground-water regions, precipitation regions, or water-balance regions. Economic or administrative regions almost never coincide with any of the physical regions, nor with each other. Perhaps the most discontinuous with the basic physical regions are the political administrative regions, or the states. The problems raised by state boundaries are now well recognized, but the articulation of the state administrative region with the other regions which make up the complex of regions concerned in water development is far from being adequately understood, or continuously treated.

The economic regions also are very important. More and more we are recognizing them as "nodal" regions oriented to the metropolitan centers. They do not correspond either to the state jurisdictional boundaries, or to the physical regions which are the engineers' beginning to water development.

It is obvious that effective planning for use of the scale economies in development needs an assessment of the different physical and economic forces stemming from the different types of region—forces which are related, but do not coincide geographically. Effective use of the scale economies calls for knowledge of the geographical extent of the different types of region, their dynamic aspects, and knowledge of the relations among the different regions. Except for the legislative authorization in the case of the Tennessee Valley, there is no organized means for providing such data and interpretations for United States water development at the present time, except on a special-project basis. This latter course notably was taken in the surveys made of the New England-New York area and the Arkansas-White-Red basins.[66] The studies of the state of Texas, sponsored by the Bureau of Reclamation and undertaken by the University of Texas, are of the same nature. The same course also now is being applied to problems of planning for further development of the Delaware River Basin in the eastern states. Permanent staffs working on general economic analysis for water development are relatively small.

All efforts of this nature have been handicapped by the fact that the economic and social data collected by the Bureau of the Census are collected and compiled in geographical forms which generally do not fit the limits of river basins. Where the political subdivisions used by the Census are territorially extensive, they present formidable problems of interpretation for the planner.

In 1957-58 a number of indications toward eventually improved treatment of the socio-economic data problem could be seen. The Bureau of the Census recently took a significant step in commencing to report industrial water use on a regional basis which is reasonably compatible with river basin boundaries. (See chapter 16, table 31 herein.) The techniques of regional study are rapidly being improved, and for the first time a sustained interdisciplinary interest in regional studies is evident.[67] Scholarly and professional analysis continues on the problems of specifically relating socio-economic data collection and interpretation, regional study, and organization for research and planning, to water development and water facilities operation.[68] The general

[66] Arkansas-White-Red Basins Inter-Agency Committee, *Arkansas-White-Red River Basins* (17 vols.), 1955; and New England–New York Inter-Agency Committee, *The Resources of the New England–New York Region* (39 vols.), 1955. The Northeastern Resources Committee, a joint state-federal agency voluntary committee, is continuing examination of the development problems of the New England area.

[67] Particularly in the activities of the professional students participating in the Regional Science Association. See for example, Regional Science Association, *Papers and Proceedings*, Vol. 3, Philadelphia, 1957 and later volumes.

[68] See for example, Joseph L. Fisher, "Notes on the Regional Analysis and Programming of Resources Developments," *Water Resources and Economic Development of the West*, Conference Proceedings Committee on the Economics of Water Resources Development of the Western Agricultural Economics Research Council, Report No. 4, Pullman, Wash., 1955, pp. 1-18; and Roy Bessey, "Needs and Alternatives in Regional Organization for an Integrated Development of Water Resources," *ibid.*, pp. 61-80.

acceptance of planning methods which demand use of more comprehensive and more penetrating socio-economic study than in the past would seem inevitable.

Data on the interrelation of the several purposes in development. To those who view multiple-purpose development as a recent phenomenon, the United States has had a surprisingly long history of concern for the relation of the different purposes in water development.[69] The first statutory concerns expressed for the effect of one type of development upon another related to navigation. The 1890 River and Harbor Act prohibited the construction of any barrier to the navigable capacity of waters under United States jurisdiction.[70] Administration was placed under the Secretary of War (now Secretary of the Army), with the implied powers of collecting data to determine the effect of any proposed structure on navigation. This was especially true after the 1899 decision of the Supreme Court which affirmed that the control applied not only to navigable waters, but "to any obstruction to navigable capacity, wherever or however done."[71] This was followed by further provision for considering the interest of navigation in the Act of March 3, 1899,[72] and in the 1906 General Dam Act.[73] The latter provided for the installation of navigation facilities at hydro power dams.

The General Dam Act of 1906 also provided for the installation of fish-passage facilities at hydro power dams, in which the interest of the United States was to be the concern of the Secretary of Commerce.[74] In the same year Congress provided for the consideration of hydro power development in connection with Reclamation Act irrigation projects. [75] Each of these duties had the implied obligation of collecting and interpreting data on the interrelation of the specified water development purposes. In 1909 Congress provided for consideration of the "development and utilization of water power for industrial and commercial purposes," and "other subjects as may be properly connected

[69] See pp. 185-94 and 235-50 for discussions of two techniques of interrelating different functions in development and operation.

[70] President's Water Resources Policy Commission, *Water Resources Law* (Vol. 3 of Report), 1951, p. 390.

[71] *Ibid.*

[72] Sec. 10, 30 Stat. 1121, 1151, 33 U. S. C. 403.

[73] Act of June 21, 1906, 34 Stat. 386.

[74] President's Water Resources Policy Commission, *Water Resources Law* (Vol. 3 of Report), p. 391.

[75] Act of April 16, 1906, Sec. 5, 34 Stat. 116, 117, 43 U. S. C. 522.

with such project" (principally limited to navigation projects).[76] The 1910 General Dam Act mentioned for the first time "a comprehensive plan for the improvement of [a] waterway . . . with a view to . . . the full development of water power."[77] The relation of other purposes in development of flood control was recognized first in 1917 flood control legislation, which provided that "all examinations and surveys of projects relating to flood control shall include a comprehensive study of the watershed," and that reports thereon shall include data on the relations of "such other uses as may be properly related to or coordinated with the project."[78] In the directions given to the Corps of Engineers in 1925 for the prosecution of the "308 reports," Congress directed that these surveys be made with the objective of improving streams "for purposes of navigation . . . in combination with development for power, flood control, and irrigation."[79] Thus the need for recognizing interrelations among the several purposes in development is clearly recognized in the instructions for these reports. Interrelation of power with other purposes also was recognized in the Federal Power Act of 1920, and in the Flood Control Act of 1938, with its penstocks provision.[80] The interrelation of several purposes with flood control was recognized in the Flood Control Act of 1944[81] and later years, where recreation, power, and irrigation are mentioned. Similar recognition was given in the Reclamation Project Act of 1939 to flood control, power, navigation, municipal water supply, and "other miscellaneous purposes"[82] where irrigation was the primary objective.

Although fishery interests had been recognized in public construction and in public regulation of private construction on streams in some way since 1888,[83] the first recognition of the effects of water development upon wildlife was in 1938.[84] At that time it was specified that Corps of Engineers investigations and improvements of waterways must include

---

[76] Act of March 3, 1909, Sec. 13, 35 Stat. 815, 822.

[77] Act of June 23, 1910, Sec. 1, 36 Stat. 593.

[78] Act of March 1, 1917, Sec. 3, 39 Stat. 950, 33 U. S. C. 701.

[79] President's Water Resources Policy Commission, *Water Resources Law* (Vol. 3 of Report), p. 409. Also Act of March 3, 1925, Sec. 3, 43 Stat. 1186, 1190.

[80] Act of June 28, 1938, Sec. 4, 52 Stat. 1215, 1216, 33 U. S. C. 701j.

[81] Act of December 22, 1944, Sec. 1, 58 Stat. 887.

[82] Act of August 4, 1939, Sec. 9(a), 53 Stat. 1187, 1193, 43 U. S. C. 485h (a).

[83] Act of August 11, 1888, Sec. 11, 25 Stat. 400, 425, 33 U. S. C. 608.

[84] Act of June 20, 1938, Sec. 1, 52 Stat. 802, 33 U. S. C. 540.

"a due regard for wildlife conservation."[85] Particularly since 1946, fish and wildlife have been among the most consistently treated subjects in water development, as far as developmental relations are concerned. The Act of August 14, 1946 gave the Fish and Wildlife Service broad powers of consultation where alteration of natural stream conditions is contemplated.[86] The basic authorizations of the 1946 Act have been supplemented since by specific legislation toward the same end. Thus the Small Reclamation Projects Act of 1956 includes a section confirming the application of the 1946 legislation to future small reclamation projects.[87] Another important law with specific application is the Colorado River Storage Project Act of 1956, which contains broad authorization for mitigation of losses and improvement of conditions for fish and wildlife as the development of the Colorado Basin proceeds further. [88]

Similarly, but with a view toward determining the effects of one function upon all other development purposes, the United States Public Health Service in 1948 was given broad authorization for the investigation of the relation of stream waste-carrying to other stream uses, including "public water supplies, propagation of fish and aquatic life, recreational purposes, and agricultural, industrial, and other legitimate uses."[89]

At least two further authorizations are of interest in the interrelation of multiple purposes. The 1939 Reclamation Act authorizes the Secretary of Interior to enter into contracts to furnish water for "municipal water supply or miscellaneous purposes."[90] The 1944 Flood Control Act grants a somewhat similar authorization to the Secretary of the Army for sale of surplus water for domestic and industrial purposes.[91] The interrelation of domestic and industrial water supply thus was entered into the official process of planning for water development. The 1944 Flood Control Act in addition authorizes the Chief of Engineers

[85] President's Water Resources Policy Commission, *Water Resources Law* (Vol. 3 of Report), p. 328.

[86] 60 Stat. 1080, 16 U. S. C. 661 *et seq.*

[87] Act of August 6, 1956, Sec. 8, 70 Stat. 1044. 43 U. S. C. Suppl. IV 422a-422k. The first loan to be made under this act (Cameron County Water Control and Improvement District No 1, Texas) was approved in March 1958.

[88] Act of April 11, 1956, Sec. 8, 70 Stat. 105. 43 U. S. C., Suppl. IV, 620-620o.

[89] Act of June 30, 1948, Sec. 2 (a), 62 Stat. 1155, 33 U. S. C. 466a(a) (Suppl. III).

[90] Act of August 4, 1939, Sec. 9 (c), 53 Stat. 1187, 1194, 43 U. S. C. 485h(c).

[91] Act of December 22, 1944, Sec. 6, 58 Stat. 887, 890, 33 U. S. C. 708.

to construct, maintain, and operate recreational facilities related to the reservoirs that may be developed by the Corps.[92] The National Park Service has co-operated with the Corps in the operation of some of these facilities. Furthermore, it has been the practice in recent years, as needed, to authorize and appropriate for a number of related data collection or research activities concerned with water development, through the Corps or the Bureau of Reclamation programs. This has been notable in the basin programs like the Missouri projects, where a variety of such activities have been supported through the major agency appropriations. Important parts of the Geological Survey programs in this field, for example, have been supported in this manner.

The authorizations described are not all that have been issued for the use of federal agencies in water development. Others relate to stockwatering supply, water quality, sediment and salinity control, and still other items. Important enabling legislation for basic data collection and analysis on water quality was contained in the 1956 Water Pollution Control Act. [93] A program for the development of continuing analytical data on sanitary water quality for major United States streams was started in 1957, following this legislation.[94] In addition some data collection and research operations are carried on by the states within their territories, and by privately maintained associations like the Izaak Walton League. The states have been particularly active in the analysis of waste-carrying effects on stream quality. However, these illustrations suffice to show the fragmented nature of responsibilities and interests in this aspect of research and data collection for large-scale development. While each of the statutes mentioned has had a beneficial effect upon the support of analysis, each also reflects a special-purpose point of view. It is a reasonable conclusion that the achievement of a truly comprehensive data background for any given basin or area under this system presents formidable problems of *ad hoc* organization. This has been provided for some of the large basins in the recent past through the creation of voluntary federal interagency committees and associated interstate organizations. The New England–New York and the Arkansas-White-Red Interagency surveys did stress interrelations on a broad basis. An exception again has been the Tennessee Valley. With the broad authorization contained in the 1933 Act, TVA has been able to under-

[92] *Ibid.*, Sec. 4, 58 Stat. 887, 889, as amended, 16 U. S. C. 460d.
[93] Act of July 9, 1956, 70 Stat. 498.
[94] Office of Statistical Standards, U. S. Bureau of the Budget, *Statistical Reporter,* October 1957, p. 174.

take a variety of essential studies of the interrelations of multiple water uses on a continuing basis. Out of these studies have come extensive sets of data on the hydrology and biology of waters under extensive artificial regulation.[95]

The need for recognizing interrelations among development purposes, and providing means of analyzing them, has also affected the organization of the special-purpose agencies. The civil functions professional staff of the Corps of Engineers now includes specialists on fisheries, recreation, industrial water supply, and electric power development, along with personnel to treat its more traditional responsibilities in navigation, flood control, and harbor construction and maintenance. The Bureau of Reclamation has shown a similar trend in broadening its professional outlook, as has the Department of Agriculture. Better organizational provision for this aspect of research related to the scale economies thus has come into existence. Both agencies also have had the co-operation of specialized agencies, like the Fish and Wildlife Service, the Forest Service, the National Park Service, and other agencies, in these tasks. However, any plan for the development of a given basin still must contend with an intricate and time-consuming operation to obtain all of the information on multipurpose relations which is demanded in the best use of the scale economies.

THE ORGANIZATION OF BASIN-WIDE OR REGION-WIDE PLANNING

Two further aspects of planning and the scale economies remain for discussion. These are: (1) organizational provision for basin-wide, or region-wide planning; and (2) interbasin planning or planning for the transfer of water, or the products or services thereof, from one basin to another. The organizational background for both has been forecast somewhat in the discussion of research and data collection, for the nature of planning organization inevitably is foretold by provision made for the research on which planning must always be based. This suggests that United States organization in this direction is far from complete.

As has been noted in the discussion of research and data collection related to planning, the ideal in planning large-scale development is

[95] Some publications on this research are listed in *A Bibliography for the TVA Program*, Knoxville, 1952. Extensive unpublished data also have been collected and used in the design and operation of the TVA program.

treatment of a hydrologic unit within the context of the economic region and the administrative units that cover the same area. Achievement of this ideal has not been approached for the country as a whole. However, steps have been taken in this direction under a number of programs. The reach of these steps, the difficulties confronting such planning, and the possibility of succeeding steps for basin-wide planning of water development are the subjects of this section. Two subjects must be considered: (1) planning of water control, water regulation, and the production of water-derived services; and (2) the disposal of water-derived services, illustrated by the planning of electric power distribution.

Administrative organization to plan for water development on a basin-wide basis dates at least from the creation of the Mississippi River Commission in 1879.[96] The Powell surveys of irrigation in western United States soon afterward added further thought and experiment with a basin-wide approach, although still from the point of view of a single purpose or a few purposes.[97] The ideas of the Powell surveys were followed to a degree in the provisions made in the Reclamation Act of 1902.[98] The latter, of course, was part of the program of President Theodore Roosevelt, who was much interested in the comprehensive development of water resources, and under whose administration many useful ideas for the organization of water development first were advanced. Further steps toward basin-wide planning were taken in the Federal Power Act of 1920 (see above, page 477), and in the directions for the Corps of Engineers "308" reports, notably those relating to Mississippi River flood control.[99] The Tennessee Valley Authority Act followed in 1933. This Act contained probably the most workable authorizations for integrated basin planning produced in this country, even to this day. The Tennessee has been planned and developed on a completely integrated, basin-wide basis from the point of view of hydraulics and land uses related to river development.

Since the 1930's the plans of the major federal agencies, and even of some state agencies (e.g., the State of California Water Resources

---

[96] Act of June 28, 1879, 21 Stat. 37, as amended, 33 U. S. C. 647. See also President's Water Resources Policy Commission, *Water Resources Law* (Vol. 3 of Report), p. 98.

[97] Wallace Stegner, *Beyond the Hundredth Meridian* (Boston: Little, Brown, 1954), p. 438.

[98] Act of June 17, 1902, 32 Stat. 388, 43 U. S. C. 391 *et seq.*

[99] Act of May 15, 1928, 45 Stat. 534, 33 U. S. C. 702-702m.

Board, and the current California Department of Water Resources), have tended to approach a basin-wide framework.[100] However, as late as 1943 both of the major centralized agencies, the Corps of Engineers and the Bureau of Reclamation, presented plans for the development of the Missouri River which were not basin wide.[101]

Even in 1958, basin-wide planning had not been achieved on a major international stream, the Columbia, despite an intensive survey of, and considerable construction on, the United States section of the main stream and its tributaries. This failing, however, was due more to the international character of the stream than any other reason. The 1948 review report on the basin came close to presenting a comprehensive plan for the *American section* of the basin.[102] By 1957 the Missouri Basin plans had been co-ordinated; the Colorado Basin had been investigated and planned for as a unit by the Bureau of Reclamation;[103] the Central Valley of California had been planned for and developed on a geographically integrated basis, as well as the above mentioned works in the Tennessee Valley. From the days of the 1907-18 Inland Waterways Commission[104] intermittent intellectual pressure advanced the basin-wide planning idea, and pushed it more and more into the everyday practice of the major development agencies. Included were the activities of the National Resources Board, the National Resources Committee, the National Resources Planning Board (the three agencies 1934-43), the President's Water Resources Policy Commission (1950), the first Hoover Commission (1948-49), the Missouri Basin Survey Commission (1951-53), and the President's Advisory Committee on Water Resources (1955-56). However, as of 1958 the

[100] As distinguished from a comprehensive framework. Not all basin-wide planning is comprehensive. See below, pp. 501-6.

[101] See Commission on the Organization of the Executive Branch, *Report of the Natural Resources Task Force* (Washington: U. S. Government Printing Office, 1949), p. 244.

[102] Recent Canadian appraisal and planning may eventually lead to a fully comprehensive program for the basin, although several difficult politico-economic problems appear to stand in the way, as of 1958. The United States program also was in process of a new review in 1958, with the original "308" survey under complete restudy.

[103] Because important areas of demand for Colorado River water lie outside that basin, even integrated basin-wide planning may be inadequate for the Colorado.

[104] President Theodore Roosevelt in transmitting the report of the Inland Waterways Commission to Congress said: "Each river system, from its headwaters in the forest to its mouth on the coast, is a single unit and should be treated as such." (Senate Doc. No. 325, 60th Cong., 1st Session, p. iv.)

two major federal development agencies did not have general statutory directives to prepare plans for full development of river basins. Only TVA has such a directive in its Act; the Corps of Engineers, the Bureau of Reclamation, and even the President of the United States do not. Working agreements among the agencies for co-ordinated planning thus have had no statutory support.[105]

It is noteworthy that even some private development was being planned in 1956 on a basin-wide basis, as in the case of the Coosa River plan of the Alabama Power Company. Privately managed development on the Wisconsin River previously had been undertaken with multiple-purpose objectives. Where possible, as in the state of California, some state plans long had reflected acceptance of basin-wide treatment. State attention to basin-wide planning was given early in the state of Ohio, where the Miami and the Muskingum conservancy districts operated on basin-wide plans.

The principle of the hydraulic unity of a river basin therefore seemed accepted, and a real effort was being made in the federal government to develop administrative organization for basin-wide planning of river regulation. Except for the Tennessee Valley Authority, this was mainly within the administrative framework of the several responsible centralized agencies, and through the interagency river basin committees, like those on the Missouri and the Columbia. Much of the initiative had been taken by, and presumably still rested with, the federal government, because the jurisdictions of the forty-eight states in few individual instances covered an entire river basin of importance. An interesting experimental exception in 1958 was the Delaware River Basin Advisory Committee, an interstate agency[106] devoted to advisory assistance in planning for further development of that river.

Four noteworthy problems remained in 1958 for the organization of water development planning on a geographically integrated basis.

1) The technique of relating the physical planning of river regulation to the economic environment within which the stream is located was little more than elementary. Even in the case of Tennessee Valley Authority planning, the full economic framework within which physical development occurred was never completely analyzed or made part of the planning process of the agency. This was more true of most other regions, with the exception of the Pacific Northwest and the possible

---

[105] Charles McKinley, letter to the authors, December 1957.

[106] Represented on the Committee: states of New York, New Jersey, Pennsylvania, and Delaware; cities of New York and Philadelphia.

exception of California. Interesting attempts in this direction were made
in the New England–New York, and the Arkansas-White-Red
surveys, authorized in the Flood Control Act of 1950[107] and completed in 1955. While each contributed useful information for the
planning of works within the two regions, both were handicapped by
lack of data on the economic articulation among the regions of the
United States and Canada. The same may be said for the equally interesting efforts of the Columbia Basin Interagency Committee in the
Pacific Northwest.[108] Further research on the nature of regions in the
United States will be needed before this part of basin planning can be
added as a useful part of the process.

2) Where several programs are administered by as many agencies
for the same river basin, as is the rule in the United States, there is difficulty in organizing a joint program which maximizes benefits and
minimizes costs. The compromises achieved through the device of the
interagency committee do not yet give assurance of the best comprehensive development. This has been very clearly described by an experienced participant in interagency committee proceedings:

> The "voluntary" federal interagency committees and their as-
> sociated interstate organizations have been handicapped because they
> always started out with plans which had been made by still other
> agencies, usually federal, which had less than a comprehensive, basin-
> wide orientation. The programs of the various agencies have all been
> oriented differently, without a central core to hold them together, and
> without any central guidance. The tug-and-haul among the various
> agencies and their individual plans has failed to weigh adequately the
> relative merits of the various purposes and uses; consequently, some
> received little consideration or evaluation. In Missouri River basin
> planning, for instance, navigation has received rather high priority
> without any evaluation of its worth as contrasted with that of hydro-
> electric power. Power was not protected by an O'Mahoney-Millikin
> Amendment.
>
> These committees are "voluntary" only in the sense that they were
> not created by law. Membership is not voluntary on the part of the
> members; they are told to serve. Neither the chairmen nor other
> members of these "voluntary" federal committees can afford to take
> an objective view of all the resources of the basin and try to plan for
> the maximizing of benefits from the development of all the water (and
> related) resources for the good of the populace. Each man is em-
> ployed and paid by a special-interest agency with limited authoriza-

[107] Act of May 17, 1950, Sec. 205, 64 Stat. 163.
[108] See Columbia Basin Interagency Committee, *Plan for Development of
Natural Resources of the Pacific Northwest*, Portland, Ore., 1952.

tions and functions far short of optimum, comprehensive development.[109]

3) Absence of statutory directive for full development to the major federal planning and construction agencies often predisposes these agencies to special points of view in planning and may lead to difficulties of co-ordination. In the absence of such directives from the Congress the Federal Power Commission, through its licensing function, can become a key agency in determining the character of a given basin plan. On occasion the Commission has done so, although it is mainly a regulatory, and not a development agency. The statutory situation thus can contribute to a confusion of objectives.

4) A fourth difficulty concerns an understanding of the size of basin to which the scale economies apply, and provision for the planned treatment of river basins of several levels of size. Development within the United States presents greater problems for use of the scale economies than in many other countries because of the existence of the great Mississippi Basin. Hydraulically and economically, this great valley has some elements of unity deserving of careful attention within a basin-wide framework. Yet the only noteworthy attempt to view the basin as a whole planning unit came under the Mississippi Valley Committee of the Public Works Administration in 1934.[110] Recent evidence of the effects of tributary storage on the quality of water in the lower Mississippi suggests that there may be many profitable opportunities in planning for the whole Mississippi Basin.[111] The only consistent basin-wide planning within the Mississippi system has been for the major tributaries, like the Tennessee and the Missouri. At the same time, problems have arisen which suggest that a unit the size of the Missouri may have been too large for the planning techniques used on it without being supplemented by tributary basin planning. The yet unsolved problems of the Kansas River Basin offer evidence that the application of basin-wide planning must make allowance for planning rivers at

[109] Personal communication to the authors, 1957.

[110] U. S. Public Works Administration, Mississippi Valley Committee, *Report of the Mississippi Valley Committee of the Public Works Administration* (Washington: U. S. Government Printing Office, 1934).

[111] See testimony of Generals Hardin and Potter on the control of salinity in the Mississippi River at New Orleans (House of Representatives Committee on Government Operations, *Hearings before a special subcommittee of the Committee,* re: Commission on Organization of the Executive Branch of the Government, Water Resources and Power Report, 84th Congress, 1st Session, Washington, 1956, pp. 1037 ff).

several levels of basin integration.[112] The relation of the Snake River Basin to the remainder of the Columbia Basin in the Pacific Northwest, an indirect issue in the Hells Canyon case, also suggests the need for multilevel treatment of the large river basins. However, by far the most challenging and difficult problem of this kind is the treatment of the Mississippi Basin. Integrated treatment of other basins is simple by comparison. For all large basins the important principle is one of balance: planning must be undertaken for the larger unit, but that planning must not be mistaken as adequate for the tributaries also. In fact, different types of planning and development may be associated with basins of different size levels. For example, the details of basic data, and interrelated land-use planning are much greater for the small tributary than for the large basin.

The appropriate geographical unit for planning may differ by function. While most water control functions are inevitably treated by watershed, the watershed unit may not always be the best for land treatment programs, and it is almost certain not to correspond to the most effective service area for a power system. Thus for a large river basin there is a question not only as to the relation of tributary basin planning to the whole, but also the reconciliation of the appropriate areas for the several purposes served in development. Whatever the technique or practice of delimiting planning areas, however, it is essential to have a broadly conceived plan for a basin as a whole. Such a plan is always needed as a basis for judging the compatibility of more detailed sub-basin plans, or related special-purpose plans.

Regional or basin-wide integration of water control or water regulation planning is more advanced in the United States than is the regional organization of planning the distribution of products or services. Yet it is obvious that the scale economies can apply also to the distribution of these services. The subject has perhaps received most attention in the organization of systems for the distribution of electricity, where the scale economies are readily realizable in the operation of a large system with diverse load and diverse hydraulic characteristics. (See chapter 9, pages 259-66; and chapter 12, pages 298-319.) The value of such scale organization was seen relatively early, as was illustrated in the "giant power" movement (*circa* 1910-25).[113] The utility holding com-

---

[112] University of Kansas, *op cit., The Kansas Basin, Pilot Study of a Watershed.*

[113]See W. S. Murray *et al., A Superpower System for the Region between Boston and Washington* (U. S. Geological Survey Professional Paper 123), Washington, 1921.

pany operations of the twenties were a reflection of the possibilities of scale economies in private utility enterprise. However, a few of the early envisioned possibilities have been realized, although "giant power" possibilities were under restudy in some quarters in 1958.[114]

In part because of the history of privately owned electric utility growth, and in part because of the consequences of the Public Utility Holding Company Act of 1935,[115] a large majority of the operating electric utility systems in the United States do not have extensive enough service areas to take advantage of the scale economies possible from integrated operation within a geographical area of suitable size for such integration. Some of the privately owned systems, like the Pacific Gas and Electric Company in California, are exceptions, however. (See figure 38, map, "United States Electric Power Utility Service Areas," map pocket.) Sixty per cent of all United States utility operating companies were reported in 1957 to have individual peak loads of less than 125,000 kilowatts. Only 14 per cent were reported to have peak loads in excess of 500,000 kilowatts.[116] Relatively few systems therefore were individually suited to the construction and use of large-scale hydro or steam generating facilities.

Among the publicly owned systems, the Tennessee Valley Authority system has managed to achieve a geographical size and extent of load which permits it to realize scale economies. Even though its hydro sites are not particularly low-cost by comparison with sites in other parts of the country, TVA has low delivered costs on its electricity partly because of the scale economies realized in a large system. To a lesser degree and under more difficult service demands, the Bonneville Power Administration's system in the Pacific Northwest also permitted the realization of some scale economies in its planning and operation.

[114] E.g. Leland Olds, "Giant Power in the Legislative Program," paper presented at National Rural Electric Cooperative Association Annual Meeting, Dallas, Texas, 1958; *idem,* "Giant Power for the American People," address before Western States Water and Power Conference, Salt Lake City, Utah, May 11, 1958.

[115] Act of August 26, 1935, 49 Stat. 303. A number of electric utility companies were under holding company control previous to 1935. Although some companies, like American Gas and Electric Corporation, have chosen to remain under jurisdiction of the Act, most of them have divested themselves of holdings in such a way that former subsidiaries now are exempted from federal regulation, and are entirely under state jurisdiction. Since state boundaries make no more logical electric service areas than they do water control regions, the optimum geographical extent of the service area is achieved only accidentally.

[116] National Association of Electric Companies, *Legislative Abstract,* Vol. II, No. 16, October 28, 1937, p. 1.

This was supplemented during the second world war by the organization of the Northwest Power Pool. The Pool has continued, and still includes all major utilities in the northwestern states. Under private direction, the New England Electric System has attempted to do for New England somewhat the same as the Bonneville Power Administration did in the Pacific Northwest.

Congress in 1935 assigned to the Federal Power Commission the responsibility of dividing the country into regions for voluntary interconnection and co-ordination of electric power transmission, production, and consumption.[117] This is "for the purpose of assuring an abundant supply of electric energy throughout the United States with the greatest possible economy and with regard to the proper utilization and conservation of natural resources. . . ." Within and between these regions it is the "duty of the Commission to promote and encourage such interconnection and co-ordination."[118] Under certain limited circumstances the Commission may even order interconnections. The progress achieved under these provisions has not been reported upon by the Commission recently. As of 1950 little had been done.[119] The slow progress in part may be ascribed to the unwillingness of a number of private utilities now exempt from Federal Power Commission regulation to place themselves under such regulation. It may be noted further that interconnection among utility service areas is hardly more than a beginning to a geographically integrated system.

Electric power at present is economically the most important of the products or services derived from water development to which the scale economies derived from regional integration of transport and distribution might be applied. While both public and private management in several cases has proved the value of such integration, it must be concluded that the administrative organization for electric power distribution within the United States has not yet fully reflected the opportunities given by technology in the scale economies.

PLANNING FOR INTERBASIN USE OF WATER
AND WATER-DERIVED SERVICES

Considering the great contrasts between the moist and the dry regions

[117] Act of August 26, 1935, Sec. 202(a), 49 Stat. 838, 16 U. S. C. 824a(a).
[118] President's Water Resources Policy Commission, *Water Resources Law* (Vol. 3 of Report), p. 287.
[119] *Idem.*

and sections of the United States, the impending concentrations of population in some areas, and the techniques available for transporting water, more widespread interbasin planning and development of water resources is indicated for the future. This is a stage beyond integrated development of a single basin, in which planning and development take into account the matching of supply and demand for water and water-derived services, whatever the relative geographical location of the two. This is most commonly done now in the case of electric power produced at hydro sites. Hydroelectricity is distributed freely across watershed boundaries, and indeed with little regard for them. The TVA power service area is a good illustration, covering parts of five watersheds other than the Tennessee.[120] In the case of water itself, a few interbasin diversions also have been made. Many are on a relatively small scale, like the diversion of Delaware River water for the New York City water supply; the Ware River diversion from the Connecticut Valley to the Boston metropolitan area; or the diversion of Lake Michigan waters into the Chicago drainage canal. In eastern United States most interbasin diversions are relatively localized projects, undertaken mainly to supply municipal needs. In the West, however, interbasin diversions and planning on a multiple-basin basis is more advanced. Among the federal agencies, the Bureau of Reclamation has been particularly interested in interbasin plans and development. Among the Bureau's interbasin projects are the completed Colorado–Big Thompson project in Colorado, and the Colorado River–Imperial Valley diversion in California. Still others, like the San Juan–Chama diversion in New Mexico, are planned. Planning for the interbasin use of waters, however, has reached its most complete application in the state of California, where the Sacramento–San Joaquin diversions of the Central Valley project now seem only a beginning to an ambitious and well co-ordinated project using the surplus waters of northern California for the benefit of the entire state.[121] At one time, the Bureau of Reclamation, as part of its

[120] Cumberland, Ohio, lower Mississippi, Warrior-Tombigbee, and Coosa River watersheds.

[121] The California State Water Plan envisions extensive transfer and exchange within the state, generally moving water from the "surplus" basins of northern California, like the Eel and Trinity rivers, to the deficit areas of southern California. The authorized Feather River Project is the first unit of the Plan. See State of California Department of Water Resources, *Program for Financing and Constructing the Feather River Project as the Initial Unit of the California Water Plan,* Sacramento, Calif., 1955; and State of California, Department of Water Resources, *The California Water Plan* (Bulletin No. 3), Sacramento, 1957.

"United Western Project," planned for the diversion of Columbia River
water into southern California and the Southwest by a system of exchanges. While feasible as an engineering plan, it met political opposition from the Pacific Northwest. The United Western Project probably
has been superseded by the more recent California State Water Plan.

On the whole, interbasin planning of water use is very much in its
youth in the United States. The development and planning that has appeared up to the present time has been mainly a response to intense
local needs, like those of southern California, rather than a response to
the opportunity presented by the scale economy techniques. Formidable
institutional obstacles, of course, exist for any extension of multiple-
basin planning within the future. The fragmentation of basins by state
lines, and the division of waters by compact, are likely to handicap most
proposals to apply these techniques indefinitely. The situation of the
state of California, with its extensive territory, and both surplus and
deficit regions within its borders, is unusual in the United States. Nonetheless it shows interesting potentialities for the application of multiple-
basin planning in other parts of the United States, should institutions
permit and should needs arise which would encourage such planning
and development. Beyond states like California or Texas, this type of
development patently is the concern of an interstate organization, or
of the federal government. Even California has benefitted from interstate interbasin diversion (Colorado River water); it could benefit
further, as in electric power interconnection with the Pacific Northwest.

Interbasin diversions promise eventually to be matters of international
concern for the United States. The effects of the proposed upper
Columbia River water diversion into the Fraser River within Canada
already have been a matter of concern to the United States. The possible large benefits from the diversion of upper Yukon waters into
southern Alaska also are economically attractive, and are likely to be
the basis of further interbasin proposals.

Planning Use Extensions of a Given Water Supply

1) MULTIPURPOSE PLANNING

The outstanding example of extending the use of given water supplies has been in employment of multiple-purpose planning and development. As noted above in connection with the discussion of research and

data collection, multipurpose planning has developed slowly from beginnings in the nineteenth century. The establishment of the Mississippi River Commission in 1879 in a degree marked the beginning of multipurpose water use in the United States, just as it marked the beginning of basin-wide planning. (See above, page 499.)[122] The act authorizing the establishment of the Commission mentions navigation, flood control, and bank protection among the responsibilities of the Commission.[123] An 1888 statute recognized the relation of irrigation and flood control in river development;[124] another 1888 statute mentions the relation of navigation, power development, and fishery uses;[125] and the 1893 provision for the California Debris Commission instructed it to reconcile the interests of hydraulic mining and navigation, to store sediment, and to provide flood relief.[126] From this period onward multipurpose planning for stream development has received increasing attention in federal legislation, as reflected in the 1902 Reclamation Act, the 1906 and 1910 General Dam acts, and the 1910 and 1912 River and Harbor acts; in 1917 flood control legislation, the Federal Water Power Act of 1920, the "308" Report directions (1927), the 1928 Boulder Canyon Project Act, and the 1933 Tennessee Valley Authority Act; in the 1938 Flood Control Act, and later federal flood control and river and harbor acts.[127] Each piece of legislation helped to extend the scope of multipurpose planning in one manner or another, culminating in the broad authorizations of the TVA Act, which permitted one of the most complete uses of this scheduling technique to date.

Multipurpose planning also was carried on on a wide scale under the auspices of the National Resources Board and its successors, between 1934 and 1943. The organization of almost every federal agency concerned with water development in the United States now reflects

[122] Multipurpose planning and geographically integrated basin planning often have been closely connected in succeeding years, and now are so much parts of the same plans that it generally is assumed that the two are inseparable. It is evident, however, that the multipurpose idea may apply to a single project no less than to a basin-wide plan. This was reflected in much of the legislation treating multipurpose planning between 1893 and 1933. For this reason they are considered separate techniques.

[123] Act of June 28, 1879, Sec. 4.

[124] Act of October 2, 1888, 25 Stat. 505, 526.

[125] Act of August 11, 1888, Sec. 1, 25 Stat. 400, 417; Sec. 11, 25 Stat. 425.

[126] Act of March 1, 1893, 27 Stat. 507, 33 U. S. C. 661, *et seq.*

[127] See President's Water Resources Policy Commission, *Water Resources Law* (Vol. 3 of Report), pp. 388-410, 408, 460-63.

the desirability and the need for multipurpose development of streams. While the machinery of administration has been criticized as cumbersome because it involves a number of special-purpose agencies,[128] the multipurpose technique has been widely recognized in the United States. Consistent efforts also have been made to fit it to the operational responsibilities of the special-purpose agencies, as well as to agencies with broad regional jurisdiction, like the TVA or the State of California Department of Water Resources and its predecessors.

2) OTHER TECHNIQUES EXTENDING THE SERVICES OF A GIVEN SUPPLY, OR INCREASING AVAILABILITY

Still other techniques may increase the availability of water, either through extension of the services of a given unit of supply, or through regulation and adjustment of use. Among the former are underground storage of water (chapter 8, pages 194-98), beneficiation as in water softening (chapter 8, pages 210-17), evaporation control of stored water (chapter 16, pages 451-57), and water-conserving land treatment (chapter 8, 201-10). Among the latter are pairing of low-quality supply and tolerant use (chapter 7, pages 148-61, and chapter 13, pages 349-55), recycling (chapter 16, pages 408-41), design and maintenance of use restrictions, and the water budget techniques (chapter 16, pages 442-51).

Organization to make use of these techniques has been much less formalized than that for multipurpose planning. With the exception of use restrictions, their application for the most part has been left to the initiative of individuals, private corporations, and local governmental units (municipalities, counties, public districts). Their employment has been determined mainly by the exigencies of individual industrial or farming operations, and by the emergencies sometimes created by rapidly rising demands at a time of unforeseen natural decreases in supply. The pairing of tolerant uses and low-quality supply has been employed industrially in New Jersey, in Texas, and in California, and agriculturally on some western irrigation projects; reuse and recycling have received attention by municipalities and corporations in a number

---

[128] See reports of Commission on the Organization of the Executive Branch, Natural Resources Task Force (First Hoover Commission), Missouri River Basin Survey Commission, and the President's Water Resources Policy Commission.

of localities throughout the arid West and Southwest; and mixing of low-quality and high-quality water supplies has been employed by Dallas, Los Angeles, and other municipalities. Water budgeting techniques have been demonstrated and applied privately, and by some agricultural experiment stations.

Probably the best organized attempt to incorporate these manifold devices into co-ordinated use in an integrated plan has been under the California State Water Plan. Because the possibility of excessive dissolved solids content is present in much of the San Joaquin Valley and southern California, it has been necessary to pay special attention to the maintenance of specific standards of water quality in the water transfers and exchanges contemplated under the plan. A Board of Water Quality Consultants was set up for the purpose of recommending for quality maintenance under the plan.[129] Pairing of low-quality of supply and tolerant use, mixing of high and low-quality supplies, and supervision of reuse are all contemplated under the plan, along with construction of storage and delivery facilities.[130]

Elsewhere little attempt has been made to fit these useful devices into the scheme of over-all planning for efficient water use, along with multipurpose planning, basin integration, and the other devices which now are part of the kit of the water development planner. Eventually public administrative organization may find a place for considering them on something more than an isolated local basis.

The history of land treatment measures is somewhat different. Through the activities of the federal Soil Conservation Service, the state extension services, and experiment stations, and the local soil conservation districts, a great deal of organized initiative has been devoted to water-conserving[131] types of cultivation, and land use. The final application of these measures necessarily has been in the hands of the private entrepreneur, the farmer. However, the long-continued attention of the federal and state agencies to these problems has created an organization which has promoted joint consideration of land treatment programs in most recent comprehensive planning for water development. Department of Agriculture agencies, including the Soil Conservation Service,

[129] State of California, Department of Water Resources, *The California Water Plan* (Bulletin No. 3), Sacramento, 1957, p. 34.

[130] Blair Bower, communication to the authors, 1957.

[131] "Conserve" is used here in the sense of promoting a time distribution of water availability adjusted as well as possible to the incidence of crop or other water requirements.

have been represented on all deliberations of the Interagency Committee on Water Resources, its predecessor, and associated field committees. Water conservation is specifically mentioned in the basic legislation which established the Soil Conservation Service, which is the heart of technically improved land treatment measures on private agricultural lands.[132]

There also are important land treatment programs applying to the public lands. Both the Forest Service and the Bureau of Land Management of the Department of the Interior administer activities directed toward watershed protection. They are especially significant in the eleven westernmost states, where more than half of the total land area is federally owned. The Bureau of Land Management in particular has lacked adequate financial means for carrying out its program of watershed protection, but some statutory and administrative provision for such protection on the Taylor Grazing Act lands does exist.[133] Watershed protection directives are more explicit in the national forest legislation. "Securing favorable conditions of water flows" is mentioned as early as the 1897 statute, which provided for national forest administration.[134]

Public administrative provision for land treatment on private forest lands does not parallel that for private agricultural lands and on the public domain. Charles McKinley considers that the machinery of the Soil Conservation Districts probably needs supplementation by state forest and debris-siltation measures before effective land treatment for water-quality conservation on private forest lands becomes a reality.[135] However, a number of the larger timberland owners now have effective long-range forestry practices.

Use restrictions have received attention particularly in the subject of pollution control. Concern for the effects of waste carrying as a stream function has brought about enabling legislation for the United States Public Health Service and for a number of state public health agencies in this field. Administrative provision thus has been and is being made for the application of use restrictions to streams where important waste-carrying functions conflict with other uses for the same water. Planning

[132] The Soil Conservation and Domestic Allotment Act of 1935 (Act of April 27, 1935, 49 Stat. 163, as amended, 16 U. S. C. 590a *et seq.*).

[133] President's Water Resources Policy Commission, *Water Resources Law* (Vol. 3 of Report), p. 365.

[134] *Ibid.,* p. 355.

[135] Charles McKinley, *op. cit.,* December 1957.

for the application of this method of regulation is advancing, with varying degrees of control exercised by the several states, and with the federal government increasingly involved in planning for pollution control.[136]

A major organizational difficulty in applying this type of use restriction is traceable to the lack of correspondence between the extent of river basins and the territorial jurisdiction of the several states. The pollution problems of many a downstream state have their origins partly in the industrial or other sources in an upstream state. Limited territorial jurisdiction has meant limited interest on the part of the upstream states. To unify these interests on a basin-wide pattern, interstate compact commissions have been tried, like the Ohio River Valley Water Sanitation Commission and the Interstate Commission on the Potomac River Basin. While such agencies unquestionably are improvements over fragmented administration of use restrictions for pollution control, there has been serious question as to their demonstrated capacity to keep pace with the growing problems of interstate pollution. For this reason the place of federal administrative organization for this purpose continues to be an issue of importance in water management.

Use restrictions also are recognized legislatively in the preferences assigned to water uses in the western states. Certain use preferences have been expressly stated in western states' water law for many years.[137] These preferences generally give recognition to the life-sustenance purposes, giving first place to domestic water supply, followed by stock water and irrigation in that order. Industrial water supply, electric power,[138] and navigation definitely are of a lower order of preference within these states, in case of competition for short supplies. Use preferences generally are superseded by priorities in time, being most effective at the time when appropriations are made. Typically, if two filings are made within a state at the same time with different use preferences, the water is awarded to the one with higher use preference. As a rule a high

---

[136] Through the Water Pollution Control Act of 1948 (Act of June 30, 1948, 62 Stat. 1155, as amended, 33 U. S. C. 466) and subsequent legislation (Act of July 9, 1956, 70 Stat. 498).

[137] See, for example, Oregon Comp. Laws Ann., Sec. 116-601; Vernon's Texas Civ. Stats., art. 7471; Utah Code Ann., 1943, Sec. 100-3-21; and Wyoming Comp. Stats., 1945, Sec. 71-402.

[138] Except for the purpose of steam use in thermal generation. Texas, for instance, places steam generation just after domestic, municipal, and stock water uses in the list of preferences; Wyoming has a similar order of preference.

preference filing does not take precedence over a low preference appropriation which was allowed previously.[139] However, there are exceptions, and in practice some states show a degree of flexibility in administering water rights both as to use preference and time priority. Thus industrial water supply may be secured by purchase of the land to which irrigation water rights are attached. Where competition arises between municipalities and irrigated farms for the same water supply, the same solution may be used for the problem of competition, although the municipal demand may have been given preference in the legal system. This approach is typified in the law of Idaho.[140] A somewhat different arrangement is found in the state of Washington, where state law makes it the responsibility of the courts to determine which use is the higher in the event of conflict between two or more different types of use. Such determination will be made on the basis of the public interest in each case examined.[141]

In 1944 and subsequently, these use preferences were recognized in federal law in the so-called O'Mahoney-Millikin amendment to the Flood Control Act of that year. In this and in subsequent legislation the priority of "beneficial consumptive" uses is recognized over that of navigation.[142] In this respect use restriction is now a part of federal water development planning, just as pollution control is also entering planning in the same way.

Use restrictions also have been experimented with in other situations. An important phase of the interagency comprehensive plan for the Columbia River and its tributaries was the reservation of the lower tributaries (below the mouths of the Deschutes and the Klickitat rivers) for salmon and steelhead runs, and such other uses as did not conflict with fish spawning. This meant no future storage or power dam construction, a measure agreed to by the major federal construction agencies, all fishery agencies, and implemented by the Washington State legislature. The conception was one of achieving multiple purpose development in part through dedication of specific tributaries of the

[139] See Wells A. Hutchins, *The California Law of Water Rights* (Sacramento: State of California Printing Division, 1956). The suggestions of Stephen C. Smith on this point also have been helpful.

[140] Wells A. Hutchins, *The Idaho Law of Water Rights* (Boise, Idaho, Idaho State Department of Reclamation, 1956), pp. 56-58.

[141] S. V. Ciriacy-Wantrup, "Some Economic Issues in Water Rights," *Journal of Farm Economics,* 37 (1955), 885.

[142] Act of December 22, 1944, Sec. 1, 58 Stat. 887; repeated in Act of March 2, 1945, Sec. 1, 59 Stat. 10 (1945 River and Harbor Act). Since made applicable in each River and Harbor and Flood Control Act.

system to their most economic use within a comprehensive plan. The plan ultimately failed because the Federal Power Commission, in discharging its licensing and other functions, interpreted comprehensive development in a different way than the major planning, construction, and operational agencies in the basin. In licensing the Mayfield and Mossy Rock projects proposed by Tacoma City Light on the Cowlitz River, the Commission in 1951 assumed that fishery and power uses could be complementary on the lower tributary streams, a premise not accepted in the Interagency Committee plans.[148] In the 1952 licensing of the Pelton project on the Deschutes River to the Portland General Electric Company, the commission adopted a similar assumption.

While finally not applied in the case of the Columbia, use restriction as an instrument of comprehensive multiple-purpose planning still is to be considered a technique of potential importance. In general, the organized application of use restrictions is only partly understood. Administrative provision for planning their application, and that of the other techniques extending services or increasing availability (except multipurpose planning), has not been methodically approached either in the federal establishment or elsewhere in the United States administrative organization for water development.

Extending the Physical Range of Water Recovery

There are two principal ways to which water development planning might have taken account of extensions of the physical range of water recovery: (1) through techniques assessing the availability and promoting the recovery of ground water (chapter 10, pages 270-84, and chapter 15, pages 384-406; and (2) through appraisal of the long-range availability of surface and gravitational water supplies (chapter 4, pages 81-83). Neither way is simple, and the two differ greatly in the approaches which promise results. Furthermore, neither has yet been cultivated to any degree of completeness.

There are three aspects to the ground-water phase of this problem: accurate delineation of shallow-lying aquifers and appraisal with known

---

[148] The plans finally adopted for the Mayfield Dam include elaborate ladder, tank transfer, and flume arrangements for movement of fish upstream and downstream around the dam and power tunnel. (See Carl Pflugmacher, "Mayfield Dam Fish Facilities," *Pacific Northwest Public Power Bulletin*, March 1957, pp. 7-10.)

techniques of measurement; development of techniques for delineating and measuring the extent of deep-lying aquifers; and development of techniques for extending the economic range of recovery from the deep-lying aquifers. From a public point of view these matters have been mostly in the hands of the United States Geological Survey and those of the individual state geological surveys or water resources agencies. Within their budgetary limitations these agencies have gradually unfolded a picture of the water present within the upper part of the mantle rock. While there are many local details to be known, this part of the task has been methodically approached and an organization has been maintained for prosecuting the work.

On the other hand, the development of techniques for extending the reach of water measurement within the earth, and of techniques for deepening the economic range of water recovery underground, has not been methodically treated as a problem from a public point of view. The major developments in this field generally have been ancillary to other objectives or other pursuits. Thus the development of geophysical techniques for charting the nature of underlying formations was particularly associated with the petroleum industry; the same may be said for deep-drilling methods. As new devices have appeared, electronic or otherwise, they have been made use of as they fitted the programs of public agencies in the field. However, in spite of some desire within the Geological Survey to do so, it cannot be said that a conscious program has been directed toward the broader development of these techniques. This is in contrast to the application of those techniques which fortuitously have appeared. Considered on a national scale, pressure for the use of ground-water resources possibly has not yet reached a critical stage.

Appraisal of the long-range availability of surface and gravitational water supplies is intimately tied to progress in the science of climatology. Within the United States the advance of this science has depended upon the efforts of the United States Weather Bureau; university departments of meteorology, climatology, and geography; and, more recently, a few private institutions like the Laboratory of Climatology at Centerton, N. J. and the Munitalp Foundation of New York City. While precipitation most obviously is an important part of meteorological analysis and climatological compilation, again methodical organization has not been realized to make use of the devices of meteorology and climatology for appraising long-range water availability. The problem of drought, of course, has figured prominently in the studies of the federal Weather Bureau and others who have undertaken climatological studies. However, the problem of supporting long-range precipitation analysis has not been viewed consistently from the point of view of its potential contribution to the longer range development of water supplies.

In brief, water development planning in the United States has not included consistent and timely support for the basic climatological studies that are increasingly known to be essential to proper appraisal of the long-range availability of water in any given region. The history of basic data collection in river basin development illustrates that it has been difficult enough to plan for adequate stream flow records, not to speak of basic climatic investigations over century-long or longer periods.[144] Many of those who have been concerned with past river basin planning in the United States probably would have regarded study of the surface of the sun,[145] or dendrochronology,[146] or geomorphology,[147] as having only a visionary and academic relation to water development planning. Yet it is from such subjects as these, and esoteric studies of the circulation of the atmosphere, that we must hope to obtain the truly basic information on the chronological frequency pattern of water availability within the large basins. Thus far, water development planning has depended upon the by-product of investigations with other objectives. Yet as the nature of basic information in water development is analyzed more systematically, it would seem that support of long-range climatic research may deserve a place in water development planning. Full exploitation of the technical and scientific means of obtaining long-range water supply data will be needed, and may be arrived at by proper planning.

Appraisal of Influences Accelerating the Rate of Exploitation

In its procedure for assessing water and water-services demand, water development organization in the United States has been unsystematic

---

[144] See House Document No. 706, 81st Cong., 2nd Sess., *A Program to Strengthen the Scientific Foundations in Natural Resources.*

[145] Like that presently being undertaken by the High Altitude Observatory of the University of Colorado at Climax, Colo. (See Director's Quarterly Reports of this Institution.)

[146] Study of tree rings. The University of Arizona has been a center for development of the subject; studies have been made in several sections of the West.

[147] Literally, the science of earth forms. Geomorphological research particularly is concerned with natural erosional and tectonic processes which produce the physical landscape of the earth.

and multidirectional. In spite of the "water consciousness" of this nation, the nature of water and water-service demands is as yet poorly understood within it. The general demand premises for development of water facilities in recent years have been three: (1) the development of a resource, or the construction of facilities to make use of a resource, automatically creates a demand for water or water-derived services; (2) political expression of demand is adequate indication of the social validity of the demand; (3) for private development, the premise has proceeded from a more conservative base—development can be undertaken where amortization of capital expenditures and an operational profit can be forecast from projections of past consumption pattern changes and reasonable population projections.

The first demand premise has dominated the policy of the Bureau of Reclamation for its irrigation development, and the Tennessee Valley Authority in its development of electric power generating and transmission facilities. The second has been an important premise for the Corps of Engineers and the Bureau of Reclamation. It also has lain behind municipal provision of urban water supplies, and many local public power movements. The third has been typified in the operations of the privately owned electric power utilities. However, a few privately owned utilities have had an element of the first premise in their policies, i.e., that development creates its own demand, and they have used promotional rates toward that end.

Within this structure certain allowances have been made for the influence of technology as it has operated upon demand for water and water-derived services. The changes in electricity system load characteristics which might be brought about by electric space heating and by air conditioning were a part of TVA planning at a time when large demand increases on this basis were far from an accepted premise in utility circles throughout the United States. (See chapter 6, pages 140-45.) At the present time other utilities also are commencing to plan their development programs on this basis. However, throughout the country as a whole, the demand effects of technical changes are only slowly assessed in some phases of electricity distribution.[148] In other respects, appraisal of the effects of technological changes has been somewhat easier and better attended to. The impact of the electro-

[148] See statement by Gordon W. Evans, president of Kansas Gas and Electric Co., as reported in *New York Times,* December 7, 1956. Also Gordon R. Clapp, *The TVA, an Approach to the Development of a Region* (Chicago: University of Chicago Press, 1955), p. 206.

metallurgical industries, because of the comparatively large blocks of power which increments to their productive capacity demand, has been much more closely followed. Yet on the whole the impact of technology upon demands for electric power production, and its associated effect upon water development, is not a subject which has been methodically examined in the sense of having administrative provision to make it part of water development planning. And electricity production and transmission, among all the water-derived services, consistently has been among the most sensitive to technical change.

Appraisal of technical influences accelerating the rate of exploitation has been made more difficult and more important by the increasing range of economic alternatives that are available for the production of services and the use of resources. This may be illustrated by the case of irrigation and food production. The traditional manner of estimating the food component of demand for increased agricultural production has depended upon population projections (including composition) with factor adjustments for consumption trends, nationally and regionally. This in turn has been related to land capability and productivity indexes. Out of the comparison have come interpretations of the relation of forecast food demands to the need for irrigation water requirements. The interpretations of the Department of Agriculture specialists and the Bureau of Reclamation not infrequently have been in disagreement, but generally these forecasts have been used in support of the need for developing further irrigation water supplies in western United States.[149]

Technology in the past has enhanced irrigation's place in fulfilling these demands as they have arisen. (See chapter 6, pages 107-19, 132-37, and chapter 16, pages 442-51.) Transportational and agronomic changes have brought the West into favorable relation to the centers of demand. Some more recent technical innovations may have operated in the other direction. The development of lightweight metal and plastic piping, efficient small pumps and small prime movers with low operating costs, and the water budget technique have changed the potentialities of irrigation agriculture throughout the humid and sub-humid sections of the United States. The potentialities of humid land agriculture also have been changed profoundly by developments in fertilizer production and application. All these are items to which some study has been devoted within the structure of specialized agencies

---

[149] See, for example, U. S. Congress, House Committee on Interior and Insular Affairs, *Colorado River Storage Project* (Washington: U. S. Government Printing Office, 1954), p. 909.

of the federal government. At the same time, administrative organization within the federal government and elsewhere does not provide
machinery for *weighing the balance of these factors* when water development planning is undertaken. In the instance of food production
increments, some machinery is afforded through the evaluations of the
Department of Agriculture and the medium of the federal Interagency
Committee on Water Resources.[150]

The special commissions which periodically have been created on an
ad hoc basis also may be mentioned in this connection. The President's
Materials Policy Commission, for example, considered the problem of
balancing agricultural product demands against means of likely agricultural production over the period 1951-75. Its staff conclusions
pointed toward considerably less requirement for reclamation of new
land than reclamation-oriented interests would recommend.[151] On the
other hand, the President's Water Resources Policy Commission
previously came to somewhat contradictory conclusions.[152] Organizational treatment of this problem obviously is not an easy matter,
particularly in view of the fact that reclamation is an issue with many
political overtones.

Evidence of the inadequate professional appraisal of the technical
influences at work within the country's food production plant continues
to appear. This organizational shortcoming has been illustrated most
frequently in the hearings for authorization of the largest water development proposed within the decade for future federal construction, the
Upper Colorado River Basin project.[153] Under present federal administrative arrangements, the client-agency relations of all the major federal
construction agencies can obscure objective technical appraisals of
development. Excellent and accurate appraisals can be made within
the framework of agency-client special interest (e.g., Bureau of Recla-

[150] By presidential letter dated May 26, 1954, the prior Federal Interagency
River Basin Committee was replaced by the Interagency Committee on Water
Resources. In this letter, the President approved the "Inter-Agency Agreement
on the Coordination of Water and Related Land Resources Activities," a joint
agency statement containing the co-ordination procedure to be followed by the
major water resource agencies.

[151] President's Materials Policy Commission, *Resources for Freedom*, Vol. 5
(Washington: U. S. Government Printing Office, 1952), p. 72.

[152] President's Water Resources Policy Commission, *A Water Policy for the
American People* (Vol. 1 of Report), p. 165.

[153] House Committee on Interior and Insular Affairs, *Colorado River Storage
Project, op. cit.*

mation and western reclamation interests), but comparable technical organization representing the full public interest is available only in the very small staff of the United States Bureau of the Budget. The inclusion of an adequate over-all technical appraisal of development within the federal organization, not identified with group interests and compatible with the political process, is still an unsolved administrative problem, for both high and low levels.

Agriculture is but one area within which technical influences on the location and form of basic economic activities are increasing. Many of them are appearing in manufactural industry. It is possible, for instance, that in the future the background of industrial location may be much more important than agriculture to use of water in the West. Some consideration recently has been given to these influences in the Bureau of Reclamation's programs and in those of the Corps of Engineers.[154] In the main, however, these evidences of planning are related mainly to the addition of municipal and industrial water to multipurpose storage capacity. But questions surrounding industrial water supply now are much more complex than planning for capacity alone. More and more, industrial demands for water must specify water of special quality, and it is within this area that technical changes have had the most profound influence. For instance the very large water withdrawals made for the thermal generation of electric power require a component of increasingly pure water. The latest techniques prescribe both very high pressures and high temperatures (i.e., supercritical temperature and pressure)[155] for the most efficient equipment. Boiler feed requirements for one recent plant are such that water of 99.99998 per cent purity must be supplied.[156] Future adequate anticipation of industrial demands therefore may be a much more complex part of planning than hitherto. Organization for large-scale development of water in the United States has not yet entered this phase as an articulated part of planning for development.

[154] As, for instance, in the Upper Colorado River Basin Project, authorized by the Congress in 1956, and in the Congressional consideration of the 1956 River and Harbor Act (vetoed).

[155] 374 degrees C. and 3,200 pounds per square inch are the temperature and pressure at which water will change continuously into steam without need for overcoming latent heat. Temperatures and pressures beyond these are "supercritical." Thermal efficiency of a plant increases as both operating pressure and temperature is increased. (See chap. 12, pp. 305-6.)

[156] Paul Cootner, "Water Requirements for Steam-Electric Power Generation, Outline of Projected Research," Resources for the Future, Inc., ms., 1956, p. 10.

Promoting an Environment Favoring Application
of Further Technical Improvements

In the water development efforts of the nation as a whole, a majority of the most important technical influences have entered the scene after they had been promoted by or had appeared in other fields. This situation is entirely natural and normal. As metallurgy progresses—or refrigeration engineering, genetics and plant breeding, agronomy, statistical design, chemical engineering, nuclear research, or many other scientific and technical fields—effects in water development planning inevitably must be felt. The nature of the scientific process and the structure of our society make windfall technical benefits important for any specialized work like water development. An administrative structure alert to the application of these windfalls would seem to be a highly desirable part of planning, both for ascertaining the nature of demand and the capacity of supply. In what degree have we had such an administrative structure?

Conscious organization for the generation of technical change within the field also should be an important part of planning. It is unlikely, for instance, that the technique of multipurpose use could have found a better testing ground than water management. How well has United States administrative structure provided for the generation of technical innovation?

Provision for these functions particularly is to be found within the federal government of the United States. The first function, application of the technical windfalls, would appear to be somewhat better provided for than the second, conscious organization for innovation. However, in recent years particularly, there has been much activity within both these fields in the federal government. Every major agency concerned with water development in the Executive Branch provides some organization for these aspects of planning. Contributions have also been made by state agencies, like the Land Grant Colleges, and by private industry, including the utilities.

Examples of administrative provision for studying the application of windfall changes are to be found in the work of several federal agencies. The Denver Laboratories of the Bureau of Reclamation offer one illustration in their fruitful study of the application of materials to irrigation development. Another example is found in the hydrologic studies of the Geological Survey. A recent review of the research interests of the Survey included, among other things, mention of the following: "thermodynamics of induced infiltration,[157] . . . automatic computation of stream flow records, equipment for making quality of water analysis, methods of measuring evapotranspiration of water from land surfaces, and ultrasonic measurement of the mean velocity of streams."[158] Further examples might be taken from the work of the United States Public Health Service's Robert A. Taft Sanitary Engineering Center in Cincinnati, that of the Beltsville station of the Department of Agriculture, of the Vicksburg, Mississippi, station of the Corps of Engineers, and of the state agricultural experiment stations. From private industry, experiment with and application of water budget techniques by the Seabrook Farms, Centerton, N. J., is a further example. Much more widely applied has been the application of digital computer techniques to system load studies in improving the operational efficiency of private electric-utility generating, transmission, and distribution systems.

It is to be noted that most of the effort applied to study of the windfall effects of technology in water development concerns the development of supply. Appraisal of the effects on demand is much less widely provided for.

The generation of technical innovation within the field also has been organizationally provided for within the agencies mentioned above, as well as in others. The Tennessee Valley Authority work illustrates particularly the scale economies and the scheduling techniques. The Corps of Engineers also has advanced the techniques of water development through the program of its Vicksburg, Mississippi, experiment station, in its fisheries-dam reconciliation studies on the Columbia River, and in other works. Again, the contributions of private industry should not be overlooked, such as the present program of the Texas Public Service Corporation to desalt brackish water for industrial use.

Thus American water development agencies include within their organization a number of provisions for appraising windfall technical innovations, and for stimulating the generation of others. There is, however, one outstanding gap in the administrative pattern. Nowhere has there been provision for a *co-ordinated view* of technical influences

---

[157] A study prompted by the increasing volume of heated water returned to the ground from factory process cooling and from air conditioners.

[158] John C. Reed, "Recent Trends in the Minerals and Water Resources Programs of the Geological Survey," statement presented before the Mid-Atlantic Division, Association of American Geographers, meeting December 1, 1956, pp. 6, 7.

bearing on development, nor co-ordination of the many different ways in which technical innovations may influence the field. Most of the administrative provision is for special-purpose views on the application of technology.

THE IMPORTANCE OF QUALIFIED PERSONNEL

Provision for the co-ordinated view of technical influences brings into focus other and still broader considerations that influence the character and efficiency of government administration. It is fully realized that administrative organization alone is not a unique nor always an entirely reliable solution to the broader problems of planning in a complex, rapidly changing industrial society. In the final analysis foresighted planning, well conceived and well carried out, will depend on men of high competence—trained generalists as well as experienced specialists. Without broadly trained men as bureau heads, division chiefs, planning or secretarial staff officers, and in other positions, administrative organization cannot of itself demonstrate alertness to new problems, and new developments. With such men, many variations in organization can be made adaptable as needed. It is highly important, therefore, that public agency personnel policies be so designed in selection, advancement, and compensation that these indispensable people are continually recruited and retained in the public service. Just as important is the psychological "climate" encouraged by sympathetic legislative and executive procedure, and the esteem of the general public. Without these many of the best men will not be encouraged to remain in service. These problems clearly take precedence over administrative organization; but it is equally clear that the existence of organization which gives adequate channels to competent men can make public service careers more attractive to them, and hence contribute toward better co-ordinated and alert planning.

REVIEW OF EXPERIENCE

For all except recently established agencies, an essential part of planning is the review of past experience. The larger the scope and complexity of the operation, and the greater the responsibility of the agency or agencies, the more vital the review functions become. Particularly within large public bureaus, the past can crystallize a mold unsuited to present or future if a vigorous policy of review is not followed. Provision for such review, indeed, is important in giving competent leadership the range needed for foresighted planning.

Almost every large resource development agency in the federal government, some state agencies, and most large private corporations have staff provision for review functions of this kind. Particularly after the widespread application of linear programming during the Second World War, with its concepts of "control" and "feedback," review of experience with consequent organizational revision has received increasing attention.[159]

Each operating agency has differing provisions for experience review, from continuing formalized staff provisions, to *ad hoc* committee or inspection team arrangement. Treatment of the problem may depend on the functions of the agency, the level of administration, and on political policy.

One of the earliest major provisions for experience review was made within the federal Department of Agriculture. During the late 1930's and the early 1940's the Administrative Council and the Program Planning Board of the Department assisted in development and evaluation of the departmental programs through well-conceived professional administrative analysis. Both of these departmental policy units had the research staff assistance of the Bureau of Agricultural Economics, which was in a position to supply background data for review from the level of local details to that of the most comprehensive national agricultural problem. Even before such review and planning had been developed on a departmental basis, the Forest Service had had its own provisions for program review and administrative planning for that bureau.

In some situations the review function is a partial responsibility of an advisory staff which has co-ordinating responsibilities. This has been the custom for the Department of the Interior, where review functions are the responsibility of the Technical Review Staff in the Office of

---

[159] Review of experience functions long preceded linear programming, but they were sharpened and popularized where linear programming techniques were applied. Linear programming generally separates planning and control or feedback functions, but in resource development the relation is obviously very close. Separation of the two is not considered justified for the purpose of this book. For a description of the linear programming concepts of organization, and their application to business, see Melvin E. Salveson, "High-Speed Operations Research," *Harvard Business Review*, 35 (July-August 1957), 89-99.

the Secretary (previous to 1953 the Program Staff). In other instances a principal administrative officer will be assigned review functions along with other planning responsibilities. TVA treated its review problem for the agency as a whole in this manner, through the Assistant General Manager (Program Analysis), during 1952-55.

Perhaps the clearest and most consistent organizational treatment of this function within public agencies occurs in the Department of Defense. Like other Defense agencies, the Corps of Engineers has an inspector general (the Engineer Inspector General) who has broad powers for organizational and procedural review within the Corps. His work is supplemented by specially organized temporary "Command Inspections" appointed as necessary by the Chief of Engineers to review important organizational or procedural problems. The reports of the Inspector General, or of the special Command Inspections, become the basis for action by the Chief of Engineers (if accepted by him) or by his subordinate officers. This is a good example of a very clearly defined channel for "feedback" in organization.

The most consistent and best staffed review functions in water development, aside from the review units in private corporations, have been at the agency or bureau level of government. There remains the important superagency level, represented in the White House staff, the Bureau of the Budget, and the Congress. The only agency maintaining a permanent staff with responsibilities of this kind is the Bureau of the Budget, which has a small unit in its Office of Management and Organization maintaining an interest in water development. The Office is described as providing "guidance and co-ordination in Bureau activities toward better agency management and organization; conducts organizational studies; co-ordinates the Bureau's management improvement efforts; and conducts work to improve governmentwide management and service practices and procedures."[160] However, it must be stressed that the Bureau of the Budget staff is small in proportion to the size of the review function at this level, if it is considered fully. Furthermore, the admixture of fiscal and administrative review has never been adequately supported by technical review within the Bureau. There is a reasonable question as to whether or not broadly conceived technical review and planning can be undertaken in such close conjunction with budgeting and fiscal planning and review.

---

[160] *United States Government Organization Manual 1956-57* (Washington: U. S. Government Printing Office, 1956), p. 61.

Since August 1955 an additional "superdepartmental" review unit has operated within the Executive Branch of the federal government. This is the office of the Public Works Coordinator, which maintains a very small staff of technical personnel to assist in the drafting of counter-cyclical public works programs. Water development projects are included.

Additional reviews of a detailed nature have been supplied from time to time by *ad hoc* commissions, already mentioned in other connections. The two Hoover Commissions [first (1949) and second (1955) Commission on the Organization of the Executive Branch of the Government] and the President's Water Resources Policy Commission all treated the review function in detail and with care. Within recent years, the disposition of the federal government has been toward discharging the review function in this broadest and very important context by temporary agency, with little direct feedback connection to the operating, construction, and planning agencies. The continuing staff devoted to these problems, while entirely competent, has been hopelessly small.

The absence of searching, continuing technical and administrative review on a superdepartmental level in part may be responsible for the relatively small amount of critical professional debate in the United States on water development projects and programs. While there is consistent critical public examination of water development on politico-economic grounds, and there are consistent discussions of administrative organization, these interests are not matched by professional engineering and other technical debate on projects and project planning.[161] Only when one of the sporadic commission investigations is under way is there widespread disposition toward technical controversy. This must be considered a loss for the cause of efficiency in water development.

---

[161] One noted water development engineer stated to the authors that, in his view, ". . . professional debate on . . . water expenditures in this country has almost disappeared since the discussion on the Mississippi Valley flood control following the proposal of the Jadwin plan."

CHAPTER 19

The Relation of Technologic Advance
to Realization of Development

Securing or Supplying Entrepreneurship—provision of suitable construction agencies; construction on a multiple-use basis; balancing water development with the development of other resources; provision for meeting dispersed demand; reducing local intensity of demand. Provisions for Financing Water Resource Development—use of private funds; first reasons for financing from public funds; public appropriations, the scale economies, and the scheduling techniques; financing of other activities related to water development; financing of research and scientific experiment; construction and engineering organization. Allocation Devices. Application of Mass Production Techniques and Interchangeability to the Management of Construction. Summary Comment.

The next steps after the completion of planning (or preconstruction planning), concern the physical construction of facilities, establishment of the essential legal structure for management of the resource, and establishment of the essential financial structure for support of the development. Ideally they are commenced even before the completion of planning. As indicated in chapter 17, these steps are: securing or supplying enterpreneurship for development and later operational management; securing or supplying financing, and administering the disbursement of investment funds; securing the necessary legal authorizations for construction and other developmental facilities; and management and supervision of construction or other developmental activities.

Some of the needs which have appeared in connection with these steps in the past have become continuing problems. Among them are: (1) provision of construction agencies capable of accepting responsibility for large works and geographically extended construction operations; (2) construction on a multiple-use basis; (3) balancing water development with the development of other resources; (4) provision for meeting dispersed demands; (5) organized attention to reducing the local intensity of demand, or dispersing it within socially desirable limits; (6) provisions for large-scale financing; (7) provision of allocation devices which match technical possibilities; (8) application of mass production techniques and interchangeability to the management of construction. It is interesting that the first five of these are needs or problems of entrepreneurship.

Securing or Supplying Entrepreneurship

The classic definitions of "entrepreneurship" in the process of economic production must be interpreted somewhat flexibly when applied to the process of water resource development. Indeed, the term is here used for want of a more precise one which would describe the essential step of supplying initiative and accepting responsibility for the organization and management of the developmental process. In simple economic definition, an entrepreneur is a person (actual or juridical) who assumes the responsibilities of risk, innovation, organization, and management decision in production. The risk is assumed with a view toward acquiring a profit. Actual practice in today's United States economy often separates risk-taking from innovation, organization, and management. Corporation stockholders take the risk, but as a group have little direct control over organization and management, which is in the hands of a relatively few professional managers. Modern entrepreneurship in fact is more a matter of organization and organization-building ability than any other single attribute.[1]

For water development, the "profit" often is one of social benefit, measured by its indirect effects upon the economy or society. At the same time, profit-making in the business sense may be a very definite consideration, as it is in electric power production and sale. Because social benefit is at least a correlative goal, public entrepreneurship is required; and in this one finds an almost complete separation of risk-taking and management. A government agency and its special-interest

---

[1] Frederick Harbison, "Entrepreneurial Organization as a Factor in Economic Development," *Quarterly Journal of Economics,* LXX (1956), 364-79.

clientele may supply the initiative; the agency undertakes organization and management. The risk is assumed by the taxpayers at large, the great majority of whom receive very little in the way of direct benefits.

Whatever the objective, social benefit or profit, initiative is needed, and so is acceptance of responsibility for the management of construction and other activities connected with development. Even if it is supplied by the federal goverment in a manner in which risk-taking is divorced from initiative and management, the entrepreneurial function must be attended to if development proceeds.

PROVISION OF SUITABLE CONSTRUCTION AGENCIES

Application of the scale economies to the development of American surface waters has been undertaken mainly by the federal government. Through its construction agencies the federal government has supplied a large share of the initiative for the construction of large works and the responsibility for geographically extended operation. The reasons for this appear very clearly in the history of water development in this country. Before the appearance of techniques enabling the achievement of the scale economies, there had been five sources of initiative in development: private enterprise, the states, municipalities or other communities, specialized local public bodies,[2] and the federal government. Their initiative was applied in the following directions: state development of canals and other navigation aids; private enterprise development of canals; private development of relatively small-scale, single-purpose power (first water and later hydroelectric) projects; private enterprise development of community or industrial water supply districts for irrigation, drainage, and flood control; community development of community or industrial water supply; and federal government development for special reasons of social benefit.

It is of interest that an important part of the earliest initiative for development of United States waters was supplied by private enterprise. This is true even in the development of navigation: a number of canal companies commenced operation in this field in the eighteenth century. Legislators' interest in navigation of course had been manifest from the first Congress. As early as 1820 Congress decided that all of the initiative needed for facilitating commerce among the states and for promoting western expansion was not being supplied by private enterprise. In that year appropriations were made for a survey of parts of

[2] Viz. districts.

the Ohio and Mississippi rivers and certain of their tributaries.[3] Extension of the 1820 survey came in 1824, together with appropriation for removal of obstructions from the Ohio and the Mississippi rivers.[4] From that time onward the federal government increasingly supplied the initiative for development of navigation on the nation's streams. By 1882, more than $100 million already had been appropriated by the Congress for rivers and harbors.[5] Federal government initiative in undertaking navigation improvements was soon followed by federal initiative in flood control work, which commenced in the Swamp Land Acts of 1849 and 1850.[6] Although interest and some activity in flood control continued from these dates, large-scale flood-control initiative did not come until 1917, after the Mississippi and Sacramento river floods of 1915 and 1916.[7] Community, state, and even individual enterprise supplied much of the initiative in flood control during the 1850-1917 period. However, it is significant that this water problem, like that of providing navigation, increasingly was recognized as one appropriate to federal initiative.

A somewhat similar history, broadly considered, is seen in the case of irrigation. Here again much of the early initiative was supplied by individual, local community, or local public district enterprise, as is illustrated by the history of the Spanish settlement in the Southwest, Mormon settlement in Utah, and the later establishment of hundreds of irrigation districts throughout the arid West. Early federal development legislation on this subject consisted largely of attempts to reinforce the position of such enterprise.[8] However, the initiative supplied in this manner was not sufficient to develop the extensive public lands of the West[9] as rapidly as the Congress considered desirable. Accordingly, the

[3] Act of April 14, 1820, 3 Stat. 562, 563. A report on this survey by the Board of Engineers was sent to the Congress in 1823 by President Monroe. J. D. Richardson, *Messages and Papers of the Presidents*, Vol. 2 (Washington: U. S. Government Printing Office, 1896), p. 199.

[4] Act of May 24, 1824, 4 Stat. 32.

[5] President's Water Resources Policy Commission, *Water Resources Law*, Vol. 3 of Report (Washington: U. S. Government Printing Office, 1950), p. 89.

[6] Act of March 2, 1849, 9 Stat. 352; Act of September 28, 1850, 9 Stat. 519.

[7] Act of March 1, 1917, 39 Stat. 948.

[8] As for instance, the Desert Land Act of 1877 (Act of March 3, 1877, 19 Stat. 377, as amended, 43 U. S. C. 321); and the 1894 Carey Act (Act of August 18, 1894, 28 Stat. 372, 422, 43 U. S. C. 641 *et seq.*). The same might have been said for early federal legislation on both navigation and flood control.

[9] The effect of failures under the homesteading laws and private irrigation difficulties also must be taken into account.

Reclamation Act of 1902 was passed,[10] "and with its passage, Congress established irrigation in the West as a national policy."[11] From 1902 onward the federal government became an increasingly important entrepreneur in irrigation development, largely to meet the challenging problems placed before the nation in its ambition to develop the West.

It is clear, therefore, that the need for federal initiative in water development was felt strongly long before the appearance of technical development favoring the scale economies. As these needs were felt increasingly, agencies grew within the Executive Branch which supplied the initiative for development—mainly the United States Army Corps of Engineers and the Department of the Interior Bureau of Reclamation.[12] Thus, by the time that understanding came as to the widespread possibilities of using scale economies in water development, the organizations in the best position to be entrepreneurs for projects using those economies were in the federal government. This occurred between the late 1920's and early 1930's, when mechanical earth moving, improved concrete construction techniques, improved electrical generating, and transmission equipment, multipurpose water control, and other technical changes came into use.[13] Accustomed as they were to treating development problems on a national or half-national scale, responsible for whole regions or river basins, pressed for the solution of a number of problems within some basins, these federal agencies turned naturally to the use of the scale economies. The same was true of the TVA, which was materially assisted in its early days by imaginative personnel from the older agencies.

Over the period during which the important federal agencies were developing their interests and capacities, other sources of initiative or "entrepreneurship" did not manifest a similar capacity or interest in accepting geographically widespread responsibility. The objective of nearly all private entrepreneurship was service to a very limited area.

[10] Act of June 17, 1902, 32 Stat. 388, 43 U. S. C. 391 *et seq.*

[11] President's Water Resources Policy Commission, *Water Resources Law* (Vol. 3), p. 181.

[12] In principle, initiative is supplied only by the Congress or the President, as public statements of Corps officials often have repeated. In practice, however, both the Corps and the Bureau certainly share the function of supplying initiative for development. No responsible, *technically competent* agency exists above the Departmental level.

[13] Some large structures were built a decade or so earlier, but with techniques which did not favor large-scale construction as later conceived. The rubble-masonry Roosevelt Dam on the Salt River, Arizona, completed in 1911, is an example.

The same certainly was true of community initiative. The concept of geographically extended service[14] did not prevail at any time in private navigation efforts, private irrigation on a district basis, private or district efforts toward flood control, domestic and municipal water supply, or even in the privately owned electric utilities. Extended financial integration of the electric utilities in the 1920's did not mean extended service objectives. With a very few exceptions, like the Pacific Gas and Electric Company in northern California, company policies tended to concentrate upon service to high-density markets and mainly urban areas. A notable exception of nonfederal flood-control efforts was the organization of the Miami Conservancy District in Ohio.

Additional reasons for nonfederal lack of interest in geographically extended entrepreneurship, previous to 1930, were the size of the river basins, the nonreimbursable nature (or tradition) of several water development purposes, and the production alternatives for the principal vendible service derived from water, i.e., electricity. The privately owned utilities did not consider themselves as water development agencies on more than an incidental basis. In the populous parts of the country steam sources of electricity rather than hydro often were more satisfactory for a profit-making organization.

Thus the initiative for applying the scale economies to water development, which started in the late 1920's, fell almost by default to the federal government of the United States. Circumstances may have made this mandatory. In any event, the large organizations needed to take advantage of the scale economies have been supplied since that time by the federal government. They have been the Corps of Engineers, the Bureau of Reclamation, and the TVA. In a more limited way the International Boundary and Water Commission, United States and Mexico which has jurisdiction over the Rio Grande below Fort Quitman, Texas, and the recent St. Lawrence Seaway Development Corporation also should be included. These agencies have been the main vehicles for supplying "entrepreneurship" for use of the scale economies. Noteworthy past projects were Boulder Dam (Colorado River, Bureau of Reclamation, dam completed 1935), Grand Coulee (Columbia River, Bureau of Reclamation, dam completed 1941), the Central Valley Project (California, Bureau of Reclamation, commenced 1935),[15]

---

[14] "Geographically extended" indicates an area larger than a locality or group of adjacent localities.

[15] This project had been planned by the state of California, but the Bureau of Reclamation came in as "entrepreneur."

Fort Peck (Missouri River, Corps of Engineers, completed 1939), Bonneville Dam (Corps of Engineers, Columbia River, completed 1938), and the series of sixteen dams on the Tennessee and its tributaries completed by the TVA between 1936 and 1944.

It is principally since 1950 that any notable works which make use of the scale economies have been undertaken by agencies other than those of the federal government. This change has followed the unavailability of federal funds for financing and federal government policies encouraging nonfederal entrepreneurship since 1953. The plans and construction operations of several of the Public Utility District in the Pacific Northwest for dams on the Columbia River or its tributaries, the New York State Power Authority operations, and Alabama Power Company plans for the integrated development of the Coosa River in Alabama are examples. The operations of the Pacific Gas and Electric Company in California, effective long before 1950, should be listed as another example. In its transmission system, and in the integration of a number of hydro plants located in the California valleys, this company did make use of scale economy principles. However, the largest of its hydro projects were of only moderate size,[16] and were not so located as to be part of an integrated water control system. The same can be said for the system of the Southern California Edison Company, which also took advantage of scale economies in its system before 1950. Since that time these companies and others, like the American Gas and Electric Company, increasingly have applied scale economies within the limits of the opportunities available to them. However, for many utilities the size of service area still is a constraining factor.

An interesting example of the relation of size of agency to the undertaking of large works has occurred since 1950 in the Hells Canyon case. In this, a relatively small agency, the Idaho Power Company, was licensed to develop a site on the Snake River (Idaho-Oregon), which might have been developed as a major storage work and hydro plant— the proposed Hells Canyon installation. The company, which had a limited generating capacity as late as 1956, a limited service area, and no multipurpose interests, proposes to develop the reservoir site with three dams. The construction of one of these three dams on the Hells Canyon reservoir site, now under way, does not fit an integrated basin

[16] Largest hydro plant of Pacific Gas and Electric Company in 1950: Pit No. 5, Pit River, California, 128,000 kilowatts installed capacity. See President's Water Resources Policy Commission, *Ten Rivers in America's Future* (Vol. 2), p. 92.

plan for the Columbia, and it does not make the best use of the scale economies in developing the site.[17] Lack of size and scope on the part of the entrepreneur in this case clearly discouraged full attention to the scale economies in development.

Thus in 1958 the federal government continued as the principal entrepreneur undertaking large water development works. Evidences of this were to be seen in the actual operations and plans of the Corps of Engineers (as on the Columbia and Missouri rivers), the St. Lawrence Seaway Development Corporation, and the Bureau of Reclamation (Upper Colorado River Storage Project).

CONSTRUCTION ON A MULTIPLE-USE BASIS

Responsibility for initiative on multiple-use development has had a history in the United States somewhat parallel to that of employing the scale economies. Here again federal "entrepreneurship" has become prominent. The reasons are similar to those applying to the emergence of the federal government as entrepreneur for large or geographically extended systems of works. The nonreimbursability of many of the purposes in development have made them unattractive or prohibitive to agencies not possessing taxing power. Among the eight distinctive uses of water, four have been almost entirely nonreimbursable (in the sense of the entrepreneur recovering his investment with return, plus operation and maintenance charges, from the users or beneficiaries). They are navigation, fish and wildlife, recreation, and waste carrying. Flood control and drainage are largely nonreimbursable. Irrigation has become increasingly nonreimbursable.[18] Only electric power production from hydro, industrial water supply, and municipal water supply "pay their way" in the manner of business practice. The stage thus is set for

[17] J. V. Krutilla and Otto Eckstein, *Multiple Purpose River Development—Studies in Applied Economic Analysis* (Baltimore: The Johns Hopkins Press for Resources for the Future, 1958), chap. v.

[18] The fact that payment for recent western irrigation projects has been tied to electric power earnings from the power facilities associated with the irrigation projects does not make irrigation itself reimbursable. Where power earnings are used for payment, the principal difference from the position of flood control is that the immediate burden of payment does not fall on the general tax receipts. Where the interest component on the power investment has been used to repay irrigation investment, however, the burden of payment does fall on tax receipts. Another difference from flood control is that direct beneficiaries (water users) do make some contractual repayment for the facilities constructed.

a public agency as the likely multiple-use entrepreneur in water development. This is strengthened by the fact that multiple uses are likely to be encountered in greater number and in more intricate relation in some rough proportion to the size of area considered. The larger the basin or other development area, the greater the total benefit likely to be realized from multiple-use development. The same relation prevails where water use is intensive, as in a densely settled industrialized basin.[19] The emergence of public agencies as the entrepreneurs for multiple-use developments thus has been more or less foreordained.

If the position of private enterprise or community initiative is compared with that of the federal government, the more favorable position of the federal government should be clear. The position of the individual state governments vis-à-vis the federal government is less sharply defined. Individual states obviously have multiple-use interests in water often covering the entire range of uses. Many of the states have agencies which express these interests officially. The size of American river basins and the long history of federal geographically extended operations have favored the federal government again, but need not have made its agencies the dominant multiple-use entrepreneurs that they have become.

Even though centered in the federal government structure, organization for realizing multiple-use development in the United States has been cumbersome. Each of the major federal agencies with national responsibilities in the field has had a special point of view, special-interest clienteles and political ties, and functionally limited responsibilities. Entrepreneurship for development thus has appeared with one of two characteristics: (a) elaborate co-ordinating mechanisms to carry out the multipurpose view in construction and other development;[20] or (b) development undertaken with a dominant point of view of one of the major specialized agencies, with an ancillary position for the uses not within the province of the entrepreneurial agency. The elaborate mechanism of the federal interagency river basin committees (see chapter 18, pages 501-3) has been necessary not only in the plan-

[19] Suggested by Blair Bower, letter to the authors, January 1958.

[20] Thirty-nine major federal bureaus or independent agencies were listed in 1950 as having interests in water development on a national basis. Four more were concerned where some special regions were concerned, like international boundary basins. President's Water Resources Policy Commission, *Water Resources Law* [(Vol. 3 of Report) p. 430]. This has been reduced to thirty-seven since 1953.

ning of river basin work, but in co-ordinating the construction and other development on a multi-use plan. In the maze of negotiations necessary to maintain this kind of entrepreneurship, it readily can be seen that this technique must have operated with some delay and friction within the United States. Since the establishment of the "partnership" policy in 1953, the number of agencies to be considered in multiple-use entrepreneurship has grown, and the problems of co-ordination have grown accordingly. These problems have been well illustrated in the history of development within the Columbia Basin since 1953.

The principal case that departs from this pattern has been the TVA, which was empowered in its Act to supply initiative on a broad multi-use pattern in its responsibility for the Tennessee River. While the development of the Tennessee has not been without problems of co-ordination (as between TVA and the Department of Agriculture), the simplified arrangement under a single management has given the multiple-use technique a fair test of its potentialities.

Trends in the federal government organization to meet the needs of multiple-use development may perhaps be discerned in the gradual addition of new interests on the part of the major development agencies. As has been observed previously, the recent civil works history of the Corps of Engineers has shown a gradual broadening of its functional interests. To the original navigation supervision, flood control and hydro power production have been added. The 1944 Flood Control Act enabled the Corps to provide for recreational facilities at projects, and to contract for the sale of surplus water for domestic and industrial purposes.[21] It seems likely that in the near future the Corps will further be enabled to construct for industrial and municipal water supply on a basis coequal with other purposes.[22] In addition, organizational provision has been made within the Corps to treat recreation and fish and wildlife on a professional basis, although the co-operative assistance of the Fish and Wildlife Service and the National Park Service has been available to the Corps for some time. Through its long-standing responsibility for the lower Mississippi Valley, it is entirely possible that eventually the Corps also may be drawn into irrigation

[21] Act of December 22, 1944, Sec. 6, 58 Stat. 887, 890, 33 U. S. C. 708. For recreation: *idem,* as amended, 16 U. S. C. 460d.

[22] The low flow regulation suggested in the vetoed 1955 flood control act would have this effect.

development.[23] A similar broadening of interest, although with somewhat different manifestations, may be traced within the Bureau of
Reclamation. Authorization for broadened responsibility was particularly notable in the Reclamation Project Act of 1939.[24] Thus it seems
that United States water development organization slowly is adjusting
to the use of multiple-purpose development.

BALANCING WATER DEVELOPMENT WITH THE
DEVELOPMENT OF OTHER RESOURCES

The next step beyond multiple use in development is the relation
of water to other resources in their development. The interrelation of
water and other resources is frequent and apparent. The very term
"irrigation" implies a relation to the land resource. There also is a close
relation to both forest and mineral resources in their use. Among the
many interesting aspects of this topic only one needs brief treatment
here. It concerns those instances in which technologic change has
altered a relation in such a manner that some reflection on administrative organization is indicated.

Many, if not most, of the developmental relations among resources
are determined by the mechanism of the marketplace. As already noted,
the nonreimbursable nature of many water-derived services in the
United States makes important parts of American water development
unsuitable to the market mechanism. Consequently, some problems
have arisen for the administrative organization of development which
concern the relation of water to the development of other resources.
Two illustrations will be used: relation of irrigation development to
national agricultural productive capacity, and the relation of water
development to the use of coal resources.

Irrigation and the land resource. The development of irrigation works
in the United States for long has been undertaken under four forms
of entrepreneurship: private, local district, state, and federal government
organization. Privately organized works have brought major additions

[23] No official statements of the Corps of Engineers support this conjecture.
However, the past position of the Corps as the lower Mississippi development
agency, the regional position of the principal irrigation agency, the Bureau of
Reclamation, the recently commenced intensive development survey of the lower
Mississippi may be interpreted as pointing in this direction. Studies of irriga-
tion under Corps auspices also are under way in the Arkansas, Potomac, and
Delaware river basins.

[24] Act of August 4, 1939, 53 Stat. 1187, 1194, 43 U. S. C. 485h.

to irrigated acreage in the West within the last fifteen years.[25] Private enterprise also has been active in the extension of irrigation within eastern United States. Local district programs have been of great importance in the past development of the West. State development programs have been of a lesser order, but in some states, like Montana and Nebraska, they are significant. In part, state and local entrepreneurship has held an important place in irrigation because small and moderate-sized developments frequently have been feasible. However, the recent plans of the state of California for a $1.6 billion implementation of the California State Water Plan may place state entrepreneurship in an entirely different size class in the future. Finally there is the federal government, which has acted as entrepreneur in this field at least since the 1902 Reclamation Act.

The succeeding discussion is concerned with the place of the federal government in accepting entrepreneurship, as it relates to technical developments. In view of the fact that private, local district, and state government entrepreneurship has continued, and has given recent signs of vigorous activity, it should be assumed that any federal government activity might best be undertaken with a distinctly national point of view. As far as irrigation entrepreneurship is concerned, this national point of view might be expressed in three different ways: (1) undertaking works designed to add needed productive capacity to the nation's agricultural lands at lowest apparent costs; (2) inclusion of irrigation facilities in large-scale multiple-purpose projects justified by other purposes, but promising irrigation benefits economically justified on an incremental basis; and (3) providing an entrepreneur of long-range view for sections of the country whose longer range potentialities are not matched by immediate attractions to the other sources of entrepreneurship in irrigation. The latter function might be termed a frontier function of the federal government in the past, although regional situations in the United States have shown that this type of need by no means is limited to the frontier.

The Bureau of Reclamation and its predecessors have been the principal agents of the federal government as entrepreneur in the field of irrigation. Their policies and practice have included an element of all three functions throughout the history of federal government entrepreneurship in constructing irrigation works. At first, from the

---

[25] The High Plains Underground Water Conservation District (Lubbock, Texas) reports that about 3 million acres have been brought under irrigation on the High Plains of Texas within this period by private enterprise.

days of Major Powell and the initial Reclamation Act, the first and third functions were probably of equal importance in the federal government operations. The West then was the principal hope for added farming opportunities, to take care of coming immigrants and those who, to a degree, had worn out some of the lands of the East. The hand of the federal government also was necessary to cope with the problems of this difficult and unusual frontier. Thus the irrigation entrepreneurship built around the Bureau of Reclamation came into existence, and thus it has operated.

In the period since 1902, however, striking changes in the techniques of agricultural production have taken place. Some of these, like techniques of large-scale construction, have been to the benefit of western irrigation, and have helped to vindicate the judgment of those who saw opportunities of national importance in the lands of the West. At the same time the major twentieth century impact of technical changes in agriculture has been on the lands of the East, where the greatest over-all cropping potential[26] of the country lies. Within this period have come significant improvements in producing, distributing, and using commercial fertilizers, dozens of improvements in plant and animal stock including hybridized grains, mechanized operation of farms, and efficient techniques of humid land (or "supplemental") irrigation. Each of these new elements has helped to change the status of eastern agriculture as a vehicle for the further development of the nation's agricultural productive capacity. We are now beginning to realize that proper provision for the application of water to the land is as important in capturing the full productive potential of eastern agricultural lands as it was for the capture of the great natural fertility of the arid lands of the West. Thus while some of the post-1902 technical changes have also been applicable in the West, by far the greatest improvement in potential productivity has been experienced on humid lands.

Administrative provision for federal entrepreneurship in irrigation development has not reflected these technical changes fully. The principal agent for federal enterprise in the field is still the Secretary of the Interior,[27] for whom the Bureau of Reclamation acts. His jurisdiction

---

[26] This does not necessarily mean the highest cropping potential per unit area. Where irrigation water is available, some western lands have very high production potentials per unit area, as in the Imperial Valley of California.

[27] The United States Department of Agriculture has some irrigation interests through its Soil Conservation Service activities, and the Corps of Engineers has

is still limited in these affairs by the provisions of the 1902 Act, which applied to sixteen western states[28] and to Texas.[29] There is, accordingly, no federal entrepreneurship which takes the interest of the entire United States into account as far as irrigation is concerned. Furthermore, there is little provision for considering irrigation and other techniques (fertilizers, seeds, etc.) of adding to agricultural production on a simultaneous, coequal basis. Irrigation is only one of several techniques, the application of which may add needed increments to agricultural production. Efficient administration would provide means for reviewing the problems of increasing agricultural production which would take into account all technical possibilities. In view of these facts, there is at least reasonable doubt that administrative provision has been made for one of the national functions in this field: that of adding needed agricultural capacity to the national plant at the lowest apparent cost, and of undertaking irrigation works as part of a general program for such capacity. Federal entrepreneurship for water development thus is not matched to land development in the manner that technical developments of the last decades might suggest.

The reasons for this failure probably are deeply imbedded in the political process. The disposition and practice of direct federal agency relations with individuals and local units no doubt has encouraged decision-making on the basis of immediate sectional politics rather than on long-range national economics. In this instance the bypassing of the state in Bureau of Reclamation development may have caused the maintenance of a much more provincial point of view in the national government administration than is appropriate to it.[30] In any event, there is still ample opportunity for profitable employment of co-ordinated land use improvement, including irrigation facility development.

assisted irrigation development incidental to some of its projects, as in the St. Francis Basin (Arkansas-Missouri). They are minor in comparison with the Bureau's operations. The Department of Agriculture also has operated under the Water Facilities Act (Act of August 28, 1937, 50 Stat. 869, 16 U. S. C. 590r-590x), with projects limited to $100,000 federal expenditure on any single project. The Bureau of Indian Affairs, U. S. Department of the Interior, also has been an irrigation entrepreneur on behalf of reservation Indians.

[28] Sec. 1, 32 Stat. 388, 43 U. S. C. 391. The states are Arizona, California, Colorado, Idaho, Kansas, Montana, Nebraska, Nevada, New Mexico, North Dakota, Oklahoma, Oregon, South Dakota, Utah, Washington, and Wyoming.

[29] The Reclamation Act of 1902 was made applicable to the entire state of Texas by the Act of June 12, 1906 (34 Stat. 259, 43 U. S. C. 391).

[30] See Henry M. Hart, *The Dark Missouri* (Madison: University of Wisconsin Press, 1957), pp. 98-119, 170-83.

Relation of water development and coal resources. Another facet of the relation of water to other resources is best seen through the field of electric energy supply and demand. Although hydroelectricity is not competitive with other energy sources for many uses, there are many and expanding uses for electricity. Low-cost and moderate-cost hydro have very useful positions in our national energy supply, particularly where demand for a large-capacity system has emerged, or where other energy supplies regionally are deficient. Even high-cost hydro may be useful for capacity in some special circumstances.

It is particularly in the areas which have other energy resources than falling water, and which have developed (or are likely to develop) needs for a large-capacity system, that the question of the relation of water to other energy sources arises.

Within the region where abundant coal resources lie and where the coal industry developed in the past, it often has been assumed that water development which included hydro was not in the best interest of the coal industry.[31] Since steam electric generation was well suited to single-purpose needs in the coal-producing areas, entrepreneurship for water development which included hydroelectric generation has in some instances found little support. It is probable that the development of the great Ohio River Basin has suffered because of this supposed competition.[32] Some other streams in eastern United States bear similar histories. An assumed competitive relation between coal and hydro discouraged entrepreneurship from multiple-purpose development of the coal area streams, thus shutting out not only hydro development, but the cheaper development of navigation channels under multiple-purpose construction. Navigation facilities are of definite advantage in moving coal for thermal generation.[33]

The extent of electricity demand assumed in the coal-hydro competition theory, however, appears to have been much underestimated. The experience of the Tennessee Valley has been one where extensive hydro development on the river was accompanied by an increase in coal

[31] This attitude has been found in several parts of eastern United States, but has been best illustrated in Ohio, Pennsylvania, and West Virginia.

[32] See President's Water Resources Policy Commission, *Ten Rivers in America's Future* (Vol. 2 of Report), pp. 672-73.

[33] For testimony on this point see Committee on Government Operations, House of Representatives, *Hearings before a Special Subcommittee on Commission on Organization of the Executive Branch of the Government (Water Resources and Power Report)*, Part 7, Milwaukee, Wis., November 7-8, 1955 (Washington: U. S. Government Printing Office, 1956), p. 1407.

production far above national indices.[34] The value of hydro for peaking purposes where the base load is served by large-capacity steam units is now recognized, not only in the operation of the TVA system, but for some other large systems as well. The flexibility of hydro generation can make it much more economical as stand-by capacity or as a "spinning reserve" than large or small steam units.[35] Under present technology, and indeed under that of the past fifteen years, the ideal large-capacity system for many parts of eastern United States would make judicious use of both coal and hydro in a carefully planned combined use of the two resources for producing electricty, with hydro to provide needed peak energy, and steam plants for the base load. This has come to be particularly true as steam generating units have increased in size and their fuel use has become more efficient. A national organization for water development, prepared to make alert and responsive adjustment to changing technology in electricity generation, should have provided entrepreneurship capable of implementing this point of view. In this instance, as in that of an area with growing urbanization and industrialization, early recognition of the hydro needs was important because of the danger of site foreclosure by other land uses. This was provided in the TVA system, and to a degree it has been provided in the areas where some privately owned utility systems have operated, like the Duke Power Company's service area in North Carolina, and in California. In general, however, the limited extent of most eastern United States electric utility service areas, the nonreimbursability of several water development purposes, and the disinclination to consider the position of both coal and hydro in the same development efforts, have not produced entrepreneurship balancing these two resources in the manner that technology for some time has indicated they profitably can be balanced. It is of interest that alert management is constructing steam generating facilities based on coal, even where moderately large federal hydro facilities are being built at present. This is happening in North Dakota, eastern Montana, and Minnesota.[36]

[34] Coal production in Tennessee and Kentucky averaged 42,192,000 short tons per year from 1931 to 1935; two decades later this figure had increased to 72,256,000 tons. TVA's coal consumption for electricity production in 1958 is expected to be 18 million tons.

[35] Assuming a reasonable initial investment. These relations of course are determined by individual plant costs. However, in some situations even high-cost hydro can have considerable value for service to peak loads.

[36] Edwin Wilson, personal communication to the authors, 1957.

One of the more interesting problems attached to the field of water development, or indeed to resource development generally, is the problem of meeting geographically dispersed demand. Technologic improvements often have had a way of concentrating their benefits in the districts or localities where people are most concentrated. The entrepreneur or entrepreneurs are always most concerned about meeting the needs of the area of most concentrated demand first. Where the market mechanism applies, supply of course will adjust eventually to meet all valid demands, no matter how dispersed they are. However, this adjustment may be very prolonged. In addition, where public entrepreneurship must be depended on, as for some phases of water development, public organization must provide for meeting the dispersed demand.

One of the best examples of the problem of meeting dispersed demand occurred in the retail distribution of electricity. From its beginning in New York City, electrical service spread without difficulty to include all cities and most towns in the United States. Entrepreneurship was readily provided for the needed construction and service to meet the more concentrated demand. At the same time, the dispersed needs of rural areas for electrical service had not been met over the greater part of the country. As late as 1935, only about 12 per cent of the country's farms had electrical service,[37] and many of these were in the vicinity of towns or cities. Clearly entrepreneurship for the provision of rural electrical service was lacking over a long period, in spite of a definitely expressed demand for such service. For this reason hundreds of rural electric cooperatives were organized throughout the country. They were backed by the lending powers of the federal Rural Electrification Administration (see page 557). The problem of entrepreneurship was met finally by public administrative organization on a large scale, and after a few years was supplemented by private utility efforts in the same direction. By 1956 the problem of dispersed electricity demand could be described as nearly nonexistent throughout the country (figure 38).

The same type of problem, although less widespread, may be found in other kinds of demand served wholly or in part by water facilities. Problems of providing for flood control, irrigation, drainage, domestic water supply and stock watering, pollution abatement, recreation, and still

[37] G. C. Clapp, *The TVA, an Approach to the Development of a Region* (Chicago: University of Chicago Press, 1955).

other needs may be recognized for areas of dispersed demand even while they have been successfully met for more peopled districts or localities. In most instances organization for entrepreneurship is not yet effectively provided in the United States.

A striking illustration of the existence of this problem comes from the Tennessee Valley. After twenty years of operation, including the installation of almost complete control works for the Tennessee River and its tributaries, it became apparent to the TVA Board that the development of the river and works associated with it was having a differential impact upon the counties of the region. Scattered throughout the Valley and the TVA power service area were districts which had lagged behind the Valley as a whole, and particularly behind the more favorably located farming and urban areas. Benefits from the TVA program, insofar as they could be tied directly to income and other measurements of economic progress in the region, had been unevenly distributed. While both geographers and economists will recognize in this situation the old problem of differential development influenced by differences in both resources and space relations, the application of new technology (multiple-purpose development) did magnify the differences. A problem of entrepreneurship for more equitable development thereby arose. The lagging districts were called "Areas of Special Need" by the TVA, and a program was set up within the organization to provide some entrepreneurship for meeting the development problems of these parts of the Valley.[38]

Similar problems have emerged elsewhere in the United States as the larger water development programs have proceeded. The need for small watershed flood-prevention and flood-control measures has been apparent for some years since the attention of the Corps of Engineers has fallen mainly on the large volume of higher priority needs. It was in part to meet this need that Congress empowered the Secretary of the Army in 1948 to use a small part of appropriated flood-control funds for the construction of unspecified small flood-control projects.[39] This was the first national program designed to provide entrepreneurship for the dispersed local flood-control demands. It was also in part to meet this need

---

[38] *Annual Report of the Tennessee Valley Authority, 1953* (Washington: U. S. Government Printing Office, 1953), pp. 48-49; and Tennessee Valley Authority, *Working with Areas of Special Need,* Knoxville, 1953.

[39] Act of June 30, 1948, Sec. 205, 62 Stat. 1171, 1182, 33 U. S. C. 701s (Supp. III). The maximum set for total expenditures in any one year was $2 million. This sum in 1950 was increased to $3 million.

that the Soil Conservation Service in the Department of Agriculture was authorized in the Watershed Protection and Flood Prevention Act of 1954[40] to undertake a program of small watershed development, including flood control. TVA also established a Local Flood Relations Branch within its organization. However, the function of the TVA office was much more one of encouraging the growth of local entrepreneurship in this field than of supplying TVA entrepreneurship. In addition, a few states maintain organizations which might be interpreted as exercising some degree of initiative in small-project flood control. Fifteen states have administrative provision within their organization for considering flood-control matters.[41] However, only six are reported to have programs which involve financial expenditure by the state itself.[42] In general, state initiative is probably more concerned with relations to the Corps of Engineers' program than with independent flood-control programs.

The situation has been similar for small-project irrigation.[43] While private enterprise has supplied the greater part of the needed entrepreneurship for past small-project irrigation, enough special needs have been felt to bring public initiative into the field for at least twenty years. The first federal provision for such initiative was in the Water Facilities Act of 1937.[44] Administration of the program was assigned to the Farmers Home Administration. Since 1949 this program has applied to projects with a maximum federal expenditure of $100,000.[45] Since 1956

[40] As amended in 1956 (Act of August 8, 1956, 70 Stat. 1088). The Department of Agriculture previously had been authorized to undertake an *upstream* flood control program by the Flood Control Act of 1936 (Act of June 22, 1936, 49 Stat. 1570, 33 U. S. C. 701a). However, relatively little work was done under the terms of this authorization until 1953. Since 1953, sixty-two "demonstration" watersheds have been established which include upstream flood control.

[41] California, Connecticut, Florida, Illinois, Indiana, Iowa, Kentucky, Louisiana, Michigan, Minnesota, New York, North Carolina, Pennsylvania, Vermont, and Washington.

[42] California, Florida, Illinois, Michigan, Pennsylvania, and Washington. (The Council of State Governments, *State Administration of Water Resources* [Partial Report], Chicago, 1956 [mimeographed], p. 16.)

[43] The National Resources Committee maintained an Advisory Committee on Small Water Projects during part of its existence, in recognition of lack of attention to this as well as other small-project problems. One example of more recent Congressional concern with the problem occurred in the activities of the Honorable Reva Beck Bosone, U. S. Representative from Utah, during 1951 and 1952. (See for example House Document No. 408, 82nd Congress, 2nd Session, 1952.)

[44] Act of August 28, 1937, 50 Stat. 869, 16 U. S. C. 590r-590x.

[45] Act of June 10, 1949, 63 Stat. 171, 16 U. S. C. 590z-5 (Supp. III).

the Bureau of Reclamation also has made administrative provision for the construction of small projects.[46] The problem also has been recognized by at least two state governments. The Montana Water Conservation Board has administered a small irrigation projects development program within the state since 1934. A similar program recently was started in Wyoming. The North Dakota Water Conservation Commission also is responsible for a modest program of small dam maintenance and construction within that state.[47]

One special need of this type has been the irrigation of Indian lands in the West. Administrative provision has been made for this enterprise within the federal government since 1867, through the activities of the Bureau of Indian Affairs in the Department of the Interior.[48]

REDUCING LOCAL INTENSITY OF DEMAND

The water development demands created by the concentration of people and their economic activities within favored spots scattered over the country generally have been viewed as problems for the manipulation of supply. Theoretically, however, attention toward directing the location of demands might also help to match supply and demand. Within the American economic system, this is assumed to be taken care of in cost-price relations. Redirection of demand through public organization therefore has had a minor place in this country. To a degree, past efforts at city planning and regional planning have had some influence on the location of water demands. However, in such planning other considerations usually have overshadowed water demand location; the effects upon the location of water demand have been incidental. Even at the expense of ignoring the cost-price relations of the marketplace, American policy has favored meeting demand by the manipulation of supply. An example has been the federal reclamation policy for western irrigation. Generally, dispersal or relocation of demand has not been so

[46] Small Reclamation Projects Act of 1956 (Act of August 6, 1956, 70 Stat. 1044). This is basically a loan and grant program, but in practice a degree of leadership may be supplied by the Bureau for initiation of works.

[47] The Council of State Governments, *State Administration of Water Resources,* Chicago, 1957, p. 27.

[48] The first project was the construction of a canal for irrigation of the Colorado River Reservation in Arizona. (President's Water Resources Policy Commission, *Water Resources Law,* Vol. 3 of Report, p. 246.)

favored that it has received any administrative provision in public organization for water development.

Two exceptions may be cited, although they are only partial exceptions for the country as a whole. They relate to flood prevention and to stream waste carrying. As pollution of streams has increased in recent years, the conflict of industrial, municipal, and residential waste carrying with other water uses has become so evident that public organization for the dispersal, reduction, or elimination of waste-carrying demands now is found in many states. This usually is a function of a water pollution control board or commission, or an executive bureau or department within the state. However, the federal government increasingly has demonstrated its concern.[49] Reduction in the intensity of the waste-carrying need (through sewage treatment, or industrial waste treatment) has been the most common objective of these activities, but indirectly they may have affected the location of some industry with heavy waste-disposal problems. (Chapter 7, pages 169-83.) For this and other reasons, the presence of chemical industries with objectionable wastes has been discouraged in the immediate vicinity of some urban areas. Stream classification also has been a method used to protect water supplies. Some streams may be assigned primarily waste-carrying functions, while others have legal recreational or withdrawal uses.[50]

Flood prevention also has emerged as a problem susceptible to manipulation of demand. If occupancy of flood-plain sites can be limited to uses that are not likely to incur heavy damage in the event of flooding, the costs of flood prevention in many places can be materially reduced. In general, the American policy and practice for flood prevention has been to supply flood control works wherever flood-plain occupancy had reached an intensity that made levees, flood walls, or detention reservoirs economically warranted.

Some municipalities, like Cincinnati and Milwaukee, have sought to control flood damage by zoning flood plain occupancy.[51] However, in recent years a more general practice among settlements susceptible to

[49] See the Water Pollution Control Act Amendments of 1956 (Act of July 9, 1956, 70 Stat. 498).

[50] The Council of State Governments, *op. cit.*, pp. 33-35. See also James J. Flannery, *Water Pollution Control: Development of State and National Policy* (doctoral dissertation), University Microfilms, University of Michigan, Ann Arbor, 1956.

[51] R. E. Behrens, "Zoning Against Floods in Milwaukee County," *The American City*, 67 (1952), 112-13.

flood damage has been that of seeking Corps of Engineers' assistance in the construction of prevention works. Within the Tennessee Valley states, zoning of flood-susceptible lands has been encouraged through the community planning assistance and local flood relations program of TVA, and through the planning agencies of the states, like the Tennessee State Planning Commission. Since the Flood Control Act of 1954, federal provision in the direction of encouraging zoning also has included the Corps of Engineers' programs.[52] While the effects of the Federal Flood Insurance Act of 1956 are yet to be ascertained, it is possible that the dispersal of demand for flood prevention through zoning may be stimulated by the Act's provisions.[53] Commencing in the fiscal year 1959, public zoning restrictions can be required as a condition of insurance coverage under the Act. Where not already available, a state enabling act for zoning is required if communities within the state are to qualify for federal insurance aid.

In general, however, administrative provision for the redirection of water demand through the techniques of regional and city planning is of minor importance in this country's capacity to shape water development at this time.

Provisions for Financing Water Resource Development

Once the entrepreneurship is provided, provisions for financing projects still must be made. In practice, questions of entrepreneurship and financing may be closely related, and the nature of one at least partly dependent upon the other. Thus public agency entrepreneurship is likely to mean financing from public funds, and private enterprise is likely to mean financing from private sources. But there are significant exceptions.

There have been seven major direct sources of funds for financing water development facilities and operations: (1) appropriations from tax funds; (2) sale of securities on the private bond market; (3) participation of private equity capital; (4) private bank loans; (5) public credit corporation loans; (6) funds acquired from the sale of public

---

[52] Act of Sept. 3, 1954, Sec. 209, 88 Stat. 1256. Also see Gilbert F. White *et al., Changes in Urban Occupance of Flood Plains in the United States* for a description of flood prevention problems related to zoning (Chicago: University of Chicago, Department of Geography Research, Paper No. 57, 1958).

[53] Act of August 7, 1956, Sec. 13 (d), 70 Stat. 1078.

assets; and (7) operational revenues over and above capital costs, operation, maintenance, and replacement.

From the point of view of the problems posed for water development, the principal distinctions among these seven sources of funds are between the public and private sources of financing. The important problems brought by technology have been the financing needs for large-scale integrated development; the meeting of dispersed demands; and response to the possibilities of the scheduling techniques, technical development, and scientific experiment.

USE OF PRIVATE FUNDS

Private financing has played a very important, if not a dominant, role in the development of water facilities in the United States, considering the extent and productivity of the facilities alone. It has been employed wherever vendible products or services could result from development. Some early navigation improvements were privately financed. Most early irrigation development capital also came from private funds, and a substantial volume of financing for the smaller scale irrigation facilities is still from private sources.[54] A very high percentage of the total invested in domestic and industrial water supply facilities has been from private sources, whether the entrepreneurship was private company, corporation, municipality, or district. Many single-purpose hydro-electric projects throughout the country have been financed through the private bond market and the participation of private equity capital. All of the private utility hydro projects have been financed in this manner,[55] and some public agency hydro projects have been financed by revenue bond sales.[56] Private financing also has played a major role in swamp- or bog-land drainage in the Midwest and Southeast, which contributed

[54] Examples: the 3-million-acre High Plains irrigation development in Texas; sprinkler irrigation developments within humid areas, and all measures and equipment required directly on any irrigated farm. It also should be remembered that banks play a part in financing the establishment of irrigation farms on federal or state projects, through equipment, building, and other loans.

[55] Exceptions would occur in those instances in recent years where accelerated tax amortization has been granted on hydro projects. This in effect is a contribution to the cost of the project from public funds.

[56] For example, the Public Utility District projects in Washington State, or the projects of public power municipalities like Seattle and Tacoma, Wash.

about a quarter of the nation's crop land.[57] Finally, private investment also figures in the financing of facilities which are needed for suitable disposal of wastes in streams by both industries and municipalities, and in the peripheral developments which assist the use of water recreation facilities.

FIRST REASONS FOR FINANCING FROM PUBLIC FUNDS

The relation of the use of public funds for financing water development projects to technology can best be understood in connection with the history of the nonreimbursable water development purposes. The basic reasons for such financing are the constitutionally expressed concern of the federal government for the general welfare, the promotion of commerce, and the conduct of the national defense.

The earliest federal investments were in navigation. In 1828 and 1846 Congress made land grants to Alabama, Iowa, and Wisconsin to encourage river improvements in those states.[58] Although Congress was apparently hesitant to assume sole responsibility for developing navigation works in the nation, its appropriations even before 1885 were generous.[59] After 1879 its policies increasingly favored free passage on all waterways, and it used improvement appropriations,[60] donations,[61] condemnation,[62] and purchase[63] of private and state facilities toward that end. By the mid-1920's the principle was firmly established that in the interests of promoting the nation's commerce, navigation facilities were to be open to free passage, and that the federal government had an abiding interest in providing adequate investment funds for that purpose. Thus a major water development purpose became nonreimbursable.

[57] J. R. Whitaker and E. A. Ackerman, *American Resources* (New York: Harcourt, Brace & Co., 1951), p. 145.

[58] President's Water Resources Policy Commission, *Water Resources Law* (Vol. 3 of Report), p. 89; B. H. Hibbard, *A History of the Public Land Policies,* 1924, pp. 240-41.

[59] *Idem.*

[60] Act of March 3, 1879, 20 Stat. 363, 371; Act of August 2, 1882, 22 Stat. 180, 189; and others following.

[61] Act of June 14, 1880, 21 Stat. 180, 189 (St. Mary's Falls Canal, Mich.); and others following.

[62] Act of June 3, 1896, 29 Stat. 202, 217 (improvements on the Monongahela River).

[63] Act of July 25, 1912, 37 Stat. 201, 206 (Chesapeake and Albemarle Canal); and others following.

A similar but less extended history brought forward the same policy for flood control. The first federal interests in flood control, like those in navigation, were expressed in land grants.[64] However, until 1917 a large percentage of the investment made in flood-control facilities came from private funds, local levies, or state appropriations. In 1917 came the first large appropriation specifically for flood control.[65] It followed the impressive and disastrous floods of 1915 and 1916 on the lower Mississippi River. The huge size of the works required, the engineering complexity of a secure system of protection, and the number of states concerned (eleven on the lower Mississippi and Ohio rivers alone) disclosed the necessity of federal financial participation in flood prevention and management soon after the occupancy of the lower Mississippi made it economically important to the nation. In fact, the existence of a drainage basin of the Mississippi's size predisposed the nation to the degree of federal participation in flood control and mangement which came late. The 1917 appropriation was followed by a flood-control appropriation of similar size in 1923.[66] After the huge appropriation of 1928 ($325 million)[67] there was little question that flood control would be firmly established also as a nonreimbursable purpose in water development. While some provision for local contribution has been continued through all federal flood-control legislation, the principle of federal responsibility for the greater part of flood-control investment in the country on a nonreimbursable basis was quickly established. This was spelled out in the 1936 Flood Control Act, the basis for subsequent national policy.[68] Congress considered this policy desirable in the interest of the national welfare, although the impracticality of recovering costs from downstream beneficiaries undoubtedly also influenced policy formation.

One other major water development purpose, irrigation, had a degree of nonreimbursability for investment in it long before the 1920's.

---

[64] The Swamp Land Acts of 1849 and 1850. (Act of March 2, 1849, 9 Stat. 352; Act of September 28, 1850, 9 Stat. 519.)

[65] Act of March 1, 1917, 39 Stat. 948. Appropriation of $50 million.

[66] Act of March 4, 1923, 42 Stat. 1505. Appropriation of $60 million.

[67] Act of May 15, 1928, 45 Stat. 534.

[68] ". . . the Federal Government should improve or participate in the improvement of navigable waters or their tributaries, including watersheds thereof, for flood-control purposes if the benefits to whomsoever they may accrue are in excess of the estimated costs, and if the lives and social security of the people are otherwise adversely affected." (Act of June 22, 1936, Sec. 1, 49 Stat. 1570, 33 U. S. C. 701a.)

Federal contributions to irrigation development date from the Desert Land Act of 1877,[69] under which 640 acres of the public lands would be sold to any person who would irrigate the land within three years from the date of sale. The 1894 Carey Act authorized gifts to each public-land state of desert lands therein up to a million acres each, to aid in reclamation and settlement.[70] The 1902 Reclamation Act[71] continued this pattern, and that of initial federal assistance in other water development purposes, with the establishment of the Reclamation Fund. This fund was established and maintained initially with the monies received from the sale of public lands in sixteen western states. While repayment of construction costs was required of direct beneficiaries on irrigation projects, payment from the outset was without interest.[72] This, in effect, amounted to some contribution to the cost of the project (as compared to costs under private entrepreneurship) by the federal government. While the initial repayment period of ten years was short, it was extended in 1914 to twenty years,[73] and in 1926 to forty years.[74] Each extension of the repayment period increased the value of the interest contribution, which presumably was a recognition of the need of partial nonreimbursability to promote irrigation development. This need was borne out by the actual record of repayment from some of the irrigation projects.[75] Thus by 1930 the irrigation development policy of the federal government included some nonreimbursability.

Thirty-three irrigation projects were authorized, placed under construction, or completed before 1930, under the repayment provisions of the 1902 Act, the 1914 Act, and the 1926 Act. The total costs of projects allocable to irrigation under the first were $309 million; under the second, $30 million; and under the third, $89 million.[76] If an

[69] Act of March 3, 1877, 19 Stat. 377, as amended, 43 U. S. C. 321.

[70] Act of August 18, 1894, 28 Stat. 372, 422, U. S. C. 641 *et seq.*

[71] Act of June 17, 1902, 32 Stat. 388, 389, 43 U. S. C. 391 *et seq.*

[72] *Report of the President's Water Resources Policy Commission,* Vol. 1, p. 70.

[73] Act of August 13, 1914, Sec. 1, 38 Stat. 686, 43 U. S. C. 472.

[74] Act of May 25, 1926, Sec. 46, 44 Stat. 636, 649, 43 U. S. C. 423e.

[75] Repayments from settlers on Bureau of Reclamation projects were so small that they partly accounted for a dearth of federal construction between 1907 and 1920. Only one new project was constructed in this period. (President's Water Resources Policy Commission, *A Water Policy for the American People,* Vol. 1 of Report, p. 151.)

[76] Commission on the Organization of the Executive Branch, *Task Force Report on Water Resources and Power* (Washington: U. S. Government Printing Office, 1955), Appendix E following p. 716.

average interest rate of 3 per cent had been assumed for the entire period, the nonreimbursable element in irrigation construction would have been about 11.46 per cent of the total initial investment under the ten-year repayment provisions of the 1902 Act; about 36.4 per cent of the total initial investment under the 1914 Act; and about 88 per cent under the 1926 Act. In actual practice, required repayment by the water users was less than the allocation of costs to irrigation. Thus, of the total irrigation investment costs of projects authorized, under construction, or completed by 1930 ($429 million), approximately $63 million was receivable from other sources than water users' repayment.[77] If the total interest component at 3 per cent ($122 million) is added to these reimbursements from other sources, the total element of nonreimbursability either contemplated in legislation or experienced by 1930 was on the order of one-third of the combined initial investment costs and interest component thereon allocable to irrigation facilities.

While the degree of federal contribution was considerably less than for flood control and navigation, an element of nonreimbursability still was conspicuously present in national reclamation policy. On the eve of the period when the scale economies commenced to be applicable to water development, three important water development purposes had been well established in national policy as wholly or partly in the domain of investment from public funds, repayment being made through tax levies. Only hydroelectric power, just then coming into its own, still remained fully reimbursable among the important purposes of development.

PUBLIC APPROPRIATIONS, THE SCALE ECONOMIES,
AND THE SCHEDULING TECHNIQUES

The national policy developed over the years, and the local expectations nourished by that policy almost predestined the use of the scale economies and scheduling techniques to public financing. With the normal market procedure inoperable for three of the four major purposes, private interest in the financing of integrated multiple-purpose river basin development was relatively small when these techniques

---

[77] Power revenues, Congressional write-offs, contributions, and miscellaneous revenues (*idem*).

commenced to be applied to water development. Two further circumstances favored the financing from public sources of large-scale works and integrated multiple-purpose river basin development. One was the economic depression of the 1930's, and the other was the somewhat pessimistic forecast by private entrepreneurship of the market for large blocks of electric power.[78] Neither favored the assumption of responsibility by private financing during the 1930's. Thus many of the works associated with the application of the scale economies and the scheduling techniques to water development have been undertaken with public appropriations. They include the TVA water control and power systems, Hoover (Boulder) Dam on the Colorado, Grand Coulee and other works on the Columbia, the Central Valley project, and the Missouri River Basin project. Public financing in all of the major instances meant use of federal appropriations.

Private financing also has figured in multiple-purpose development, at first mainly through the entrepreneurship of states. For example, the state of Texas in 1929 established the Brazos River Conservation and Reclamation District, and in 1934, the Lower Colorado River Authority. In 1935 Oklahoma created the Grand River Dam Authority.[79] While the works of these agencies were more modest than the large-scale undertakings of the federal government in this period, they were multiple-purpose ventures of regional importance. All of these agencies were financed in part by the sale of revenue bonds on the private market. However, all of them also received federal grant and loan assistance in their important beginning stages. The Lower Colorado River Authority and the Grand River Dam Authority have particularly benefited from federal grants and loans. Of the three, the most complete multiple-purpose development would appear to have been carried out by the Lower Colorado River Authority.[80] Some irrigation districts also have undertaken modest multi-purpose operations.

As conditions have changed, including private estimates of the electric power market and the continued high level of economic activity, the interest of private institutions in financing multiple-purpose integrated

[78] For example, see Clapp, *op. cit.*, p. 99.
[79] The Council of State Governments, *op. cit.*, pp. 27-28. Other examples also might be found, like the Santee-Cooper project in South Carolina.
[80] Comer Clay, "The Lower Colorado River Authority," in E. S. Redford, *Public Administration and Policy Formation* (Austin: University of Texas Press, 1956), pp. 224-25. Discusses the financing of the Lower Colorado River Authority in particular.

river basin development has grown. However, with one possible exception, their interest has concerned participation in financing, rather than sole responsibility, for a multiple-purpose development. Thus private funds could be found to finance those costs of a multiple-purpose project which are directly attributable to electric power, and the share of the joint costs allocable to power. As in the cases of the southwestern state river authorities, private funds also could be found where the taxing power of the state stood back of the payment of capital charges and the amortization of the investment (general obligation bonds).

Particularly since 1953, national policy has encouraged the participation of private financing in what has been termed the "partnership" program of multiple-purpose river basin development in the United States.[81] Implementation of this policy since 1953 has been particularly active in the Pacific Northwest, although some projects also have been planned or undertaken in other sections of the country. At least twenty-eight nonfederal hydroelectric projects were reported as either licensed or under construction within the Columbia Basin or on other streams of western Washington and western Oregon in 1958.[82] One notable eastern project was the St. Lawrence Seaway Development, of which the United States section was jointly undertaken by the St. Lawrence Seaway Development Corporation, a federal agency, and the New York State Power Authority. Although financing in all instances was sought and obtained or was to be sought from private sources, entrepreneurship for some of these projects was through a public agency.[83] Private financing also was contemplated in implementing the California State Water Plan.[84]

---

[81] See, for example, Presidential Advisory Committee, *Water Resources Policy* (Washington: U. S. Government Printing Office, 1953), pp. 31-35; also periodic statements of Executive Branch policy, like the Presidential Message accompanying the presentation of the federal budget for the fiscal year 1958. Federal-nonfederal "partnership" actually has existed for a number of years, as in the Hoover Dam power plant operations.

[82] John Nuveen and Co., *More Power in the Pacific Northwest*, 1957, Exhibits 5 and 7.

[83] Thirteen of the twenty-eight Pacific Northwest nonfederal projects had public agency entrepreneurship (*idem*).

[84] State of California, Department of Water Resources, *The California Water Plan*, pp. 225-26. Public, and particularly federal funds were being sought to finance costs attributable to nonreimbursable functions of projects contained in the Plan.

FINANCING OF OTHER ACTIVITIES RELATED
TO WATER DEVELOPMENT

Three further aspects of water development deserve brief mention insofar as technology may have given rise to problems of financing. They are the means of meeting dispersed demands, the financing of research and scientific experiment, the development of equipment, and construction and engineering organization. For each of them a combination of financing from private and public sources has existed, although one form usually has dominated in each.

The problem of meeting dispersed demands has been well exemplified in rural electrification systems. When technical developments made possible the economic rural distribution of electricity and its advantageous farm use, entrepreneurship and financial support for rural electrical systems did not keep pace with either the apparent demand in the United States or the rate of rural system formation in other countries.[85] Accordingly, federal government financing of rural electrification was considered necessary by the Congress during the 1930's, and the Rural Electrification Administration was established for this purpose in 1936.[86] Nearly all of the rural electric cooperatives created in the United States since the mid-thirties have been financed with REA loans, which bear an interest rate of 2 per cent and a maximum amortization period of thirty-five years. Financing from public sources thus has borne an important share of the load in establishing rural electric systems in this country.

Private financing, however, also has shared in the support of rural electrification. On the eve of the establishment of the REA about 11 or 12 per cent of the farms of the country already were electrified, and a much larger share of the farms of three specific regions (the Pacific Coast states, New England, and the New York–Pennsylvania–New Jersey area). About 46 per cent of the farms of the three Pacific Coast states were electrified by 1934, for instance.[87] This had been undertaken almost entirely with privately obtained financing and under private entrepreneurship. After the success of REA had proved the financial soundness of rural electrification for the remainder of the

---

[85] Clapp, *op. cit.*, pp. 103-4.
[86] Act of May 20, 1936, 49 Stat. 1363, 7 U. S. C. 901, *et. seq.*
[87] President's Water Resources Policy Commission, *A Water Policy for the American People* (Vol. 1 of Report), p. 232.

nation, private financing of further rural electrification was accelerated throughout the country. The now nearly complete rural electric systems of the United States[88] thus have been financed from both public and private sources of funds.

It is easy to see that a parallel problem of meeting dispersed demand exists for rural domestic water supply. Particularly since the development of new pumps and lower cost forms of piping, the practicality of rural water systems has been apparent in the more densely settled rural areas. The possibility of much improvement in many individual rural water supplies also has been apparent for many years. However, the principal extensions of water distribution systems for domestic use have continued to be associated with the urban fringe. Only a few districts, like the vicinity of Johnson City and Elizabethton, Tennessee, and Clarke County, Washington, have undertaken the creation of rural water distribution systems. Private sources have supplied their financing. In 1957 an area-wide water supply system was reported to be under consideration in northern New Jersey.[89] Although this is a highly urbanized section, some rural areas are included within it. Improvement of individual unit, rural, domestic water supply facilities has been more common, but even in that respect a very large number of unimproved supplies are to be found on individual farms in several parts of the country.[90] In this field, also, little public attention has been paid to the financing of dispersed domestic water supply and distribution. As yet there has been no parallel to the policy and administrative treatment of rural electrification in the field of rural domestic water supply.

FINANCING OF RESEARCH AND SCIENTIFIC EXPERIMENT

As the illustrations in chapters 5 through 16 indicate, a large variety of activities is touched in considering water development. Almost all now have some associated research and experimental aspects. Financing of research and experiment as a whole is shared by all of the sources for funds in the country. The proportion customarily assumed by public or private, state or federal, or other entity may vary greatly

[88] About 95 per cent (94.2 per cent as of June 30, 1956) of the total number of farms in the country were reported by the Rural Electrification Administration to be electrified in 1956.

[89] Maynard Hufschmidt, communication to the authors, 1957.

[90] See *Annual Report of the Tennessee Valley Authority, 1956.*

among the major classes of research activities. Those activities which
center upon agriculture generally have depended heavily upon publicly
appropriated tax funds for their financing. Both state and federal
governments have contributed toward the mass effort of research and
experiment in agriculture which bears upon water use, the former
through the many Department of Agriculture programs, those of the
Bureau of Reclamation and the Tennessee Valley Authority, and some
less prominent programs.[91] State support of research and experiment
in this field has come through the state universities, the agricultural
experiment stations, and in other ways. Privately obtained funds have
figured in the biological and other research of the privately endowed
universities, projects supported by research foundations, some trade
association research, and some commercial corporations.

The distribution of the financing function for other phases of research and development related to water development in the United
States shows a similar pattern to that of agriculture-related research.
Generally speaking, a large share of the "basic" research upon which
future development may depend is publicly supported, either through
federal bureaus or through universities or other state institutions. Examples are to be found in the hydrologic and physiographic research
projects of the United States Geological Survey, the Vicksburg laboratory
of the Corps of Engineers, the Office of Saline Water and the Bureau
of Reclamation laboratories in the Department of the Interior, the
United States Weather Bureau, and in the meteorological, geographical,
geological, and engineering departments of many publicly supported
universities throughout the country. However, very important contributions have been made by research in privately supported institutions.
While they are fewer than the publicly supported institutions, privately
financed universities continue to share the significant research in these
fields. Mention of the meteorological and climatological research done
at Harvard, Massachusetts Institute of Technology, California Institute of Technology, the University of Chicago; attention to sanitary
engineering at Johns Hopkins and Harvard; and physiography at
Harvard, Columbia, and Chicago may serve to illustrate the place of
these private institutions. In addition, increasingly valuable research
and experiment by commercial companies add another dimension.
This is illustrated in the research and experiment by Langmuir, Schaefer,
Vonnegut, and others at the General Electric Company laboratories

---

[91] Such as the basic research programs of the Department of Defense.

pioneering in weather modification in this country; and the work on saline water conversion conducted by Ionics, Inc., and others.[92] Thus the financing of these aspects of research related to water development again would seem to be a web with both public and private participation.

Equipment and materials development. More than in any other aspect of research and experiment, the development of materials and equipment for water facilities construction and operation holds a "by-product" position. The same materials and equipment are likely to be useful in some or many other places of the country's economy. In part because of this multifunctional position, privately supported research and experiment is dominant in this field. The development of earth-moving equipment, alloyed metals, higher strength cement and concrete, hydraulic turbines, electric generators, transforming and transmission equipment, plastics, and water treatment equipment illustrates, in all instances, the important place of the American commercial structure in the great improvement of materials and equipment experienced within the last fifty years. However, even here the dual structure of financing research and experiment appears. At least some of the research carried on by private corporations must be considered as supported by public funds, through defense and other contracts, or through a share in the benefits to private corporations of accelerated amortization for tax purposes. In addition, publicly supported research of importance is carried on by public institutions. The state universities again require mention, as do the federal Bureau of Standards, the Denver laboratories of the Bureau of Reclamation, and a few other federal offices.

Pattern for research and experiment as a whole. The pattern of financing research and experiment related to water development is one which has evolved partly from the nature of our economic institutions and partly from historical accident. In those sectors of the economy where large-scale private enterprise has been possible, private financing of important research and experiment has reached a dominant position. Elsewhere, financing from public sources has appeared to bear the main burden. No important phase of research and experiment, however, has been exclusively the domain of public or private financing; the degree of progress achieved notably has resulted from a dual system.

---

[92] The examples of universities and commercial companies here mentioned are illustrative only; other equally significant examples might have been chosen. Notation also must be made of the fact that significant amounts of research at private institutions are supported by public funds.

CONSTRUCTION AND ENGINEERING ORGANIZATION

Mention of the place of private financing in water development would not be complete without inclusion of the private contracting, construction, and engineering organizations that are found throughout the country, some on a very large scale. Private firms of engineering consultants have a role in the engineering phases of planning, which are as wide as the field of water development and management. These private concerns are employed, as needed, on every governmentally managed and every privately managed undertaking of any size. Even the TVA, noted for its force account policy[93] and its highly competent engineering staff, employs private consulting assistance of this type.

Much larger sized operations are found in the contracting organizations that undertake the actual construction. Most large construction projects under government or private auspices during peace time are undertaken by contract. They include projects of the largest scale, like the Grand Coulee Dam, the St. Lawrence Seaway Project, and the Kitimat-Kemano development (British Columbia). The large organizations, like Merritt-Chapman-Scott, Bechtel Corporation, and Morrison-Knudsen Company, engage in a wide variety of engineering and construction operations on a near world-wide basis. It is probable that their existence permits a flexibility and rapidity in development which otherwise would not be possible. They, and the private financing under which they maintain their organizations, are an important part of the developmental structure of the country, even for governmentally supported projects in the field.

Allocation Devices

One of the most troublesome problems appearing in the wake of changes in water development techniques for the United States has been the problem of allocating benefits. Benefit allocation usually has been

---

[93] A force account system means management and supervision of the construction work directly by the entrepreneurial agency. The agency in effect acts as its own contractor, by contrast with the system of letting a contract to an outside organization after competitive bidding. The comparative merits of the two systems and their best individual application are deserving of further study.

achieved through the allocation of water or water-derived services. This problem must be solved for each program or facility before a water development may be considered fully realized. Essentially, the problem may be outlined with the question: To what degree does the allocation of benefits among the beneficiaries of development permit or prevent realization of all the economies that technically are possible? In other words, do allocation devices in the United States match technical possibilities?

The problem is best illustrated in the use of the scale economies in water development. As previously noted (pages 492, 501), extremely few important river basins are entirely within the political jurisdiction of one, or even two, states. Also, where water demands have long exceeded supply, the allocation of water resources and benefits therefrom to individual citizens (or juridical persons) has been the province of the state, as distinct from the federal government. Throughout the western states the unit for allocation of benefits on the major basins, therefore, has become the state. In the West the prevailing method of apportioning water has been prior appropriation—"first in time, first in right." It thus has been within the seventeen western states that the country has had to face the allocation problem in a practical manner.

The means hitherto generally employed for dividing the benefits of development within an interstate basin has been a physical division of the water itself. Such allocation has been by court decision,[94] or by compact agreement.[95] While compacts have been negotiated and court decisions rendered on specific water development purposes,[96] the simple division of water decisions and agreements have come closest to touching multiple-purpose integrated river basin development, with its scale economies. With the exception of the Central Valley basins of California, the system of water division which thus has grown is of deep significance to the use of the scale economies in western water development. Because of the impending consideration of prior appropriation

[94] As in the 1907 Supreme Court decision on the case of *Kansas vs. Colorado* (206 U. S. 46 [1907]) and subsequent decisions.

[95] For example, the Colorado River compact agreement among Arizona, California, Colorado, Nevada, New Mexico, Utah, and Wyoming, which divided the water among the "upper basin" states and the "lower basin" states. (U. S. Department of the Interior, Bureau of Reclamation, *Hoover Dam Power and Water Contracts and Related Data,* 1950, pp. 5, 6.)

[96] Especially navigation, pollution, and flood control. See President's Water Resources Policy Commission, *Water Resources Law* (Vol. 3 of Report), pp. 65-67.

or other allocative water rights legislation in at least some of the thirty-one eastern states,[97] the possibility of interstate water division there is of equal import.

A physical division of water among several states of a river basin does not result *a priori* in a situation wherein the maximum benefits may be derived from the development of the basin. Theoretically, there could be a compatibility between physical division of water and the most efficient possible development of a basin if a full engineering plan of optimum development had first been laid out, with consideration given to geographical location of water-related resources and future population. Allocation of water among several state jurisdictions then might be made in a manner which allowed approximation to the economically and socially optimum plan, provided the states of the basin could agree on such a division. In practice, however, such plans have not been available for water allocation negotiations or decisions, nor have they generally been considered the necessary foundation of court decision or compact agreement.

The Supreme Court of the United States, as final arbiter in interstate litigation, has based its decisions in these cases "upon the principles of equitable apportionment," fitting the individual decision to "the facts of the controversy, without adherence to any particular formula."[98] Compact agreements, in their turn, have often had as their basis "a trading compromise of conflicting claims."[99] The Court's view on the proper procedure for the allocation of benefits (actually the apportioning of water) among contending state jurisdictions is revealing as to the principles lying behind the present administration of allocation in the United States. In the 1943 decision on *Colorado vs. Kansas,*[100] the Court had this to say:

> . . . such disputes . . . involve the interests of quasi-sovereigns, present complicated and delicate questions, and, due to the possibility of future change of conditions, necessitate expert administration rather

[97] The state of Mississippi passed prior appropriation water rights legislation in 1956. Studies examining the water rights situation within the state had been undertaken in a majority of eastern states by 1957.

[98] President's Water Resources Policy Commission, *Water Resources Law* (Vol. 3 of Report), p. 59.

[99] *Ibid.,* p. 67. The quotation is from the United States Supreme Court in the decision on the case of *Hinderlider vs. La Plata River and Cherry Creek Ditch Co.* (304 U. S. 92 [1938]).

[100] 320 U. S. 383, 392 (1943).

than judicial imposition of a hard and fast rule. Such controversies may appropriately be composed by negotiation and agreement, pursuant to the compact clause of the federal Constitution. We say of this case, as the court has said of interstate differences of like nature, that such mutual accommodation and agreement should, if possible, be the medium of settlement. . . .

Thus the administrative structure and underlying water policy of the United States, with its "quasi-sovereign" forty-nine states, places the individual interests of the states sharing a water resource in the foreground of any allocation of benefits. Our policy and administrative structure have not placed consideration of an optimum plan of development[101] in a coequal, or even a prominent position, as far as allocation of benefits from a scarce resource is concerned. The only instance in which an optimum plan of development has been carried through occurred within the eastern section of the country, where until recently surpluses of water were apparent to development planners. This was on the Tennessee. It is interesting to speculate as to whether the Tennessee development might have been carried through in its present form if water had been considered a scarce resource by the states of the Tennessee Valley at the time the major phases of TVA development were undertaken.

The outstanding case in eastern United States in which water allocation and comprehensive development have been an issue, concerns the Delaware River. The basin of the Delaware is in the midst of one of the most densely settled parts of the country, including the Philadelphia metropolitan area, and adjacent to the huge New York metropolitan area. At least since the early 1920's its waters have been the subject of independent development objectives, negotiation, and litigation on the part of the three principal states and two major cities concerned.[102] It has been subject to intensive study by federal, state, and city agencies, and at least two comprehensive river basin plans have been suggested for the basin, one by the Corps of Engineers in 1934,[103] and another by

---

[101] Optimum plan of development is here defined as that combination of projects or facilities and nonstructural measures which (1) provides the highest net return in economic benefits compatible with the other social benefits desired by the sponsoring group; and (2) maximizes the social benefits (other than economic) within the limits of the expenditures which the sponsoring group is willing to appropriate for that purpose.

[102] New York, Pennsylvania, and New Jersey; Philadelphia and New York City.

[103] U. S. Army Corps of Engineers, *Delaware River and Tributaries,* 73rd Congress, 2nd Session, House Document No. 179, Washington, 1934.

the Interstate Commission on the Delaware River Basin in 1950.[104] Another comprehensive plan was in preparation in 1957-58 by the Corps of Engineers, in co-operation with the states and cities concerned. Yet the major implementation of planning has been based on Supreme Court decisions, one in 1931, and a second in 1954.[105] These decisions have been necessitated by the desire of New York City to divert water from the basin for its municipal supply in plans which have been contested by the downstream states. Although the 1954 decree allows for a major diversion (800 million gallons per day) for New York City, the judicial process in the case did not include consideration of a comprehensive basin plan, even though the results of two previous comprehensive studies were available.[106] Hufschmidt's perceptive study of the case notes that the Court decision method of planning, while it has the virtue of clarity, also has the disadvantages of being piecemeal, static in concept, and open to more litigation. He notes further that such decisions need not ignore comprehensive planning and development, but he also notes that the judicial process inherently is attracted to the piecemeal approach.[107] Adaptation to and favoring of sound, comprehensive basin planning is a significant problem for future United States jurisprudence.

One other aspect of American administrative structure bears mention in connection with the allocation of benefits. As has been noted, the principal vendible product from water development in recent years (in terms of value) has been electric power. The benefits of electric power production and distribution are channeled through individual power systems, which are both privately and publicly managed. Many of them are geographically small, the creatures of commercial accident and of legal environment. (See pages 505-6.) Many also are vertically integrated, in the sense that they produce their power as well as distribute it at retail. Federal agency systems, however, distribute very little power at retail.

These management units are like the states in that they are usually of smaller size than the river basins of the country, and often are of a

[104] *Report on the Utilization of the Waters of the Delaware River* (prepared for the Commission by the consulting firm of Malcolm Pirnie and Albright and Friel, Inc.).

[105] Maynard M. Hufschmidt, *The Supreme Court and Interstate Water Problems—The Delaware Basin Example*, Littauer Center, Harvard University, 1957 (dittoed).

[106] *Ibid.*, pp. 69-70.

[107] *Ibid.*, pp. 65-73.

capacity so small that they cannot readily absorb the large units of supply possible in integrated development. Nor do they encourage the scale economies possible in a large and varied generating and distribution system managed as a unit. The effects of this structure upon the scale economies has been illustrated dramatically in the previously mentioned case of the Idaho Power Company at Hells Canyon. (See page 534.) The company, understandably seeking to preserve its own vertical integration and its independence, will fall $2.7 million short of the annual benefits which would have been realizable for two dams operated as part of the Columbia River system as a whole.[108] This is 11 per cent of the total annual costs of the three-dam system. While few other effects of this administrative structure have been so dramatically displayed as in the Hells Canyon case, there can be little doubt that the existing structure for distributing power contains so many small independent systems that they have been a negative rather than a positive influence on the use of the scale economies in river basin development. These systems include both public and private management.

It has been suggested that there also may be unrealized scale economies in the irrigation district pattern. Particularly where separately organized small-scale developments have taken place on contiguous territory, consolidation of management and joint use of some facilities might yield economies. The California State Water Plan, as previously noted (pages 507-8), embodies some important principles of scale organization. However, there may be other possibilities which involve considerably less capital investment. A recent Reed College study of the Hermiston and Deschutes projects, Oregon, suggests that this may offer an interesting field of investigation elsewhere.[109]

The composite water and electric power policies of the federal government, the states, and private enterprise thus have not produced an organizational structure for allocating benefits which is notably favorable to, or even compatible with, use of the scale economies in integrated, multiple-purpose river basin development. Other considerations appear to have dominated our methods of allocating benefits.

The present gap between administrative organization for allocation and distribution of benefits in water development and the employment of the scale economies may be widened in the future. Increasingly severe problems, and danger of increasing deviation of economic actuality from physical potentiality may be expected as (1) pressure

[108] See Krutilla and Eckstein, *op. cit.*, p. 158.
[109] Charles McKinley, personal communication to the authors, 1957.

of demand upon supply grows in the West, with mounting need for interbasin diversions; (2) public attitudes and legal provisions in eastern states become oriented toward water as a scarce commodity; and (3) mounting electric energy demands increase the value of hydro in meeting peak loads. The provision of allocation devices which match technical possibilities therefore may be set forth as a problem which time alone is not likely to solve. In 1958 it was a problem for which intensive professional research clearly could be considered necessary. Better understanding both of the nature and methods of ascertaining the incidence of benefits was called for.

Application of Mass Production Techniques and Interchangeability to the Management of Construction

Each river basin, like each geographical region on earth, is unique. So is every dam site. The regulation of a river to produce the optimum multiple-purpose benefits therefore has an element of art in its most successful prosecution. While formal programming and mathematical analysis with computer use hold promise for stream regulation, as yet the new computer research techniques have found only limited application to these problems.[110] Nevertheless, it has been shown that some of the machinery and ideas of standardized mass production are applicable to the design of river control facilities. Each of the large government construction agencies has developed standardized design to some extent. This has been possible because under the United States system the design of structures and facilities generally has been kept in the hands of the entrepreneurial public agency, as contrasted with the European system, under which design as well as construction is open to competitive bidding. Standardization and interchangeability probably were carried farthest in the TVA construction schedule, which operated within a single river system, on a force account system, and with a well-maintained construction schedule over a period of twenty years. Not only were procedures and smaller equipment standardized wherever possible, but even the housing for construction workers, and the com-

---

[110] Principally in some experiments by the Corps of Engineers planning for the operational management of the main stem reservoir system of the Missouri River. They have been most frequently employed in electric power operation analyses, on both privately owned and publicly managed systems.

ponent parts of dams. In part to take advantage of standardized procedure and interchangeability, TVA used a single type of dam, the
gravity dam, throughout its water control system.

Other water development projects in the country have not lent
themselves to standardized procedure to quite the same degree as the
development of the Tennessee. While the practice of designing within
the agency has contributed toward standardization, as mentioned above,
the division of interest and responsibility for a given basin—among the
Corp of Engineers, the Bureau of Reclamation, municipalities, public
utility districts, irrigation districts, and private companies in the West;
and among the Corps, municipalities, and private companies in the
East—has not permitted the kind of standardized procedure and
attendant economies practiced by TVA.

A degree of standardization also is provided in the operations of the
contractors who directly manage much of the construction for which
either public agency or private company is entrepreneur. The skillful
management of a large construction organization like the Morrison-
Knudsen Company or the Bechtel Corporation could not fail to produce
under competitive conditions standardized procedures and equipment
productive of economies. Assuming that the economies are reflected in
the accepted bid prices, they of course are then available in the construction operations which customarily are carried on by contract.

The scale of development in the country as a whole in recent years
also has afforded suppliers of equipment a home-market opportunity
for mass production and standardized design not encountered elsewhere
in the world for such things as earth-moving equipment, pumps, construction steel, wire, transformers, switchyard equipment, and a variety
of other components of the facilities for regulating a river and the
distribution of its products and services.

In conclusion, the present and immediately past organization of construction operations for water development in this country to some
extent has permitted the employment of standardized design and
procedure, and mass production economies. These economies perhaps
have been most frequently encountered for equipment usable for a wide
variety of purposes beyond water development alone. Such economies
also have been employed in the actual construction of river regulation
facilities, both through the management techniques developed by large
construction contractors, and through the application of force account
construction by an agency entrepreneur working a single basin on a
consistent schedule. However, the multiplicity of construction re-

sponsibilities characteristic of the country as a whole has discouraged full achievement of the economies presently known.

Summary Comment

The problems presented by technical progress for the realization of water development have been met with varying degrees of success and attention in the United States. The most varied and pervasive challenge has been presented by the techniques promoting the scale economies. The problem of entrepreneurship and financing for large-scale, geographically extended construction projects has been successfully met. Application of mass production techniques and standardization to construction operations also has shown marked achievement. On the other hand, the promotion of the scale economies by adequate procedures of resource or benefit allocation has shown decidedly less adaptability in American practice. So has the interrelation of water resources and other resources, like land and minerals, in the development process.

Still other problems presented by technology have drawn differing adaptations. Attention to multiple use, for example, has had a long history, but practice in 1958 still fell short of the technical possibilities, although notable strides had been made. Attention to the technique of reducing demand likewise has had a slow growth in spite of the presence of some critical problems, like stream pollution. On the other hand, the challenge presented by dispersed demand had been met very completely in at least one important function, electricity distribution.

While all levels of organization participated in the achievement which could be recorded, the initiative in several prominent ways has fallen to federal government entrepreneurship and financing. This has been true for the construction activities connected with many large scale water development programs. Federal initiative also has been notable in promoting multiple use, and in meeting the geographically dispersed demands throughout the country. Other types of organization also have responded to technology, notably private organizations, but they often have been handicapped in meeting the problems presented by scale opportunities because of limits in jurisdiction or special-interest objectives. As of 1958 the position of the United States in this respect was one of numerous further opportunities for giving scope to technical possibilities by administrative change.

CHAPTER 20

The Organization of Operational Management

The Operational Management Complex in Present-day United States. The Opportunities and Problems for Operational Organization Presented by Technologic Change. Three Opportunities and Organizational Response—(1) water control flow; (2) conservation and quality control; (3) distributional organization and the scale economies. Four Organizational Problems Presented by Technology—(1) relation of the different levels of administrative organization; (2) relations of planning, construction, and management agencies; (3) flexibility in administrative organization; (4) relations of operational management of water resources and other resources.

Once river regulation facilities have been constructed or other water development undertaken, the problems of administrative organization are far from diminishing. In a sense they have only begun. Major construction may be completed within a decade. The need for operational management of the dams, locks, levees, canals, generators, transmission lines, and other works installed may last for forty or fifty years, or even longer. The forms of operational management influence the planning and development administration, and they should be kept in mind by the planning and development agencies, whatever they may be. Technical advance has presented certain opportunities for the organization of operational management of water control and related facilities, and it also has brought some interesting, if not serious, problems for governmental and private administration in this country. The nature of these problems and opportunities is the subject of this chapter.

The Operational Management Complex in Present-day United States

There are five principal phases of operational management of water regulation or water use facilities: (1) the necessary physical and

570

mechanical operation, including maintenance; (2) the sale of products and services, or supervision of the use thereof; (3) debt service and other disbursement; (4) co-ordination of operations among organizations, agencies, or individuals having overlapping or impinging interests; and (5) regulation of operational agencies.[1]

Responsibility for these five aspects of operational management in the United States is shared by a complex of responsible agencies. The pattern of operational management differs from region to region, and even from locality to locality within a given region. Included among the operational agencies are public and private organizations, federal, state, municipal, and other local governmental entities. Among them are both public and private corporations, public agencies with commercial objectives, public agencies with social objectives, and public agencies which have a mixture of social and commercial responsibilities. The array of agencies is larger in some regions than others, but in each region multiple responsibilities are to be found. A high degree of development does not appear to have brought with it any special pattern of management.

Five regions, the Columbia River Valley, the Central Valley of California, New England, the Lower Mississippi Valley, and the Tennessee and Cumberland valley region are briefly described on table 35 as to their 1956 operational management responsibilities for water control and use. The Tennessee-Cumberland valley region and the lower Mississippi Valley have patterns which are unique, the first because of the existence of TVA, and the latter because of the singular physical characteristics of the lower Mississippi and the long history of Corps of Engineers' work in the region. The basic patterns of the other three regions, the Columbia Valley, the Central Valley, and New England, however, are repeated in other regions of the country. As is to be expected from previous description of planning and construction responsibilities, West (Central Valley and Columbia Valley) shows a different pattern from East (New England).

Essentially the pattern in the West is one in which the large federal executive agencies, the Corps of Engineers and the Bureau of Reclama-

---

[1] Regulatory agencies are included in this list because they, in reality, share the responsibility for shaping the exact character of debt service, and the distribution of products and services by a number of agencies. This is true for the distribution of electricity and, in the West, also for water distribution. Regulatory agencies also have a close relation to physical operations in waste carrying (pollution control).

TABLE 35. *Agencies of Operational Management for Water Facilities, Five United States Regions*

	Columbia River Valley	Central Valley of California	New England	Lower Mississippi Valley	Tennessee and Cumberland Valleys
1. Physical and mechanical operations					
a. Reservoir operation	Corps of Engineers, Bureau of Reclamation, Municipalities, Private utilities, Public utility districts	Corps of Engineers, Bureau of Reclamation, Municipalities, Private utilities, irrigation districts	Private utilities and other corporations, Corps of Engineers	Corps of Engineers	TVA, Corps of Engineers, private corporations
b. Irrigation facilities	Bureau of Reclamation, Irrigation districts	Bureau of Reclamation, Irrigation districts, Private corporations, State Department of Water Resources	N.A.	Drainage and other special districts, Private companies	N.A.
c. Drainage facilities	Irrigation districts	Irrigation districts, Private companies	N.A.	Drainage, Flood control and levee districts	N.A.
d. Levees and floodwalls	Corps of Engineers, Municipalities, Flood control districts	Corps of Engineers, Municipalities, Levee districts	Municipalities	Corps of Engineers, Flood control and levee districts, municipalities	Corps of Engineers, Municipalities
e. Electric generation	Bureau of Reclamation, Corps of Engineers, Municipalities, Private utilities, Public utility districts	Bureau of Reclamation, Private utilities, Municipalities	Private utilities	Private utilities	TVA, Corps of Engineers, Private corporations
f. Domestic, industrial and municipal water supply	Municipalities, Irrigation districts, Private companies, Public utility districts	Municipalities, Private companies, Irrigation districts	Municipalities, Private companies	Municipalities, Private companies	TVA, Municipalities, Private companies

Facilities, Five United States Regions—Continued

	Columbia River Valley	Central Valley of California	New England	Lower Mississippi Valley	Tennessee and Cumberland Valleys
g. Recreation, fish and wildlife	State fish, game and parks agencies; National Park Service; U. S. Bureau of Sport Fisheries and Wildlife,[1] County and municipal park agencies, Private organizations, U. S. Forest Service, and U. S. Bureau of Land Management	State Division of Beaches and Parks, State Department of Fish and Game, U. S. Forest Service, U. S. Bureau of Sport Fisheries and Wildlife, National Park Service, Private organizations	State recreation, fish, game, natural resources and parks agencies, U. S. Bureau of Sport Fisheries and Wildlife, U. S. Forest Service, National Park Service, Private organizations	U. S. Bureau of Sport Fisheries and Wildlife, Corps of Engineers, State conservation, fish and game agencies, Private organizations	TVA, U. S. Bureau of Sport Fisheries and Wildlife, State conservation or parks agencies, National Park Service
h. Waste carrying[2]	Municipalities, Private companies	Private companies, Municipalities	Private companies, Municipalities	Private companies, Municipalities	Private companies, Municipalities
i. Locks and other navigation facilities	Corps of Engineers	Corps of Engineers	Corps of Engineers	Corps of Engineers	Corps of Engineers, TVA
2. Sale of products and services, or supervision of user thereof					
a. Electricity	Private utilities, Bonneville Power Administration, Bureau of Reclamation, Municipalities, Public utility districts, Electric co-operatives, Irrigation districts	Private utilities, Bureau of Reclamation, Municipalities, Electric co-operatives, Irrigation districts	Private utilities, Municipalities, Electric co-operatives	Private utilities, Electric co-operatives	TVA, Southeastern Power Administration, Municipalities, Rural Electric co-operatives, Private companies

TABLE 35. *Agencies of Operational Management for Water Facilities, Five United States Regions—Continued*

	Columbia River Valley	Central Valley of California	New England	Lower Mississippi Valley	Tennessee and Cumberland Valleys
b. Irrigation water	Bureau of Reclamation, Irrigation districts	Bureau of Reclamation, Irrigation districts, Private companies	N.A.	Drainage and other special districts, Private companies	N.A.
c. Domestic, industrial and municipal water supply	Municipalities, Public utility districts, Private companies	Municipalities, Irrigation districts, Private companies	Municipalities, Private companies	Municipalities, Private companies	Municipalities, Water districts
d. Recreation, fish and wildlife	State agencies, National Park Service, County and municipal park agencies, Private organizations, U. S. Forest Service, U. S. Bureau of Sport Fisheries and Wildlife, and U. S. Bureau of Land Management	State Division of Beaches and Parks, State Department of Fish and Game, National Park Service, Municipalities, U. S. Forest Service, U. S. Bureau of Sport Fisheries and Wildlife, Private organizations	State agencies, U. S. Forest Service, U. S. Bureau of Sport Fisheries and Wildlife, Private organizations	State conservation, fish and game agencies; U. S. Bureau of Sport Fisheries and Wildlife, Private organizations	State agencies, TVA, U. S. Bureau of Sport Fisheries and Wildlife, National Park Service
3. Debt service and other disbursement					
a. Carrying charges	As for 2a,[3] plus Federal Treasury for nonreimbursable costs	As for 2a,[3] plus Federal Treasury for nonreimbursable costs	Private utilities, Municipalities, Private companies (other than utilities), Federal Treasury	Federal Treasury, Private utilities, Municipalities, Electric co-operatives	As for 2a,[3] plus Federal Treasury for nonreimbursable costs
b. Amortization	As for 3a	As for 3a	As for 3a	As for 3a	As for 3a

Facilities, Five United States Regions—Continued

	Columbia River Valley	Central Valley of California	New England	Lower Mississippi Valley	Tennessee and Cumberland Valleys
4. Co-ordination[4] of operations among organizations, agencies, or individuals having impinging interests	Columbia River Basin Interagency Committee and affiliated groups, Bonneville Power Administration	Bureau of Reclamation	None[5]	Mississippi River Commission, Corps of Engineers	TVA, State planning and development agencies
5. Regulation of operational agencies					
a. Electricity	FPC,[6] State commissions	FPC,[6] State Commission	FPC,[6] State commissions	State commissions	FPC,[6] TVA,[7] State Commissions
b. Waste-carrying	State health or sanitary agencies	State Water Pollution Control Board	State health or sanitation agencies	State health or sanitation agencies	State health or sanitation agencies
c. Water use or appropriation	State engineer, State departments of conservation or reclamation, County and state courts (Montana)	State Water Rights Board	None	State Board of Water Commissioners (Mississippi only)	State Board of Water Commissioners (Mississippi only)

N.A. = not applicable.

[1] Part of Fish and Wildlife Service, U. S. Department of the Interior.
[2] Except reservoir operation for flow regulation.
[3] Minor operations not included.
[4] Only those agencies included which have a positive objective of co-ordinating multiple-agency operations.
[5] In 1958 the Northeastern Resources Committee, a state-federal voluntary interagency group, existed. However, its activities at that time mainly were planning efforts.
[6] FPC—Federal Power Commission.
[7] Through resale provisions of contracts.

tion, are dominant in operation of major works like dams and the associated reservoirs. However, some major works are under the management of municipalities, public utility districts, private utilities, irrigation districts, water conservation districts, and flood control districts. State regulatory agencies have an important place in supervising the distribution and use of water. The sale of water or water-derived services includes the entire range of the several types of agencies having operational responsibilities in the field. The sale and distribution of electricity, for instance, is undertaken by private utilities, the Bonneville Power Administration, the Bureau of Reclamation, municipalities, public utility districts, electric co-operatives, and irrigation districts.[2] Nonreimbursable services are supervised by federal and state agencies and by municipalities. Debt service is undertaken by an equally wide variety of agencies, corresponding to the physical or commercial responsibility which has been assumed or granted. In addition the federal treasury is involved insofar as there is some nonreimbursable investment which should be entered in the long-term national debt. Co-ordination of functions among the several operational agencies has been attempted mainly by the nonofficial federal interagency committees, or by affiliated interstate committees and by intra-departmental "field committees" (as in the Department of the Interior). In some specialized ways, single agencies have undertaken a degree of co-ordination, like the Bonneville Power Administration in the Columbia Valley, and the Bureau of Reclamation in the Central Valley.

Within eastern United States, the pattern differs in the absence of the Bureau of Reclamation, the dominant position of the Corps of Engineers in managing major works (except the Tennessee Valley), the dominant position of private utilities in distributing electricity (except in the TVA power service area), the paucity of organizations having irrigation responsibilities, the paucity of supervised or regulated water use and distribution (except for pollution control), and the minor position of co-ordinating efforts (except in the Tennessee Valley, on the lower Mississippi River, and by the Delaware River water master).

Whatever its strengths and weaknesses, the pattern of operational management for water and water-derived services in the United States at this time may be described as having a high degree of local individuality. However, throughout the West state regulatory agencies hold strategic positions by virtue of their administration of water rights laws,

[2] E.g. the Salt River Valley Water Users' Association (Arizona), and the Modesto Irrigation District (California).

and throughout the country federal agencies hold strategic positions in their management of major control works on surface streams.

The Opportunities and Problems for Operational Organization Presented by Technologic Change

The present administrative structure for water management in part has been the product of response to technical opportunities. To see how large or how small the response has been, three specific illustrations are briefly examined. Since the subjects already have been introduced in the previous discussions on planning and the realization of development, presentations here are brief. The three opportunities are: unified operational management of water control facilities within a basin; comprehensive management for water conservation and quality control; and the creation and maintenance of distributional organizations sufficiently large to take advantage of economies of scale.

As distinct from opportunities presented by technical change, some organizational difficulties have been presented by technology. The opportunities may be regarded as desirable ends from the point of view of economically efficient operation; on the other hand, the difficulties are mainly those of articulating the administrative structure. The answers to these problems still are not clear, and the effects upon efficiency of organizational change to meet the problems are moot questions within the nation. Four problems are suggested in this connection: (1) the relations of the different levels of administrative organization, including balance among them and the degree of centralizing necessary; (2) the relations of operational agencies with planning and development agencies; (3) the degree of flexibility desirable in organization; and (4) the relation of the operation of water facilities to other resource management operations.

Three Opportunities and Organizational Response

1) WATER CONTROL FLOW

The advantages of close co-ordination of operating facilities using and reusing the same water in a river basin should be obvious. The manner of operation in fact is the critical aspect of the entire technique of integrated river basin development. Whatever the facts of the physical and legal environment within which the use of water must take place, closely co-ordinated operation of multiple-purpose facilities can yield benefits unobtainable under independent operation of individual projects. This is true even in those situations within the West where the larger rivers have been "compartmentalized" by compact agreements or judgments apportioning the available water into parcels for development within specific states. In brief, this is the opportunity which technology has brought to development. Its scope may have been reduced somewhat by systems of water allocation, but on every watershed of any size an opportunity for economies in co-ordinated operation remains. This has been recognized by almost every major operational agency in the United States.

Unified operational management of a basin's water theoretically might include flow control on regulatory structures on the basin's streams, propagation and enforcement of measures designed to conserve waste water for useful purposes, and unified quality control (including attention to paired use and supply).

Three organizational answers to the opportunity presented by technology for unified flow control have been experimented with in the United States. One has been the single water master, in which all regulation through major works for whatever purpose is directed from a single agency. This is perhaps best exemplified by the TVA water management on the Tennessee, in which releases and flow for navigation, flood control, and electric power[3] are directed from TVA, even for privately owned structures within the basin.

A second type of management for flow control, and an older form, is single-agency management for a single purpose. The Corps of Engineers' responsibility for control of flood flows on the entire Mississippi river system is an example.[4]

The third type has been that of voluntary co-ordination of several agencies for multiple purposes. Wherever the Corps of Engineers and

[3] These are the "statutory" purposes specified in the TVA Act. Recreation and low flow regulation for sanitary purposes also receive attention, but not on a priority basis.

[4] The Corps has similar responsibilities in other basins. The Tennessee Basin is a part of the Mississippi system. Water control of the Tennessee is a TVA responsibility. However, during times of Mississippi or Ohio floods the discharge of the Tennessee at its mouth is regulated by TVA according to Corps' stipulations.

the Bureau of Reclamation have divided responsibilities on western streams a degree of co-ordinated flow control has been sought and obtained. Within the Columbia Basin plans for an extension of the co-operative arrangements for flow control have included all agencies operating facilities within the basin, particularly those producing electric power. Similar plans have been explored for the management of structures within the Missouri Basin.

Each of these arrangements has certain advantages and handicaps. The first obviously has the potentiality and capacity for greatest efficiency for a fully developed basin, since it can operate with the simplest organizational structure and the least organizational friction. The second is well suited to the incompletely developed basin, where only the need for special-purpose management thus far has been recognized. The third gives the maximum opportunity for preserving the voice of particular interests or agencies, as they are vitally concerned in the operation of water facilities and the benefits therefrom.

As of 1958, unified management of flow control on major American streams geographically was far from prevailing. The most closely co-ordinated flow control operation was on the Tennessee, but even on the Tennessee the co-ordination of secondary purposes (e.g. recreation) and primary purposes (e.g. navigation, flood control) had so little statutory authorization that fully unified multiple purpose management was not achieved. Co-ordination of flow control for multiple purposes also was to be found or was in a late stage of planning on the Columbia, in the Central Valley, on the Missouri main stem, and on the Cumberland. Unified flow control under private management was to be found on the Wisconsin River. On most other American streams co-ordination of flow control was either very incomplete or undeveloped. Under the prevailing division of responsibilities, achievement of unified operational management was confronted with formidable problems. Where a dozen or more agencies are autonomous in their operations within a basin, multiple-purpose operation must always have a background of different, often divergent, objectives in operation. Where conflicts occur among purposes (see table 14, page 84) or where scarcity gives rise to apportionment, co-ordination therefore must be on an *ad hoc* basis, and through cumbersome committees and other organizational devices. While earnest efforts have been made in the direction of unified flow control, satisfactory administration of flow management has yet to be achieved on most American streams.

The flow control discussed thus far has referred entirely to surface flow. Our knowledge of hydrology now is sufficient to make it clear that the connections between surface flow and ground water often are close. This is particularly true where streams flow over extensive deposits of alluvium or other unconsolidated sediments, or semiconsolidated and porous rock materials. Ultimately, fully unified flow control of basin water must contemplate operational management which considers surface waters, gravitational water, and other related moving ground water, as one.

Sections of the West have gone farthest in the direction of such unified management of surface and ground water. Knowledge of the value of return flow in irrigated areas has stimulated points of view in local water management which now are sweeping in their inclusion of all moving water within a basin. The unified management of surface- and ground-water flows has been given its most complete attention in the Santa Clara Valley, the San Joaquin Valley, and the southern coastal plain of California. The integrated use of ground- and surface-water supplies has been proposed in every state-wide water plan in California since 1931. Local operations of this kind have been undertaken for a number of years by the Turlock and Modesto irrigation districts, the Kern County Land Company (Bakersfield) in the San Joaquin Valley, by the United Water Conservation District (Santa Clara River, southern California), the Santa Clara Valley Water Conservation District (San Francisco Bay area), the Los Angeles County Flood Control District, and the Orange County Flood Control District.[5] Recent proposals for revised methods of managing the Salt River watershed in Arizona offer an interesting example of extending intensive unified ground-surface management to another area.[6] Similar views are held for some of the irrigated basins of Utah, and in the Platte River Valley of Nebraska. The principle now also is specifically recognized in New Mexico water rights administration. However, knowledge of the movement of ground water for most of the nation still is inadequate to permit joint consideration of ground and surface flow on more than a localized basis. Except in California, unified flow management which includes both ground and surface water has not been applied or considered for any except relatively small tributary watersheds.

[5] Information from Blair Bower, Delaware Basin Advisory Committee, 1957, and Stephen C. Smith, University of California, 1958.

[6] George H. Barr and Associates, *Recovering Rainfall*, Part I (Tucson, Ariz.: Arizona Watershed Program, 1956).

2) CONSERVATION AND QUALITY CONTROL

Ultimately, unified operational management for multiple purposes within a river basin (or other chosen administrative area) can include attention to conservation of water in use and to quality control, including the pairing of tolerant uses and mineralized supplies, as well as to conservational storage.

Attention has been given to all of these subjects in the operational management of water resources in the country, although again mainly on a localized, or on a single-purpose, basis.

Water conservation measures have been administered mainly on an emergency basis. However, some continuous programs of an educational nature have been developed, and cost analysis has encouraged industrial establishments to make some notable conservational efforts in scarcity areas. Examples of the emergency application of water conservation are most frequently found in urban areas. Many cities have had to apply use restrictions (like the use of water at certain hours, or prohibitions against luxury use) during seasons of low supply. Such restrictions were applied in the drought area of the Southwest for several years during the 1950's, and in 1953 were applied in an estimated third of municipalities of more than 50,000 throughout the country.[7] An example of emergency water conservation applied to a multiple-purpose system as a whole occurred on the Tennessee River in 1954, when recreational craft could move through locks only on a limited basis during the low-water season.

An example of a continuous educational program for water conservation is shown in the activities of the High Plains Underground Water Conservation District, with headquarters in Lubbock, Texas. Foreseeing the possible fate of an irrigated area making heavy withdrawals on a ground-water deposit with little recharge, this organization for several years has campaigned consistently for voluntary attention by users toward waste elimination.[8]

However, as of 1958, no major basin or other major area has developed an organization for systematic attention to conservational use

[7] K. A. MacKichan and J. B. Graham, *Public Water Supply Shortages, 1953* (Water Resources Review Supplement No. 3), U. S. Department of the Interior, Geological Survey, Washington, 1954), p. 2.

[8] See *The Cross Section,* a monthly publication of the High Plains Underground Water Conservation District No. 1, Lubbock, Tex,

of water as part of the efficient use of waters in a basin. This may be true because pressures for water use nowhere have reached a point where such organization is considered necessary. On the other hand, there is some reason to believe that the allocation system which prevails throughout the scarcity regions of the country may not be conducive to carefully planned and executed conservational use.[9]

In recent years quality control has received increasing attention in operational management of water facilities in the United States. The large majority of efforts have centered on use restrictions which concern municipal sewage effluent and industrial wastes. Enforcement actions are mainly in the hands of state regulatory agencies, while effluent treatment is the responsibility of the municipality or industry originating the waste. In addition, a degree of quality control is achieved both automatically and consciously by the operation of reservoir storage projects or systems of projects for minimum low flow discharges. Salinity control is an objective of management on some western projects, as for the facilities constructed by the Bureau of Reclamation on the Central Valley Project.[10] Sediment control has received much wider attention, as it has been authorized on a number of Corps of Engineers and Bureau of Reclamation projects, and in the work of the California Debris Commission and the Soil Conservation Service.[11] Sediment control authorizations for water development agencies, however, are mainly directed toward the construction and maintenance of storage space for impounding sediment. A notable exception occurs in the agricultural and forestry programs of TVA, which were partly designed to reduce the production of sediment on Tennessee Valley watersheds. The land treatment programs initiated by the Soil Conservation Service and by other Department of Agriculture agencies aid on a wider scale in sediment reduction. At the same time some important and unsolved sediment problems remain and have

---

[9] Most prior appropriation laws specify that the water appropriated must continue to be used beneficially, or the right to it lapses. Nonuse for a prescribed period of years usually results in forfeiture of the right. See, for example, Montana Rev. Codes 1947, Ann., secs. 89-802. As previously noted (pp. 513-15), most western state water legislation also provides for preference among different types of use. However, conservational use otherwise is a question generally left to the influence of economics upon entrepreneurs' decisions.

[10] Authorized in the Act of August 26, 1937, sec. 2, 50 Stat. 844, 850.

[11] See President's Water Resources Policy Commission, *Water Resources Law*, Vol. 3 of Report (Washington: U. S. Government Printing Office, 1950), pp. 337-38.

regional importance. Such are the production of sediment and other debris from logging operations in the Pacific Northwest, and the control of sediment movement in irrigation water on western irrigation projects.[12]

Water quality control has not received the comprehensive attention technically possible. No administrative machinery has been in motion which examines all aspects of water quality on a basin-wide scale and applies measures of quality control, including paired use plans, on a unified basis to the watershed.

3) DISTRIBUTIONAL ORGANIZATION AND THE SCALE ECONOMIES

Application of the scale economies to water facilities development may depend in part upon the scale and coverage of related distribution facilities under co-ordinated management. The size and diversity of the market served influence both the scale on which works can be constructed and the opportunity for a variety of incidental operational economies. This is most pertinent to electric power distribution systems, but it also applies to irrigation facilities, to navigation, and to the operational management of flood-control facilities.

The largest scale co-ordinated management of benefit distribution in the United States is the Corps of Engineers' flood-control management on the Mississippi River system. The routing of flood waters in co-ordinated manner for the entire drainage basin, including the major headwater tributaries, results in construction economies and a scale of benefits which otherwise would be unachievable in the middle and lower Mississippi Valley. Were each major tributary (or other tributary drainage unit) operated independently, the probability of higher flood crests on the lower river would be increased.[13] Single management also permits more effective planning and location of flood control

[12] Charles McKinley, personal communication to the authors.

[13] R. K. Linsley, M. A. Kohler, and J. L. H. Paulhus, *Applied Hydrology* (New York: McGraw-Hill, 1949), pp. 612-13, discusses the general benefits and disadvantages of system operation. However, the management of benefit distribution and benefit maximization on the Mississippi system has not yet received independent appraisal. While appraisals made for Corps of Engineers' program and project planning permit some general conclusions like those suggested in this text paragraph, there remains an opportunity for further study of Mississippi water control benefits which can be of great importance in future American water development.

works, since a basin-wide flood-control plan can be devised and adhered to.

The Corps of Engineers has a similar position in its management and operation of inland navigation facilities. However, only a few of the possible operational economies have been realized (except within the Great Lakes system) because several vital interbasin connections have not been constructed.[14] In the absence of these interconnections, the Corps has not yet had a large-scale system connecting the centers of origin and destination of heavy traffic in the country.

In the field of irrigation, the Bureau of Reclamation has organized large-scale distribution facilities in both the Columbia Basin and the Central Valley projects. When and if completed, some of the water distribution facilities within the Missouri Basin, like the James River diversion, also will be on a large scale. Still others, like the Texas coastal project,[15] are in the planning stage. The Bureau thus definitely takes advantage of organizing large-scale distribution facilities where they physically are possible. However, it has not yet approached the consolidation of established irrigation districts of small scale to produce similar economies.

Federal government agencies also have organized relatively large-scale distribution of electric power. The generating and transmission system of TVA, covering an 80,000-square-mile service area, is such an operation. The transmission system of the Bonneville Power Administration, tying together the subregional systems of the three northwestern states and the federal generating plants on the Columbia, is another. Some privately owned distribution systems, like that of the Pacific Gas and Electric Company of California and the Montana Power Company, also have geographically extended coverage. The major systems in large cities—Consolidated Edison Company of New York, Philadelphia Electric Company, and the Commonwealth Edison Company of Chicago—also are large-scale distributors, even though their coverage is geographically limited. In at least one case integrated operation of private and publicly owned systems with adjoining service territories has been successfully undertaken. This is by the Puget

[14] For example, the Lake Erie–Ohio River canal, the Delaware–Hudson canal, a deepened Lake Ontario–Hudson River canal, the Tennessee–Tombigbee interconnection, and others. This statement is not intended as a judgment on the economics of water transport in comparison to rail, highway, or other overland transport.

[15] Senate Document 57, 83rd Congress, 1st Session, *Water Supply and the Texas Economy*, Washington, 1953.

Sound Utilities Council in the state of Washington. The Council's member utilities are the Puget Sound Power and Light Company, the Chelan County Public Utility District, the Snohomish County Public Utility District, Seattle City Light, and Tacoma City Light.

On the whole, however, distribution of electric power within the United States tends to be undertaken by small-scale units, typically with a peak load of less than 125,000 kilowatts (see page 505). Geographical coverage generally is intrastate, and often is made up of a single urban area and its surrounding service area. In some sections of the country there are overlapping privately operated and publicly operated systems. (See figure 38, map pocket.) Since most of the systems are vertically integrated (i.e. generating, transmission, and retail distribution), loads are not diverse, and small or moderate sized generating plants are the rule. This pattern has its extreme in southern New England. While there are some privately owned generating and transmission companies of regional scope affording some scale advantages, like the New England Power Company, the many small distribution systems have not favored scale economies in electric power production. The country still has taken only a few steps toward the super-power systems foreseen by some, with their indicated efficiency.[16] While the optimal size of a closely co-ordinated distribution system may be no more than regional, few of the country's systems in 1958 reached this scale. The Federal Power Commission's responsibility to secure interconnection and co-ordination of electric facilities[17] applies only to interstate utilities, and hence is not applicable to many of the existing systems. It also appears that this authorization has not been vigorously administered by the Commission.[18]

A recent legislative attempt to encourage the construction of large-scale facilities for private company operation appeared in the 1957 "Bricker Bill" (S. 2552, 85th Congress) which sought to amend the Holding Company Act of 1935. This would grant exemptions from the terms of the Act to utilities owning more than 10 per cent of the voting

[16] U. S. Department of the Interior, Bureau of Reclamation, *A Study of Future Power Transmission for the West*, Washington, 1952. Also Leland Olds, "Giant Power in the Legislative Program," paper presented at National Rural Electric Cooperative Association Annual Meeting, Dallas, Texas, 1958; and *idem*, "Giant Power for the American People," address before Western States Water and Power Conference, Salt Lake City, Utah, May 11, 1958.

[17] Act of August 26, 1935, Sec. 202(a), 49 Stat. 838, U. S. C. 824 a(a).

[18] See President's Water Resources Policy Commission, *Water Resources Law* (Vol. 3 of Report), pp. 287-88.

stock of any company engaged solely in the generation and transmission of electric energy and not distributing such energy. It is thought that proposals for joint company entrepreneurship in large-scale facilities, like the Pacific Northwest Power Company's million-kilowatt Mountain Sheep project, on the Snake River, thereby would be facilitated.[19]

Organization of electricity distribution which responded to the opportunity given by technology thus would seem to have been in a beginning stage in the United States of 1958.

Four Organizational Problems Presented by Technology

1) RELATION OF THE DIFFERENT LEVELS
OF ADMINISTRATIVE ORGANIZATION

One of the most troublesome of the organizational problems raised by technology has been the relation of the several levels of public administration of water facilities operation. This question is present in both planning and development, but it reaches its crisis in arrangements for operational management.

Relations of federal, state, municipal, private corporation (under franchise), county, co-operative, and district administrative entities have many wellsprings other than those occurring in technical change. Such are the governmental concerns for the general welfare, defense and security, promotion of commerce, and other public responsibilities. The entire complex of those relations should be considered if a description of the problem of water resource administration is to be complete. This is beyond the scope of this study. Nevertheless, *some aspects of these relations have predisposing conditions in technology. The succeeding description is limited to these conditions,* and is intended only to outline the problem observed.

There can be little doubt that technology has strengthened the federal hand in the joint responsibility for operational management of water facilities in the United States. The rivers of the United States have been suited to a scale of works which, when technically feasible, were also suited to financing by an agency of national or regional size. Since it

[19] National Association of Electric Companies, *Legislative Abstract*, Vol. II, No. 16 (October 28, 1957), p. 1.

undertook financial responsibility for construction of these works, and since the works often could be instruments of national political policy, operation by the federal government also followed naturally. Because of the division of the basins among several states, the application of integrated basin management, insofar as it has been undertaken, also fell naturally to the federal government. Thus, both the financing requirements of individual works and the integration of works within a major basin (or a major tributary basin) opened a field of operational management for federal attention which no existing agency at another level immediately cared to challenge.

During the two and one-half decades since these technical factors commenced to be felt in the development of water facilities, the most striking change in management responsibilities resulting from the application of the scale economies has been in the position of the states and state organizations. The most consistent, and still operationally unsolved, problem has been the place of the state government agency, the quasi-state agency[20] and state government in the management of these facilities. Localized organizations have generally expanded and have extended their operational responsibilities, along with the federal agencies. Rural electric co-operatives, municipal organizations, irrigation districts,[21] and soil conservation districts are examples. While the states and their related agencies have not retrogressed in the acceptance of responsibilities, they generally do not have the same *relative* place in water facilities administration that they had in the 1920's. At least in part this may be traced to the financial incapacity of the states, which have had many other increasing obligations within the last three decades. The outstanding feature of the period within which the application of the scale economies and the scheduling techniques has taken place has been the joint growth of local and federal institutions in their operational responsibilities.

The position of the states (the quasi-sovereign units in the federal system) accordingly merits some special notation. As of 1958, their position in the operational management of water facilities and uses might be summarized thus:

[20] Quasi-state agencies consist mainly of those private utilities operating under franchise on a state-wide basis, or over a sufficiently large part of a state, so that they are of more than local or district significance.

[21] The classification of irrigation districts as local agencies is not always definite. In California, for example, irrigation districts are state agencies, subject to state supervision, even though they are under local management.

a) The states have retained, strengthened somewhat, and expanded appreciably their responsibilities for administration of regulatory functions.[22] Water rights administration has been strengthened further in the West, has been extended to one eastern state, and is likely to be instituted in other eastern states. Ground-water regulation has appeared in some states, although it has not yet met some of the challenging questions raised by technology.[23] Regulation of the waste-carrying functions of streams, on the other hand, has been taken in hand much more firmly by state administration.

b) Only one state, California, thus far has shown inclination to plan realistically for operational management of large-scale water facilities on a basin-wide or an interbasin basis. This is in the California State Water Plan.[24] Delay elsewhere generally appears to have stemmed from a reluctance or incapacity on the part of individual states to assume financial responsibility for construction. The prevailing attitude of interests representing the states has been one of encouraging federal financing and construction, with later assumption of operational responsibility by the state. This has not been realized because federal and state political objectives often have been divergent as to the distribution of benefits from water development.[25] Financing and operational responsibility for some secondary facilities, however, has been undertaken in a number of states (e.g. New York State Power Authority electrical transmission facilities).

The principal exception to this pattern has been the willingness of some private utilities to assume responsibility for the financing, and hence operation, of the electric power facilities of the federally constructed large multiple-purpose projects, like those on the Columbia River. This has been realized in only a few instances because of divergent political objectives between Congress, the national executive, and state administrations. Private ownership has not been in a position to stress the social development objectives considered essential by some recent administrations.

[22] The activities of the Oregon Water Resources Board and the Oregon Hydroelectric Commission are examples.

[23] Such as the future of areas depending on "mined" ground-water deposits, and waste elimination within them.

[24] The Feather River project, one unit in the California State Water Plan, had been approved, appropriations made, and construction started in 1957.

[25] Such as those on the "160-acre" provision of federal reclamation law, and the "preference clause" favoring public bodies and co-operatives in electric power distribution.

c) No effective interstate agency has been formed to act as entrepreneur in inland water development, or to undertake operational management of a system employing the scale economies and the scheduling techniques.[26] Regulatory agencies for administering apportionment of water have been formed and are operating. In addition compact organizations with ambitions toward basin management have been explored, as in the Columbia Basin, but had not yet been agreed upon in 1958. On the contrary, some notable attempts toward a basin-wide management organization, like that of the Interstate Commission on the Delaware River Basin, had been abandoned.

d) Even though limited financial capacity and limited jurisdiction have been important reasons for lack of state attention to the scale economies, state organization has not generally responded to some opportunities of more modest financial requirements and territorial coverage. Such are the problems of meeting dispersed demands, as in the case of rural electricity and rural domestic water supply. Other examples are to be found in further application of water use restrictions.

2) RELATIONS OF PLANNING, CONSTRUCTION, AND MANAGEMENT AGENCIES

The problem of relations among planning, construction, and management agencies has arisen since large-scale development became common in the United States. Because much of this development has been with public agency entrepreneurship and public financing, it also is a problem peculiarly associated with public agencies. However, the entire complex of public and private agencies interested in development has exhibited, in places, some facet of the problem of these relations. The problem is most clearly and sharply manifest in multiple-purpose, integrated basin development and management.

During the years before basin-wide multiple-purpose planning and construction, the problem was scarcely visible because planning, construction and operational management most commonly were undertaken within the same organization, or by closely related organizations. Thus the Corps of Engineers planned, constructed, and operated (where necessary) facilities for navigation improvement; privately

---

[26] The New York Port Authority may be cited as an interstate agency concerned with marine development.

owned electric utilities planned, constructed, and operated hydro facilities; and the Bureau of Reclamation planned, constructed, and operated irrigation facilities. Moreover, development generally was carried out within discrete local areas on a single (or perhaps dual) purpose basis, where conflict of interest generally was less likely.

A long-standing question on flood-water management

One pre-1930 illustration of the problem occurred in floodwater management. Where flood-control works were federally constructed along the Mississippi and elsewhere, maintenance of the works (excepting reservoirs) on occasion was left to local bodies. These arrangements for maintenance failed in enough instances for them to be considered unsatisfactory.[27] However, the problem hardly could be considered serious since Congress seemed disposed to further use of federal funds where neglect of maintenance contributed to a flood danger or a flood emergency.[28]

A somewhat more serious problem of planning, construction, and management relations was the general absence of community, county, and state restraint upon flood-plain occupancy in flood-susceptible localities. Zoning restrictions or other land-use regulation rested in the authority of states, counties, or communities. As planning of flood control and construction of flood-control facilities became more and more a federal function, the considerable power of local regulation to lessen flood damage dangers by zoning rarely was used effectively. While some progress has been made within the last decade, this problem still continues. It is illustrated by the situation of Chattanooga, Tennessee. Chattanooga has been provided a substantial amount of flood protection through the federally operated TVA reservoir system. As late as 1958, however, the city had failed to establish zoning restrictions concerning the occupancy of flood-susceptible areas remaining within the city, in spite of accurate technical appraisal of the remaining danger there.[29] The same story may be repeated for many other flood plains.

[27] W. G. Hoyt and W. B. Langbein, *Floods* (Princeton, N. J.: Princeton University Press, 1955), pp. 172-74.

[28] *Ibid.*

[29] *Annual Report of the Tennessee Valley Authority, 1956* (Washington: U. S. Government Printing Office, 1956), pp. 18-19. See also Gilbert F. White et al., *Changes in Urban Occupance of Flood Plains in the United States* (Chicago: University of Chicago, Department of Geography Research Paper No. 57, 1958).

Four present-day questions on the relation of planning, construction, and management

As the opportunity came for applying the scheduling techniques and the scale economies to harnessing a river system, the problem has taken on a much broader meaning than the relatively simple one illustrated in federal-local relations on flood damage prevention. The kind of organizational issues raised are suggested in four questions: How far should water development go in the direction of what might be called "vertical integration" of planning, construction, and management? How are planning and operational management to be co-ordinated in basins of geographically divided agency or national jurisdiction? How much should general regional planning shape project planning and development? How closely are operational interests for the secondary purposes to be brought into planning, and how are they to be reconciled with the primary purposes in case of conflict?

The question of vertically integrated organization. There is no complete vertical integration of water resource development in the United States, in the sense of a single agency undertaking multiple-purpose planning, constructing and operating multiple-purpose facilities, and distributing product and services to the user. Privately owned electric utilities have the farthest reaching integration, since they have planned, constructed, and operated dams, and sold electricity from their plants to the retail consumer. But this is on a single-purpose basis. The Corps of Engineers also exhibits such integration for navigation, flood control, and other flow regulation. The TVA and the Bureau of Reclamation, which have had the highest degree of multiple-purpose integration among the public agencies treating water development as a whole, both undertake planning, construction, and operation of major facilities. However, both rely on independent, smaller, localized organizations to care for final distribution of products and services.[30] This

---

[30] The Bureau of Reclamation sells water to or contracts with water users' associations and irrigation districts for irrigation services rather than individual farmers or other water users. It sells power over its own transmission lines to co-operatives, municipalities, and private companies, or to separate federal power systems, like the Bonneville Power Administration or the Southwestern Power Administration, which maintain the transmission lines and sell power at wholesale to industrial corporations and utilities. The TVA sells electric power to municipalities, co-operatives, and private companies for final distribution; TVA-produced fertilizers are distributed by co-operatives, co-operative associations, or private firms for test-demonstration purposes.

is the rule for nearly all of the operations conducted by the federal government, which thus far has been the principal entrepreneur employing the scale economies and the scheduling techniques in water development. On a smaller scale some irrigation districts, like the Modesto Irrigation District of California, have vertically integrated operations on a multipurpose basis.

Thus the first question generally has been answered in United States practice by stopping far short of full vertical integration. Complete integration has never been a very seriously recognized objective. The question instead has centered on the precise intermediate point at which the consolidation of planning, construction, and operational functions does end.

The three major federal agencies that carry on large-scale multiple purpose operational management—the Corps of Engineers, the Bureau of Reclamation, and TVA—all plan, construct, and operate.[31] When they reach the operational step, however, the three differ in the stage to which operational responsibility is carried. As their geographical jurisdiction permits, all three operate facilities on the principle of co-ordinated water control on a stream. Planning therefore is toward that end. Water storage and releases for navigation, irrigation, and flood control now are the result of a process where planning, development, and operation were closely tied. The principal difference centers on the relation of planning to the distribution of electric power, the chief vendible service. The TVA operates a closely integrated primary transmission system, designed to take advantage of all the economies possible from operating a large system with diverse generating characteristics with a large marketing area of diverse load. The Corps of Engineers does not do this, relying on other agencies for dispatching, transmission, primary and retail distributions of power. The Bureau of Reclamation falls between the two. Its electrical operations in some regions (e.g. Hoover Dam) are on the pattern of the Corps. In other areas, like the Missouri Basin and the Central Valley of California, the agency also maintains primary transmission systems.

Where the Corps pattern prevails, co-ordination of planning and development with the needs (and opportunities) of operational management requires special care. Perhaps the best documented illustration

[31] Other federal and state agencies working on a smaller scale have a similar range of practice. Among them are the International Joint Commission, the dam or river authorities in Texas and Oklahoma, and the Indian Service.

has been that of the installation of electric generators at McNary Dam on the Columbia River.

According to testimony offered in the Senate and the House of Representatives on the several Columbia Valley Administration bills in 1949, the Corps of Engineers had chosen high-reactance generators for installation at the McNary generating plant. As compared to low-reactance generating equipment of the same installed capacity, an investment saving of approximately $3 million was possible in the Corps' choice. However, the high-reactance generators would have required one more circuit in transmission than the low-reactance type, a fact which soon became apparent to the Bonneville Power Administration, which was responsible for the distribution of power from the McNary plant. Additional transmission equipment investment costs for the use of the high-reactance generators would have been $10 million, a net additional cost for generating and transmission equipment combined, of $7 million. The preference of the Corps on one hand and of the Bonneville Power Administration on the other required approximately eighteen months of negotiation for final settlement between April 1947 and October 1948. The Federal Power Commission, because of its responsibility for the design of generators of the Corps-built dams, also was involved in the negotiations. The low-reactance type of generator was finally chosen.[32]

Similar problems are known to have arisen elsewhere, although they are less a matter of the public record. For example, electric power plants on Corps of Engineers projects are often logically designed for peaking operation and consequent integration into regional power systems. At the same time they may be called upon to operate under the "preference clause" of the Flood Control Act of 1944,[33] by which public bodies and co-operatives are given prior consideration over other agencies in the delivery of electric power. The demands often are not for peaking power, but for base load power from these sources. While the difficulty of foreseeing eventual demands and even the discovery of the appropriate "operational voice" must be freely admitted in this instance, the need for that voice in planning would seem clear.

[32] 81st Congress, 1st Session, Senate Committee on Public Works, *Hearings on S. 1595, S. 1631, S. 1632, and S. 1645* (Columbia Valley Administration), 1949; testimony of Secretary of the Interior J. A. Krug, pp. 278-79. House Committee on Public Works, *Hearings on H. R. 4286 and H. R. 4287,* 1949; testimony of Representative Henry M. Jackson, Secretary Krug, and Assistant Secretary of the Interior C. Girard Davidson, pp. 39, 41, 60, 77.

[33] Act of December 22, 1944, Sec. 5, 58 Stat. 887, 890, 16 U. S. C. 825s.

Such problems of course are not insurmountable if detected in time. However, the separation of development and operational management does require the introduction of operational voices in the planning of water development facilities.

Problems introduced by geographically divided jurisdiction. Where scale economies are possible through unified water control within a basin, and through an extended primary electricity transmission system under one direction, divided geographical jurisdiction of agencies may create a co-ordination problem in relating operational management and planning. This might be called a problem of horizontal integration. The critical division is between upstream and downstream territory. The type of development undertaken upstream obviously has an effect on the operational results of downstream works. For instance, not infrequently water retained in a major upstream storage facility may be more valuable in power production at downstream plants than at the site of the upstream reservoir itself.[34] Such problems of co-ordination have arisen particularly on the western streams that have been shared by the Bureau of Reclamation and the Corps of Engineers, and on international streams. The Columbia, which has both an international and an agency division of jurisdiction, offers a good illustration of the problems involved.[35] The vast Missouri Basin development, orginally planned in two sections by the Corps (downstream) and the Bureau (upstream), also has an interesting history in this respect. Initial agency independence was gradually replaced by elaborate organizational structure and co-ordinating procedure for relating operational needs downstream to planning and construction upstream (and for other purposes also).[36]

While the best known jurisdictionally caused problems of relating planning and operation have arisen on the major national agencies and international streams, they are not absent in other quarters. An illustration occurs in the development of the Kansas Basin, where the Corps' development for Kansas City flood control has omitted con-

[34] Illustrated by the Hungry Horse reservoir, on the Flathead River in Montana, and by the Albeni Falls development on the Pend d'Oreille River, Idaho.

[35] See President's Water Resources Policy Commission, *Ten Rivers in America's Future* (Vol. 2 of Report), pp. 1-77, and especially pp. 60-67.

[36] *Ibid.*, pp. 161-281, and especially pp. 245-48. Missouri Basin Survey Commission, *Report* (Washington: U. S. Government Printing Office, 1953). Commission on the Organization of the Executive Branch, *Natural Resources Task Force Report* (Washington: U. S. Government Printing Office, 1949), p. 244.

sideration of the upstream agricultural and municipal needs.[37] Here the development encouraged in the interest of scale economies, and on a basin-wide plan, would seem to have leaped over the more localized needs of the smaller tributary basin. It is a different sort of operational management-planning co-ordination problem, in which downstream effects on upstream facilities raise the question. Still another situation in which the co-ordination of two or more units becomes a problem is that occurring where interbasin diversions exist, as from the Delaware River.

Two types of planning. Problems of planning-development-management co-ordination often have been created because there are two kinds of planning customarily carried on in distinctly different organizations. Both, however, may influence the course of development. While the two merge into each other in practice, the core of each activity is distinct enough to warrant the separation here suggested.

One type of planning is that essential to any development agency: the selection of development sites, planning of the relation of single facilities to system of works in a basin, and determination of the scale of works economically warranted. This kind of planning is conducted as part of the normal activity of each development agency, major or secondary, multipurpose or single purpose. It is found alike in publicly sponsored agencies and private corporations. We might term it program and project planning.

A second type is planning for the economic development of a region as a whole. This may be called regional planning. In it economic needs are anticipated, economic potentialities explored, social targets may be identified and set. It is essential if the most effective physical planning is to be undertaken.[38]

In this country regional planning has been undertaken in the past by the National Resources Planning Board,[39] by state planning boards and planning commissions, by *ad hoc* federal committees like the Columbia Basin Interagency Committee, New England–New York Interagency Committee, and the Arkansas-White-Red Interagency Committee, by the TVA and other federal development agencies, by regional affiliates of the privately sponsored National Planning As-

[37] See University of Kansas, *The Kansas Basin, Pilot Study of a Watershed,* Lawrence, Kan., 1956, p. 103.

[38] This is true, whatever the political beliefs and structure, and whatever the general societal framework of the community or larger region.

[39] Includes predecessors and affiliates of the Board.

sociation, by private utilities, by chambers of commerce, by railroad companies, by quasi-public organizations like the New England Council, and still other groups.

The scope of this type of planning obviously is much broader than water development, and in our society it usually is carried out as data collection and educational and advisory activities. Yet such planning may be as closely related to operational management of resources as program and project planning, because it is an attempt to anticipate the future environment in which operation of water facilities and other factors of production will be undertaken. It also may be an effort to shape that environment. Thus where water is scarce, regional planning might attempt to shape the future environment by determining and publicizing the least costly pattern of urban settlement, and the economically most profitable "mix" of water uses. These and other possibilities all should concern operational management of water resources and the design of water resource systems.

It is clearly to the advantage of operational management of water resources to have in advance as much as possible of the data of regional planning and regional planning analysis. The needs of operational management thereupon may be entered into program and project planning more realistically and more effectively. However, the connections between regional planning, operational management, and program and project planning often have been either disregarded or tenuous. There have been a number of reasons for this. For one thing, good regional planning requires highly skilled professional practitioners. There have been far too few trained superior men to meet the challenge of a most exacting technical and diplomatic task. For another reason, such planning has been beyond the scope of most water development agencies, sometimes actually discouraged by the Congress.[40] A third has been a quiet assumption by program and project planners that a region and its economy would adjust to the project without bothering about the complicated process of region planning analysis.[41]

Still other reasons have been the location of development in sparsely settled regions with few organizational resources, lack of common objectives between public and private agencies, single-purpose objec-

---

[40] For example, the regional studies and the government relations and economics budgets of TVA.

[41] Illustrated in the early plans for the Missouri Basin development. See Senate Document No. 191, 78th Congress, 2nd Session, 1944. Also Commission on the Organization of the Executive Branch (1949), *op. cit.*

tives, naturally vague delineation of some regions, and rigid concepts of planning by professional planners. The latter in the past often have failed to present the possibilities of regional planning, or its vital place, in terms which would be understood by project planner and Congressional committee member. Even the joining of regional planning analysis and program planning in a single agency has not always successfully co-ordinated professional regional planning and program planning.[42]

It should be noted that some successful co-ordination between the economic phases of regional planning and program planning for resource development has been carried on by private electric utilities within their service areas. Organizations like the Georgia Power Company and the Texas Electric Service Company have had a consistent interest in stimulating the development of their service areas in a planned program. Many of these organizations, however, have been hampered by the small size and the almost accidental boundaries of their service areas. The utilities' history would seem to show that the deeper the interest of a development agency in operational management, the more attention is likely to be paid regional planning action.

Within recent years the two types of planning have been drawn somewhat closer in water development. Both development agencies and their program planners have shown an increasing respect for the usefulness of regional planning data and for broad anticipation of the problems of operational management. The reports of the New England–New York and Arkansas-White-Red interagency committees; the activities of the more permanent (but still voluntary) federal interagency river basin committees in other regions; the Department of the Interior and the Department of Agriculture field committees, and increasingly sophisticated concepts as to essential planning data by the development agencies, all point to some improvement of the past poor connections between regional planning analysis, operational management, and program planning. Aside from the TVA, the most sustained and comprehensive effort has been in the Pacific Northwest, where wide consciousness of regional development objectives appears to be present.

An interesting experiment in combining the two types of planning on the Delaware River was in progress in 1958. The Delaware River Basin Advisory Committee was attempting to marshal data on

[42] Illustrated in the prewar TVA, where relatively rigid professional concepts of planning did not meet sympathetic response on the part of program planners.

the regional economy of the basin and adjacent territory, and fit it into
the program of survey and planning for further physical development
in the basin being conducted by the Corps of Engineers and other
federal agencies. Such data also could be used in working out recommendations for administrative organization to manage the later operation of facilities, an analytical responsibility undertaken by the Committee.

The "secondary" purposes. Even if co-ordination of planning, construction, and operational management were perfect within our present
administrative structure for development, the problem of the so-called
secondary purposes would remain. All of the major public development
agencies have definite statutory assignments to certain "primary" purposes. The major private development agencies have such interests by
choice. Yet the facilities constructed for navigation, flood control,
irrigation, and electric power may and do affect vitally the statutory
"secondary" purposes.[43] These are illustrated by fish and wildlife use,
by waste carrying, and by recreation.

The most recurring illustration of this problem in the United States
has been the management needs of fish and wildlife as they impinge on
plans for major regulation works. Many although not all of the most
difficult questions concern the provisions for anadromous fish, like
salmon, shad, steelhead, and sturgeon. Project planning long solved
this problem simply by excluding the fish; that is, their way barred by
a dam, they ceased to be a part of life in the river upstream from the
barrier. This solution was used even after the construction of large-
scale facilities commenced, as at Grand Coulee Dam, Washington.[44]

The Columbia River, on which Grand Coulee Dam is located, may
be used as an illustration of the manner in which fishery needs were
handled in the planning of the last twenty years. In time, a combination of commercial fishery interests, privately organized game fishing
and conservation groups, state and federal fish and wildlife agencies,
mobilized public opinion in such a way as to make careful consideration of this secondary purpose essential in program and project planning.

[43] The use of the term "secondary" here refers to the past statutory position of
these purposes. It is not intended to suggest that they should be considered
secondary in efficient comprehensive development.

[44] After completion of Grand Coulee Dam, arrangement was made for trapping
at Rock Island Dam, Washington, all spawning anadromous fish bound upstream. These fish are taken to hatcheries for spawning, and the fry are later
released in tributaries open to fish movement.

From the middle 1930's onward the salmon and steelhead preservation problem was very much in the minds of the Corps, Bureau of Reclamation, and other planners responsible for future major regulation works on the Columbia. By the late 1940's, when planning for development on the river reached its peak, extended conference and negotiation among the interested agencies had produced a plan for preservation of the fish runs on the river. As reported, it included "removal of obstructions, abatement of pollution, screening of diversions, fishery (fishway) construction, transplantation of runs, extension of artificial propagation, and establishment of fish refuges."[45] This plan was developed through the offices of the Columbia Basin Interagency Committee and its co-operating agencies—state, federal, and private. In it the operational needs of the fishery received fully as careful consideration in planning as the primary development purposes.

Implementation of the plan then proceeded, through the construction of fishways in the dams otherwise barring runs on streams designated for fishery preservation, through hatchery propagation and other measures. However, full implementation of the plan has been shadowed in some doubt, because the effects of a series of major dams upon the mortality of descending smolt was unknown and unascertainable without extensive research. The capital value to be attached to a resource theoretically available in perpetuity also was not evaluable. Full implementation also encountered difficulty because some development agencies had sensitivity more to local than to regional public opinion, and were unwilling to sacrifice their interest for the fishery plan. Such a conflict arose in the previously cited plans for the Mayfield and Mossyrock projects on the Cowlitz River, a lower tributary which was a key spawning ground in the fishery plan.[46] Thus the fishery problem on the Columbia continued to intrude in project planning as late as 1958. It was judged to be of sufficient importance to warrant support of a major interagency fishery research program on the river through funds appropriated to the Corps of Engineers, a major entrepreneur on the river. Provision also was made for permanent professional fishery staff assistance within the Corps itself.[47]

While the lower Columbia fisheries plan has encountered dif-

[45] President's Water Resources Policy Commission, *Ten Rivers in America's Future* (Vol. 2 of Report), p. 45.

[46] *Ibid.*, p. 47.

[47] A Fisheries Section is included within the office of the Division Engineer, Portland, Oregon. The Corps also has the active co-operation of the Fish and Wildlife Service.

ficulties it also has produced some results. In general, hatchery propagation, obstruction removal, and other measures appear to have improved the anadromous fishery habitat on some lower basin tributaries. The accomplishments were considered encouraging enough so that the fisheries program was extended to important Idaho tributaries, like the Salmon and the Clearwater rivers in the federal Civil Functions Appropriation Act of 1957.[48]

The Columbia case of relating a secondary operational purpose in water development to program and project planning is of interest, because adequate "feedback" of operational needs for the fishery into water project planning was first stimulated by an alerted and militant set of special-interest groups. Once pressed, the major development agencies proved to be very responsive in bringing this secondary operational need into planning, and in co-operating with the federal Fish and Wildlife Service and state agencies toward this end. However, the more local agencies with development ambitions on the river were somewhat less responsive. Their position was strengthened by the views of the Federal Power Commission, which considered a different fishery preservation plan to be valid and at the same time compatible with local power development interests. (See pages 514-15.)

Few other problems of relating the operational needs of secondary purposes to project planning have been as sharply raised as the fishery problems. That they exist, however, is illustrated by the recreational use issues in comprehensive planning for the Potomac Valley.[49] This aspect of relating planning, construction, and operational management is suggested as important for future consideration particularly because of the increasing importance of recreational water use and waste carrying. The rapidly growing population, the even more rapid urbanization of the American population, changes in work habits (including earlier retirement), and longer life span in the United States point in the direction of public recreational development on a scale hitherto unanticipated. Urbanization, wider distribution of industry, more extensive concentrations of industry, and use of nuclear products and processes, all suggest waste-carrying problems on a scale hitherto unexperienced. Con-

[48] For a description of some plans in the new fisheries program, see Fishery Steering Committee, Columbia Basin Interagency Committee, *Columbia River Fishery Program*—Part I: Comprehensive research program for development of the fishery resource; and Part II: Inventory of streams and proposed improvements for development of the fishery resource (Portland, Ore., 1957).

[49] President's Water Resources Policy Commission, *Ten Rivers in America's Future* (Vol. 2 of Report), pp. 603-8.

sidering this outlook, it is pertinent that the administrative means for relating the secondary purposes to planning for water development often have been the product of emergency situations detected by special interests appealing to the public at large.

An interesting and unusual example of the relation of "secondary" purposes in planning and operation is to be found on the Marias River in north central Montana. Here the Tiber Dam and Reservoir were constructed by the Bureau of Reclamation primarily for irrigation and flood control. However, the region which the reservoir was constructed to serve has not adjusted its economy to the capacities of the project in the manner expected by the project planners. Contrary to earlier expressed plans, relatively few farmers of the area wished to irrigate after completion of the dam and reservoir, so no distributional canals were built.[50] Farmers' personal preferences for dry-land farming, the short growing season, farm product surpluses, and other factors may delay irrigation development as part of the Tiber project for a decade or more. On the other hand, the recreational use of the reservoir (swimming, fishing, boating, camping, picnicking) has far exceeded initial expectations and has become the principal use of the reservoir.[51]

Another interesting example of the evolution of the "secondary" purposes occurs in the case of Pacific Northwest salmon and steelhead fishing. Coincident with declining anadromous fish runs in Northwestern streams has come a greatly increased interest in sports fishing of these species. On a number of streams salmon and steelhead now are game fish only, and an important sports fishery is found also on most streams that are commercially fished. Every coastal stream in Oregon by 1958 had been closed to commercial fishing of salmon.[52] It is probable that more people in the Pacific Northwest now are concerned about the anadromous fisheries for sporting purposes than for commercial fishing. Their proper maintenance now may touch the interest of ten game fishermen where one commercial fisherman was concerned thirty years ago.

These developments, and other experiences with the recreational

[50] E. G. Nielsen, address before the Montana Reclamation Association, Great Falls, Montana, November 18, 1957 (U. S. Department of the Interior, Information Service release, Nov. 18, 1957), pp. 3-4.

[51] Edwin Wilson, Billings, Montana, personal communication to the authors, 1957.

[52] Charles McKinley, personal communication to the authors, January 1958.

values of multiple-purpose reservoirs, certainly raise questions as to whether the statutory secondary purposes in our publicly supported water developments in fact should continue to be so. If even a farming population, with its access to the out-of-doors, finds recreational value in a water development unsuspected by project planners working within their statutory limitations, how then should we interpret the recreational needs of the great urban population now certain in the future of the country? At the very least, a review of the position of all secondary purposes in our water planning and development appears to be needed. There may no longer be nationally preferred purposes for development, or the preferred purposes of the past may be socially and economically anachronistic.

3) FLEXIBILITY IN ADMINISTRATIVE ORGANIZATION

Water development has many facets and, both physically and economically, is a maze of complex interrelations. Technical change affecting water development within the last three decades had been rapid. Viewed together, these two facts speak for a flexible administrative organization in planning, construction, and operational management. They also speak for administrative experiment and the habit of analytical review of such experiment. In practice, the problem is one of maintaining administrative flexibility without encouraging cumbersome procedure, friction, and confusion. It also is one of encouraging orderly progress in administrative organization.

Several questions therefore are appropriate for brief consideration: How flexible has our administrative organization for water development been? Is physical design compatible with administrative flexibility? How much opportunity for experiment have we permitted? What has been the price of flexibility? Has the nation a procedure for analytical review and eventual further application, where appropriate, of successful experiment?

Considered on a national basis over the last fifty years, our administrative organization for water development and management has been flexible, although there have been some points of rigidity. The points of rigidity are important, and bear more than passing mention. One such point has been our water allocation system in areas of scarcity, related as it is to state territorial jurisdictions and prior appropriation rights. Another is the number of small-scale power service areas, unresponsive to the scale economies of larger distribution systems because of entrepreneurial inclination and statutory discouragement. These examples will not be elaborated further because of previous reference in this study.

A third point of rigidity is created by the manner of design of some of the works built for stream regulation. The same rapid-moving technology which speaks for flexibility in administrative organization also suggests the desirability of building flexibility into the design of multiunit, multiple-purpose systems of stream regulation works. Suitability to flexible use and adaptability to changed methods of operation to conform to a changing economic, technical, and social environment would seem to be sound planning. While many extenuating circumstances may excuse the planners of past projects, the fact remains that few of the works constructed may be considered to have the desired flexibility. The previously cited "penstocks provision" of the flood control acts (see page 477) is a lonely and minor example of a principle of design which needs to be considered much more carefully. Contrasted with this principle is the practice represented in Miami River and Muskingum River (Ohio) flood control planning, which commits system facilities almost irrevocably to a very narrow range of operational change. Until it is so considered, the works constructed in the past will themselves offer some encouragement to rigidities in the administrative organization for development and operation.

Bower and Thomas recently have undertaken a brief but perceptive analysis of the reasons for and technical feasibility of flexible design in river basin development. Among the reasons cited for incorporating the maximum flexibility in design are: population changes in the service area, changes in the level and components of economic development, technologic changes affecting demand, institutional changes (e.g. water allocation devices), hydrometeorologic and geomorphic changes, changes in system boundaries, changes in engineering design and practice and other technical changes affecting supply. Among the methods of incorporating flexibility into design the following are illustrative: detailed use of statistical probability studies, design for future shifting of use among all feasible purposes, design of units in a system to operate over a wide range of conditions, provision for step construction, use of numbers of replicate units rather than a few large units, and interchangeability of operating parts.[53]

[53] Blair Bower and Harold A. Thomas, Jr., "Flexibility in River Basin System Design," Cambridge, Mass., 1957, mimeo draft, 16 pp.

However, these technical possibilities will be faced with certain rigidities in the American institutional structure. A most important rigidity is the attachment of specific social (or "special") interests around the existing administrative and Congressional committee structure. Such are the irrigation development interests of the West, the river and harbor groups which center on the Mississippi Valley, the private power group, the public power groups, and others. Rearrangement of administrative structure becomes particularly difficult because these groups may be unwilling to exchange the certainty of established means of communication for the uncertainties associated with adaptation to a newer administrative organization, even if it is rational and sound from a national point of view. The weight of these influences has been amply displayed in the events which have attended efforts to implement the water development organization recommendations of the first Commission on the Organization of the Executive Branch, the President's Water Resources Policy Commission, and the Missouri Basin Survey Commission. The net administrative change from the manifold carefully considered recommendations of all these commissions has been very slight, in part because they were viewed suspiciously by established social interests. The effects have also been shown in authorizations for major development work, as in a now well known 1949 action of the Congress. At that time the joint Corps of Engineers–Bureau of Reclamation plan for the Columbia Basin was split in authorization. The Corps' part of the program was authorized; that of the Bureau of Reclamation was rejected.

Finally, an element of rigidity has been introduced into the administrative structure for water development by the position of the agencies operating on a national basis, especially since the mid-1930's. After the establishment of the TVA, any experiment in regional administration of large-scale development which did not include strategically situated participation by one or more of the major nationally interested development agencies[54] was destined for failure. Thus definite limits appeared to be set within which experiment would be tolerated, but beyond those limits there actually was little administrative flexibility.

The points of rigidity in the United States system nonetheless should not obscure the general nature of our national administrative procedure. In the main it shows flexibility in organizational form, and alertness to experiment. Although there has been an increasing tendency toward

[54] Corps of Engineers, Bureau of Reclamation, U. S. Department of Agriculture.

centralization in recent years, as the Corps of Engineers' and other national programs of development gathered momentum, the number of development agencies still has been large, and their points of view diverse. State, municipal, other public, and private organizations have found their place along with federal agencies. For whatever reason, various forms of administrative experiment have been sought and encouraged. The TVA was designed, tried, and continued in one region to meet broad social needs, but it did so by creating a type of management organization alert to technical opportunity. Instead of reproducing it elsewhere, still other forms of co-ordination, like the interagency river basin committees, were tried. When existing private development organizations failed to meet the opportunities presented for large-scale public development and the geographically dispersed demands for water-derived services, public agencies moved into the openings. More recently, under the Republican national executive, the "partnership" approach has been tried, and continues. The partnership approach, as it has been implemented, applies particularly to the hydroelectric generation purpose in water development. This policy stresses the point of view that "the primary responsibility for supplying power needs of an area rests with the people locally."[55] Within any given area it also emphasizes "the best development of natural resources,"[56] development responsibility to be shared by states, local communities, private citizens and the federal government. The principal unanswered question in the partnership policy, as it relates to technology, is the application of scale economies in the organization of electrical generation, transmission, and distribution. Emphasis on local responsibility easily leads to a fragmented operational structure, which does not favor the organization of large scale systems or the generating units which they can support.

Generally, as new technical opportunities have appeared, some agency also has appeared to cultivate them, apply them to water resource development, and enter them into the economy. Two striking illustrations of this may be drawn from private entrepreneurship, in the case of applying improved pump irrigation to the aquifers of the Great Plains and the Southwest, and in the case of commercial rainmakers experimenting with known weather modification techniques throughout the western states.

[55] U. S. Department of the Interior, Information Service release, August 18, 1953.

[56] *Idem.*

Viewed from a national standpoint, the price of the flexibility, alertness to new opportunities, and willingness to experiment, has been a
degree of administrative confusion. In many ways the evolution of
our administrative structure for water development has not been
orderly. Divergent objectives among the many planners and entrepreneurs, and abrupt changes of direction for reasons quite unrelated to
technical opportunities, have produced situations in which organizational evolution was thoroughly obscured. Such was the situation on
the Columbia River in 1957, when 7.5 million kilowatts of hydro
generating capacity and eleven major multi-purpose works were
planned and authorized or under construction by two federal agencies,
five private electric utilities, two public utility districts, and two municipalities. This was within the United States part of the basin. There were
in addition far-reaching plans by wholly different agencies within the
important Canadian headwaters of the Columbia. The best that could
be said for the Columbia of 1957 was that the nature and form of
needed future operational co-ordination was very much in doubt.
Even the status of some individual projects under construction was in
doubt for years because of divergent objectives among operational interests.[57] Furthermore, co-ordination had been a cumbersome, time-consuming procedure.

No attempt will be made here to assign reasons for the obscure
evolutionary pattern of administrative organization. Beyond the
fluidity favored by rapidly changing technology and the numerous
entrepreneurial interests, however, the national procedure for analytical
review of water administration may also have contributed. It merits
brief mention as the means for taking advantage of the flexible organization to produce a more orderly evolution of administration.

Analytical review of administrative responses to new problems and
opportunities, like those presented by technology, has been achieved in
two ways within the United States. One has been through the normal
functioning of the interested planning, development, and operational
agencies as they meet recognized responsibilities, seek new responsibilities, and set their strategy thereon. This has been a steady process
within the Corps of Engineers, the Department of Agriculture, the
Bureau of Reclamation, other federal agencies, the private utilities
and other private organizations, and state agencies like the California

---

[57] For example, the Idaho Power Company Brownlee and Oxbow dams under
construction within the reservoir site of the proposed federal Hells Canyon Dam,
until 1958.

Department of Water Resources. Results of this process would appear to be a continually widening multiple-purpose interest on the part of the federal operational and construction agencies,[58] a recognition of the need for adjustment to the scheduling techniques for multiple purposes by single-purpose agencies like the privately owned utilities, and expansion of basic and applied research related to technical improvement.

Another manner of analytical review of administrative organization and structure has been through *ad hoc* agencies, designed to review and report as of a specific date. A succession of these agencies has appeared within this century. The Inland Waterways Commission of 1907, the National Waterways Commission of 1912, the Mississippi Valley Committee of 1934, the President's Committee on Administrative Management in 1937, the Commission on the Organization of the Executive Branch of 1948, the President's Water Resources Policy Commission of 1950, the Missouri Basin Survey Commission of 1953, the Commission on the Organization of the Executive Branch of 1955,[59] all reviewed and recommended in varying degrees of detail upon the administrative structure for operation, among other things. In addition, a number of hearings have been held by Congressional committees upon legislation under consideration, some involving intensive investigation and review.[60]

Last, some provision for continuous review has been made in the Federal Power Commission authorizations, and in the fiscal review and budget-making functions of the Bureau of the Budget. For the period between 1935 and 1943 the National Resources Planning Board also afforded some facilities for such review. For effectiveness in this

[58] Illustrated by the concern of the Corps of Engineers within the 1950's for low flow regulation (related to municipal and industrial water supply and waste carrying), the concern of the Bureau of Reclamation for municipal water supply, and the entry of the Department of Agriculture into flood prevention work.

[59] Dates are given as the year of report or recommendations presented by the agency. The federal Advisory Committee on Weather Control was another recent *ad hoc* agency with specialized responsibilities which may relate to later water development administration.

[60] The hearings before the House Committee on Public Works on H. R. 4286 and H. R. 4287, and before the Senate Committee on Public Works on S. 1645 in 1949 (81st Congress, 1st Session) were of this type. These hearings related to the proposed Columbia Valley Administration. See President's Water Resources Policy Commission, *Water Resources Law* (Vol. 3 of Report), p. 429, for a list of others.

particular function the Planning Board suffered from a massive, all-embracing view of its scope; the two other agencies have always had relatively small staffs. The federal agencies giving this function continuous attention thus have never been in the position to furnish adequate professional review of national administrative response to a changing environment and technology. There has been nothing on the order of the Federal Reserve Board's review of the financial structure, or the Council of Economic Advisers' pulse-taking on the economy as a whole. Yet the problems of organizing resource development and water facility operation are not without some parallel in their need for review.

As of 1958, then, the problem of guiding an orderly evolving administrative structure for water facility operation had been met mainly on a sporadic basis, often by *ad hoc* bodies committed to a special predetermined objective. No methodical professional review had been provided of the problems and opportunities brought by technical change, the administrative results of organizational experiment, and the applicability of administrative devices under alternative political policies. Separation of the political and the technical under this practice has been difficult or impossible. Moreover, the best employment of the technical in reaching political decisions has not been possible. The *ad hoc* investigations and review almost inevitably take on political overtones, sometimes confusing the technical reviewer's function with political objectives.

4) RELATIONS OF OPERATIONAL MANAGEMENT OF WATER RESOURCES AND OTHER RESOURCES

The nature of the relations between water and other natural resources has been suggested in preceding chapters (especially chapters 3, 4, and 19). It concerns principally forest lands, grazing lands, cultivated lands, and mineral deposits in addition to water.

The location, quality, amount, and nature of use of other resources affect demand for water and water-derived services. They also affect water supply. Conversely, water use and availability affects the productivity or attractiveness (in the case of recreation) of other resources, and water regulation and use may actually make other resources physically unavailable.

Examples of the effect upon demand are well illustrated in the demand for irrigation water which arises in the presence of cultivable lands, and in the demand for processing water, electric energy production, navigation facilities, and waste carrying, which accompany the growth and maintenance of forest and mineral-based industry.

The use of other resources may affect the quality of water, as in the discharge or seepage of acid wastes from coal mines, the discharge of compounds in processing metallic ores, and the intrusion of salt water from petroleum-bearing strata into fresh-water aquifers or streams during oil exploitation. Land use for cultivation and grazing may affect water quality through accelerated movement of sediment from soils. Water quantity also may be affected by the nature of use of other resources. For example, the manner of forest exploitation may affect the seasonal distribution of runoff, and it may even affect the total amount of runoff available for use.[61] Substitution of other resources for water can affect quantities available, where multiple demands are experienced. Substitutions of thermal electricity production for hydro; nonwater forms of recreation for water recreation facilities; land transport for navigation, and waste treatment and nonriverine disposal of stream waste carrying may affect either local or basinwide water availability.

The manner of water use, on the other hand, may enhance (or detract from) productivity or attractiveness of other resources as they are used. This is exemplified in the use of grazing lands, minerals, nonwater recreational resources, and lands where cultivation is dependent upon soil moisture. Water use may actually make some resources physically unavailable, as where reservoirs drown mineral deposits or natural (scenic or otherwise) recreational sites. The latter problem has been especially recurrent in relation to sites of recreational value. Hetch-Hetchy Dam (California), the Tehepite and Cedar Grove exclusions from Kings Canyon National Park (California), the proposed reservoirs in the Potomac Gorge (Maryland-Virginia), the proposed Echo Park dam (Utah-Colorado), and the proposed Glacier View dam (Montana) all have offered bitter political expressions of this inherent conflict.

These potential and actual relations of water and other resources in use create a need for the following functions in administrative organization for water development: (*a*) ascertaining and anticipating demands for water arising from other resource use; (*b*) regulating or conditioning the use of resources other than water, or undertaking their

---

[61] See George H. Barr *et al., Recovering Rainfall* (Tucson, Ariz., Arizona Watershed Program, 1956), Parts I and II.

exploitation for the purpose of deriving water benefits; (c) operating water resource facilities so as to maximize total benefits from all resources in a given area, including water.

It therefore is apparent that the fullest, most efficient water resource management is not the multiple-purpose management of water resources alone, it is the multiple-purpose, co-ordinated management of water and resources related to water in their use. The manner in which this has been treated administratively in the United States accordingly is of interest, insofar as technologic change may have intensified or created problems in such relations.

Ascertaining and anticipating water demand stemming from the use of other natural resources

United States administrative organization to care for the anticipation of water demand from the use of other resources has been loose and informal when the whole is considered. Results or progress mainly have been derivative from the action of many independent groups of limited interest, leavened by appropriate legislation from time to time. In this, perhaps more than in the other phases of water development and management, the characteristic functioning of a free enterprise economy has been displayed. A large part of the initial determination (as distinct from professional evaluation) of need for irrigation water, for flood protection, for navigation facilities, for municipal, domestic, and industrial water supply, has rested with communities, localities, or special-interest groups. The determination of need for electric energy production, and its form, has rested with the utility in charge of a given power service area. At times this has not led to the soundest long-term development, as in the case of the development of the Snake River and its relations to the phosphate lands of Idaho, Montana, and Wyoming. In this the long-term development of the region as a whole would not appear to have been fully considered.

A degree of regional cohesion has been given to these determinations in a variety of ways. For the western states, the Bureau of Reclamation has given an over-all view of water needs derived from land development possibilities. For the East particularly, the Corps of Engineers has preserved a broad view, relating navigational requirements to the development of other resources. The private Mississippi Valley Association has given a regional political view of the flood control and other water regulation needs of that important region. The TVA has maintained an analytical view of the water development needs of its valley, in terms of the entire resource complex of the area. In a somewhat more limited way, the Bonneville Power Administration in the Pacific Northwest brought multiple-resource views on a regional scale to its area between 1943 and 1953. It had been preceded by the Pacific Northwest Regional Planning Commission. The New England Council, in a wholly different manner, has brought forward such views in New England. A number of other cases of efforts to provide a cohesive regional multiple-resource view, like the temporary New England and the Arkansas-White-Red Interagency committees, and the recent Northeastern Resources Committee, also could be cited. Within its state the California Department of Water Resources also has directed effort toward multipurpose regionally co-ordinated development.

In general, this multiple-agency, multiple-resource, multiple-purpose process has covered this function, although at times with delayed effect. Its most notable deficiencies have appeared in unsatisfactory (or delayed) resolution of the place of inland waterway navigation in eastern United States; the incomplete program of development in some major coal-bearing valleys (like the Ohio), where the operational relation of hydro and coal-based thermal power was misjudged or incompletely understood; and in the hesitant manner in which the potentialities of humid-land irrigation have been approached.

Regulating the use of or treating other resources, or developing them for water benefits

The relations of other resource use to water supply may be treated organizationally in four ways: regulating use or imposing use restrictions; substituting the use of other resources for some water functions; conditioning the resource; and altering the nuisance capacity of a resource.

a) Regulation. The imposition of use restrictions, in line with the governmental location of many regulatory functions, has generally been the responsibility of state agencies. Thus Texas, Oklahoma, and other petroleum-producing states regulate the disposal of salt waters lifted in the pumping of oil. Such regulation is administered by state agency. All states have regulation against the disposal into streams or employed aquifers of especially toxic inorganic wastes. Some, like New Jersey, have regulations against the disposal of organic wastes, and administrative supervision of the regulation.

Regulation, especially of waste carrying, implies or requires the substitution of other resources for water. The achievement of such substitution has been in the hands of the operating industry, like the individual canning or food-preserving firm, or the pulp-manufacturing corporation. This is characteristic of the administration of this particular type of substitution in the United States, and of other measures of substitution as well.

b) Substitution. Substitution probably has been carried farthest in the field of transportation. While pipeline, highway, railroad, and now conveyor belts[62] may be substitutes for water transportation, the relations between rail transport and navigation have been closest in national policy and practice. After the construction of some modern waterways in the early part of this century, it was evident that water-borne traffic might offer serious competition to established railways over some important routes in the eastern half of the country. The relation of these two forms of transportation has been made the closer by statutory provisions for rail traffic. Under Section 4 of the Interstate Commerce Act[63] railroads are allowed to establish water-competitive freight rates along lines competitive with water routes, upon approval by the Interstate Commerce Commission. This prerogative has been freely used by rail companies in the United States, placing their lines on an equal, and perhaps superior, competitive basis as compared with inland water carriers.[64] Although waterway use has grown over the years, land transportation facilities thus have been substituted in a very important way for the use of waterways in the United States. Initiative for this has rested with the management of rail companies, as they chose to make use of the provisions of Section 4(e) relief under the supervision of the Interstate Commerce Commission.

Another field within which other resources have been substituted for water has been energy production. Particularly in sections of the country with adequate and well-exploited coal resources, electric energy production has tended to be based on a thermal, rather than hydro system. This choice again has been in the hands of dozens of operating

[62] J. F. Lawrence, "Fuel Flow," *The Wall Street Journal*, February 19, 1957, pp. 1, 18; M. J. Barlow, "The Second Transport Revolution," *Harper's Magazine*, 214 (March 1957), 37-43.

[63] Act of February 4, 1887, 24 Stat. 379, Sec. 4 amended by Act of June 18, 1910, 36 Stat. 539, 548.

[64] Equal as far as freight rates are concerned, but superior in the territorial coverage and speed of movement.

companies or agencies.[65] Where exercised, it has been a distinctly nongovernmental function.

Most substitution of other resources for performing services which might be derived from use of water have not been with the objective of producing water benefits. Indeed, the case of transportation in the "4(e)" relationship between navigation and rail transport, exhibits an inverse stimulus. The development of waterways may be sought by communities or regions because the existence of the waterway may serve to lower rail rates, even though the creation of the waterway in places may detract from achievement of maximum water use benefits.[66] Only in the recently advancing function of pollution control has resource substitution been undertaken mainly with the objective of water benefit. Most other cases of substitution have rested on short-term considerations of institutional, community, or regional economic benefit.

  c) Treatment of other resources for water benefits. Conditioning other resources for the production of water benefits most frequently is encountered in the vegetational treatment of upper watersheds, where the state of forest planting or maintenance is known to affect the seasonal availability and flow on a given watershed. In some environments it is possible that the total flow of water may be affected by the manner in which the vegetation on water source areas is managed.[67] Administrative overseeing of this relation has been a peculiarly public agency function in the United States, as in all lands at all times in history. Where water supplies have been short, as in the West, vegetational management of the upper watersheds has been considered an important task. This may have been a partial reason for the establishment of the National Forests, and of the Forest Service to administer

---

[65] This could be considered a choice only where the potential hydro was incompletely developed, as in the previously mentioned Ohio Valley area.

[66] For example, navigation on the middle and upper Missouri, and on the Arkansas River.

[67] Where winter snowfall on mountains is an important source of summer stream flow to western bolsons, or alluvial valleys, experiment has suggested that tree spacing on the watershed may be a relation to water production. H. G. Wilm, *Effect of Timber Cutting on Water Available from a Lodgepole Pine Forest,* U. S. Department of Agriculture Technical Bulletin No. 968, Washington, 1948. *Idem,* "Timber Cutting and Water Yields," pp. 593-602 *Trees, The Yearbook of Agriculture, 1949,* U. S. Department of Agriculture, Washington. While the connection between vegetation and water production in other situations, like that of the Southwest, is being examined, the connection of vegetational management and total water production within a basin is not yet proven.

them.[68] It has been an important consideration in several federal laws subsequently enacted, which now form the basis of the very influential federal government activity in this field.[69] Although state forest lands have not been as extensive as those under federal agency management, some state forest reserves exist in almost every state. Every state also has a state forest agency, which influences the management of private lands for watershed purposes through its own and co-operative federal programs. In recent years, since the establishment of large-scale forestry programs by private corporations, like those of the Weyerhaeuser Timber Company, the International Paper Company, The Crossett Company, the Long-Bell Lumber Company, and others, watershed benefits have been produced incidental to forest management for wood production. However, the management of forests or other vegetation primarily or secondarily for water production or regulation is most typically a publicly administered function in the United States.

d) Altering nuisance capacity. Alteration of the nuisance capacity of another resource is outstandingly illustrated in soil erosion control. Erosion control in itself may be incidental to other objectives in land management, and water benefits may be a secondary purpose in erosion control. From the point of view of the farm, the operational manage-

[68] Although the Act of March 3, 1891, which set up the national forest reserves (16 U. S. C. 471), and the Act of June 4, 1897 (16 U. S. C. 473, 475-82, 551), which provided for their administration, are ambiguous on the relation of the forest reserves to water supply benefits, the subject was referred to in Congressional debate preceding enactment of the 1891 law. (See President's Water Resources Policy Commission, *Water Resources Law,* Vol. 3 of Report, p. 355.)

[69] For example, the Act of March 1, 1911 (36 Stat. 961, as amended, 16 U. S. C. 480, 500, 513-19, 521, 552, 563), which provided for federal acquisition of lands in watersheds of navigable streams over the country as a whole. (The previous national forests were only reservations from the public domain.) The 1911 Act is also known as the Weeks Law. Also the Clarke-McNary Act (Act of June 7, 1924, 43 Stat. 653, 16 U. S. C. 564) that provided for co-operation with state agencies in a program of fire prevention and suppression. Also the McSweeney-McNary Act (Act of May 22, 1928, 45 Stat. 699, as amended, 16 U. S. C. 581-581i) providing for investigation and experiment upon the maintenance of "favorable conditions of water flow." Still other legislation with watershed management objectives has been that on the "O and C" lands (Act of August 28, 1937, 50 Stat. 874), the Sustained-Yield Forest Management Act (Act of March 29, 1944, 58 Stat. 132, 16 U. S. C. 583-583i), the Co-operative Forest Management Act (Act of August 25, 1950, 64 Stat. 473), the Forest Pest Control Act (Act of June 25, 1947, 61 Stat. 177, 16 U. S. C. 594-1 [Supp. III] *et seq.*), and other acts, as for instance, the TVA Act (Act of May 18, 1933).

ment unit for erosion control in the United States, the reduction of sedimentation and other water benefits are usually incidental to more immediate farm management objectives, that is, production of crops and livestock. That they are secondary and incidental in such management makes them no less real; erosion control often produces water benefits.

For these reasons, responsibility for starting the process of erosion control for water benefit has been almost entirely a public agency function. Thus the Soil Conservation and Domestic Allotment Act specifically mentions water benefits among the objectives of federal soil conservation activity.[70] Succeeding federal legislation on activities relating to soil conservation has given increasing emphasis to water benefits, as in the Watershed Protection and Flood Prevention Act of 1954.[71] Thus the Soil Conservation Service of the Department of Agriculture as it works with the 2,701 Soil Conservation Districts in the United States, is at least partly an operational agency for water benefits. The intimate relation between hydraulics and one phase of the erosional process has kept the professional orientation of the Service always sensitive to water problems. Because the territorial unit of farm management responsibility is always relatively small, water benefits from erosion control usually are geographically removed from the site of erosion control work. Thus the creation of public agencies with this operational view was natural.

While the federal Soil Conservation Service and its associated administrative units[72] form the principal public organization for the managed treatment of soil erosion in the United States, other agencies also have a part. The state Agricultural Extension services, with federal co-operation, have included erosion control in their program. Extension Service activities have been related most clearly to water benefits in the Tennessee Valley, where the TVA–Extension Service–Co-operative test demonstration fertilizer program has been a part of

[70] Act of April 27, 1935, Sec. 1, 49 Stat. 163, as amended, 16 U. S. C. 590a-590f. ". . . it is declared to be the policy of Congress to provide permanently for the control and prevention of soil erosion and thereby to preserve natural resources, control floods, prevent impairment of reservoirs, and maintain the navigability of rivers and harbors . . ." (Sec. 1). This Act is the basic federal legislation enabling soil conservation activity by federal executive agency.

[71] Act of August 4, 1954, 68 Stat. 666. Act of August 8, 1956, 70 Stat. 1088.

[72] The District Board of Supervisors of the Soil Conservation District and the State Soil Conservation Committees.

the multiple resource program enabled by the TVA Act[73] and conducted by the TVA.

Still other operational organizations concerned with soil erosion control include the agencies with public lands under their care, like the Forest Service of the Department of Agriculture, the Bureau of Land Management of the Department of the Interior, the National Park Service, state park departments, and other state land management agencies. Locally, irrigation districts or water users' associations also may concern themselves with erosion control activities.

Finally, incentive programs conducted under other administrative units within the Department of Agriculture also have been directed toward the achievement of soil conservation. Such was the Agricultural Conservation Program, with its now well-known "ACP payments." This program first was authorized in 1936, and included water benefits among its purposes.[74] For a long period this was under the administration of the Production and Marketing Administration of the Department of Agriculture. Since 1955 the program has been administered by the Department's Agricultural Conservation Program Service. The major federal expenditures for soil conservation during the last twenty years have been made through this administrative route. However, because of the spread of purposes within the authorizing legislation,[75] the water benefits must be considered more incidental than purposeful as the program has proceeded. The "Soil Bank" act of 1956[76] is a newer version of the ideas contained within the older ACP program with somewhat different procedural objectives. It is under the administration of the Commodity Stabilization Service.

Operationally, the production of water benefits by altering the nuisance capacity of the land resource is in the hands of farmers, graziers, and forest land operators. Under the American system the effective operation therefore overwhelmingly is under privately organized supervision. The erosion and sedimentation problem, as influenced by land-use practices, has been predominantly on lands used for cropping, although geographically lesser nuisances have had their origin in

[73] Act of May 18, 1933, Sec. 5(d), 48 Stat. 58.

[74] Act of February 29, 1936, Sec. 7 (a), 49 Stat. 1148, as amended, 16 U. S. C. 590g (a). The stated purposes include "(4) the protection of rivers and harbors against the results of soil erosion in aid of maintaining the navigability of waters and water courses and in aid of flood control . . ."

[75] Farm income stabilization and re-establishment of farm purchasing power were also purposes. (*Idem.*)

[76] Public Law 540, 84th Congress, Chap. 327, 2nd Session.

grazing lands in the West, and on sloping forest lands. In the past both forest-land exploitation and cropping have been adversely influenced (from the point of view of erosion and sediment production) by the mechanization of operations favored by technical advance.[77] Some of the widespread soil conservation problems raised or intensified by mechanized agriculture and logging have been on public land reserves, and in public operational management of them. On the increasingly important national forest lands the Forest Service retains the power of supervising cutting and haulage operations by private contractors who undertake timber cutting. The erosion problem incidental to mechanized operation would appear to be most nearly under control on these lands.

Special and now well-known problems have arisen in the mechanized operations of the mining industry in several parts of the country. The earliest of these came in the placer mining of the Sierra Nevada in California, and led to the establishment of the California Debris Commission, as previously mentioned. Later and much more extensive operations are characteristic of the coal industry, particularly in the mechanized strip or surface mining. Pennsylvania and other coal states now regulate the disposal of coal wastes into streams, a substantial source of sediment at one time. The important strip-mining states, like Illinois, also prescribe for mining methods and final treatment of strip-coal lands for several reasons, including sediment control.

A number of specialized government agencies have been created, whose responsibility it is to install and extend erosion control as a part of American land use. This they do by incentive payment, by education, by technical services, and physical aid. In every instance they act through the private owner, farm-land operator or forest-land operator, who is the final arbiter as to the action taken. Erosion control itself may be incidental to other purposes in the incentive payment programs. In most programs the alteration of the nuisance capacity of lands for excessive sediment production is incidental to other objectives in land use. The small watershed activities of the Soil Conservation Service and the TVA have come closest to administration of conserva-

---

[77] W. A. Raney, T. W. Edminster and W. H. Allaway, "Current Status of Research in Soil Compaction," *Soil Science Society of America, Proceedings,* 19 (1955), 423-28; G. R. Free, "Compaction as a Factor in Soil Conservation," *Soil Science Society of America, Proceedings,* 17 (1953), 68-70; S. Weitzman and G. R. Trimble, "Skid-road Erosion Can Be Reduced," *Journal of Soil and Water Conservation,* 7 (1952), 122-24.

tional activities for joint land and water benefits. Some co-ordination of operational objectives also has been reached through interagency committee and conference. For the most part, however, this part of a multiresource view on water management has had a by-product relation, rather than being an integral part of water management itself.

Operating water facilities so as to maximize the productivity of other resources

The obverse of the problem of altering the nuisance capacity of other resources, and production of water benefits thereby, is the management of water facilities to add to the productivity of other resources. This statement of problem is not intended as "an elaboration of the obvious," as might be inferred from known primary purposes of water development. Irrigation, as noted above, is undertaken to maximize the productivity of land resources. A substantial part of electricity production is used for ore reduction. Improvement of inland navigation may bring fuel resources and industrial raw materials within effective range of each other. These are so much a part of planning for and operating water facilities that they have been implicit, as well as explicit, in the entire preceding discussion.

At the same time, the scale economy techniques and the scheduling techniques have raised problems of operational management of water facilities which bear mention. They bring the question of optimum benefit not only on a multiple-purpose but on a multiple-resource basis.

Water facilities operation generally is on a preferred-purpose basis (see page 598), either because of economic preference (in the case of private and some public agencies) or because of statutory instruction (for all public agencies). In large measure the result has been management of the water resource for achievement of productivity in other resources. In fact, achievement of productivity in other resources would seem far more a part of water management than the management of land resources for water benefits.

The questions raised in the course of past and present operational management generally are those of the optimum "mix" of purposes to which management of the water of a given basin should be devoted. The growing centralized agency operation, as well as preferred-purpose operational schedules, have brought an administrative structure which sometimes appears unfortunately rigid in the face of operational problems in the management of resources that may be related to water.

This problem recently was illustrated in the operation of the TVA reservoir system, which suffered four years of extremely low water flow during the period 1953-56. Upper tributary reservoirs, like Fontana on the Little Tennessee River, were drained to very low levels to maintain the navigation channel of the Tennessee and produce incidental electric power. Reservoirs were so low that they were not only of little recreational use, but also were unsightly, detracting in some instances from the scenic attractiveness of mountain areas in which they were located. It is possible that operation of the reservoir system which stressed maximum benefits from water and all related resources, including the extensive recreational lands of the Great Smoky Mountains, may have resulted in the same operational pattern as that between 1953 and 1956 on the TVA water control system. But it might also have been otherwise. The important point is that the TVA statutory requirements for preferred purposes did not give the agency flexibility to vary its mix of operational purposes in response to new situations in the economic environment of the agency.[78] Water management on the Tennessee could not be undertaken with the objective of maximizing total benefits from water and all related resources.

If such a problem can arise for an agency distinguished in the United States for its multiple-purpose and multiple-resource view, it follows that similar problems are to be expected elsewhere in the country on other basins. A privately owned single-purpose agency like an electric utility finds it difficult even to recognize the problems, unless means are at hand to compensate monetarily for multiple-resource operation. Other public agencies likewise bring the problem into relief at intervals. It is especially likely to appear in connection with recognition of the operational voice in planning. Thus the Corps of Engineers' flood prevention program in the Kansas River Basin has encountered opposition because the operational needs of the land resource were not fully anticipated. A state-inspired view of the Kansas Basin water development consequently stressed the operational needs of land resources.[79] Likewise the Bureau of Reclamation, within the

[78] "The Board is hereby directed in the operation of any dam or reservoir in its possession and control to regulate the stream flow primarily for the purposes of promoting navigation and controlling floods. So far as may be consistent with such purposes, the Board is authorized to provide and operate facilities for the generation of electric energy at any such dam . . ." (Act of May 18, 1933, Sec. 9(a) as amended, 49 Stat. 1076, 16 U. S. C. 831 h-1.)

[79] University of Kansas, *op. cit.*

area of its territorial jurisdiction, has found some difficulty in stressing the development and management of water resources with multiple resource production objectives.[80]

Perhaps the most persistent problem of this type has been that of the anadromous fisheries (salmon, smelt, shad and similar fishes) on many of the streams which empty into the Atlantic and Pacific oceans. For many years the principal manner of treating the problem was to ignore the needs of stream passage and spawning for these fish in favor of other stream uses, like electric power production or waste carrying.[81] Only after the anadromous runs were reduced to a very few streams did the relation of stream use to the marine resource become an important part of planning for water development and facility operation on streams. As previously illustrated for the Columbia River (see pages 598-600), it is still an incompletely solved problem of operational management, although now consistently considered.

The desirability of maximizing the productivity of other resources emphasizes the need for planning and construction in a manner whereby flexibility and room for adjustment may be introduced into the operation of water regulation systems and the other works and operations associated with them. The difficulty of foreseeing the future demands for water services is compounded by the added difficulty of foreseeing the demands which may also be placed on other resources. If facilities and administrative organization are cast in a rigid mold, only slowly adaptable to the changes inevitably introduced by technology, time, and new social demands, efficiency and social benefit are both the poorer for it.

---

[80] For example, the mineral resources of the upper Rio Grande Basin, and local industrial use thereof.

[81] See E. A. Ackerman, *New England's Fishing Industry* (Chicago: University of Chicago Press, 1941), pp. 29-31, for a brief account of the former extent of these fisheries in one region as compared to their subsequent decline.

CHAPTER 21

Summary: Adaptation of Administrative Structure for Water Development to Problems Presented by Technology

The details given in the preceding chapters suggest that the administrative structure for water development in the United States is in the midst of evolution. The pattern of evolution is not yet a clear one, but it is certain that the pressures and challenges of technological change are causing administrative organization to adjust and adapt. Some needs have been met very well, some adequately, some partly, and others not at all. Because of the complexity of functions which characterize water development, and the large number of agencies concerned with it, a summary of the apparent 1958 pattern may be of interest. In each of the three major functions, planning, realization of development, and operational management, differences appear in the degree of organization and administrative attention to pertinent activities.

In describing the pattern of administrative evolution there is an important distinction between the activities in which a methodical approach is developing, and those treated incidentally or episodically, if at all. Administrative provision for continuing attention characterizes the former, but not the latter. This is the only distinction that will be made here in summarizing the planning, realization of development, and operational management observable at this time.

Within the field of *planning,* examples of methodical approach to the following activities may be found in the United States, with some administrative provision for continuing attention: site selection for major and secondary works, routine collection of basic physical data, basin-wide water-control planning, planning for multiple use, and promotion of an environment favoring further technical improvements.

The following planning activities appear to be episodically approached: forestalling the economic and physical encumbrance of sites for major works; adequate scheduling of basic data collection in relation to development needs; precise observation of the effects of development for one use upon the quantity and quality of water available for another; methodical provision for scheduling techniques other than multiple-purpose use (pairing, recycling, water budget calculation, use restrictions, etc.); appraisal of the long-range availability of surface and gravitational water supply (as part of water planning); appraisal of influences accelerating demand; planning interbasin use; the co-ordinated analysis of all types of pertinent basic data; and co-ordinated formation of broad policy and program planning.

As far as the *realization of development* is concerned, some provision of administrative structure for meeting the following needs has been made: entrepreneurship for large-scale, multiple-use works; meeting dispersed demands for energy; balancing the use of water and other resources; financing economically feasible works; application of standardized techniques in construction; and allocation of water among contending users (in the West).

These potential functions of entrepreneurship have not been met methodically: redirection or relocation of water demands, and meeting the dispersed domestic water supply demands. The problem of allocating water has been considered, but not yet met administratively in the eastern states.

Among the functions or attributes of *operational management,* some methodical administrative provision is found for: regulation of stream flow; federal operation of water facilities on a geographically extended basis; relation of water use to the productivity of other resources; substituting the use of other resources for water; and organizational flexibility for administrative experiment.

The following functions or attributes of operational management have been more incidentally approached in United States administration of water facilities: realization of the scale economies in distributional organization; nonfederal operation of water facilities on a geographically extended basis; conservation (in the narrow sense) and quality control of water; introduction of the operational voice into planning; operation for the so-called secondary purposes; management of other resources for water benefits; and the analytical review of administrative experiment and experience for the purpose of further application in the orderly evolution of the administrative structure.

The analytical dissection of the numerous functions of planning, realization of development, and operation should not obscure the fact that both ideally and in practice these three steps in water development and use are closely associated. The web which runs from planning to operational management and back again is so close that it is appropriate to end on a few observations which include all three steps.

1) Improved techniques of development and other aspects of technical change have introduced a wide range of new functions into water planning, development, and management. Fully efficient water use will not be obtained until each is approached methodically, and as part of the basic problem of matching supply and demand.

2) A fragmentation of user interest and developmental responsibility is characteristic of United States water administration. Even in the smaller river basins this is true. Within the Delaware River Basin, for instance, more than one hundred governmental entities—federal, state, municipal—are concerned in the planning and eventual operational management of water facilities in the basin.[1] A large number of private corporations have use interests in the same water, and a few of them have operational interests.

3) The characteristic fragmentation of interest and responsibility and the multiplicity of functions introduced by technology indicate that:

a) Efficient development requires a sophisticated professional study of the economic and demographic environment in which project and water program planning is to take place, and that water project and program planning may be advanced by the existence of competent regional planning analysis.

b) If optimum efficiency is sought, very careful attention to integration must be a part of planning, development, and management of water facilities.

c) Relatively small units of integrated planning and development must eventually take their place in the structure of development. It must be a place compatible with the needs for scale economies and

[1] Walter Phillips, "Administrative Aspects of Delaware River Basin Water Resource Planning and Development," address before the American Society for Public Administration, January 29, 1957.

the broader regional and national interests in development. Administrative organization responsive to the multiplicity of functions on a local level will be needed, but one in which the responsibilities are clearly defined vis-à-vis the state and national administrative agencies for these functions.

4) Recent experience has shown that the pattern of operational responsibility has considerable power to determine the entire course of planning and development, and the opportunity for the application of known techniques. This has been well illustrated in the field of electric power distribution. Critical examination of the complex of operational responsibilities, and their relation to the scale economies in particular, may be considered a part of good planning from a national point of view.

5) An important share of water development is undertaken for the purpose of improving the productivity of other natural resources. The co-ordination of planning and management on a multiple resource basis therefore would seem to be a part of efficient administrative procedure.

6) Within the last thirty years technology has given certain advantages to federal government administration of planning, construction, and operational management, which has grown to dominating proportions in the application of the new techniques of water development. Further accretions of function appear to be in progress, in spite of federal executive policy declarations favoring "de-concentrated" planning, development, and management, increasing the role of states, communities, and private enterprise.

The growth of federal administration of water planning, facility construction, and operational management in part may be traced to the creation over several decades of nonreimbursable development purposes supported by tax-derived funds. Special impetus came during the 1930's, when Federal funds became very important for creating employment during depression conditions, and when states' financial capacities were low. In part, however, it also may be traced to the size of investment funds required by the large-scale works which technical development made feasible, and to the advantages given to federal organization by highly specialized research and engineering personnel. The move toward federal administration has gained strength because of the limited technical capacity and the limited jurisdiction of other governmental units and private organizations in relation to the size of river basins in the United States. Finally, federal administration has continued to grow because federal administrative and policy leadership conceived agency responsibilities in terms of broad operational responsibility rather than on a strategic project, demonstration pattern.

7) There is a trend toward the centrally administered federal agency. At the same time provision has not been made for adequate and continuing professional analytical review of administrative experience in the light of technical change and needed responses in the administrative structure of water development. Among the obvious undetermined questions are: the efficiency of the centrally administered federal agency in coping with regional problems; the means of methodically treating the now incidentally treated functions; the place of the federal-regional or other regional agency in the administrative structure; dependence to be placed on state, community, and private initiative for known but uncultivated functions; the manner of maintaining operational flexibility; and others. A number of experienced observers, including men within the federal bureaucracy, have concluded that the central staff of a nationally centralized agency cannot efficiently discharge the responsibilities which it has. The scope of responsibilities in these central staffs, and the pressure of always immediate problems precludes the deliberate, sustained, methodical study of regional and local situations so necessary to the best decisions. Without a bureaucracy much larger than it is, centralized staffs cannot be expected to be intimately acquainted with regional and local economies and with the functioning of state and local resource agencies. It follows, in this view, that they do not have the necessary basis for technologic and administrative decisions which they must make. The degree of correctness in this view, and necessary measures to improve the administrative structure toward better articulation at the four levels, national, regional, state, and local, must be considered an outstanding problem of the day.

The existence of the problem is emphasized by the views in a succession of professional studies of resource administration. Both the first Hoover Commission Task Force on Natural Resources (1948) and the President's Water Resources Policy Commission emphasized the need for further centralization within the federal establishment, with at least a Central Technical Review Board above the existing developmental and operational agencies. Charles McKinley also has analyzed in detail the possible structure of a unified federal water development agency, and has concluded that it is not incompatible with decentralization of important development and operational functions toward regional federal administrative units.[2]

8) Five major groups of technical change have influenced the development of water in this country within the recent past: techniques increasing the rate of water demand; techniques decreasing the rate of demand; techniques extending the services afforded by a given unit of supply; technical improvements promoting the scale economies; and techniques extending the physical range of water recovery. While no summary answer will apply to every phase of these groups, some distinctive features of past administrative response to the appearance of these techniques would seem to be:

a) The influences accelerating demand: Important administrative problems which have arisen here are the monitoring of total effects, research on and development of solutions to problems of dislocating effect, and the regulation of conflicting or monopolizing use (e.g. excessive waste discharge). In general response has been slow, and often the product of crisis or emergency situations. In the western states, where systems of prior rights are familiar and in effect, administrative structure as a whole seems better organized to cope with management and development problems arising from these sources, than in the eastern states. Eastern administrative structure has not yet been able to care adequately for the two most important problems which recently have arisen—rapidly mounting volumes of waste discharge, and extension of crop irrigation. For the nation as a whole, better organized monitoring, strengthened research and development on technical and administrative solutions, and improved regulatory structure seem indicated whenever they are politically possible.

b) Techniques decreasing demand and those extending the services afforded by a given unit of supply: The principal administrative problems for these groups have been the organization of research to develop or refine the techniques, and the development of a framework for their application. Response to the multiple-purpose scheduling technique also has been somewhat slow, but adaptation of the administrative structure to it has been progressive. On the other hand, organized application of the pairing techniques and the water conserving and treatment techniques has not been approached in

---

[2] Charles McKinley, *Uncle Sam in the Pacific Northwest* (Berkeley: University of California Press, 1952), pp. 567-617.

terms of their net total contribution, or of the administrative framework within which they might be given methodical attention as a group. A large volume of research has been carried forward on techniques within this group. An exception is research pertaining to the interrelation of these techniques as they might be applied, or are applied together.

c) *Technical improvements promoting the scale economies:* The principal problem in this instance has been the adjustment of administrative jurisdiction, both territorially and functionally, to realize these economies in planning, construction, and management. While some notable regional adjustments have been made, they have been slow for the country as a whole, and suffer from confused and sometimes contradictory objectives.

d) *Techniques extending the physical range of water recovery:* In the main, these have given rise to problems of administering research and development. Long approached on a haphazard basis, they have profited from a much more methodical approach since the end of World War II. Administrative and policy responses in this instance would seem very satisfactory, considering federal, state, and local public and private efforts in the several fields.

CHAPTER 22

The Emerging Technology and Administrative Organization for Water Development

Background Assumptions. Regional Incidence of the Emerging Techniques. Effects of the Emerging Techniques on Existing Problems and Opportunities in Administering Water Development. Centralization, Federal-State-Local-Private Enterprise Relations, and the Co-ordinated View in Water Development—what emerging technology suggests about centralization; distribution of responsibility among different levels and types of administration in the future; conclusions.

It is evident that both the opportunities and the problems presented by technologic change in the past have met less than full response in American administrative organization for water development. A long list of problems and opportunities introduced by technology may be considered to be episodically rather than methodically treated in present-day United States organization (chapter 21). These range from precise observation of the effects of use upon supply through realization of the scale economies, to the desirability of multiple resource management. Most significantly, the administrative means are missing for weighing and making use in a co-ordinated manner of the many technical approaches to matching demand and supply in the most efficient manner.

Background Assumptions

The emerging technology affecting water use and supply not only reintroduces the questions raised by adaptation to past technical progress, but also raises questions as to necessary new adjustments

628

which may be in prospect. This assumes a desire on the part of the nation as a whole to plan, develop, and operate its water facilities as efficiently as science, engineering technique, and administrative technique permit.

To postulate a desire on the part of the nation as a whole to plan, develop, and operate as efficiently as possible is a policy assumption. This study has been limited to an examination and interpretation of the relation between technology and administrative organization. It therefore treats but one facet of the national problems of matching water demands with facilities for dependable supply. So as to have a clearly understood background for conclusion, several other assumptions also are appropriate.

1) The United States will continue as a federation of quasi-sovereign states.

2) Water resource development policy will include other elements than efficiency, as it has in the past. It is assumed that pursuit of such objectives as the general welfare and national defense will require compromises for a policy which includes economic efficiency. However, it is also assumed that the extent of the compromise should be the subject of careful review in each instance where it is required.

3) Private enterprise will continue to have a major entrepreneurial and operational management function within the economy of the nation.

4) Public bodies at the three levels, federal, state, and local-community, will all continue to maintain active interest in planning, developing, and operating facilities for meeting water requirements. The past intense interest of private enterprise in these problems will be continued. No assumption is made as to the degree of financial responsibility likely to be attached to any of the four in development and operation.

5) Population will continue to grow rapidly, will be at least as mobile as at present, and will continue to concentrate in the urban areas of the country.

6) Uncertain international relations will continue, with a constant risk of rapid alteration of the problems of operating the American economy. In view of the importance of water-deficient sections of the country to defense activities, avoidance of rigidity in water resource planning, development, and management is especially desirable. So is avoidance of investment commitments to facilities with risk of obsolescence before amortization, and to service of submarginal land resources.

Continued international tension is likely to be of particular significance in maintaining or increasing the complexity of relations between private and public organization for resource development. As defense elements enter into all basic production, the interweaving of public interest and the private entrepreneurial function becomes more intricate. The common interest of all levels of public administration in the action taken by one level also becomes more evident. The past tendency for private entrepreneurs often has been to translate the element of public interest into public financial support, as in the case of accelerated tax amortization for the construction of electric generating equipment during the 1950's, and for other capital investment. The past tendency, furthermore, has been one of increasing public participation in economic decisions, with accepted restrictions on private decision making in the most essential resource fields. Pressure for clarity in administrative organization to cope with continuing complexity of relations among all levels of interest thus may be expected.

It is further assumed that the time period of interest for statement of conclusion is approximately the remainder of this century.

Regional Incidence of the Emerging Techniques

The regional incidence of emerging technical changes is likely to differ considerably. Several changes are of special interest in those parts of the country where present water use is pressing upon the physical limits of known available supply. Each technique bearing upon a major flow or withdrawal use is likely to have regional differentiation in its impact. Electric power production techniques, for example, promise to have some sharp regional distinctions. The same is true for all techniques bearing upon waste-disposal requirements or techniques of waste disposal. Finally, all of the techniques which make additional water available through discovery, precipitation, quality improvement, or reuse appear to have selective regional affinities.

Regional and local cost relations among them therefore are important, as well as the physical possibility of their application.

Weather modification appears to be the cheapest source of possible moisture increase wherever the technique becomes applicable. (Chapter 14, pages 356-83.) Extremely high benefit-cost ratios seem likely *where results can be obtained,* high enough to make this source of water desirable even in comparison to present sources. However, the area of probable near-term application of this technique is sharply limited, with promising results only in the Pacific Coast and Rocky Mountain states. Benefit-cost ratios are high, in part because of the high flow value of water falling on upper watersheds for hydro power generation and in part because the water is capturable for withdrawal uses as runoff or as soil moisture.

In whatever process it is usable, *untreated* salt water is the second cheapest source of water supply. (Chapter 13, pages 349-55.) However, its employment is obviously limited to sea coast or estuarine locations, and possibly to locations near salt-water aquifers. Functionally it appears usable principally for industry or low-grade municipal use.

The equivalent of cheap fresh water also may be obtainable through reuse of industrial and municipal supplies (chapter 16, pages 408-41), through the water budget (pages 442-51), and through evaporation control (pages 451-57).

As already illustrated by regional interest, the techniques of reuse and of evaporation control appear destined for most complete application in the arid and subhumid sections of the country. Reuse techniques also promise to be applicable in other sections of the country where there are large urban populations and concentrations of industry. California and the manufacturing belt of the northeast and midwest most nearly answer this description in this country. The water budget technique, on the other hand, appears destined for humid region use more than for arid region attention. Although the costs of the water made available by the application of these techniques will differ according to local situation, water obtained through extending the use of a given supply *generally* may be considered low-cost water. These techniques accordingly may be examined as one of the first marginal sources where additional water is needed.

Where water transfers are possible, the next cheapest water would appear to be that made available through long-distance transportation. Costs of storage and moving water 750 miles from northern to southern California are reported to be from $0.15 to $0.23 per thousand gallons, depending on the interest rate used in calculation.[1] These

[1] J. C. DeHaven and Jack Hirshleifer, "Feather River Water for Southern California," paper presented at Econometric Society meeting, Cleveland, December 1956 (mimeo.), p. 11. The lower estimate assumes a 2.7 per cent interest rate, and the higher a 5 per cent rate. Actual costs also will depend upon the allocations made to the individual purposes, or to user groups of all jointly used

costs appear to be below the immediate goal of desalting experiments, about $0.38 per thousand gallons, the next cheapest prospective source.[2] (Chapter 13, pages 335-49.) The latter are production costs, with no component for transportation or storage.

The most likely geographical area of interest in connection with desalting accordingly appears to be the localities in southern California near the seacoast, but not in a position to be served with water transported from other parts of the state. Similar situations are found along the Texas Gulf Coast. Desalting also may be of local interest in the interior West where surface waters have accumulated undesirable concentrations of salts. On the whole, however, large-scale desalting presently appears to have a rather narrowly limited area of application geographically.

Some sections of the West, and probably a high proportion of the total area west of the 98th meridian, must consider ground water the most likely marginal source. On the average this is likely to be "mined" water, and it probably will be the highest cost source of marginal water in the country. But it is water, and it will be turned to further use, even as it has been used already in areas like the Texas High Plains.

In the light of the present evidence it seems reasonable to expect that the emerging techniques, and others undescribed or unknown, will be selective in their regional application. This is a fact of substantial importance in anticipating the problems of efficient administrative organization for water development.

Effects of the Emerging Techniques on Existing Problems and Opportunities in Administering Water Development

Recognizing that the effects of the developing technology are likely to differ in regional incidence, it is next of interest to analyze the possible effects of these techniques upon the problems and opportunities previously outlined in this study (chapter 21). In some instances the

structures, like dams, the aqueduct, pumping plants, or other. See also J. W. Milliman, "An Economist Looks at State Water Planning," paper presented to the Water and Power Committee, Los Angeles Chamber of Commerce, November 1, 1957 (mimeo.).

[2] U. S. Department of the Interior, *Third Annual Report on Saline Water Conversion* (1954), Washington, January 1955.

newer techniques will intensify or magnify problems that have arisen in the past; in other cases they may relieve or lessen problems. In some cases wholly new problems may be raised; others will not be altered one way or the other.

The most obvious contributions of the emerging techniques, of course, are the increments which they may add to scarce water supplies. Beyond this, some troublesome or potentially troublesome situations may be mitigated as the newer techniques come into wider use. The techniques of industrial reuse and recirculation are of particular importance in this connection. Where they become widely applied they may relieve significantly the problem of industrial waste carrying in surface streams. Demands for flow use of surface water in this manner thus may be lessened at the same time that withdrawal demands for cooling and process water are lessened. Reuse thus promises to be a major conservational technique, and one which will make the problem of conscious quality control of streams much easier[3] than it otherwise might be. At the same time it promises to lessen the problems now anticipated for water allocation within the industrialized areas of the eastern states. As suggested, evaporation control, weather modification, salt water use, desalting, and added ground-water exploitation also promise increments to the available supply. Where they are applicable, their effects upon troublesome situations seated in scarce supply, like waste carrying and quality control, will be similar to those from recycling and reuse.

These are substantial contributions which will eventually yield a large sum of total benefits. However, the newer techniques intensify some problems already recognized as having arisen from past technical improvements. They are a long list. Furthermore they raise a few previously minor problems to a level of significance which as yet they have not had. There follows a brief description of twelve administrative problems which the new techniques promise to intensify, or which they raise to significant proportion for the first time.

1) They intensify the problem and the opportunity presented in applying the scale economies to water control and to the distribution of electricity.

a) The great new thermal electric generating units that are possible, and the very conspicuous economies that they effect, may make small-scale electricity distribution systems technically even

---

[3] Quality control of water within a plant, however, will be more exacting.

more out-of-date than they now seem. If geographically extended, large-scale transmission and distribution systems are accepted as desirable, then co-ordinated water control for hydro within a given system's generating and service area likewise is highly desirable. The problem is provision of the means of encouraging, promoting, or even requiring such scale operations in electricity generating, distribution, and related water regulation. In some regions such scale organization appropriately will include several river basins.

b) Weather modification, where it is successfully applied, also is likely to intensify the need for basinwide, co-ordinated water regulation where it does not already exist. Both in its potential for benefits (added water supply) and disbenefits (possible flooding or overseeding) weather modification will have basin-wide ramifications. This is assured by the location of its most frequent practice, the higher sections of the watershed to which cloud seeding is applied. In the basins where weather modification is applied periodically, pressure will be increased for some form of unified water-administering body which crosses local governmental jurisdictions. Where the effects are of interstate scope, the need for compact designed or federal administrative mechanisms is likely to be felt. Whatever the form to be used, the need for some intergovernmental legal jurisdiction is certain to accompany application of this technique.

2) The new techniques of power generation and weather modification reopen the question of the proper design of multiple-purpose storage facilities. The value of water storage for hydro generation in systems using the new techniques would seem to be enhanced by comparison with the past, particularly for peaking or reserve to meet steam unit outage. Consideration of hydro generation therefore would appear prudent in all future comprehensive water development, although its pattern of use may vary according to the specific system. The problem in this instance is one of achieving adequate recognition in planning for the new situation derived from the newer generating techniques. Parts of the problem lie in: (a) vested interest in specific patterns of allocation or distribution, or competition for benefits, as between public and privately owned electricity distribution systems; (b) professional inertia or opposition to systems of accurate data collection and interpretation which pave the way for the new techniques, or opposition to the techniques themselves.

Water receipts from weather modification undoubtedly will have to be stored in some cases, if they are to be of greatest economic benefit.

Storage facility design and construction eventually may have to provide for receipt of water from this origin.

3) New techniques of water freight movement open a question of the best integrated basin planning and development, and a question of co-ordination of inland waterway development with the land transportation facilities of region and nation. A number of recently applicable techniques may change the cost-benefit outlook for inland waterway development if adopted as a part of development and operation. Among them are the following: new techniques of earth movement and channel improvement, increase in lock size, motive power improvements yielding increased speed and economy; flexibility of operation and adaptability of equipment (interchangeable marine and inland use); terminal and port layout design and unloading practice improvements, decrease of cargo space requirements, through compression, concentration or packaging techniques; marine architectural design improvements; interchangeability of containers as between land and water movement; and extension of the range of services through new refrigeration equipment and methods.[4]

4) There are many major unsolved administrative problems and uncertainties in the development of atomic energy. Among these are the relative position of government and private industry with respect to atomic power generation; supply of fuels; reprocessing of spent fuels; relative costs of atomic power and conventional energy sources; and conservation of conventional fuels for chemicals, transport and other indispensable uses. Planning and administration of water resources is equally important to sound planning in nuclear energy development. The provision for public safety in water supply systems and in the control of radioactive waste disposal is vital. At the present time, there is complete federal control of contamination hazard through the licensing requirements of the Atomic Energy Commission. Will this type of control prove most effective, or will a degree of local control also be desirable?

In addition to safeguarding atomic installations insofar as possible from the hazard of massive contamination of water supplies, prepara-

---

[4] These and other technical aspects of water freight movement possibility might have been included in a case description in Part III of this book. They were not because of space limitations within the design of the study. However, the authors consider them of sufficient general importance to warrant notation in a list of probable administrative problems. The authors acknowledge with gratitude the suggestions of Roy Bessey on this point.

tions must be made in case of accidental contamination. Acceptable methods of chemical treatment, diversion, suitable warning systems for water users, decontamination procedures, provision of substitute supplies during purification of the contaminated water and works, and other services will need to be established.

Still further questions pertain to the liabilities of various parties in the event of radioactive pollution either from a routine operation or from an accident. As yet, there is too little experience with nuclear reactors for adequate insurance programs to be established. There is, however, a need for protection from heavy financial loss, both for industrial firms developing atomic power and for the water users who might be injured financially by contamination from accidents.

Finally, public control of radioactivity in streams and ground water will have to be effectively established. The degree to which water resources agencies may have to co-operate with authorities concerned with all types of radioactive contamination, as in the atmosphere, will have to be clearly established. Co-ordination within the country as a whole will be essential. Indeed, co-ordination on an international basis may prove to be vital in the face of the very far-reaching impact of small quantities of radioactive wastes.

5) The new techniques intensify the need for precise, accurate, and comprehensive basic data collection, and for the fundamental science upon which such collection must be based. Included are precise observation of the effects of use, professional appraisal of the influences accelerating demand, and accurate appraisal of the long-term characteristics of regional water receipts from the atmosphere.

a) Appraisal of long-term water receipts is of particular interest for the West, both in the areas where weather modification may be applicable and where it is not. The long-range chronological pattern of moisture receipts is of interest in the weather modification areas because artificial precipitation may increase the range in total moisture receipts during periods of maximum as compared to periods of minimum precipitation. It is of even greater interest in semi-arid sections where weather modification is not applicable, and where understanding of the pattern of moisture receipts is the only key to rationally planned land use. The problem is one of achieving a much more extensive observation and deeper understanding of the entire geographical environment within which water development and management take place.

The importance of long-term appraisal of water receipts is

Plate 27. Modern lock, barges, and tug; U. S. Army Corps of Engineers Lock No. 2, Monongahela River, Pennsylvania.

emphasized by dominant influences on water policy formation and management practice within the United States. The influence of crises of water surplus or water deficit upon our water development history is very evident. To mitigate the crisis influence in the future, accurate measurement of the availability of water over long periods of time must be sought. Only then can the country achieve the range of development planning now permitted by technology.

b) Appraisal of the influences accelerating demand is of vital interest in monitoring the effects of applying new techniques of electric energy production, both those based on the nuclear fuels and those related to the use of hydrocarbon fuels in electricity production. The problem of monitoring the effects and anticipating prospective effects of water use is likely to require organization and accuracy far beyond the system to which the United States has been accustomed in the past.

6) The technique of weather modification adds to the problem of administering the allocation of available water. For instance, within the states that maintain the right of prior appropriation wholly or in part as the water law, the question of low flow allocation is certain to appear. If flow during a dry season is not adequate to meet the established rights of all users, can the additional flow claimed to result from cloud seeding be claimed by the supporter of the cloud seeding, even if he holds only low priority rights? If he cannot, the prospect for support of seeding would not seem as bright as if water could be capturable by the entrepreneur. If he can capture the water, a complicated problem of continuous evaluation of results arises.

7) The management of "mined" ground water almost certainly will be of much wider importance in the future than in the past, considering the newer techniques of exploration and evaluation, and greater demands in specific areas for the use of ground water.

Although relatively little is known about the movement of water within aquifers, it is probable that problems associated with joint exploitation of an aquifer will increase. Already experienced as intrastate problems in Texas, Arizona, New Mexico, California, and elsewhere, joint interest in underground water may well lead to some interstate management problems, just as in the case of surface water. Some productive known aquifers cross state boundaries, and others of a similar nature are likely to be proven as demand for ground water rises. An efficient pattern of administering such deposits has yet to be designed.

8) The newer techniques raise problems of adjustment to the risks of

obsolescence for publicly supported development. This problem is present for private entrepreneurship and management also, but it particularly is a problem in public management and development because of recent tendencies to extend amortization and repayment periods of reimbursable construction, and to extend the list of nonreimbursables. Under these conditions, obsolescence risk is a much smaller consideration in planning and development than under conditions where the price of misjudging the risk is financial loss, as it is for private management. The emerging techniques introduce possibilities of rapid change for which prudent public administration will have to make allowances beyond the conservative estimates of capacity change based on present evidence. The desirability of building greater flexibility for operation into the design of multiple-purpose systems of facilities is likely to become more apparent as the risk of obsolescence rises. Water allocations and other legal rigidities, however, confront any administration based upon a principle of flexible operation. Careful study may be required of the need for rigid legal provisions and their costs in American water development and management.

9) Further application of the emerging techniques suggests an intensified need for bringing "the operational voice" more closely and more responsibly into planning of construction and development. The new pattern of power system operation and of weather modification are both examples of technical changes which can be of profound importance in proper planning for further development of water facilities.

10) The newer techniques as a group, in their present capacities, are likely to increase the difference between the western states and the eastern states in problems of administering water development and facility operation. National programs accordingly are certain to be faced with differences in program needs, which must be reflected in administrative organization if they are to be treated effectively.

11) As water recovery becomes more three-dimensional than ever before, and as the activities which may influence the availability or quality of water become more numerous, the opportunities facing co-ordinated or integrated planning for the employment of these techniques become much more attractive. The often unrealized and unappreciated benefits possible from co-ordination in the past would seem minor by comparison with the benefits possible from technically comprehensive matching of water supply and water requirements. This raises a corollary question as to how far the administration of water-related research, development planning, and operational management should be part of programs with other broader and often unrelated objectives.

12) Both the differential regional pattern of applicable techniques and their wide range raise questions as to the desirable degree of centralization in water facilities research, planning, construction, and operational management for the nation. Present federal tendencies are clearly in the direction of greater department-level centralization, stimulated partly by the superior technical resources of federal government agencies, but perhaps much more inspired by the prospect of increasing nonreimbursable appropriations for an increasing number of water development purposes. Do the emerging techniques suggest anything as to the desirability or undesirability of this tendency?

Among these problems for organizing the administration of water development and management none is more important than the last. Indeed the national capacity to make good use of opportunities presented by technology, and the capacity to meet highly specialized, radically new problems may depend on the balance achieved between national, state, local, and private agencies in research, planning, construction, and operational responsibilities. This subject, and the problem of co-ordinating the use of a number of applicable techniques, deserve special mention in concluding.

Centralization, Federal-State-Local-Private Enterprise
Relations, and the Co-ordinated View in Water Development

Centralization of administration in water development here is taken to mean the concentration of responsibility for planning, entrepreneurship, financing, and operational management in the hands of department-level federal government agencies, as they are supervised by the Chief Executive and provided with tax funds and policy direction by the Congress of the United States. Centralized administration occurs when national agencies are so organized as regularly to require central staff decisions for treatment of administrative problems that are primarily of local or regional application and importance. While field organization for such a national agency may exist, field autonomy is not stressed, and co-ordination of different field functions of regional scope tends to proceed through a central staff. At the other extreme from centralized federal agency activity are the planning, entrepreneurship, and operational management of private enterprise, co-operatives, and local public bodies. State organizations and regional organizations of interstate scope are intermediate administrative units. The latter theoretically could be a regional federal agency, or a nonfederal entity.[5]

The balance of responsibility as among these units of administration may be affected, *perhaps most importantly, by considerations other than technical change.* Among such factors are social motivation and both national and local political policy. These are excluded from the determination of any conclusions that are drawn from the foregoing discussion, because the discussion has been restricted to relationships, between the apparent state of technology associated with water development, its inferred effects on economic efficiency, related geographical factors, and the meaning for administrative organization. *These conclusions therefore must be tested in the light of other factors, as well as technology, before they may be considered valid patterns for action in our society.*

WHAT EMERGING TECHNOLOGY SUGGESTS ABOUT CENTRALIZATION

There can be little doubt that centralized responsibility and administration recently has been accepted popularly as the easy answer to water development planning, construction, financing, and (to some degree) operational management problems of the present and the future. Centralized department-level federal agencies early accepted and, to some extent, sought for the opportunities which past technology provided for large-scale development and integrated management. The trend toward centralization has been strengthened by policies of subsidization and nonreimbursability[6] for certain development purposes, which now have an imminent prospect of further broadening.[7] Does the emerging technology add to pressure for centralization, as past technology seems to have done?

In some respects it appears to do so, although the case is by no means as clear as in past development. Certain regulatory functions, the monitoring of changes in water use and quality deriving from

[5] The suggestions of Charles McKinley have helped the authors in drafting this paragraph.

[6] Nonreimbursability is taken to mean the practice of requiring no payment from direct and indirect beneficiaries of a project or program, payment being made from the general tax funds collected by the United States.

[7] As for example in the small watersheds program of the Department of Agriculture, and in "low flow regulation" on Corps of Engineers constructed reservoirs.

technical changes in industry, support of highly specialized fundamental research, integrated basic data collection, and the need for advance planning of large scope, will cause further pressure for centralized attention on the part of the federal government.

On the other hand, other aspects of the emerging situation speak just as eloquently of further difficulties for centralization. First, the complexity of planning, development, and management will be increased further. It already has been illustrated that even present-day planning and management is confronted with a highly complex field. If federal efforts are diffuse and incomplete, they are so with some reason. The addition of still further complexities will bring the danger of further inattention and friction in the administrative process if centralized organization is extended.

Second, increasing complexity calls attention to the mechanical problem of sheer size. Where complexity is great the encouragement of a competitive series of alternatives in planning and appraisal becomes very important. Yet under a centralized department-level approach the encouragement of such competitive thinking in any effective manner is increasingly difficult. It may be almost impossible where the central direction of a large scale agency is closely identified with special-interest pressures. The necessary central organization should be so constructed as to decide wisely and expeditiously among alternatives, not to minimize their generation and presentation.[8]

Third, emerging technology promises to accentuate, rather than smooth over, differences among the several regions of the United States. The difference between the water management problems of East and West may be even sharper in the future than now.

The Pacific Northwest will continue to present a unique set of water resource problems. The difference between the Great Plains and other regions shows no promise of lessening. In brief, regional differences promise to be heightened by technical changes, and these differences favor specialized regional planning, development, and managements.

Fourth, the emerging technology suggests as high a risk of rapid obsolescence of prevailing techniques and existing facilities as ever before. Perhaps it will be higher. Yet centralized responsibility for development has produced increasingly less disposition to consider obsolescence, as demonstrated in the lengthening repayment periods sought for reimbursable facilities constructed by the federal government,

[8] The suggestions of Abel Wolman on these points have been helpful.

and by the lengthening list of nonreimbursable works.[9] If centralized federal responsibility, nonreimbursable works, and high risk of obsolescence before amortization continue to show the correlation exhibited in the past, emerging technology does not strengthen the case for further centralization.

DISTRIBUTION OF RESPONSIBILITY AMONG DIFFERENT LEVELS AND TYPES OF ADMINISTRATION IN THE FUTURE

If doubts may be cast on the desirability of highly centralized responsibility for water development in the future, a second question arises as to what other form of administrative structure in the country might be more compatible with the technology which is foreseen. The assumptions stated at the beginning of this chapter, including the continuance of the federal system and the important entrepreneurial function of private enterprise, continue as background to the succeeding discussion.

Among the many functions connected with development of a resource some are adapted to administration at the most local level, including that of locally controlled private enterprise.[10] Others with broader territorial coverage, or with broader implications of the public interest, are suited to administration at the level of the state. Still others are interstate, but not national in scope, and therefore adapted to regional organization. Finally come those which are most efficiently conducted by a centralized national organization.

One of the most difficult and important of the questions that have

---

[9] It is recognized that groups with a vested interest in policies which minimize the importance of obsolescence risk are politically responsible for continuance of such policies. Yet the centralized agency pattern has given these groups effective technical assistance in promoting increasingly great separation of the incidence of benefits and the incidence of costs. Where the immediate incidence of benefits falls to a special group (as in flood prevention, irrigation, small watershed development), and the costs upon the public at large, concern with obsolescence rarely is raised for a serious hearing.

[10] A distinction is made here between locally controlled private enterprise and private enterprise operating with more distant financial or administrative direction. This distinction is of practical importance principally among the private electric utilities, some of which respond to administrative or policy direction from New York or another financial-administrative center for corporate operations. Where such policy or administrative control is exercised, a utility is not to be equated with local communities or local public bodies.

arisen, on the location of responsibility and decision making, has concerned the regional functions. It is noteworthy that federal responsibility can be compatible with discharging the regional functions at an interstate level. That federal administration does not necessarily mean centralized administration has been illustrated in the case of the Forest Service as well as that of the Tennessee Valley Authority. As Charles McKinley and others have noted, the Forest Service very successfully has combined decentralized management at three different field levels with centralized determination of basic policies and technical standards.[11] The TVA, of course, has not been a test of reconciling centralized staff and regional function, but it has illustrated the feasibility of federal agency administration of regional functions in a decentralized manner. Federal administration and centralized responsibility therefore are not necessarily synonymous.

The question here is not one of contrasting the federal administration with all other levels, and determining what can be detached from federal responsibility. Rather it is one of examining the functions to be performed in their territorial scope and in their degree of application to the public interest, and then determining what pattern of responsibility allocation is indicated.

An answer may commence at the level of planning, entrepreneurship, and management farthest from centralized federal government administration. This is with private enterprise, co-operatives, and local public bodies—i.e., all units of management below the state level. These entities are and will continue to be most intimately connected with the management of other productive factors used in combination with water. The fact that a large share of the production and distribution machinery of the country is managed from this level is of far-reaching importance in considering the outlines of a well-adjusted pattern of responsibility for administration of development.

Locally controlled private enterprise, co-operatives, and local public bodies. In the past, locally controlled private enterprise, co-operatives, and local public bodies[12] have conducted and successfully assumed responsibility for the following activities related to the planning, development, and operational management of water resources:

Industrial use of water, and supply therefor.

Agricultural use of water, and supply therefor.

[11] Charles McKinley, *Uncle Sam in the Pacific Northwest*, pp. 283-87.
[12] For example, municipality, irrigation district, levee district, or other entity legally empowered to levy specific assessments.

Municipal water supply and distribution, including rate regulation.

Generation, transmission, and distribution of electricity.

Flood control.

Water recreation facilities.

Mosquito control.

Design, development, and manufacture of equipment.

Weather modification.

Waste disposal.

Research and experiment related to the above activities. In the course of these activities locally controlled private enterprise and other local bodies have developed and managed both surface- and ground-water supplies, and have experimented with atmospheric sources of moisture. They have contributed to research and the development of techniques. In the important class of technical advances with windfall benefits to water development they have been dominant. In some of these activities responsibilities are recognized as having been inadequately met, as for instance in management related to the waste-carrying function of streams, and in the exploitation of ground-water deposits. In the main, however, a heavy share of responsibility for past progress in these areas has been at this level of administrative attention.

There is nothing about the emerging techniques that would suggest that similar responsibilities cannot be carried by locally controlled private enterprise, co-operatives, and local public bodies in the future. On the contrary, the intimate connection of these organizations with the practical management of the nation's productive machinery suggests that planning for a well-adjusted pattern of responsibility in water resource development should receive much initiative from them. In some directions—as in the application of industrial recirculation and reuse techniques, in the application of better scheduling techniques to irrigation, in the substitution of other resources for water, in recognition of obsolescence, and in other responsibilities or opportunities—major reliance will have to be placed on implementation at this level of administration. Successful application of the scale economies in the production and distribution of electricity throughout the country, and desalinization efforts also will be greatly aided by initiative, action, and co-operation from this level.

The special case of electricity generation and distribution. The pattern of adjustment in the administration of electricity generation, and transmission, and distribution is likely to be one of the most difficult.

Now that rural electrification is almost complete, territorial involvement in this function is nearly nationwide. Furthermore, the pattern of responsibility has become most intricate. Privately owned companies, co-operatives, municipalities, public utility districts, federal agencies, and state agencies undertake distribution.[13] Privately owned companies furthermore may be divided into those with locally controlled policy, and those responding to more distant corporate policy determination. Generation and transmission may be in the hands of the same range of agencies, although often distribution may be undertaken by one type of agency, and generation and transmission by another. Thus generation and transmission by a private company may be for its own customers, for municipal distribution, for co-operative distribution, or for sale to public utility districts. Federal agency generation may have a similar range of distributors, even though its sales are more to "preference customers" (municipalities, co-operatives, and other public bodies) than to private companies. This complicated pattern of responsibility is made further intricate by the fragmented geographical pattern of service areas.

There can be little doubt that pressure for taking advantage of the scale economies will be felt on this administrative complex. Increasing consciousness of the economies of scale, and of the new relation of hydro to thermal generation, are likely to exert such pressure. Some responses already have become regionally important, as in the TVA system, the municipal–utility district–private company co-operation in the Puget Sound Utilities Council, and combinations of private utilities for large-scale construction and operation. Other organizational possibilities for large-scale development exist, as in state agencies. The New York State Power Authority and the proposed state power agency for Oregon are examples. In general, the future performance of the devices for scale management suggested from private and local public sources will be very important in determining the involvement of other levels of public administration in this vital set of functions. In this area the difficulties of obtaining a clear distinction among the responsibilities of the several levels of government and private enterprise are patent.

Activities that have not been undertaken by private enterprise, co-operatives, and local public bodies. In spite of the wide range of activities which traditionally have been assumed by private enterprise,

---

[13] Federal and state agencies may and do distribute directly to industrial or other customers with heavy load requirements.

co-operatives, and local public bodies in the United States, there are some responsibilities in water planning, development, and management that have not been successfully met at that level of administration. Included among them are:

Most regulatory functions, e.g. use restrictions, water allocation, dam site use, conservation measures; rate regulation; geographical extent of operations; ground-water management and production regulations; maintenance of quality standards for discharge waters; etc.

Financing of large-scale multiple-purpose works.

Financing of works nonreimbursable by direct beneficiaries.

Water regulation on integrated basin-wide basis.

Leadership in providing a co-ordinated view for planning of construction and management; planning for the material needs of the nation.

Demonstration of improved techniques.

Leadership in applying the scale economies and multiple-purpose techniques to development and management.

Leadership in fundamental and applied science.

Interbasin development and long-distance transportation of water.

There is little in emerging technology to suggest that these activities, or necessities, may be transferable in any important way to the first level of administrative responsibility. The capacity of private enterprise to finance some large-scale works has been demonstrated, and in this direction some change may be possible. However, the manner of financing large-scale multiple-purpose works is likely to remain a subject for state, regional, or federal attention so long as there are significant subsidized or nonreimbursable purposes included in the development plan.

State, regional, and federal responsibilities in the balance. Those responsibilities that have not been or are not likely to be successfully treated on the level of private enterprise, cooperatives, or local public bodies, must be assumed by state, regional, or federal organization. Considering the possible demands from emerging technology, the assumptions stated at the beginning of this chapter, and the inherited pattern of organization, the following might be a compatible division of responsibility.[14]

[14] These observations are not intended as a blueprint to a division of responsibilities among federal, state, and regional organization, but as an indication of the type of activity which logically might reside at a given level, considering the problems and opportunities possible from technical change.

Activities appropriate to state organizations. Even though the nature of state territorial jurisdictions places the state at some disadvantage in comprehensive planning, development, and management of water resources, certain responsibilities still logically are appropriate to organizations at the state level. Among them one may cite:

Certain regulatory functions:
    water allocation among direct users, and other use restrictions;
    zoning of land use;
    rate regulation of utilities;
    management of ground-water exploitation (except interstate aquifers);
    water disposal regulation.
Planning, development, management of wholly intrastate watersheds.
Meeting dispersed water supply demands.
Administration of programs propagating certain scheduling techniques, viz.
    paired use;
    water budget irrigation.
Redirection of water demands, where necessary, through planned location of demand.
Basic data collection on local demand and supply.
Technical assistance to local public bodies, co-operatives and private interests.

Advisory participation in regional interstate, or federal planning and development. Even though these responsibilities are appropriate to state administration, certain problems arise in determining clear lines of division as between the state and federal government organization for some purposes. This may be illustrated by the case of intrastate watershed development in the eleven westernmost states. More than 53 per cent of the total land area of these states is under some form of federal administration, particularly as national forest or grazing district.[15] Within three-fifths to two-thirds of the total area of these states it would be very difficult to find an intrastate watershed which does not include some federally administered land. Thus, in addition to consideration of the water programs of the Corps of Engineers, the Soil Conservation Service, and the Bureau of Reclamation, co-operative planning and operational arrangements with the federal land manage-

[15] Marion Clawson and Burnell Held, *The Federal Lands: Their Use and Management* (Baltimore: The Johns Hopkins Press for Resources for the Future, 1957), p. 403.

ment agencies must be worked out for these areas. The Forest Service and the Bureau of Land Management are most likely to be involved and, less frequently, the National Park Service. Finally, as state agencies themselves may depend in part on federal aid funds for their support, other federal agencies are drawn into the complex necessary to complete administrative co-ordination. Such federal aid funds are received and used by state fish and game agencies, forestry departments, public health or sanitation agencies, and highways agencies, among others. The federal interest in direct land management is less in the midwestern and eastern states than it is in the West, but other federal interests are nationwide. The opportunity for state administration of an intrastate watershed program, therefore, at least initially will be one of leadership and initiative in a co-operative program which must necessarily include federal agencies no less than local communities and private interests. This pattern is indicated especially in the eleven western states. While the pattern of interrelation is best illustrated by small watershed development, other appropriate state functions may give rise to similar problems of co-ordination.

Activities appropriate to federal or nonfederal regional organization. Because of the nonconformity between large river basins and state territorial jurisdiction, some planning, development, and management functions cannot be carried on within the limits of individual state jurisdiction. At the same time they fall short of being functions which must be administered from a centralized national agency. In the past they have been successfully administered as parts of a regional program. However, a certain amount of confusion has been created on this point because centralized federal department agencies also have acted as regional agencies. This has tended to obscure the regional nature of most of these functions. The emerging technology may heighten the regional characteristics of these functions. The organization to which responsibility can be given may be federally sponsored, or nonfederal; but whichever the case, a regional organization may successfully administer the following responsibilities:

Balancing water development and other resource use.

Collection and interpretation of basic data on any discrete basin, including the monitoring of quantity and quality changes in supply.

Entrepreneurship for large-scale multiple-purpose works.

Programming and budgeting for development.

Basin-wide integrated water regulation, and the application of other scale economies in construction and management for interstate basins.

Co-ordination of operational management and planning for development.

Here again the difficulty of clearly separating responsibility between federal and nonfederal levels of administration and initiative becomes apparent. Theoretically, regional organization of a nonfederal character is possible, although the problem of administrative responsibility is present unless "a new layer is built into our federalism cake," as McKinley has expressed it.[16] However, the economic imbalance occasioned by differential capacities for investment, differing regional resource endowments, and differing stages of development may always create a national interest in regional development. This often has found political expression in American history, in the problems of both West and South. The national interest is also apparent in other broad policy considerations relating to development, like the volume of needed federal investment at any given period. At the same time this does not mean that centralized national administration of development programs are the necessary, or even the appropriate, means of administering the functions here described as basically regional. The national interest in regional development will remain, and some federal administrative interest is appropriate, but administrative devices which more clearly express regional functions are indicated, whether they are federal or nonfederal.

Activities appropriate to federal attention at a national level. It has been suggested here that some confusion has existed as between functions appropriate to the federal agency at the national level, the federal or nonfederal regional agency, and the state agency, with a tendency toward centralized federal attention in all three areas. Even allowing for a more clear separation of functions among these units of government, there remains a large and important field in which federal government attention is appropriate, and is likely to be necessary. In this field there are regulatory, research, technical development, financing, demonstration, development and advanced planning functions. They may be illustrated as follows:

Regulation:
  forestalling site encumbrance;
  weather modification practice;

[16] Charles McKinley, personal communication to the authors, 1957.

    interstate stream quality standards,

    achievement of legal environment suitable to scale economies;

    responsibility for control of potential atomic radiation hazard;

    interstate electric power systems.

Research and data collection:

    monitoring water quality and other effects of technical change on water demand,

    experiment with accurate, dependable means of collecting and evaluating physical data,

    fundamental research: e.g. weather cycle collection of data on weather and climate, and weather modification on continental or planetary scale.

Promotion of environment favoring further technical improvements.

Advanced planning:

    interbasin use and development,

    necessary redirection of demand on national basis,

    planning for adjustment to obsolescence,

    relation of water development to other resource use.

Large-scale financing.

Demonstration development for all individual functions, and integrated functions on all scales (including the regional if necessary), using experimental and research results.

Development considered essential to discharge of constitutional responsibilities (national defense, promotion of commerce, general welfare, etc.)

Management of the federal lands, and associated waters.

Both the existing and the emerging technology emphasize the importance of these functions in a manner not yet fully reflected in federal policy or administration. The complexity, depth, and pervasiveness of technical works may be expected to continue to place a premium on proper co-ordination and leadership. Furthermore, they require increasingly specialized professional attention in the area of both fundamental and applied research. They also require, if economic efficiency is sought, careful provision for co-ordination. Finally, more commanding regulatory problems of interstate extent may be anticipated as results of technical progress. These all concern functions which best may be assumed by agencies operating on a national scale from the financial position of the federal establishment.

The federal government has a pre-eminent research, technical, and co-ordinating position; it also has had and is likely to have heavy responsibility for national leadership in recurring emergencies. How is this position translated into development and management activity of most benefit to the nation? One way is the assumption of increasingly greater entrepreneurial and management functions centrally administered by the federal government, as has taken place in the past. Another way is the maintenance of demonstration activities on all scales, designed to provide guides in the translation of technical advance into actual development and management practice in this field. The choice between demonstration and centralized management exists at present, whether or not it is recognized in the present everyday operations of the federal agencies and in the policy set for them.

Entrepreneurship on a demonstration basis would seem to be the essential minimum, for without pilot and demonstration operations the advantages of the federal position (specialization, large funds, position for co-ordination) cannot effectively promote technical change to the national advantage. The favorable position of the federal establishment for undertaking advanced planning cannot be effectively used unless the problems of actual development are met and treated in the course of such planning. This should be possible under a demonstration program just as much as under a system where centralized department-level federal management prevails.

Federal entrepreneurship on a demonstration basis also can be highly useful in suggesting the possibilities and pattern of development. Such matters as design of projects and program, costs, management techniques, manner of financing, and evaluation techniques may all be taken into account and perfected under such an entrepreneurship.

Demonstration also can be an effective adjunct in coping with regulatory problems with which the federal government is certain to be faced. In the past, this effect has been illustrated in some federal electric power developments (e.g. the Southeast). Demonstration has been useful in providing factual bases for electric power rate regulation. The demonstration entrepreneurship also may be valuable in coping with the forthcoming problems of waste disposal in streams and ground water.

The question of general entrepreneurship for water development as a federal government function. If demonstration development would seem the minimum federal entrepreneurial function, a question arises as to how far beyond this, if at all, the federal establishment should go in its entrepreneurial function, and in the management function for water resources in the United States. No attempt will be made here to answer this question conclusively. The final decision in this instance must be a political one, based on evidence and pressures from several fields other than the technical, including national security and the general welfare. However, evidence on a corollary question will be given. The question is: Does the emerging and the existing technology favor further extension of federal entrepreneurial and water management functions in the United States?

The following evidence may be considered in examining the question:

1) True comprehensive development—the efficient matching of water demand for all purposes with water supply—is a far more complex matter than integrated river basin development, which hitherto has been the most complex planning, entrepreneurial, and management task undertaken for water resources in the United States. Well coordinated, fully comprehensive treatment of water problems with the use of all the technical "tools" available has never been undertaken.

2) The emerging technology suggests that both regulatory problems and the need for technical leadership may be more intensive for the federal government than in the past. Flexibility in organization, opportunity for experiment, and a co-ordinated view of the technical front in this field may be needed more than ever before.

3) The past dominant interest of federal agencies has been in accretions to the entrepreneurial or management functions, or both, overshadowing important federal technical functions.

4) Centralized federal agencies have mixed functions appropriate to national, regional, and even local levels of administration in their entrepreneurship and operations. With a few exceptions, no sharp differentiation of national, regional, state, and local functions has been sought.

5) Large-scale or integrated basin surface-water development is faced with problems of indivisibility at the economic margin which favors public entrepreneurship in many instances.[17] However, there is nothing in this characteristic which suggests the need for construction and operational agencies of national scope, as contrasted with regional or even (in some cases) state organizations. At the same time centralized advanced planning and performance review are essential to it.

6) Most state governments in the United States have a manifest bias toward rural constituencies. Opportunity is lacking for an increasing urban population to express itself fully, or even adequately,

[17] J. V. Krutilla and Otto Eckstein, *Multiple Purpose River Development— Studies in Applied Economic Analysis* (Baltimore: The Johns Hopkins Press for Resources for the Future, 1958), Chap. II.

through state institutions. State organization accordingly favors appeal of numerically preponderant groups of urban people to federal action, where their needs or ambitions for development are not otherwise attended to.

7) There is a lack of any regional polity of wider jurisdiction than an individual state within the nation. Consequently there is no politically organized institution below the federal government to which a regional resource development agency might be responsible.

Reviewing this evidence, one is left with further questions which the nation might profitably consider from all points of view. They are:

1) Is the present dominant centralized department-level federal agency interest in entrepreneurship for construction and water resource management compatible with efficient discharge of important federal functions of regulation, technical leadership, co-ordination, and advance planning? Or does it divert the attention of the entire federal establishment in a manner not conducive to the most efficient matching of demand and supply over the long term? Does it mask the real need for and nature of co-ordinated planning and review?

2) In the face of technology-caused intensification of regional differences in water development and management problems, is it efficient to have centralized department-level national agencies responsible for activities which can be, and have been, undertaken successfully at the regional or state level?

3) In view of the great technical complexity of true comprehensive planning, development, and management, what manner of federal leadership and co-ordination should be sought? Should a leadership at the national level rest more on its technical resources than on its financial resources? Are federal financial resources and centralized department-level agency entrepreneurship and management indivisible?

4) Can a viable nonfederal organization of regional scope be developed? If not, what means should be taken to ensure adequate treatment of regional resource development functions in the federal establishment? How do state and local agencies effectively organize advisory participation in federal regional development? Can centralized staff functions and regional functions be successfully separated within the federal organization?

CONCLUSIONS

In this discussion of centralization in water resource activities one meets a looming problem of modern times. Technology has led to large-scale and complex operations in economic production and resource development. Yet in the case of water development and management it should not be assumed that this means an accompanying administrative organization which seeks most of its important solutions in centralized national organization. It does mean, however, that a high degree of co-ordination and imaginative technical leadership are essential. These best can be provided at the overriding (federal) level, provided the functions most appropriate to that level are clear and not confused with the complete range of development responsibilities. From the point of view of efficiency it therefore would seem logical that the administrative scope of different organizations in this field should not reach farther into detail than is warranted by the geographical and functional scale of problems.

To achieve such a balanced separation of administrative function, one important step is necessary in the United States: study and design of, and agreement upon, effective regional administrative units for water development. This is probably the most confused key issue in American resource administration at present, with several centralized department-level national agencies filling most regional agency functions. The authors believe, on the grounds of this analysis of administrative-technical relations, that technology points strongly in the direction of the need for such regional units. This is because of the coming differentiation of water problems among the several regions; because of the varying effectiveness of the techniques as applied in different regions; because of the eminent feasibility of administering integrated water development on a regional basis; and because the centralized federal establishment may be a better technical leader if it is less of an entrepreneur. The form to be taken by such agencies is not a subject for the present study; only the need is indicated.

It is entirely possible that the apparent inflexibility of political institutions within many states, and the improbability of any future regional political institution, may make federal organization essential to the performance of the regional functions. Should this be true separation within the federal establishment of (1) administration of the centralized agency functions appropriate to the national level and (2) the administration of the regional planning, entrepreneurial and operational management functions, should be examined carefully and given further experimentation. Only one federal regional agency has been given a full trial thus far, the TVA. Elsewhere in the federal establishment there has been some response to the technologic pressure for regionalization of some functions with the centralized agencies, although it has been slow. The most determined attempt toward regionalization among these agencies was that of the Department of the Interior between 1946 and 1953.[18] The Department in that period attempted not only to experiment with a regional operating structure for its bureaus, but also to use its central staff to support the regional units, encouraging them to develop independence where it was statutorily possible. This experiment was abandoned in 1953, along with several other similarly directed resource development activities. Since 1953 the principal experiment in regional federal units has been in the subsidiary or associated basin committees of the Interagency Committee on Water Resources, like the Missouri Basin Interagency Committee, the Columbia Basin Interagency Committee, the Northeastern Resources Committee (New England), the Arkansas-White-Red Basin Interagency Committee, and the Delaware Basin Survey Co-ordinating Committee. These committees, however, have no statutory power, and can have only co-ordinating functions as accepted by the constituent agencies.

An important point in designing a balanced, better co-ordinated, technically more effective administrative organization for water development is the dominant place of private enterprise, co-operatives, and local public bodies in the direct management of the productive resources with which water is used. The technically most effective development and management of this resource is likely to be that which takes account of the manifold operations at this first level of administration. For example, full advantage of the scale economies in water development is not likely to be achieved on a national basis until either: (1) the place of the privately owned utilities in relation to them is taken into account and the utilities become willing participants achieving these economies and passing them on to consumers; or (2) small utility service areas are superseded by public organizations with adequate territorial jurisdiction.

---

[18] These remarks are limited to agencies with a primary concern in water development. Regional structure is well developed in some other resource agencies, like the Forest Service.

In other instances it may be found that the development agency may not necessarily be the appropriate management agency. Thus on projects for which the Corps of Engineers and the Bureau of Reclamation have been the entrepreneurs, operational functions have been transferred to local governmental units for both irrigation and recreational facilities. In this way large-scale entrepreneurship is taken advantage of where needed, as are the later advantages of local operational management in appropriate functions.

A second point of importance is the potential value of integrated policy and technical review above the level of the federal department for major problems of national significance. If the nation is to take advantage of the economic efficiency and the amenity of integrated development, and all the other possibilities brought by technology, clearer central policy standards for planning and operation would seem essential. The problem of determining these standards is beyond the level of any single agency with special functions and special interests, even though that agency has a national scope and responsibility. At the same time, it is a problem on which highly competent and imaginative technical guidance is essential.

As a step toward the permanent provision of such guidance, further study of the functions appropriate to the several levels of administration in the United States seems very desirable. The United States is in need of determining the effectiveness of its present administrative structure for water resource planning, development, and management. It also must determine the capacity of this structure to absorb the impact of major technical change. How might this structure be designed to make better use of all the scheduling and conservational techniques? How can it promote the scale economies further? How can it become more sensitive to turns in multipurpose demand? How can it bring the full possible range of supply-and-demand manipulation into planning? How can it provide effective water allocation compatible with economic efficiency? How can it avoid excessive obsolescence costs? How can it make dependable provisions for the public safety and convenience in water use? How can it be made responsive to the entry of still unconceived techniques?

Will these ends be met through making the states and local communities more a part of development and operation? Will it be through new forms of encouragement to private enterprise? Will it be through stronger superdepartmental co-ordination in the federal government? Would a moderate restructuring of existing agencies, decentralizing some responsibilities, have any effect? It is possible to place the political process on the roads which technology has already shown to be open?

These questions are sufficiently important to warrant an early reconsideration of national water policy and national water development organization.[19]

[19] See Irving Fox, "National Water Resources Policy Issues" (*Law and Contemporary Problems* 22 [Summer Issue, 1957] Durham, N. C., 472-509), for a review of present policies. This study is being pursued further.

Glossary

ACIDITY AND ALKALINITY. Acidity—the concentration of hydrogen ions in a solution, applied usually to solutions having sufficiently high concentrations to exhibit typical acidic characteristics such as sour taste and reactivity with alkalis and certain metals. Weakly acidic solutions (low hydrogen ion concentration) may be typified by vinegar and carbonated water; highly acidic solutions by sulfuric acid and nitric acid. Alkalinity relates to the hydroxyl ion concentration, and to the degree of reactivity with acids. A typical weakly alkaline solution is sodium bicarbonate (baking soda) whereas sodium hydroxide ("lye") has high alkalinity.

ACRE-FOOT. A volume of water sufficient to cover one acre to a depth of one foot. One acre-foot equals 325,851 gallons.

ADIABATIC. Occurring without transfer of heat, i.e., with neither addition nor removal of heat to or from the system involved. Adiabatic differences in atmospheric temperature are those dependent on pressure changes with elevation.

ANADROMOUS. Fish species which live in salt water but migrate periodically into fresh water streams for spawning (e.g. salmon, shad, etc.).

ANNULAR SPACE. The space between two concentric circles or cylinders. In wells, the space between the larger casing pipe and the smaller drill pipe.

AQUIFER. Water-bearing consolidated or unconsolidated sediments, or water-bearing fractured crystalline rock below the earth surface.

AUSTENITIC. Pertaining to a very hard, high strength steel composed of a solid solution of carbon or iron carbide in iron.

BASE LOAD. The average minimum electrical demand on an electric generating facility or system of facilities.

BIOCHEMICAL OXYGEN DEMAND (B.O.D.). The quantity of dissolved oxygen, measured in milligrams per liter (or parts per million) required during the stabilization of decomposable organic matter content by aerobic biochemical action. This quantity is determined by diluting a sample with water saturated with oxygen and measuring the dissolved oxygen in the mixture both immediately and after a five-day incubation period.

BLACK ALKALI. Dark-colored incrustation encountered on some arid-region soils containing organic materials dissolved by the carbonate solutions of

659

sodium and other metals. Evaporation of the solutions leaves the so-called black-alkali deposit.

BLOW DOWN. The portion of a recirculated water stream, usually involved in an evaporation cycle such as in a steam boiler, which is run to waste to avoid accumulation of impurities.

BOLSON. A local basin of interior drainage in an arid region.

BORROW AREA (PIT). The portion of land from which soil or rock is moved to another location, as for a dam or roadway.

CAPILLARY WATER. The water which is held in the soil above the water table or upper surface of the zone of saturation by capillary action. It includes the water content of the soil beyond its hygroscopic moisture content, which is held against the force of gravity by surface tension.

CEREMET. A chemical compound having a very high melting point, and resembling ceramics in hardness and resistance to chemical attack.

CHLOROSIS. A plant disease causing loss of the green color and chlorophyll activity.

CLAYPAN. A stratum or horizon of accumulated stiff, compact, and relatively impervious clay which is not cemented and if immersed in water can be worked to a soft mass.

COMPREHENSIVE. In planning, construction, and management, the inclusion of all socially pertinent purposes (or uses) on an interrelated basis, and all economically relevant means of providing for those purposes.

CONCENTRATED DEMAND. Intensive, localized demand, as in an urban area, an irrigation district, or a manufacturing plant having heavy consumption.

CONCENTRATED SUPPLY. Water sources capable of fulfilling concentrated demand, as surface streams, lakes, aquifers, etc.

CONTINUOUS DEMAND. Demand incident generally on a year-round basis, as contrasted with seasonal or other periodic demand.

CONVECTION. As used in meteorology, the vertical movement of air resulting from vertical instability in an air mass which is produced either by heating from below or by cooling aloft.

COOKING LIQUOR. The solution containing calcium bisulfite, sodium sulfide, or sodium hydroxide, along with other compounds, in which wood chips are digested in the manufacture of chemical wood pulps.

CORONA LOSS. The electricity discharge which appears on the surface of a conductor when the potential gradient exceeds a given amount.

COUNTER FLOW. An arrangement by which two streams of materials, usually one liquid and one solid, are handled so that each stream entering the process contacts the other stream.

CRITICAL TEMPERATURE, CRITICAL PRESSURE. For a particular pure substance, the temperature above which no liquid phase can exist, regardless of pressure. For water, this temperature is 705 degrees F., and the critical pressure (pressure exerted by steam in presence of liquid at the critical temperature) is 3,206 pounds per square inch.

CUBIC FEET PER SECOND (cfs). The number of cubic feet of fluid flowing past a given point during one second. One cubic foot per second, or "cusec," equals 449 gallons per minute or 1.98 acre-feet per day.

DEMAND. The quantity of water or water derived energy or services which will be used at given costs or prices of sale.

DISAPPEARANCE. Evaporation or transpiration of water, or its incorporation into a product (inanimate or biotic) for human use.

DISPERSED DEMAND. Demand which is distributed extensively over a wide area, the opposite of concentrated demand.

ELASTOMER. A synthetic plastic compound having some of the properties of natural rubber.

ELECTRODIALYSIS. The process of transferring ions through selective membranes by means of an applied direct current so that ions of like electrical charge move in one direction through one membrane and ions with the opposite charge move in the other direction through another membrane.

ELLIPSOID. A surface of which all plane sections are ellipses or circles.

EPHEMERAL STREAM. An intermittent stream which flows only in direct and immediate response to precipitation, and has no prolongation of flow from surface or underground storage.

EVAPOTRANSPIRATION. The natural process by which water is lost from a land area by the combined action of evaporation from free surfaces and transpiration by plants.

FAST NEUTRON. A neutral particle emitted at high speed from the nucleus of an element undergoing certain types of radioactive disintegration. Fast neutrons may in turn effect other nuclei in their immediate vicinity, or they may be converted to slow neutrons by passage through atoms of other elements known as moderators.

FIELD CAPACITY. The moisture content of the soil, expressed as a moisture percentage on a dry-weight basis, after excess gravitational water from a thorough wetting of the soil profile by rain or irrigation water has drained away.

FOGGARAS. Underground infiltration galleries or tunnels constructed by connecting shafts sunk in the detrital fans of piedmont regions in order to concentrate water moving underground into definite channels. They are known as foggaras in North Africa and as kanats or karezes in the Near East.

FOSSIL FUELS. Combustible gases, liquids, and solids found in the earth's crust, resulting from the metamorphosis of plants and animals living in past geologic ages.

FUSION. As applied to nuclear reactions, the conversion of two or more atoms to form fewer atoms of higher atomic weight and the release of large quantities of energy. The fusion of hydrogen atoms to form helium is the important example.

GAMMA RAY. A quantum of electromagnetic radiation emitted by a nucleus, each such photon being emitted as the result of a quantum transition between two energy levels of the nucleus.

GLAUBER'S SALT. A colorless crystalline salt composed of sodium sulphate decahydrate ($Na_2SO_4 \cdot 10H_2O$).

GRAVITATIONAL WATER. Subsurface water in excess of absorption water and pellicular water (film adhering to soil particles) which responds to the force of gravity and therefore percolates downward toward the water table.

HALF LIFE. A term expressing the rate of radioactive decay of an element. It is the length of time required for half the material present to undergo spontaneous conversion to other elements.

HEAT ENGINE CYCLE. The process by which heat transferred from fuel combustion products to a working fluid (steam) is partially converted to mechanical work by expansion of the steam in an engine (turbine). The useless heat energy remaining in the exhaust steam is discarded by condensing it with cooling water.

HORIZON (SOILS). Many soils are composed of layers possessing distinctive physical and chemical characteristics. These layers are known as horizons, and are commonly referred to by letter designations. Thus the "A" horizon may be weathered material nearest the surface, and the "C" horizon may be unconsolidated material with little chemical weathering lying adjacent to the underlying country rock.

HYDROGEN ION CONCENTRATION. The concentration of hydrogen ions in an aqueous solution is a function of the degree of dissociation of the liquid, and determines its character as an acid or a base. This variable is conveniently expressed in terms of the value pH, which represents the logarithm of the reciprocal of the gram equivalents per liter of the hydrogen ion. More simply expressed, the pH value is a number between 0 and 14 which represents the degree of alkalinity or acidity of the water. A pH value of 7 indicates, in the case of pure water at 25°C, exact neutrality, since it contains the same concentration of H+ ions as OH− ions, while lower values indicate a preponderance of H+ ions which are acidic and higher values indicate a basic character of the water with an excess of OH− ions which are alkaline.

HYDROLOGY. The study of the water of the earth, its properties and distribution and its transformations, combinations and movements, especially from the time of its precipitation on land until its discharge into the sea or return to the atmosphere. Analysis within this field is directed particularly to the occurrence of water on the earth, the description of the earth with respect to water, the physical effects of water on the earth, and the relation of water to life on the earth.

HYDROPHILIC. Water loving. The property of attracting or combining with water.

INDUCED RADIOACTIVITY. The radiation emitting property of a material, not naturally radioactive, resulting from conversion of a portion of it to a radioactive isotope by capture of neutrons, bombardment by radiation, or other means.

ION, CATION, ANION. An ion is an atom of an element containing one or more excess electrons or electrical charges or lacking one or more electrons. If it contains extra electrons, it carries a negative charge and is known as a cation. If deficient in electrons, it is a positively charged anion.

ISOTOPE. Most elements contain atoms of identical chemical properties but slightly different atomic weights because of differing numbers of neutrons in the nucleus. These different forms of a single element are called isotopes.

INSOLATION. The radiation from the sun received on the earth's surface.

INTEGRATED. As applied to design, planning or operational management, the design, planning or operation of a system of facilities in an interrelated pattern with view toward maximizing the economies possible in joint operation. An integrated system may be single purpose or multiple purpose, and may be both hydraulic and electrical.

INTERFACE. A surface which forms the boundary between two geologic strata, two immiscible liquids, or any two phases of matter with distinct boundaries.

INTERMITTENT STREAM. A stream which alternates between flow and dry channel. Many intermittent streams are seasonal, but flow in others is aperiodic.

KILOVOLT (kv). 1,000 volts, a volt being that unit of electromotive force which when applied continuously to a conductor with a unit resistance (one ohm) produces a current of one ampere.

LEACHING. The process of removing water-soluble material from the soil by passing water through it.

LOAD FACTOR. The average electric power generated during a designated period divided by the peak generation rate occurring in the same period.

LOGGING. The technique of determining underground geologic formations by use of various devices and methods to traverse a well or bore hole.

MEGAWATT. One thousand kilowatts.

MERCAPTAN. A family of organic chemicals containing carbon, hydrogen, and sulfur, and having an unpleasant odor.

MICROMHO. A unit of electrical conductance. Specifically, one-millionth the conductance of a body through which one ampere will flow when the potential difference is one volt.

MULCH. Any substance artificially applied to the soil surface, or incorporated into the surface layer of soil, designed to reduce evaporation of soil moisture, reduce soil temperature fluctuations, or accomplish other effects. Mulches often are organic substances.

OBLATE. As applied to a spheroidal object, flattened at the poles.

OROGRAPHIC PRECIPITATION. The precipitation resulting from the movement of a moisture-laden air mass across mountains or another inclined surface, like the face of a plateau. If sufficient water vapor is initially present in the air mass, precipitation falls upon the high ground which the air mass crosses. Substantial amounts of such precipitation are caused by mountain ranges athwart the prevailing direction of the air mass movement.

OSMOSIS. The diffusion of a liquid, usually water, from a solution containing a higher concentration of dissolved substances, into a solution of lower concentration, through a semipermeable membrane.

PEAK LOAD. The maximum demands for electricity which a generating facility or system is called upon to supply.

PERCHED GROUND WATER DEPOSIT. Ground water occurring in an elevated aquifer or saturated zone, separated from the main body of ground water at lower elevation by unsaturated rock.

PERHUMID CLIMATE. A climate in which the supply of moisture through precipitation more or less consistently exceeds the need for water for evaporation and transpiration. C. W. Thornthwaite has defined this climatic type as one in which the moisture index, based upon a comparison of the humidity index with the aridity index, is 100 or more.

PHREATOPHYTE. Plants which are heavy consumers of water, particularly those of arid or semi-arid regions. Phreatophytes send their roots to the water table, and typically are found on the shores of water bodies.

PICKLE LIQUOR. Acid solution used in the de-oxidizing of steel prior to galvanizing or other surface treatment. Waste pickle liquor contains a portion of the original sulfuric acid and dissolved iron sulfate.

PLANKTON. The passively floating or weakly swimming plant and animal life present within a body of fresh or salt water.

PLAYA. The surfaces at lowest elevation within a basin of interior drainage in arid regions. After heavy rainfall or flash floods these areas temporarily may become inundated, becoming shallow, muddy lakes but drying out again in the absence of further water flow.

REACTANCE. In electricity, the influence of a coil of wire or a condenser on an alternating electric current, tending to "choke" or diminish the current. This property is usefully employed in many types of electrical machines.

REQUIREMENT (water). Effective demand for withdrawal.

RUSSET. As used in the specification of potatoes and other vegetables and fruits, refers to the presence of an outer skin which possesses a coarse and rough texture.

SALINE. As applied to water, containing dissolved salts in concentrations which make the water undesirable or unfit for most withdrawal uses. Saline water usually has a total dissolved solids content of 1,000 ppm or more.

SAVE-ALL FILTER. A type of rotary filter used in pulp and paper mills for removing small amounts of fiber from large volumes of process and waste water.

SCHEDULING TECHNIQUE. Design of an operational pattern for a facility or a system of facilities which provides economies through meshing two or more demands with the diurnal, seasonal, or geographical peculiarities of supply.

SCINTILLATION COUNTER. A type of sensitive radioactivity detector utilizing the fluorescence of various materials when struck by radiation.

SLURRY. A mixture of a solid suspended in a liquid, the solid usually in sufficient quantity to produce a thick or viscous mixture.

SODA ASH. Anhydrous sodium carbonate, Na_2CO_3, obtained as a white powder or in lumps.

SUBSTRATE. In agriculture, the subsoil or soil layer underlying the normally cultivated surface layer.

SUPERCOOLING. The process of cooling a liquid below its normal freezing point in such a manner that it does not freeze. In a cloud of very small water droplets, in the absence of nuclei to initiate freezing, the droplets can be supercooled over 70 degrees below the normal freezing point without crystallizing.

TANKS. In peninsular India and Ceylon, small reservoirs, ponds and lakes that have been constructed to impound water during the wet season. (In addition to common definition of tank.)

TECHNOLOGY. Any practical art utilizing scientific knowledge, as horticulture. (Webster)

TON. Short ton (2,000 pounds), unless otherwise specified.

TRANSPIRATION. The physiological process by which plant organisms exhale water vapor.

TUBE WELL. Describes any of a variety of drilled wells which are cased. In most tube wells screening is provided in the area of aquifer penetration where water is to be withdrawn.

USE. The state of being employed (Webster). Some technological literature on water adopts a meaning for "use" which is synonymous with disappearance. Because such a definition is obviously confusing, it is avoided in this study.

TRACER. A substance added to another material for purposes of determining flow characteristics of the stream. If radioactive, the tracer can be found and measured by use of simple radiation-detecting instruments.

TRITIUM. An isotope of hydrogen with an atomic weight approximately three times that of ordinary hydrogen. It is moderately radioactive, with a half life of 12.5 years.

VADOSE LAYER. The subsurface zone between the water table and the capillary fringe is known as the vadose layer. It constitutes that zone through which gravitational water percolates.

WATER BUDGET. An accounting device to maintain a constant inventory of the soil or surface moisture conditions of any given area on the earth's surface. Deductions are made from the inventory in accordance with the computed evaporation and transpiration losses, and additions are made as precipitation occurs or water is added artificially or naturally from an external source.

WET BULB TEMPERATURE. The temperature attained by a thermometer when surrounded by a moistened wick in the air stream. It is useful in determining atmospheric humidity.

WHITE WATER. Paper mill and pulp mill processing streams consisting of comparatively low concentrations of cellulose fibers in water.

WILTING POINT. The soil moisture content at which the permanent wilting of plants occurs.

WITHDRAWAL. Diversion for beneficial use.

ZONE OF AERATION. Surficial zone of unconsolidated sediment in which the interstices are filled with air, and water held or suspended by molecular forces such as cohesion, adhesion, and surface tension.

ZONE OF SATURATION. Zone in which interstices of rock or other materials are filled with water under hydrostatic pressure.

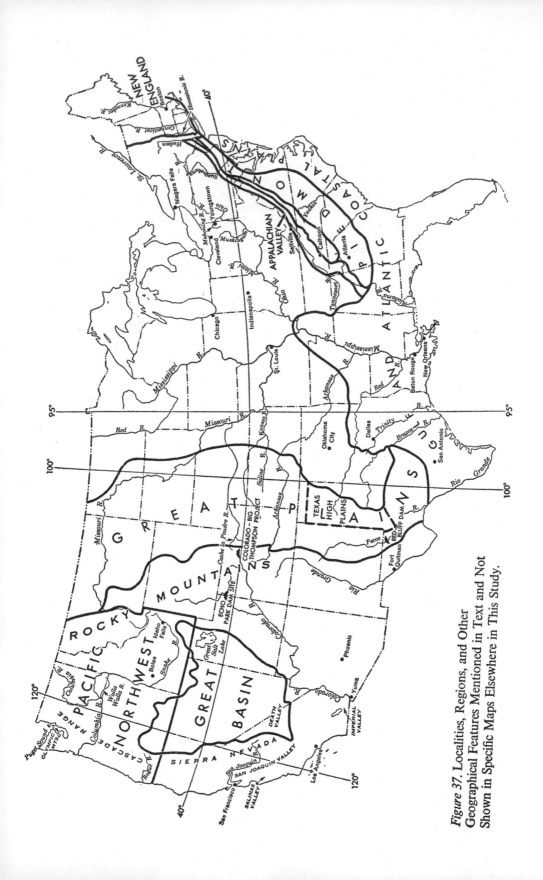

Figure 37. Localities, Regions, and Other Geographical Features Mentioned in Text and Not Shown in Specific Maps Elsewhere in This Study.

General References

A few of the more useful references applying to topics treated in this book are listed for readers' convenience.

Comprehensive Planning and Management

Commission on the Organization of the Executive Branch, Natural Resources Task Force. *Report*. Washington: U. S. Government Printing Office, 1949.

Hart, Henry C. *The Dark Missouri*. Madison: University of Wisconsin Press, 1957.

Krutilla, John V., and Eckstein, Otto. *Multiple Purpose River Development*. Baltimore: Johns Hopkins Press, 1958.

McKinley, Charles. *Uncle Sam in the Pacific Northwest*. Berkeley: University of California Press, 1952.

President's Water Resources Policy Commission. *Report*. 3 vols. Washington: U. S. Government Printing Office, 1950.

Tennessee Valley Authority. *Annual Report* (1934-58). Washington: U. S. Government Printing Office.

United Nations, Department of Economic and Social Affairs. *Integrated River Basin Development*. New York, 1958.

United Nations Scientific Conference on the Conservation and Utilization of Resources. *Proceedings*. Vol. IV, Water Resources. New York: United Nations, 1951.

U. S. Army, Corps of Engineers. Planning and survey reports for certain large river basins, like the Columbia, Missouri, Arkansas-White-Red, and other basins; usually published as U. S. Senate or U. S. House of Representatives documents.

U. S. Bureau of Reclamation. Planning reports for certain western river basin projects, like the Columbia Basin Project, Colorado River Storage Project, and the Missouri River Basin Project; usually published as U. S. Senate or U. S. House of Representatives documents.

Construction Equipment and Engineering

Black, Archibald. *The Story of Tunnels.* New York: Whittlesey House, 1937.
Fifth International Congress on Large Dams. *Proceedings.* Paris, 1955.
Houk, I. E. *Irrigation Engineering.* Vol. 1. New York: John Wiley and Sons, 1951.
Nichols, H. L. *Moving the Earth: The Workbook of Excavation.* Greenwich, Conn.: North Castle Books, 1955.
United Nations Scientific Conference on the Conservation and Utilization of Resources. *Proceedings.* Vol. IV, Water Resources. New York: United Nations, 1951.

Electricity

DEMAND

The Electrical World. New York, N. Y. Various articles appearing between 1954 and 1958.
Tennessee Valley Authority. *Annual Report* (1956). Washington: U. S. Government Printing Office.

GENERATION

Beeman, Donald, *et al. Industrial Power Systems Handbook.* New York: Mc-Graw-Hill, 1955.
The Edison Electric Institute. Statistical bulletins, and semi-annual surveys of United States electric power industry operations, since 1934. New York, various dates.
Federal Power Commission. Annual reviews of Electric Utility Statistics, since 1937.
International Conference on Peaceful Uses of Atomic Energy (Geneva, 1955). *Proceedings.* New York: United Nations, 1956.
Joint Committee on Atomic Energy. *Peaceful Uses of Atomic Energy,* Report of the Panel on the Impact of the Peaceful Uses of Atomic Energy ("McKinney Report"), Vols. I and II. Washington: U. S. Government Printing Office, January 1956.
Young, L. L. *Developed and Potential Water Power of the U. S. and Other Countries of the World.* Circular No. 367, December 1954. Washington: U. S. Geological Survey, 1955.

TRANSMISSION

Ayers, Eugene, and Scarlott, Charles A. *Energy Sources: Wealth of the World.* New York: McGraw-Hill, 1952.
Skrotzki, B. G. *Electric Transmission and Distribution.* New York: McGraw-Hill, 1954.
Sporn, Philip. "Recent and Past Progress in Power Transmission," *Electrical Engineering,* 74 (October 1955), 878-83.

Floods, Flood Management

Hoyt, W. G., and Langbein, W. B. *Floods.* Princeton, N. J.: Princeton University Press, 1955.

Leopold, Luna B., and Maddock, Thomas, Jr. *The Flood Control Controversy.* New York: The Ronald Press, 1954.

White, Gilbert F. *Human Adjustment to Floods.* Chicago: University of Chicago, Department of Geography Research Paper No. 29, 1945.

White, Gilbert F., et al., *Changes in the Urban Occupance of Flood Plains in the United States.* Chicago: University of Chicago Department of Geography Research Paper No. 57, 1958.

Ground Water

DISCOVERY

Jones, P. H., and Buford, T. B. "Electric Logging Applied to Ground Water Exploration," *Geophysics,* 16 (1951), 115-39.

Jones, P. H., and Skibitzke, H. E. "Subsurface Geophysical Methods in Ground-Water Hydrology," *Advances in Geophysics,* Vol. III. New York: Academic Press, 1957.

Vacquier, Victor. *Prospecting for Ground Water by Induced Electrical Polarization.* Socorro: New Mexico Institute of Mining and Technology, Research and Development Division, 1956.

OCCURRENCE AND USE

Bennison, E. W. *Ground Water: Its Development, Uses and Conservation.* St. Paul, Minn.: Edward E. Johnson, Inc., 1947.

Meinzer, O. E. *The Occurrence of Ground Water in the United States, with a Discussion of Principles.* U. S. Geological Survey, Water Supply Paper 489. Washington: U. S. Government Printing Office, 1923.

Picton, W. L. "The Water Picture Today—A National Summary of Ground Water Use and Projection to 1975," *Water Well Journal,* April 1956.

Sayre, A. N. "Ground Water," *Scientific American,* November 1950.

Thomas, Harold E. *The Conservation of Ground Water.* New York: McGraw-Hill, 1951.

U. S. Department of Agriculture. Yearbook 1955. *Water.* Washington: U. S. Government Printing Office, 1955. Pp. 62-78.

Hydrology, General

Kuenen, P. H. *Realms of Water.* New York: John Wiley & Sons, 1955.

Langbein, W. B., and others. *Annual Runoff in the United States.* Circular 52. Washington: U. S. Geological Survey, 1952.

Linsley, R. K., Jr., Kohler, M. A., and Paulhus, J. L. H. *Applied Hydrology.* New York: McGraw-Hill, 1949.

Meinzer, O. E. (ed.). *Hydrology.* New York: McGraw-Hill, 1942.

Thornthwaite, C. W., and Mather, J. R. *The Water Balance.* Publications in Climatology, Vol. VIII, No. 1. Centerton, N. J.: Laboratory of Climatology, 1955.

U. S. Geological Survey. Water Supply papers, published by U. S. Government Printing Office, Washington, D. C. at intervals since 1896.

Industrial and Municipal Use

Glover, J. G. (ed.). *The Development of American Industries: Their Economic Significance.* 3rd ed. New York: Prentice-Hall, 1951.

Graham, J. B., and Burrill, M. F. (eds.). *Water for Industry.* Washington: American Association for the Advancement of Science, 1956.

MacKichan, K. A. *Estimated Use of Water in the United States, 1955.* Circular 398. Washington: U. S. Geological Survey, 1957.

Mussey, O. D. *Water Requirements of the Pulp and Paper Industry.* U. S. Geological Survey, Water Supply Paper No. 1330-A. Washington: U. S. Government Printing Office, 1955.

Nordel, E. *Water Treatment for Industrial and Other Uses.* New York: Reinhold, 1951.

Shreve, R. N. *The Chemical Process Industries.* 2nd ed. New York: McGraw-Hill, 1956.

U. S. Department of Agriculture. Yearbook 1955. *Water.* Washington: U. S. Government Printing Office, 1955. Pp. 649-55.

U. S. Bureau of the Census. *U. S. Census of Manufactures: 1954.* Bulletin MC-209, Industrial Water Use. Washington, 1957.

Irrigation

Houk, I. E. *Irrigation Engineering.* Vol. 1. New York: John Wiley & Sons, 1951.

Huffman, Roy E. *Irrigation Development and Public Water Policy.* New York: Ronald Press, 1953.

Russell, E. J., and Russell, E. W. *Soil Conditions and Plant Growth.* New York: Longmans Green, 1950.

Thornthwaite, C. W. *The Place of Supplemental Irrigation in Postwar Planning.* Publications in Climatology. Vol. VI, No. 2. Seabrook, N. J.: Laboratory of Climatology, 1953. Also later publications of the same laboratory.

Thornthwaite, C. W., and Mather, J. R. *The Water Balance.* Publications in Climatology. Vol. VIII, No. 1. Centerton, N. J.: Laboratory of Climatology, 1955.

U. S. Bureau of Reclamation. *Reclamation Project Data.* Washington, D. C.: Government Printing Office, 1948. Some data revised annually, and available from the Office of the Commissioner, Bureau of Reclamation.

U. S. Department of Agriculture. *Irrigation Agriculture in the West.* Miscellaneous Publication No. 670. Washington: U. S. Government Printing Office, 1948.

U. S. Department of Agriculture. Yearbook 1955. *Water.* Washington: U. S. Government Printing Office, 1955. Pp. 247-405.

White, Gilbert F. (ed.). *The Future of Arid Lands.* Washington: American Association for the Advancement of Science, 1956.

Recycling

Fair, G. M., and Geyer, J. C. *Water Supply and Waste Water Disposal.* New York: John Wiley & Sons, 1954.

"Reuse of Water by Industry," *Industrial and Engineering Chemistry.* A collection of six papers published in Vol. 48, (December 1956) pp. 2145-71.

U. S. Bureau of the Census. *U. S. Census of Manufactures: 1954.* Bulletin MC-209. Washington, 1957.

Reservoir Design and Operation

Civil Engineering. (American Society of Civil Engineers, New York.) Numerous articles, especially those published since 1945.

Engineering News-Record. (New York.) Numerous articles, especially those published since 1945.

Fifth International Congress on Large Dams. *Proceedings.* Paris: ●55.

Houk, I. E. *Irrigation Engineering.* Vol. 1. New York: John Wiley & Sons, 1951.

Senate Committee on Interior and Insular Affairs. *Control of Evaporation Losses.* Memorandum of the Chairman. Washington, April 14, 1958.

Tennessee Valley Authority. *Engineering Data—TVA Water Control Projects and Other Major Hydro Development in the Tennessee and Cumberland Valleys.* Technical Monograph No. 55. Knoxville, Tenn., 1954.

United Nations Scientific Conference on the Conservation and Utilization of Resources. *Proceedings.* Vol. IV, Water Resources. New York: United Nations, 1951.

U. S. Public Health Service and Tennessee Valley Authority, Health and Safety Department. *Malaria Control on Impounded Waters.* Washington: U. S. Government Printing Office, 1947.

Saline Water Use, Desalting

Symposium on Saline Water Conversion (1957). *Proceedings.* Washington: National Academy of Sciences, 1958.

U. S. Department of Agriculture. *Agriculture Handbook No. 60.* Washington, 1954.

U. S. Department of Agriculture. Yearbook 1955. *Water.* Washington: U. S. Government Printing Office, 1955. Pp. 109-17.

U. S. Department of the Interior. *Demineralization of Saline Waters.* Washington, October 1952.
U. S. Department of the Interior. *Saline Water Conversion.* Report for 1957. Washington, January 1958.

Waste Disposal

Besselievre, Edmund B. *Industrial Waste Treatment.* New York: McGraw-Hill, 1952.
Fifth Annual Water Symposium. *Proceedings.* Engineering Experiment Station Bulletin No. 55. Baton Rouge: Louisiana State University, 1956.
Flannery, James J. *Water Pollution Control: Development of State and National Policy.* Ann Arbor, Mich.: University Microfilms, 1956.
Glover, J. G. (ed.). *The Development of American Industries: Their Economic Significance.* 3rd ed. New York: Prentice-Hall, 1951.
Gorman, A. E., and Theis, C. V. "The Treatment and Disposal of Wastes in the Atomic Energy Industry," *Water for Industry.* Publication No. 45. Washington: American Association for the Advancement of Science, 1956.
Rudolphs, W. (ed.). *Industrial Wastes: Their Disposal and Treatment.* New York: Reinhold, 1953.
Shreve, R. Norris. *The Chemical Process Industries.* New York: McGraw-Hill, 1945.

Watershed Treatment

U. S. Department of Agriculture. Yearbook 1955. *Water.* Washington: U. S. Government Printing Office, 1955. Pp. 121-242, 407-44.

Weather Modification

Advisory Committee on Weather Control. *Final Report of the Advisory Committee on Weather Control.* 2 vols. Washington: U. S. Government Printing Office, 1957.

Withdrawal, Disappearance

MacKichan, K. A. *Estimated Use of Water in the United States, 1955.* Circular 398. Washington: U. S. Geological Survey, 1957.
Robinson, T. W. *Phreatophytes.* U. S. Geological Survey Water Supply Paper 1423. Washington: U. S. Government Printing Office, 1958.
U. S. Department of Agriculture. Yearbook 1955. *Water.* Washington: U. S. Government Printing Office. Pp. 14-40, 219-28, 247-51, 341-45, 407-15, 615-35.
(See also Industrial and Municipal Use.)

Data Sources for Maps and Graphs, and Sources for Photographs

Figure

1. W. B. Langbein *et al., Annual Runoff in the United States,* U. S. Geological Survey Circular 52, 1952.
2. G. E. Harbeck, Jr., and W. B. Langbein, "Normals and Variations in Runoff, 1921-1945," *Water Resources Review,* Supplement No. 2, U. S. Geological Survey, 1949.

   W. B. Langbein and J. V. B. Wells, "The Water in the Rivers and Creeks," *Water,* U. S. Department of Agriculture Yearbook, 1955, pp. 52-62.
3. U. S. Department of Agriculture.
4, 5. L. Lassen, H. A. Lull, and B. Frank, *Some Plant-Soil-Water Relations in Watershed Management* (Circular No. 910), U. S. Department of Agriculture, 1952, p. 8.
6. H. E. Thomas, *The Conservation of Ground Water* (New York: McGraw-Hill, 1951), esp. maps pp. 32, 40.

   O. E. Meinzer, *Ground Water in the U. S., A Summary* (Water Supply Paper No. 836-D), U. S. Geological Survey, 1939; reprinted as Paper No. 489, 1950.

   A. K. Lobeck, *Geologic Map of the United States* (New York: Geographical Press, Columbia University, 1941).

   G. G. Parker, "The Encroachment of Salt Water into Fresh," *Water,* U. S. Department of Agriculture Yearbook, 1955, pp. 628-35.

   U. S. Congress, House of Representatives, Interior and Insular Affairs Committee, *The Physical and Economic Foundation of Natural Resources,* III—Ground Water Regions of the United States, Their Storage Facilities, 1952.

   N. M. Fenneman, *Physiography of Western United States* (New York: McGraw-Hill, 1931).
7. U. S. 81st Congress, 2nd Session, *House Document No. 706,* "A Program to Strengthen the Scientific Foundation in Natural Resources (map), November 1950, p. 91.

   A. G. Fiedler and C. L. McGuinness, *Ground-Water Problems and*

Figure

     their Relation to Army Water-Supply Installations, U. S. Geological Survey, 1948.

11. U. S. Bureau of the Census, *Statistical Abstract of the United States, 1956.*

12, 13. C. D. Harris, "The Market as a Factor in the Localization of Industry in the United States," *Annals of the Association of American Geographers,* 44 (December 1954), 315-48, esp. figs. 17, 27.

14. U. S. Bureau of the Census, *Geographic Reports* (Series GEO No. 4), December 1952.

15. New England–New York Inter-Agency Committee, *The Resources of the New England–New York Region,* Part One, "The General Report on the Comprehensive Survey," esp. p. III-12, and plates 7, 8.

     Federal Security Agency, U. S. Public Health Service Water Pollution Series, Nos. 129, 180, 223, 366, 381, 292, 450, 293, 365, 347, 294, 86, 169, 175, 234, 238 (various dates).

16. U. S. Department of Agriculture files.

17. W. Van Royen, ed., *Agricultural Atlas* (New York: Prentice-Hall, 1954), pp. 107, 201, and 241; and United States Department of Agriculture, statistical data.

18. Tennessee Valley Authority, Divisions of Engineering and Construction, *Engineering Data, TVA Water Control Projects and other Major Hydro Developments in the Tennessee and Cumberland Valleys* (Technical Monograph No. 55), 1954, p. 3-3.

19. American Water Works Association, *Water Quality and Treatment—a Manual* (2nd ed.; New York, 1950).

     U. S. Geological Survey, *Industrial Utility of Public Water Supplies in the U. S., 1952* (Water Supply Papers Nos. 1299 and 1300), 1954.

20. Tennessee Valley Authority.

21, 22. C. E. Blee, "Multiple-Purpose Reservoir Operation of Tennessee River System," *Civil Engineering,* 15 (1945), 222, 221.

23. E. J. Rutter, "Flood Control Operation of Tennessee Valley Authority Reservoirs," *Proceedings of the American Society of Civil Engineers,* 76 (May 1950), 13.

24. Tennessee Valley Authority, *Annual Report,* 1957, p. 13.

25. Louis Koenig, "Ground Water: The Impact of Recent Technological Advances in its Utilization," report prepared for Resources for the Future, Inc., December 1956. Personal communication from A. H. White, editor, *The Cross Section,* June 5, 1957.

26. Tennessee Valley Authority, Maps and Surveys Branch.

27. Department of the Interior, Office of Saline Water, *Potential Use of Converted Sea Water for Irrigation in Parts of California and Texas* (Saline Water Conversion Program Research and Development Report No. 3), 1954.

28. U. S. Geological Survey Water Supply Paper No. 1374.

29. F. H. Ludlam, "Artificial Snowfall from Mountain Clouds," *Tellus* (Sweden), 7 (1955), 283.

31. L. M. Smith, "Impact of Atomic Energy on Conservation of Natural

Figure

Resources," *Conservation is Good Business* (Washington: U. S. Chamber of Commerce, 1957), p. 17.

32. U. S. Bureau of the Census.
33. U. S. Bureau of the Census, *U. S. Census of Manufacturers: 1954.*
34, 35. C. W. Thornthwaite and J. R. Mather, *The Water Balance* (Centerton, N. J.: Laboratory of Climatology, 1955), pp. 25, 35.
36. U. S. Geological Survey, 1958. Data prepared by J. S. Meyers.
38. *Directory of Electric Utilities, 1957* (New York: McGraw-Hill, 1957).

Federal Power Commission, *Principal Electric Utility Generating Stations and Transmission Lines,* 1956; additional information supplied by state regulatory agencies.

Rural Electrification Administration, *Graphic Summary of the Rural Electrification and Rural Telephone Program,* 1953; revised and brought up to date from information available in REA files.

State regulatory agencies: published maps and official description; also correspondence with authors during 1956 and 1957.

Plate

1. U. S. Department of Agriculture, Soil Conservation Service.
2. U. S. Department of Agriculture.
3. U. S. Department of Agriculture, Soil Conservation Service. B. C. McLean.
4. U. S. Department of Agriculture, Soil Conservation Service. W. H. Von-Tebra.
5. U. S. Department of Agriculture, Soil Conservation Service.
6. Interstate Commission on the Potomac River Basin.
7. Tennessee Valley Authority.
8. U. S. Department of Agriculture. D. E. Hutchinson.
9. U. S. Army Corps of Engineers.
10. Caterpillar Tractor Company.
11. U. S. Army Corps of Engineers.
12. U. S. Department of the Interior, Bureau of Reclamation.
13. U. S. Army Corps of Engineers.
14. Tennessee Valley Authority.
15. U. S. Department of the Interior, Bureau of Reclamation.
16. U. S. Department of the Interior, Bureau of Reclamation.
17. U. S. Army Corps of Engineers.
18. U. S. Army Corps of Engineers.
19. Public Service Company of Colorado.
20. Westinghouse Electric Corporation.
21. General Electric Company.
22. Westinghouse Electric Corporation.
23. Westinghouse Electric Corporation.
24. Pacific Gas and Electric Company.
25. U. S. Department of Commerce, Weather Bureau.
26. U. S. Department of Commerce, Weather Bureau.
27. U. S. Army Corps of Engineers.

Index

Abu-Lughod, J., 410*n*, 424*n*, 434*n*
Acetate rayon plants, unit water requirements, 409
Ackerman, E. A., 133*n*, 368*n*, 383*n*, 551*n*, 620*n*
Ackerman, William C., 436*n*
Adams, Henry, 3*n*
Administrative organization (*see also* Operational management; Planning; Water development):
and adaptation to technologic problems, summary of, 621-27;
centralization of, 639-53;
distribution of responsibility among different administrative levels, 642-57;
and electricity generation and distribution, 644-45;
federal entrepreneurship, question of, 651-53;
and federal functions at national level, 649-51;
federal or nonfederal regional organization, role of, 648-49;
functions of, 465-66;
place of, in matching supply and demand, 461-65;
and political forces, 462-63;
and private enterprise, 642-46, 642*n*;
state activities, 647;
state and federal administration, problems of division concerning, 647-48;
and water technology, relation of, 7, 8-10, 99-101, 461-68, 628-57:
background assumptions, 628-30;
effects of techniques on problems and opportunities, 632-39;
regional incidence of techniques, 630-32
Advisory Committee on Weather Control, 364-65, 367, 370, 380, 382, 383, 607*n*
Aeration, zone of, 26*n*

Africa, 455; grasses from, 164, 165
Agricultural Conservation Program, 616, 616*n*
Agricultural Research Service, 486
Agriculture (*see also* Crops; Irrigation; Livestock; Refrigerated transportation; Salinity conditions; Soil moisture; Soils; Water conservation; Western agriculture): and national diet, changes in, 110; and potential effects of weather modification, 378-79; production, technical changes in, 540-41; research, financing of, 559; technology, impact of, on, 519-21
Air conditioning, 143-45
Alabama: land grants to, 551; population, withdrawal, and runoff, *Table*, 76
Alabama Power Company, 501, 534
Alabama River, 19
Albeni Falls development, 594*n*
Albuquerque, N.M., growth of, 67, 67*n*
Alcoa. *See* Aluminum Company of America
Alfalfa, 58; distribution of, 159, *Fig.*, 158., salt tolerance of, 152, 155; use of, as cover crop, 174
Alkali soils. *See under* Salinity conditions
Alkalies, output of, 120
All-American Canal, 252
Allaway, W. H., 617*n*
Allhands, J. L., 222*n*, 223*n*
Allied Chemical Corporation, 123*n*, 128
Allocation of water. *See under* Water development
Aluminum Company of America, 471, 471*n*; and TVA, agreement between 237
Aluminum Company of Canada, 267
Aluminum piping, 132-33
Amberg, H. R., 434*n*

677